Lecture Notes in Computer Science 2081

Edited by G. Goos, J. Hartmanis and J. van Leeuwen

T0216254

Springer

Berlin
Heidelberg
New York
Barcelona
Hong Kong
London
Milan
Paris
Singapore
Tokyo

Karen Aardal   Bert Gerards (Eds.)

Integer Programming and Combinatorial Optimization

8th International IPCO Conference
Utrecht, The Netherlands, June 13-15, 2001
Proceedings

 Springer

Series Editors

Gerhard Goos, Karlsruhe University, Germany
Juris Hartmanis, Cornell University, NY, USA
Jan van Leeuwen, Utrecht University, The Netherlands

Volume Editors

Karen Aardal
Universiteit Utrecht, Instituut voor Informatica en Informatiekunde
Padualaan 14, 3584 CH Utrecht, The Netherlands
E-mail: aardal@cs.uu.nl

Bert Gerards
CWI
Kruislaan 413, 1098 SJ Amsterdam, The Netherlands
E-mail: bgerards@cwi.nl

Cataloging-in-Publication Data applied for

Die Deutsche Bibliothek - CIP-Einheitsaufnahme

Integer programming and combinatorial optimization : proceedings / 8th
International IPCO Conference, Utrecht, The Netherlands, June 13 - 15,
2001. Karen Aardal ; Bert Gerards (ed.). - Berlin ; Heidelberg ; New
York ; Barcelona ; Hong Kong ; London ; Milan ; Paris ; Singapore ;
Tokyo : Springer, 2001
   (Lecture notes in computer science ; Vol. 2081)
   ISBN 3-540-42225-0

CR Subject Classification (1998): G.1.6, G.2.1, F.2.2

ISSN 0302-9743
ISBN 3-540-42225-0 Springer-Verlag Berlin Heidelberg New York

Springer-Verlag Berlin Heidelberg New York
a member of BertelsmannSpringer Science+Business Media GmbH

http://www.springer.de

© Springer-Verlag Berlin Heidelberg 2001
Printed in Germany

Typesetting: Camera-ready by author, data conversion by PTP-Berlin, Stefan Sossna
Printed on acid-free paper   SPIN: 10839281   06/3142   5 4 3 2 1 0

Preface

This volume contains the papers selected for presentation at IPCO VIII, the Eighth Conference on Integer Programming and Combinatorial Optimization, Utrecht, The Netherlands, 2001. This meeting is a forum for researchers and practitioners working on various aspects of integer programming and combinatorial optimization. The aim is to present recent developments in theory, computation, and application of integer programming and combinatorial optimization. Topics include, but are not limited to: approximation algorithms, branch and bound algorithms, computational biology, computational complexity, computational geometry, cutting plane algorithms, diophantine equations, geometry of numbers, graph and network algorithms, integer programming, matroids and submodular functions, on-line algorithms, polyhedral combinatorics, scheduling theory and algorithms, and semidefinite programs.

IPCO was established in 1988 when the first IPCO program committee was formed. The locations and years of the seven first IPCO conferences were: IPCO I, Waterloo (Canada) 1990, IPCO II, Pittsburgh (USA) 1992, IPCO III, Erice (Italy) 1993, IPCO IV, Copenhagen (Denmark) 1995, IPCO V, Vancouver (Canada) 1996, IPCO VI, Houston (USA) 1998, IPCO VII, Graz (Austria) 1999. IPCO is held every year in which no MPS (Mathematical Programming Society) International Symposium takes place. Since the MPS meeting is triennial, IPCO conferences are held twice in every three-year period. As a rule, IPCO is held somewhere in Northern America in even years, and somewhere in Europe in odd years.

In response to the call for papers for IPCO 2001, the program committee received 108 submissions, indicating a strong and growing interest in the conference. The program committee met on January 13 and 14, 2001, in Amsterdam, The Netherlands, and selected 32 contributed papers for inclusion in the scientific program of IPCO 2001. The selection was based on originality and quality, and reflects many of the current directions in integer programming and optimization research. The overall quality of the submissions was extremely high. As a result, many excellent papers could unfortunately not be chosen.

The organizing committee for IPCO 2001 consisted of Karen Aardal, Bert Gerards, Cor Hurkens, Jan Karel Lenstra, and Leen Stougie. IPCO 2001 was organized in cooperation with the Mathematical Programming Society, and was sponsored by BETA, CQM, CWI, DONET, The Netherlands Society for OR (NGB), EIDMA, ILOG, IPA, Philips Research Labs, Technische Universiteit Eindhoven, the Technology Foundation STW, and Universiteit Utrecht.

April 2001

Karen Aardal
Bert Gerards

IPCO VIII Program Committee

Table of Contents

Two $O(\log^* k)$-Approximation Algorithms for the Asymmetric $k-$Center Problem

Aaron Archer*

Operations Research Department, Cornell University, Ithaca, NY 14853
aarcher@orie.cornell.edu

Abstract. Given a set V of n points and the distances between each pair, the k-center problem asks us to choose a subset $C \subseteq V$ of size k that minimizes the maximum over all points of the distance from C to the point. This problem is NP-hard even when the distances are symmetric and satisfy the triangle inequality, and Hochbaum and Shmoys gave a best-possible 2-approximation for this case.

We consider the version where the distances are asymmetric. Panigrahy and Vishwanathan gave an $O(\log^* n)$-approximation for this case, leading many to believe that a constant approximation factor should be possible. Their approach is purely combinatorial. We show how to use a natural linear programming relaxation to define a promising new measure of progress, and use it to obtain two different $O(\log^* k)$-approximation algorithms. There is hope of obtaining further improvement from this LP, since we do not know of an instance where it has an integrality gap worse than 3.

1   Introduction

Suppose we are given a road map of a city with n buildings where we wish to offer fire protection, along with the travel times (a distance function) between each pair of buildings, and suppose we are allowed to make k of the buildings into fire stations. Informally, the asymmetric k-center problem asks how to locate the fire stations (*centers*) in order to minimize the worst case travel time to a fire (the *covering radius*). It is common to assume the distance function is symmetric and satisfies the triangle inequality. Without the triangle inequality, it is NP-hard even to decide whether the k-center optimum is finite, by a reduction from set cover. So we require the distances to satisfy the triangle inequality, which also makes sense when we think of them as travel times. But distances may be asymmetric due to one-way streets or rush-hour traffic, so in this paper we do not require symmetry.

The asymmetric k-center problem has proven to be much more difficult to understand than its symmetric counterpart. In the early 1980's, Hochbaum and Shmoys [6,7] first gave a simple 2-approximation algorithm for the symmetric

* Supported by the Fannie and John Hertz Foundation and ONR grant AASERT N0014-97-10681.

K. Aardal, B. Gerards (Eds.): IPCO 2001, LNCS 2081, pp. 1–14, 2001.

variant. Shortly thereafter, Dyer and Frieze [4] found another 2-approximation. These results essentially closed the problem because this approximation guarantee is the best possible, by an easy reduction from set cover. However, no non-trivial approximation algorithm was known for the asymmetric version until Panigrahy and Vishwanathan [14,13] gave an $O(\log^* n)$-approximation more than ten years later.

The k-center problem is one of the basic clustering problems. Uncapacitated facility location and k-median are two other prominent ones, and both of these admit constant-factor approximations in the symmetric case (see [2] for the current best factors). But in the asymmetric case, the best results yield $O(\log n)$ factors [5,10], relying on clever applications of the greedy set cover algorithm. Moreover, a natural reduction from set cover gives inapproximability results that match these bounds up to a constant [1]. In stark contrast, it is now widely believed that a constant-factor approximation algorithm should exist for the asymmetric k-center problem. This conjecture is prompted by the $O(\log^* n)$ result of [13], especially since no hardness of approximation result is known beyond the lower bound of 2 inherited from the symmetric version.

In this paper we introduce a natural linear programming relaxation that seems a promising step towards a constant-factor approximation. Our LP is essentially a set cover LP derived from an unweighted directed graph, where the optimal k-center solution corresponds to an optimal integral cover (dominating set). Whereas this LP has an $O(\log n)$ integrality gap for the set cover problem, it behaves better in our context because k-center has a different objective function. The LP objective is the number of fractional centers chosen, whereas the k-center objective is the covering radius. We present two $O(\log^* k)$-approximation algorithms for the asymmetric k-center problem, both of which use the fractional centers given by the LP solution as a guide. There is hope that our LP might be used to obtain further improvements since we do not know of any instances where it has an integrality gap worse than 3 for the asymmetric k-center problem.[1]

The $O(\log^* n)$ algorithm of [13] is entirely combinatorial. Essentially, it uses the greedy set cover algorithm to choose some (too many) centers, and then recursively covers the centers until the correct number remain. The advantage of working with our LP is that it gives us fractional center values to guide our algorithm and measure its progress. Rather than choosing centers to cover the nodes of our graph, we instead try to efficiently cover the fractional centers.

The crux of the first algorithm is our EXPANDINGFRONT routine. We select centers greedily, hoping to cover lots of fractional centers, because we are covering within two steps what these fractional centers covered within one. For this strategy to succeed, we need to somehow guarantee that the centers we choose cover many fractional centers. The key idea here is that as we choose more centers, the set A of active centers that remain to be covered shrinks, as does the number of centers necessary to cover them. We show that the amount of progress we make at each step grows substantially as the size of the optimal

---

[1] Here we use integrality gap in a slightly non-standard sense. See the remarks in Section 3 for details.

fractional cover of A shrinks, so it is important to reduce this quantity rapidly. The fractional center values from the LP allow us to enforce and measure this shrinkage.

We can also use the fractional centers to modify the algorithm of [13] to obtain the same $O(\log^* k)$ performance guarantee. Our RECURSIVECOVER routine uses them to obtain an initial cover with fewer centers than the initial cover produced in [13], then employs precisely the same recursive set cover scheme to finish.

Since achieving a constant-factor approximation seems difficult, it is natural to attempt to find a bicriterion approximation algorithm that blows up both the number of centers and the covering radius by a constant factor. This might seem to be a significantly easier task, but in fact [13] shows this would essentially give us a constant-factor unicriterion approximation. This is because we can preprocess the problem with the REDUCE routine of Section 6, and postprocess our solution by recursively covering the centers we chose, as in Section 5.

2   Formal Problem Definition and Notation

The input to the asymmetric k-center problem is a parameter k, a set V of n points, and a matrix D specifying a distance function $d : V \times V \to \mathbb{R}_+ \cup \{\infty\}$. Think of $d(u, v)$ as the distance *from* u to v. We require our distance function to obey the triangle inequality, but not symmetry. That is, $d(u, w) \leq d(u, v) + d(v, w)$ for all $u, v, w \in V$, but $d(u, v)$ may differ from $d(v, u)$.

Any set of *centers* $C \subseteq V$ with $|C| \leq k$ is a *solution* to the k-center problem. The *covering radius* of a solution C is the minimum distance R such that every point is within R of the set C. (It could be ∞.) The goal is to find the solution with the minimum covering radius R^*.

We will find it convenient to work with unweighted directed graphs on node set V instead of on the space (V, D) directly. With respect to a directed graph $G = (V, E)$, we define (for $i \in \mathbb{Z}_+$)

$$\Gamma_i^+(u) = \{v \in V : G \text{ contains a directed path from } u \text{ to } v \text{ using at most } i \text{ edges}\}.$$

Conversely, $\Gamma_i^-(u)$ is the set of nodes *from* which u can be reached by a directed path using at most i edges. We suppress the subscript when $i = 1$. Thus $\Gamma^+(u)$ is u plus its out-neighbors, and $\Gamma^-(u)$ is u plus its in-neighbors. For $S \subseteq V$ we define $\Gamma_i^+(S)$ and $\Gamma_i^-(S)$ analogously. In G, we say S *covers* T *within* i (or i-*covers* T) if $\Gamma_i^+(S) \supseteq T$. When $i = 1$, we just say S *covers* T.

For $R \geq 0$, we define the graph $G_R = (V, E_R)$, where $E_R = \{(u, v) : d(u, v) \leq R\}$. The essential connection here is that there exist k centers that cover all of G_R if and only if $R \geq R^*$. Thus we can binary search for the optimal radius R^* and work in G_{R^*}. Finding an i-cover in this graph will yield an i-approximation in the original space, since traversing each edge in the graph corresponds to moving at most R^* in the original space. Moreover, in the worst case this analysis is tight, because our given distances could be those induced by shortest directed paths in an unweighted graph.

For $y \in \mathbb{R}^S$, let $y(S)$ denote $\sum_{v \in S} y_v$. For a function g, let $g^{(i)}$ denote the function iterated i times. Finally, define $\log^* x = \min\{i : \log^{(i)} x \leq \frac{3}{2}\}$.

3   Overview of the Algorithm

We describe a polynomial time relaxed decision procedure AKC (see Figure 2), which takes as input an asymmetric k-center instance and a guess at the optimal radius R, and either outputs a solution of value $O(R \log^* k)$ or correctly reports that $R < R^*$. Since $R^* = d(u, v)$ for some $u, v \in V$, there are only $O(n^2)$ different possible values for R^*. We binary search on R, using at most $O(\log n)$ calls to AKC to yield a solution of value $O(R \log^* k)$ for some $R \leq R^*$, which gives our $O(\log^* k)$-approximation algorithm. Thus, by guessing the covering radius R, we immediately convert the optimization problem into a promise problem on G_R and thereafter think only in terms of this graph and others derived from it.

Theorem 1 *Given a parameter R and an asymmetric k-center instance (V, D, k) whose optimum is R^*, AKC (V, D, k, R) runs in polynomial time and either proves that $R < R^*$ or outputs a set C of at most k centers covering V within $R(3 \log^* k + O(1))$.*

The general framework we use was introduced by [13]. There are two substantive components – a preprocessing phase called REDUCE, and the heart of the algorithm, which we call AUGMENT. In addition, our version of AKC solves an LP. We can solve the LP in polynomial time, and it will be clear that our algorithms for REDUCE and AUGMENT run in polynomial time. Thus AKC runs in polynomial time.

We present two different ways to implement the AUGMENT phase, EXPANDINGFRONT (Section 4) and RECURSIVECOVER (Section 5), each of which improves the approximation guarantee from $O(\log^* n)$ to $O(\log^* k)$. For completeness, we also describe the REDUCE phase in Section 6, slightly sharpening the analysis given in [13] to improve the constant inside the $O(\log^* k)$ from 5 to 3. We now describe how REDUCE and AUGMENT fit together.

Assume we have a graph G, and we are promised that there exist k centers covering G. Then AUGMENT finds at most $\frac{3}{2}k$ centers that cover all of G within $\log^* k + O(1)$. How do we obtain a solution that uses only k centers? Roughly speaking, we will prove that if there are k centers that cover V, then there exist $\frac{2}{3}k$ centers 3-covering V. So we can run AUGMENT in G^3, the cube of G, to find k centers that cover all of V in G within $3 \log^* k + O(1)$.

More precisely, the REDUCE phase preprocesses the graph G by choosing an initial set C of at most k centers consisting of some special nodes, called *center capturing vertices*. These centers already cover part of V within some constant radius. The set of nodes that remain to be covered we call the *active set*, and denote it by A. We prove that there exist $p \leq \frac{2}{3}(k - |C|)$ centers 3-covering A. Then, again roughly speaking, we use the AUGMENT procedure in G^3 to augment C by at most $\frac{3}{2}p$ new centers to a set of at most k centers covering V within $\log^* p + O(1)$ in G^3.

Our LP, with respect to $G = (V, E)$ and $A \subseteq V$

---

min $y(V)$
s.t. $y(\Gamma^-(v)) \geq 1$ for all $v \in A$
     $y \geq 0$

Fig. 1. Here is our set cover-like LP, defined with respect to some graph G with nodes V and an active set A of nodes to be covered. Recall the notation $y(S) = \sum_{v \in S} y_v$.

Both EXPANDINGFRONT and RECURSIVECOVER use the linear program of Figure 1. In this LP, if we further restrict $y \in \{0, 1\}^V$, the resulting integer program asks for the smallest set of centers necessary to cover A. This is just a set cover problem where A is the set of elements to be covered, and $\{A \cap \Gamma^+(v) : v \in V\}$ is the collection of sets. We think of any solution y as identifying fractional centers, so y is a *fractional cover* of A. We note that our AUGMENT phase does not require there to exist an integral cover with p centers, nor does it require y to be an optimal fractional cover. It suffices to have any fractional cover with at most p fractional centers. We use these fractional centers to guide the AUGMENT procedure.

The following two theorems (proved in Sections 6 and 4.2, respectively) summarize the technical details. When put together with the precise description of AKC in Figure 2, Theorem 1 follows.

Theorem 2 *Given a directed graph G for which there is a cover using k centers, REDUCE (G) outputs a set of centers C and an active set $A = V \backslash \Gamma_4^+(C)$ such that there exists a 3-cover of A using at most $\frac{2}{3}(k - |C|)$ centers.*

Theorem 3 *Suppose $A = V \backslash \Gamma^+(C)$ in G, y is a fractional cover of A, and $p = y(V)$. When AUGMENT (G, A, C, y, p) is implemented with either EXPANDINGFRONT or RECURSIVECOVER, it augments C by at most $\frac{3}{2}p$ additional centers to cover A within $\log^* p + O(1)$ in G.*

We show two different ways to implement the AUGMENT phase, each producing a $(\log^* p + O(1))$-cover. The first, EXPANDINGFRONT, introduces the idea of choosing centers to cover fractional centers, and shows how to use the LP to make our future choices more efficient by reducing the number of fractional centers necessary to cover the new active set. The second, RECURSIVECOVER, shows how to combine the fractional center covering idea with the recursive set cover technique of [13] to obtain the same improvement.

Remarks. Our main contribution is in using the LP solution to define a new notion of progress based on covering fractional centers. We believe that this LP and the techniques introduced here may lead to a constant-factor approximation. Even though our LP has a $\Theta(\log n)$ integrality gap for the set cover problem,

---

AKC (V, D, k, R)

$(C, A) \leftarrow$ REDUCE (G_R)
$p \leftarrow \frac{2}{3}(k - |C|)$
$\hat{G} \leftarrow G_{3R}$ plus the edges $\{(u, v) : u \in C, v \in \Gamma_4^+(C)\}$
(where $\Gamma_4^+(C)$ is interpreted in G_R)
solve the linear program of Figure 1 defined by $\hat{G}$ and A to get y
if $y(V) > p$ then STOP and conclude $R < R^*$
else
    $p \leftarrow y(V)$
    $C \leftarrow$ AUGMENT $(\hat{G}, A, C, y, p)$
    output C

Fig. 2. Formal description of AKC.

we do not know of any examples where its integrality gap for the asymmetric k-center problem is worse than 3. That is, we are not aware of any graphs G for which there is a fractional cover using k fractional centers but there is no integral 3-cover using $\lceil k \rceil$ centers.

There is a graph G on 66 nodes that can be covered with 6 fractional centers, while the smallest integral 2-cover uses 7 centers [12]. This is the smallest example we know that establishes the LP integrality gap of 3. Probabilistic constructions yield an infinite family of graphs on n nodes having a fractional cover using k fractional centers but admitting no integral 2-cover using fewer than $\Omega(k\sqrt{\log n})$ centers.

4   Augment Phase: ExpandingFront

Recall that the AUGMENT phase takes as input a directed graph G, a set of already chosen centers C, an active set $A = V \backslash \Gamma^+(C)$ of nodes not already covered by C, and a solution y to the LP of Figure 1, with $p = y(V)$. We wish to select an additional $\frac{3}{2}p$ centers that, along with C, cover A within $\log^* p + O(1)$ in G.

4.1   Motivation and Description of ExpandingFront

For motivation, consider the case where $C = \emptyset$ so A is all of V. If there exists a cover using only q integral centers, then clearly one of these centers v must cover at least $\frac{p}{q}$ units of fractional centers. That is, $y(\Gamma^+(v)) \geq \frac{p}{q}$. It turns out that this result holds also when there are q fractional centers covering p fractional centers (see Lemma 4 below, with $z = y$), and we can apply this observation with $p = q$. Thus, there exists a node v covering a full unit of fractional centers.

It makes sense to choose such a node as a center, because it 2-covers what these fractional centers covered. If we could manage to continue covering one new unit of fractional centers with each new center we select, then we would use only p centers to cover all the fractional centers, which means we would 2-cover all of V. This would yield a 2-approximation.

The problem is that when we choose a new center v, we remove $\Gamma^+(v)$ from A, the active set of fractional centers yet to be covered. To see how many units of fractional centers our next greedily chosen center covers, we use Lemma 4. Since $y(A)$, the amount of active fractional centers, decreases, while $y(V)$ is still p, the best guarantee we can make is that each new greedily-chosen center covers at least a $\frac{1}{p}$ fraction of the remaining active fractional centers.

Lemma 4 *Let $G = (V, E)$ be a directed graph, $A \subseteq V$, and $z \in \mathbb{R}_+^A$ be any set of non-negative weights on A. If $y \in \mathbb{R}_+^V$ is a fractional cover of A, that is, y is any feasible solution to the linear program of Figure 1, then there exists $v \in V$ such that*

$$z(\Gamma^+(v) \cap A) \geq \frac{z(A)}{y(V)}. \tag{1}$$

Proof. We take a weighted average of $z(\Gamma^+(v) \cap A)$ over $v \in V$.

$$\frac{1}{y(V)} \sum_{v \in V} y_v z(\Gamma^+(v) \cap A) = \frac{1}{y(V)} \sum_{v \in V} \sum_{u \in \Gamma^+(v) \cap A} y_v z_u$$

$$= \frac{1}{y(V)} \sum_{u \in A} z_u \sum_{v \in \Gamma^-(u)} y_v$$

$$\geq \frac{1}{y(V)} \sum_{u \in A} z_u$$

The inequality follows because $z \geq 0$ and $y(\Gamma^-(u)) \geq 1$ for all $u \in A$. Since some term is at least as large as the weighted average, we know $z(\Gamma^+(v) \cap A) \geq \frac{z(A)}{y(V)}$ for some $v \in V$. □

We might be saved if we somehow knew that the present active set A had a fractional cover y_A with fewer than p fractional centers. We could then use y_A instead of y, so the denominator $y_A(V)$ of (1) would decrease along with the numerator $y(A)$. We can indeed reduce the denominator by the following observation: for any set S, the nodes in $\Gamma^+(S)$ do not help cover any of the nodes in $V \backslash \Gamma_2^+(S)$. To aid this discussion, we introduce some new notation.

With respect to a graph G and a current set of centers $C \subseteq V$, define $V_i = \Gamma_i^+(C) \backslash \Gamma_{i-1}^+(C)$, the nodes exactly i steps from from C in G, and $V_{\geq i} = V \backslash \Gamma_{i-1}^+(C)$, the nodes at least i steps from C in G. If we consider a directed breadth first search tree from the set C, then $V_{\geq i}$ consists of the nodes at and beyond the ith level, so this set shrinks as we add centers to C.

In our motivational example (where initially $A = V$), when selecting our first center v with $y(\Gamma^+(v)) \geq 1$, let us set $A \leftarrow V_{\geq 3}$ (instead of $A \leftarrow V_{\geq 2}$). See Figure 3. Since V_1 does not help cover A, projecting y onto $V_{\geq 2}$ yields a fractional

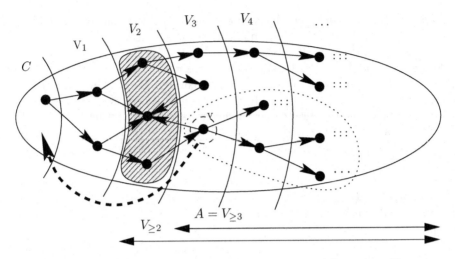

Fig. 3. This illustrates one step of the $i = 1$ phase of EXPANDINGFRONT.

cover of A, so we can replace $y(V)$ by $y(V_{\geq 2}) \leq p - 1$ in the denominator of (1) for the next greedy choice. Unfortunately, the numerator decreases to $y(A) = y(V_{\geq 3})$, which is smaller by exactly $y(V_2)$, so the next greedily chosen center may cover less than one additional unit of fractional centers.

This problem seems to be unavoidable. The difficulty is that $y(V_2)$ might be quite large, say $\frac{p}{2}$. Since these fractional centers are no longer in the active set, our chosen centers may never cover them, so they will always contribute to the denominator in (1).

The idea of EXPANDINGFRONT is to greedily choose a few centers, then "flush" the irritating fractional centers trapped at the "front" V_2, by setting $A \leftarrow V_{\geq 4}$. We can then ignore the centers in V_2 since they no longer help cover A. This expands the front (the location of the irritating nodes) to radius 3. In the bad case where $y(V_2)$ is large, we flush a lot of fractional centers. We then repeat this process. The penalty is that each time we flush, we expand the radius and so we do not get a constant-factor approximation. As it turns out, we need flush only $\log^* p + O(1)$ times. See Figure 4 for a precise description. For simplicity, the algorithm begins with phase $i = 0$, whereas this motivating discussion corresponds to the $i = 1$ phase.

Figure 3 shows part of the breadth first search tree from C at one of the steps of the $i = 1$ phase. Suppose v (circled) is the center chosen greedily from $V_{\geq 2}$. Adding it to C "pulls" it's out-tree (encircled by the dotted oval) to the left in the diagram. The shaded area is the "front," i.e. the irritating strip of nodes that cause our lower bound on $y(\Gamma^+(v) \cap A)$ from (1) to be less than one. By the end of phase 0, we had chosen $\frac{3}{4}p$ centers, reducing $y(A)$ to $\exp(-\frac{3}{4}) \cdot p \approx 0.472p$. We then moved the front one strip to the right, so we know at most $0.472p$ fractional centers are necessary to cover the current A throughout phase 1.

ExpandingFront (G, A, C, y, p)

---

for $i = 0, 1, 2, \ldots$ do ("phase" i)
    for $j = 1, 2, \ldots \lceil \frac{3}{4} \frac{p}{2^i} \rceil$ do (j is just a counter)
        if $y(A) < 1$ then STOP, output C
        else
            find $v \in V_{\geq i+1}$ that maximizes $y(\Gamma^+(v) \cap A)$
            $C \leftarrow C + v$
            $A \leftarrow A \backslash \Gamma^+_{i+1}(v)$ (equivalently, $A \leftarrow V_{\geq i+2}$)
    $A \leftarrow V_{\geq i+3}$ (expand the front)
    (now y_A is y projected onto $V_{\geq i+2}$)

Fig. 4. Formal description of ExpandingFront.

4.2 Analysis of ExpandingFront

At the beginning of each inner loop we have $A = V_{\geq i+2}$, so all inactive nodes are $(i+1)$-covered by C. If $y(A) < 1$, then since every $v \in A$ is covered by one unit of fractional centers, some node in $\Gamma^+_{i+1}(C)$ must cover v. Hence C $(i+2)$-covers all of V. So to prove Theorem 3 for ExpandingFront it suffices to prove Lemma 6 below, and to show that we do not choose too many centers (Lemma 10).

Definition 5 *Define the* tower function $T(n)$ *by* $T(0) = 1$ *and* $T(n+1) = e^{T(n)}$ *for* $n \geq 0$.

Lemma 6 ExpandingFront *terminates after at most* $\log^* p + O(1)$ *phases, where we call each outer loop a phase.*

Proof. Claims 7, 8 and 9 below establish that at the beginning of phase i we have $y(A) \leq \frac{p}{a_i}$ where a_i grows like a tower function. This establishes Lemma 6, since we stop once $a_i > p$. $\qquad\square$

In phase i, since $A = V_{\geq i+2}$, we take our fractional cover y_A to be the restriction of y to the nodes $V_{\geq i+1}$, and we work on decreasing $y(A)$, which we can upper bound using Claim 7 below. Then by expanding the front at the end of the phase, our old $y(A)$ becomes our new $y_A(V)$.

Claim 7 *If at the beginning of a phase we have* $y_A(V) = a$ *and we choose* b *centers in the phase, then we reduce* $y(A)$ *at least by a factor of* $e^{-\frac{b}{a}}$.

Proof. Since y_A covers A, each chosen center v satisfies $y(\Gamma^+(v) \cap A) \geq \frac{y(A)}{a}$ by Lemma 4. In phase i, choosing v as a center actually reduces $y(A)$ by $y(\Gamma^+_{i+1}(v) \cap A)$, but we know how to account only for the reduction due to $y(\Gamma^+(v) \cap A)$. Thus, each of the b new centers reduces $y(A)$ by a factor of $(1 - \frac{1}{a}) \leq \exp(-\frac{1}{a})$. $\qquad\square$

Claim 8 *Define $\{a_i\}$ by $a_0 = 1$ and the recursion*

$$a_{i+1} = a_i \exp\left(\frac{3}{4}\frac{a_i}{2^i}\right) \quad \text{for } i \geq 0. \tag{2}$$

Then $\frac{p}{a_i}$ is an upper bound on $y(V_{\geq i+1})$ at the beginning of phase i.

Proof. At the beginning of phase i, $y(A) = y(V_{\geq i+2}) \leq y(V_{\geq i+1}) \leq \frac{p}{a_i}$. By Claim 7, at the end of the phase we have $y(A) \leq \frac{p}{a_i} \exp(-\frac{3}{4}\frac{p}{2^i}/\frac{p}{a_i}) = \frac{p}{a_{i+1}}$. We then expand the front, and the lemma follows. □

Table 1 illustrates how our bounds progress for the first several phases.

Table 1. This table illustrates the progress of EXPANDINGFRONT. It indicates the number of centers used at each stage, the ratio between this and our upper bound on the size of our present cover for A (as used in Claim 7), our new upper bound on $y(A)$ at the end of the stage, and the a_i from (2).

stage i	centers used	upper bound on $y(\Gamma^-(A))$	ratio	cut $y(A)$ to	a_i
0	$3p/4$	p	0.75	$0.472p$	1
1	$3p/8$	$0.472p$	0.794	$0.214p$	2.12
2	$3p/16$	$0.214p$	0.878	$0.0888p$	4.68
3	$3p/32$	$0.0888p$	1.06	$0.0309p$	11.27
4	$3p/64$	$0.0309p$	1.52	$0.00676p$	32.40
5	$3p/128$	$0.00676p$	3.47	$0.000211p$	147.96
6	$3p/256$	$0.000211p$	55.6	$1.51 \cdot 10^{-28}p$	4744.26
7	$3p/512$	$1.51 \cdot 10^{-28}p$	$3.89 \cdot 10^{25}$	tiny	$6.63 \cdot 10^{27}$

Claim 9 *For $i \geq 6$, $a_i \geq \frac{4}{3}2^i T(i-4)$.*

Proof. From Table 1 we have $a_6 \approx 4744$, while $\frac{4}{3}2^6 T(2) \approx 1293$. The theorem follows by induction, using the recurrence (2). □

Lemma 10 EXPANDINGFRONT *augments C by at most $\frac{3}{2}p$ centers.*

Proof. Suppose there are l phases prior to termination. By Lemma 6, $l \leq \log^* p + O(1)$. Since in phase i we choose no more than $\frac{3}{4}\frac{p}{2^i} + 1$ centers, overall we choose at most $\frac{3}{2}p + l - \frac{3}{2}\frac{p}{2^l} \leq \frac{3}{2}p$ centers. □

5   Augment Phase: RecursiveCover

We now present RECURSIVECOVER, a second way to implement the AUGMENT phase (see Figure 5). RECURSIVECOVER uses a subroutine GREEDYSETCOVER

RecursiveCover (G, A, C, y, p)

(find an initial 2-cover using $p \ln p$ centers)
$S_0' \leftarrow \emptyset$, $A' \leftarrow A$
while $y(A') \geq 1$ do
      find $v \in V$ that maximizes $y(\Gamma^+(v) \cap A')$
      $S_0' \leftarrow S_0' + v$
      $A' \leftarrow A' \backslash \Gamma^+(v)$
$S_0 \leftarrow S_0' \cap A$
(recursively cover the centers to reduce their number but increase their covering radius)
for $i = 0, 1, 2, \ldots$ do
    if $|S_i| \leq \frac{3}{2}p$ then STOP, output $C \cup S_i$
    else
        $S_{i+1}' \leftarrow$ GREEDYSETCOVER (G, S_i)
        $S_{i+1} \leftarrow S_{i+1}' \cap A$

Fig. 5. Formal description of RECURSIVECOVER.

(G, S), which is simply the greedy set cover algorithm on the instance where $S \subseteq V$ is the set of elements to be covered, and $\{\Gamma^+(v) : v \in V\}$ is the collection of sets.

Ours is an enhanced version of the RECURSIVECOVER algorithm given in [13]. The place where our version of RECURSIVECOVER differs from that given in [13] is in our choice of S_0'. They use GREEDYSETCOVER (G, A) to get S_0' so that S_0' covers A and $|S_0'| \leq pH(\frac{n}{p})$. (Here the usual function H is extended to fractional arguments as follows: for $i \in \mathbb{Z}^+$ and $f \in [0, 1)$, define $H(i+f) = 1 + \frac{1}{2} + \frac{1}{3} + \ldots + \frac{1}{i} + \frac{f}{i+1}$.) We eliminate the dependence on n by using the LP to guide our choice of a 2-cover for A. The greedy set cover algorithm is well-studied (see [3,8,11]), and the following theorem is known.

Theorem 11 *If there exists a fractional set cover for S using p centers, then* GREEDYSETCOVER (G, S) *outputs a cover of size at most* $pH(\frac{|S|}{p})$.

This follows easily from Lemma 4 with $z \equiv 1$ and induction on $|S|$. Now the following three lemmas yield Theorem 3 for RECURSIVECOVER.

Lemma 12 $|S_0| \leq p \ln p$.

Proof. Since $y(V) = p$ and y is a fractional cover of A, Lemma 4 shows that each new node added to S_0' reduces $y(A')$ by a factor of at least $(1 - \frac{1}{p})$. Thus, within $p \ln p$ iterations we drive it below 1.   □

Lemma 13 *For $i = 0, 1, 2, \ldots$ we have* $\Gamma_{i+2}^+(S_i) \cap \Gamma_{i+3}^+(C) \supseteq A$.

Proof. At the end of the while loop, every node in $A \backslash A'$ is covered by S_0', hence either covered by S_0 or 2-covered by C. Since $y(A') < 1$, every node in A' is covered either by some node in $V \backslash A$ (hence 2-covered by C) or by some node in $A \backslash A'$ (hence lies in $\Gamma_2^+(S_0) \cup \Gamma_3^+(C)$). This establishes the claim for $i = 0$. The inductive step follows since S'_{i+1} covers S_i, so every node in S_i is either covered by S_{i+1} or 2-covered by C. ☐

Lemma 14 *For $i \geq 0$, $|S_i| \leq pH^{(i)}(\ln p)$.*

Proof. The case $i = 0$ is Lemma 12. We use Theorem 11 to induct. ☐

Since $H(x) \leq 1 + \ln x$, we see from Lemma 14 that $|S_i|$ drops below $\frac{3}{2}$ when $i = \log^* p + O(1)$. This establishes Theorem 3 for RECURSIVECOVER.

6   The Reduce Phase

Our REDUCE phase (see Figure 6) is identical to that given in [13], and our analysis is essentially the same, but we include it here for completeness. The paper [13] proves a slightly different version of Theorem 2, guaranteeing a 5-cover of A using only $\frac{1}{2}(k - |C|)$ centers. This allows AUGMENT to use $2y(V)$ new centers when run on the fifth power of G, instead of $\frac{3}{2}y(V)$ when run on the cube of G. The end effect of using our version is to decrease the constant inside the $O(\log^* k)$ guarantee of Theorem 1 from 5 to 3. Our Theorem 2 relies on Lemma 16 below, which was suggested to the author by Kleinberg [9].

Following [13], we define the notion of a center capturing vertex.

Definition 15 *In a directed graph G, v is a* center capturing vertex *(CCV) if $\Gamma^-(v) \subseteq \Gamma^+(v)$.*

Since every node has some optimal center covering it, each CCV v covers at least one center, so it 2-covers what that center covered. Thus, adding v to our set of centers and removing $\Gamma_2^+(v)$ from the active set A decreases by one the number of centers necessary to cover A. Note that in the special case where distances are symmetric, every node is a CCV, so REDUCE returns $A = \emptyset$ and C is a 2-cover. This is exactly the 2-approximation given in [6].

The following lemma and intermediate theorem establish Theorem 2.

Lemma 16 *If H is a directed graph with nodes U and no self-loops, then there exists $S \subseteq U$ such that $|S| \leq \frac{2}{3}|U|$ and S covers all nodes in U with positive in-degree.*

Proof. By induction on $|U|$. The claim is vacuously true if H has no edges. Otherwise let v be any node with positive out-degree, and let T be the directed breadth first search tree rooted at v. Let S_{even} and S_{odd} respectively be the non-root nodes at the even and odd levels of T. Let S_1 be S_{even} or S_{odd}, whichever is smaller. Then $S_1 + v$ covers T, and $|S_1| \leq \frac{2}{3}|T|$, the worst case occurring when T is a directed path on three nodes. Since removing T from H does not alter the in-degrees of the remaining nodes, we may inductively choose S_2 to cover the positive in-degree nodes of $U \backslash T$, with $|S_2| \leq \frac{2}{3}|U \backslash T|$, and take $S = S_1 \cup S_2$. ☐

Reduce (G)

$A \leftarrow V,\ C \leftarrow \emptyset$
while there is a CCV $v \in A$ do
$\qquad C \leftarrow C + v$
$\qquad A \leftarrow A \backslash \Gamma_2^+(v)$
$A \leftarrow A \backslash \Gamma_4^+(C)$
output (C, A)

Fig. 6. Formal description of REDUCE.

Theorem 17 *Let $G = (V, E)$ be a directed graph, $C \subseteq V$, and $A = V \backslash \Gamma_2^+(C)$. Suppose A has no CCV's and there exist $\bar{k}$ centers that cover A. Then there exists a set S of $\frac{2}{3}\bar{k}$ centers that 3-cover $V \backslash \Gamma_4^+(C)$.*

Proof. Let U denote the set of $\bar{k}$ centers that cover A. We call $v \in U$ a *near* center if $v \in \Gamma_3^+(C)$, and a *far* center otherwise. Then C 4-covers all of the nodes in A covered by near centers. Thus it suffices to choose S to 2-cover the far centers, so S will 3-cover the nodes they cover.

Let x be any far center. Since A contains no CCV's, there exists y such that y covers x but x does not cover y. Since $x \notin \Gamma_3^+(C)$ we have $y \notin \Gamma_2^+(C)$, so $y \in A$. Thus there exists $z \in U$ covering y, and $z \neq x$. So z 2-covers x. If we define an auxiliary graph H on the centers U with an edge from u to v if and only if u 2-covers v in G, we see that all far centers have positive in-degree. By Lemma 16, there exists $S \subseteq U$ with $|S| \leq \frac{2}{3}\bar{k}$ such that S covers the far centers in H, so S 2-covers them in G. □

Acknowledgments. The author thanks Éva Tardos for suggesting this problem and for reading many drafts of this paper, Anupam Gupta for helpful comments on the presentation, Jon Kleinberg for suggesting Lemma 16, R. Ravi for a useful conversation, and Ryan O'Donnell for sharing his thoughts on the LP integrality gap.

References

1. A. Archer, "Inapproximability of the asymmetric facility location and k-median problems," unpublished manuscript.
2. M. Charikar and S. Guha, "Improved combinatorial algorithms for the facility location and k-median problems," *Proc. 40th Annual IEEE Symp. on Foundations of Computer Science*, (1999) 378-388.
3. V. Chvatal, "A greedy heuristic for the set covering problem," *Math. Oper. Res.*, 4 (1979) 233-235.
4. M. Dyer and A. Frieze, "A simple heuristic for the p-centre problem," *Oper. Res. Lett.*, 3 (1985) 285-288.

5. D. S. Hochbaum, "Heuristics for the fixed cost median problem," *Math. Prog.*, 22 (1982) 148-162.
6. D. S. Hochbaum and D. B. Shmoys, "A best possible heuristic for the k-center problem," *Math. Oper. Res.* 10 (1985) 180-184.
7. D. S. Hochbaum and D. B. Shmoys, "A unified approach to approximation algorithms for bottleneck problems," *J. ACM*, 33 (1986) 533-550.
8. D. S. Johnson, "Approximation algorithms for combinatorial problems," *J. Computer and System Sciences*, 9 (1974) 256-278.
9. J. Kleinberg, personal communication, November 1999.
10. J.-H. Lin and J. S. Vitter, "ϵ-approximations with minimum packing constraint violation", *Proc. 24th Annual ACM Symp. on Theory of Computing*, (1992) 771-782.
11. L. Lovász, "On the ratio of optimal integral and fractional covers," *Discrete Math.*, 13 (1975) 383-390.
12. R. O'Donnell, personal communication, June 2000.
13. R. Panigrahy and S. Vishwanathan, "An $O(\log^* n)$ approximation algorithm for the asymmetric p-center problem," J. Algorithms, 27 (1998) 259-268.
14. S. Vishwanathan, "An $O(\log^* n)$ approximation algorithm for the asymmetric p-center problem," *Proc. 7th Annual ACM-SIAM Symp. on Discrete Algorithms*, (1996) 1-5.

Strongly Polynomial Algorithms for the Unsplittable Flow Problem

Yossi Azar[1] and Oded Regev[2]

[1] Dept. of Computer Science, Tel-Aviv University, Tel-Aviv, 69978, Israel.
azar@math.tau.ac.il ***
[2] Dept. of Computer Science, Tel-Aviv University, Tel-Aviv, 69978, Israel.
odedr@math.tau.ac.il

Abstract. We provide the first strongly polynomial algorithms with the best approximation ratio for all three variants of the unsplittable flow problem (UFP). In this problem we are given a (possibly directed) capacitated graph with n vertices and m edges, and a set of terminal pairs each with its own demand and profit. The objective is to connect a subset of the terminal pairs each by a single flow path as to maximize the total profit of the satisfied terminal pairs subject to the capacity constraints. Classical UFP, in which demands must be lower than edge capacities, is known to have an $O(\sqrt{m})$ approximation algorithm. We provide the same result with a strongly polynomial combinatorial algorithm. The extended UFP case is when some demands might be higher than edge capacities. For that case we both improve the current best approximation ratio and use strongly polynomial algorithms. We also use a lower bound to show that the extended case is provably harder than the classical case. The last variant is the bounded UFP where demands are at most $\frac{1}{K}$ of the minimum edge capacity. Using strongly polynomial algorithms here as well, we improve the currently best known algorithms. Specifically, for $K = 2$ our results are better than the lower bound for classical UFP thereby separating the two problems.

1 Introduction

We consider the unsplittable flow problem (UFP). We are given a directed or undirected graph $G = (V, E)$, $|V| = n$, $|E| = m$, a capacity function u on its edges and a set of l terminal pairs of vertices (s_j, t_j) with a demand d_j and profit r_j. A feasible solution is a subset S of the terminal pairs and a single flow path for each such pair such that the capacity constraints are fully met. The objective is to maximize the total profit of the satisfied terminal pairs. The well-known problem of maximum edge disjoint path, denoted EDP, is the special case where all demands, profits and capacities are equal to 1 (see [5]).

The EDP (and hence the UFP) is one of Karp's original NP-complete problems [6]. An $O(\sqrt{m})$ approximation algorithm is known for EDP [7] (for additional positive results see [12,13]). Most of the results for UFP deal with the

---

*** Research supported in part by the Israel Science Foundation and by the US-Israel Binational Science Foundation (BSF).

K. Aardal, B. Gerards (Eds.): IPCO 2001, LNCS 2081, pp. 15–29, 2001.

classical case where $d_{max} \leq u_{min}$ (the maximal demand is at most the minimal capacity). The most popular approach seems to be LP rounding [2,10,14] with the best approximation ratio being $O(\sqrt{m})$ [2]. A matching lower bound of $\Omega(m^{1/2-\epsilon})$ for any $\epsilon > 0$ is shown in [5] for directed graphs. Both before and after the $O(\sqrt{m})$ result, there were attempts to achieve the same approximation ratio using combinatorial methods. Up to now however, these were found only for restricted versions of the problem [5,10] and were not optimal. Our combinatorial algorithm not only achieves the $O(\sqrt{m})$ result for classical UFP but is also the first strongly polynomial algorithm for that problem.

The extended UFP is the case where both demands and capacities are arbitrary (specifically, some demands might be higher than some capacities). Due to its complexity, not many results addressed it. The first to attack the problem is a recent attempt by Guruswami et al. [5]. We improve the best approximation ratio through a strongly polynomial algorithm. By proving a lower bound for the extended UFP over directed graphs we infer that this case is really harder than the classical UFP. Specifically, for large demands we show that unless $P = NP$ it is impossible to approximate extended UFP better than $O(m^{1-\epsilon})$ for any $\epsilon > 0$.

Another interesting case is the bounded UFP case where $d_{max} \leq \frac{1}{K}u_{min}$ (denoted K-bounded UFP). It is a special case of classical UFP but better approximation ratios can be achieved. As a special case, it contains the half-disjoint paths problem where all the demands and profits are equal to $\frac{1}{2}$ and edge capacities are all 1 [8]. For $K \geq \log n$, a constant approximation is shown in [11] by using randomized rounding. For $K < \log n$, previous algorithms achieved an approximation ratio of $O(Kn^{\frac{1}{K-1}})$ ([2,14] by using LP rounding and [3,9] based on [1]). We improve the result to a strongly polynomial $O(Kn^{\frac{1}{K}})$ approximation algorithm which, as a special case, is a $O(\sqrt{n})$ approximation algorithm for the half disjoint case. Since this ratio is better than the lower bound for classical UFP, we achieve a separation between classical UFP and bounded UFP. The improvement is achieved by splitting the requests into a low demand set and a high demand set. The sets are treated separately by algorithms similar to those of [1] where in the case of high demands the algorithm has to be slightly modified. We would like to note that in our approximation ratios involving n, we can replace n with D where D is an upper bound on the longest path ever used (which is obviously at most n).

As a by-product of our methods, we provide online algorithms for UFP. Here, the network is known but requests arrive one by one and a decision has to be made without knowing which requests follow. We show on-line algorithms whose competitive ratio is somewhat worse than that of the off-line algorithms. We also show that one of our algorithms is optimal in the on-line setting by slightly improving a lower bound of [1].

We conclude this introduction with a short summary of the main results in this paper. We denote by d_{max} the maximum demand and by u_{min} the minimum edge capacity.

- Classical UFP ($d_{max} \leq u_{min}$) - Strongly polynomial $O(\sqrt{m})$ approximation algorithm.
- Extended UFP (arbitrary d_{max}, u_{min}) - Strongly polynomial $O(\sqrt{m}\log(2+\frac{d_{max}}{u_{min}}))$ approximation algorithm; A lower bound of $\Omega(m^{1-\epsilon})$ and of $\Omega(m^{\frac{1}{2}-\epsilon}\sqrt{\log(2+\frac{d_{max}}{u_{min}})})$ for directed graphs.
- Bounded UFP ($d_{max} \leq \frac{1}{K}u_{min}$) - Strongly polynomial $O(Kn^{\frac{1}{K}})$ approximation algorithm.

2   Notation

Let $G = (V, E)$, $|V| = n$, $|E| = m$, be a (possibly directed) graph and a capacity function $u : E \rightarrow \mathbf{R}^+$. An input request is a quadruple (s_j, t_j, d_j, r_j) where $\{s_j, t_j\}$ is the source-sink terminal pair, d_j is the demand and r_j is the profit. The input is a set of the above quadruples for $j \in T = \{1, ..., l\}$. Let D be a bound on the length of any routing path; note that D is at most n.

We denote by u_{min} (u_{max}) the minimum (maximum) edge capacity in the graph. Similarly, we define d_{min}, d_{max}, r_{min} and r_{max} to be the minimum/maximum demand/profit among all input requests. We define two functions on sets of requests, $S \subseteq T$:

$$r(S) = \sum_{j \in S} r_j \quad d(S) = \sum_{j \in S} d_j$$

A feasible solution is a subset $\mathcal{P} \subseteq T$ and a route P_j from s_j to t_j for each $j \in \mathcal{P}$ subject to the capacity constraints, i.e., the total demand routed through an edge is bounded by the its capacity. Some of our algorithms order the requests so we will usually denote by $L_j(e)$ the relative load of edge e after routing request j, that is, the sum of demands routed through e divided by $u(e)$. Without loss of generality, we assume that any single request can be routed. That is possible since we can just ignore unroutable requests. Note that this is not the $d_{max} \leq u_{min}$ assumption made in classical UFP.

Before describing the various algorithms, we begin with a simple useful lemma:

Lemma 1. *Given a sequence $\{a_1, ..., a_n\}$, a non-increasing non-negative sequence $\{b_1, ..., b_n\}$ and two sets $X, Y \subseteq \{1, ..., n\}$, let $X^i = X \cap \{1, ..., i\}$ and $Y^i = Y \cap \{1, ..., i\}$. If for every $1 \leq i \leq n$*

$$\sum_{j \in X^i} a_j > \alpha \sum_{j \in Y^i} a_j$$

then

$$\sum_{j \in X} a_j b_j > \alpha \sum_{j \in Y} a_j b_j$$

Proof. Denote $b_{n+1} = 0$. Since $b_j - b_{j+1}$ is non-negative,

$$\sum_{j \in X} a_j b_j = \sum_{i=1,\dots,n} (b_i - b_{i+1}) \sum_{j \in X^i} a_j$$

$$> \alpha \sum_{i=1,\dots,n} (b_i - b_{i+1}) \sum_{j \in Y^i} a_j = \alpha \sum_{j \in Y} a_j b_j$$

3   Algorithms for *UFP*

3.1   Algorithm for Classical *UFP*

In this section we show a simple algorithm for classical *UFP* (the case in which $d_{max} \leq u_{min}$). The algorithm's approximation ratio is the same as the best currently known algorithm. Later, we show that unlike previous algorithms, this algorithm can be easily made strongly polynomial and that it can even be used in the extended case.

We split the set of requests T into two disjoint sets. The first, T_1, consists of requests for which $d_j \leq u_{min}/2$. The rest of the requests are in T_2. For each request j and a given path P from s_j to t_j define

$$F(j, P) = \frac{r_j}{d_j \sum_{e \in P} \frac{1}{u(e)}},$$

a measure of the profit gained relative to the added network load.

Given a set of requests, we use simple bounds on the values of F. The lower bound, denoted α_{min}, is defined as $\frac{r_{min}}{n}$ and is indeed a lower bound on $F(j, P)$ since P cannot be longer than n edges and the capacity of its edges must be at least d_j. The upper bound, denoted α_{max}, is defined as $\frac{r_{max} u_{max}}{d_{min}}$ and is clearly an upper bound on $F(j, P)$.

PROUTE
run $Routine_2(T_1)$ and $Routine_2(T_2)$ and choose the better solution

$Routine_2(S)$:
foreach k from $\lfloor \log \alpha_{min} \rfloor$ to $\lceil \log \alpha_{max} \rceil$
run $Routine_1(2^k, S)$ and choose the best solution

$Routine_1(\alpha, S)$:
sort the requests in S according to a non-increasing order of r_j/d_j
foreach $j \in S$ in the above order
if $\exists$ path P from s_j to t_j s.t. $F(j, P) > \alpha$ and $\forall e \in P, L_{j-1}(e) + \frac{d_j}{u(e)} \leq 1$
then <u>route</u> the request on P and for $e \in P$ set $L_j(e) = L_{j-1}(e) + \frac{d_j}{u(e)}$
else <u>reject</u> the request

Theorem 1. *Algorithm PROUTE is an $O(\sqrt{m})$ approximation algorithm for classical UFP.*

Proof. First, we look at the running time of the algorithm. The number of iterations done in $Routine_2$ is:

$$\log \frac{\alpha_{max}}{\alpha_{min}} = \log(n \frac{r_{max}}{r_{min}} \frac{u_{max}}{d_{min}})$$

which is polynomial. $Routine_1$ looks for a non overflowing path P with $F(j, P) > \alpha$. The latter condition is equivalent to $\sum_{e \in P} \frac{1}{u(e)} < \frac{r_j}{d_j \alpha}$ and thus a shortest path algorithm can be used.

Consider an optimal solution routing requests in $\mathcal{Q} \subseteq T$. For each $j \in \mathcal{Q}$ let Q_j be the route chosen for j in the optimal solution. The total profit of either $\mathcal{Q} \cap T_1$ or $\mathcal{Q} \cap T_2$ is at least $\frac{r(\mathcal{Q})}{2}$. Denote that set by $\mathcal{Q}'$ and its index by $i' \in \{1, 2\}$, that is, $\mathcal{Q}' = \mathcal{Q} \cap T_{i'}$. Now consider the values given to α in $Routine_2$ and let $\alpha' = 2^{k'}$ be the highest such that $r(\{j \in \mathcal{Q}'|F(j, Q_j) > \alpha'\}) \geq r(\mathcal{Q})/4$. It is clear that such an α' exists. From now on we limit ourselves to $Routine_1(\alpha', i')$ and show that a good routing is obtained by it. Denote by $\mathcal{P}$ the set of requests routed by $Routine_1(\alpha', i')$ and for $j \in \mathcal{P}$ denote by P_j the path chosen for it.

Let $\mathcal{Q}'_{high} = \{j \in \mathcal{Q}'|F(j, Q_j) > \alpha'\}$ and $\mathcal{Q}'_{low} = \{j \in \mathcal{Q}'|F(j, Q_j) \leq 2\alpha'\}$ be sets of higher and lower 'quality' routes in $\mathcal{Q}'$. Note that the sets are not disjoint and that the total profit in each of them is at least $\frac{r(\mathcal{Q})}{4}$ by the choice of α'. From the definition of F,

$$r(\mathcal{Q}'_{low}) = \sum_{j \in \mathcal{Q}'_{low}} F(j, Q_j) \sum_{e \in Q_j} \frac{d_j}{u(e)} \leq 2\alpha' \sum_{j \in \mathcal{Q}'_{low}} \sum_{e \in Q_j} \frac{d_j}{u(e)}$$

$$\leq 2\alpha' \sum_{j \in \mathcal{Q}} \sum_{e \in Q_j} \frac{d_j}{u(e)}$$

$$= 2\alpha' \sum_{e} \sum_{j \in \mathcal{Q}|e \in Q_j} \frac{d_j}{u(e)}$$

$$\leq 2\alpha' \sum_{e} 1 = 2m\alpha'$$

where the last inequality is true since an optimal solution cannot overflow an edge. Therefore,

$$r(\mathcal{Q}) \leq 8m\alpha'.$$

Now let $E_{heavy} = \{e \in E | L_l(e) \geq \frac{1}{4}\}$ be a set of the heavy edges after the completion of $Routine_1(\alpha', i')$. We consider two cases. The first is when $|E_{heavy}| \geq \sqrt{m}$. According to the description of the algorithm, $F(j, P_j) > \alpha'$ for every $j \in \mathcal{P}$. Therefore,

$$r(\mathcal{P}) = \sum_{j \in \mathcal{P}} F(j, P_j) \sum_{e \in P_j} \frac{d_j}{u(e)}$$

$$\geq \alpha' \sum_{j \in \mathcal{P}} \sum_{e \in P_j} \frac{d_j}{u(e)}$$

$$= \alpha' \sum_{e} \sum_{j | e \in P_j} \frac{d_j}{u(e)}$$

$$= \alpha' \sum_{e} L_l(e) \geq \frac{1}{4}\sqrt{m}\alpha'$$

where the last inequality follows from the assumption that more than $\sqrt{m}$ edges are loaded more than fourth their capacity. By combining the two inequalities we get:

$$\frac{r(\mathcal{Q})}{r(\mathcal{P})} \leq 32\sqrt{m} = O(\sqrt{m})$$

which completes the first case.

From now on we consider the second case where $|E_{heavy}| < \sqrt{m}$. Status: RO

Denote by $R = Q'_{high} \setminus P$. We compare the profit given by our algorithm to that found in R by using Lemma 1. Since $\frac{r_j}{d_j}$ is a non increasing sequence, it is enough to bound the total demand routed in prefixes of the two sets. For that we use the notation $R^k = R \cap \{1, ..., k\}$ and $\mathcal{P}^k = \mathcal{P} \cap \{1, ..., k\}$ for $k = 1, ..., l$. For each request $j \in R^k$ the algorithm cannot find any appropriate path. In particular, the path Q_j is not chosen. Since $j \in Q'_{high}$, $F(j, Q_j) > \alpha'$ and therefore the reason the path is not chosen is that it overflows one of the edges. Denote that edge by e_j and by $E^k = \{e_j | j \in R^k\}$.

Lemma 2. $E^k \subseteq E_{heavy}$

Proof. Let $e_j \in E^k$ be an edge with $j \in R^k$, a request corresponding to it. We claim that when the algorithm fails finding a path for j, $L_j(e_j) \geq \frac{1}{4}$. For the case $i' = 1$, the claim is obvious since the demand $d_j \leq u_{min}/2$ and in particular, $d_j \leq u(e_j)/2$. Thus, the load of e_j must be higher than $u(e_j)/2$ for the path Q_j to overflow it. For the case $i' = 2$, we know that $u_{min}/2 < d_j \leq u_{min}$. In case $u(e_j) > 2u_{min}$, the only way to overflow it with demands of size at most $d_{max} \leq u_{min}$ is when the edge is loaded at least $u(e_j) - u_{min} \geq u(e_j)/2$. Otherwise, $u(e_j) \leq 2u_{min}$ and since $d_j \leq u_{min} \leq u(e)$ we know that the edge cannot be empty. Since we only route requests from T_2 the edge's load must be at least $u_{min}/2 \geq u(e_j)/4$.

Since each request in R^k is routed through an edge of E^k in the optimal solution, $d(R^k) \leq \sum_{e \in E^k} u(e)$. The highest capacity edge $f \in E^k$ is loaded more than fourth its capacity since it is in E_{heavy} and therefore $d(\mathcal{P}^K) \geq \frac{u(f)}{4}$. By Lemma 2, $|E^k| \leq |E_{heavy}| < \sqrt{m}$ and hence,

$$d(R^k) < \sqrt{m} \cdot u(f) \leq 4\sqrt{m} \cdot d(\mathcal{P}^k).$$

We use Lemma 1 by combining the inequality above on the ratio of demands and the nonincreasing sequence $\frac{r_j}{d_j}$. This yields

$$\sum_{j \in R} \frac{r_j}{d_j} d_j \leq 4\sqrt{m} \sum_{j \in \mathcal{P}} \frac{r_j}{d_j} d_j,$$

or,

$$r(R) \leq 4\sqrt{m} \cdot r(\mathcal{P}).$$

Since $\mathcal{Q}'_{high} = R \cup \mathcal{P}$,

$$r(\mathcal{Q}'_{high}) = r(R) + r(\mathcal{P}) \leq (1 + 4\sqrt{m}) r(\mathcal{P}).$$

Recall that $r(\mathcal{Q}'_{high}) \geq r(\mathcal{Q})/4$ and therefore

$$\frac{r(\mathcal{Q})}{r(\mathcal{P})} \leq 4 + 16\sqrt{m} = O(\sqrt{m})$$

3.2   Strongly Polynomial Algorithm

$Routine_1$ is strongly polynomial. $Routine_2$ however calls it $\log \frac{\alpha_{max}}{\alpha_{min}}$ times. Therefore, it is polynomial but still not strongly polynomial. We add a pre-processing step whose purpose is to bound the ratio $\frac{\alpha_{max}}{\alpha_{min}}$. Recall that l denotes the number of requests.

$SPROUTE(T)$:
   run $Routine_3(T_1)$ and $Routine_3(T_2)$ and choose the better solution

$Routine_3(S)$:
   For each edge such that $u(e) > l \cdot d_{max}$ set $u(e)$ to be $l \cdot d_{max}$.
   Throw away requests whose profit is below $\frac{r_{max}}{l}$.
   Take the better out of the following two solutions:
      Route all requests in $S_{tiny} = \{j \in S | d_j \leq \frac{u_{min}}{l}\}$ on any simple path.
      Run $Routine_2(S \setminus S_{tiny})$.

Theorem 2. *Algorithm $SPROUTE$ is a strongly polynomial $O(\sqrt{m})$ approximation algorithm for classical UFP.*

Proof. Consider an optimal solution routing requests in $\mathcal{Q} \subseteq S$. Since the demand of a single request is at most d_{max}, the total demand routed through a given edge is at most $l \cdot d_{max}$. Therefore, $\mathcal{Q}$ is still routable after the first pre-processing phase. The total profit of requests whose profit is lower than $\frac{r_{max}}{l}$ is r_{max}. In case $r(\mathcal{Q}) > 2r_{max}$, removing these requests still leaves the set $\mathcal{Q}'$ whose total profit is at least $r(\mathcal{Q}) - r_{max} \geq \frac{r(\mathcal{Q})}{2}$. Otherwise, we take $\mathcal{Q}'$ to be the set containing the request of highest profit. Then, $r(\mathcal{Q}')$ is $r_{max} \geq \frac{r(\mathcal{Q})}{2}$. All in all, after the two preprocessing phases we are left with an UFP instance for which there is a solution $\mathcal{Q}'$ whose profit is at least $\frac{r(\mathcal{Q})}{2}$.

Assume that the total profit in $\mathcal{Q}' \cap S_{tiny}$ is at least $\frac{r(\mathcal{Q})}{4}$. Since the requests in S_{tiny} have a demand of at most $\frac{u_{min}}{l}$ and there are at most l of them, they can all be routed on simple paths and the profit obtained is at least $\frac{r(\mathcal{Q})}{4}$. Otherwise, the profit in $\mathcal{Q}' \setminus S_{tiny}$ is at least $\frac{r(\mathcal{Q})}{4}$ and since algorithm $PROUTE$ is an $O(\sqrt{m})$ approximation algorithm, the profit we obtain is also within $O(\sqrt{m})$ of $r(\mathcal{Q})$.

The preprocessing phases by themselves are obviously strongly polynomial. Recall that the number of iterations performed by $Routine_2$ is $\log(n \frac{r_{max}}{r_{min}} \frac{u_{max}}{d_{min}})$. The ratio of profits is at most l by the second preprocessing phase. The first preprocessing phase limits u_{max} to $k \cdot d_{max}$. So, the number of iterations is at most $\log(n l^2 \frac{d_{max}}{d_{min}})$. In case $S = T_1$, $d_{max} \leq \frac{u_{min}}{2}$ and $d_{min} \geq \frac{u_{min}}{l}$ since tiny requests are removed. For $S = T_2$, $d_{max} \leq u_{min}$ and $d_{min} \geq u_{min}/2$. We end up with at most $O(\log n + \log l)$ iterations which is strongly polynomial.

3.3   Algorithm for Extended UFP

In this section we show that the algorithm can be used for the extended case in which demands can be higher than the lowest edge capacity.

Instead of using just two sets in $SPROUTE$, we define a partition of the set of requests T into $2 + \max\{\lceil \log d_{max}/u_{min} \rceil, 0\}$ disjoint sets. The first, T_1 consists of requests for which $d_j < u_{min}/2$. The set T_i for $i > 1$ is of requests for which $2^{i-3} u_{min} < d_j \leq 2^{i-2} u_{min}$. The algorithm is as follows:

$ESPROUTE(T)$:
for any $1 \leq i \leq 2 + \max\{\lceil \log d_{max}/u_{min} \rceil, 0\}$ such that T_i is not empty
    run $Routine_3(T_i)$ on the resulting graph
choose the best solution obtained

The proof of the following theorem is left to Appendix A.1:

Theorem 3. *Algorithm $ESPROUTE$ is a strongly polynomial $O(\sqrt{m} \log(2 + \frac{d_{max}}{u_{min}}))$ approximation algorithm for extended UFP.*

4   Algorithms for K-Bounded UFP

In the previous section we considered the classical UFP in which $d_{max} \leq u_{min}$. We also extended the discussion to extended UFP. In this section we show better algorithms for K-bounded UFP in which $d_{max} \leq \frac{1}{K} u_{min}$ where $K \geq 2$.

4.1   Algorithms for Bounded Demands

In this section we present two algorithm for bounded UFP. The first deals with the case in which the demands are in the range $[\frac{u_{min}}{K+1}, \frac{u_{min}}{K}]$. As a special case, it provides an $O(\sqrt{n})$ approximation algorithm for the half-disjoint paths problem where edge capacities are all the same and the demands are exactly half the edge capacity. The second is an algorithm for the K-bounded UFP where demands are only bounded by $\frac{u_{min}}{K}$ from above.

$EKROUTE(T)$:

$\mu \leftarrow 2D$

sort the requests in T according to a non-increasing order of r_j/d_j

foreach $j \in T$ in the above order

if $\exists$ a path P from s_j to t_j s.t.

$\sum_{e \in P}(\mu^{L_{j-1}(e)} - 1) < D$

then <u>route</u> the request on P and for $e \in P$ set $L_j(e) = L_{j-1}(e) + \frac{1}{\lfloor \frac{K \cdot u(e)}{u_{min}} \rfloor}$

else reject the request

$BKROUTE(T)$:

$\mu \leftarrow (2D)^{1+\frac{1}{K-1}}$

sort the requests in T according to a non-increasing order of r_j/d_j

foreach $j \in T_i$ in the above order

if $\exists$ a path P from s_j to t_j s.t.

$\sum_{e \in P}(\mu^{L_{j-1}(e)} - 1) < D$

then <u>route</u> the request on P and for $e \in P$ set $L_j(e) = L_{j-1}(e) + \frac{d_j}{u(e)}$

else <u>reject</u> the request

Note that algorithm $EKROUTE$ uses a slightly different definition of L. This 'virtual' relative load allows it to outperform $BKROUTE$ in instances where the demands are in the correct range.

The proof of the following theorem can be found in Appendix A.2:

Theorem 4. *Algorithm $EKROUTE$ is a strongly polynomial $O(K \cdot D^{\frac{1}{K}})$ approximation algorithm for UFP with demands in the range $[\frac{u_{min}}{K+1}, \frac{u_{min}}{K}]$. Algorithm $BKROUTE$ is a strongly polynomial $O(K \cdot D^{\frac{1}{K-1}})$ approximation algorithm for K-bounded UFP.*

4.2   A Combined Algorithm

In this section we combine the two algorithms presented in the previous section: the algorithm for demands in the range $[\frac{u_{min}}{K+1}, \frac{u_{min}}{K}]$ and the algorithm for the K-bounded UFP. The result is an algorithm for the K-bounded UFP with an approximation ratio of $O(K \cdot D^{\frac{1}{K}})$.

We define a partition of the set of requests T into two sets. The first, T_1, includes all the requests whose demand is at most $\frac{1}{K+1}$. The second, T_2, includes all the requests whose demand is more than $\frac{1}{K+1}$ and at most $\frac{1}{K}$.

$CKROUTE(T)$:

Take the best out of the following two possible solutions:

Route T_1 by using $BKROUTE$ and reject all requests in T_2

Route T_2 by using $EKROUTE$ and reject all requests in T_1

Theorem 5. *Algorithm $CKROUTE$ is a strongly polynomial $O(K \cdot D^{\frac{1}{K}})$ approximation algorithm for K-bounded UFP.*

Proof. Let Q denote an optimal solution in T. Since $BKROUTE$ is used with demands bounded by $\frac{1}{K+1}$ its approximation ratio is $O(KD^{\frac{1}{K}})$. The same approximation ratio is given by $EKROUTE$. Either T_1 or T_2 have an optimal solution whose profit is at least $\frac{r(Q)}{2}$ and therefore we obtain the claimed approximation ratio.

5   Lower Bounds

In this section we show that in cases where the demands are much larger than the minimum edge capacity UFP becomes very hard to approximate, namely, $\Omega(m^{1-\epsilon})$ for any $\epsilon > 0$. We also show how different demand values relate to the approximability of the problem. The lower bounds are for directed graphs only.

Theorem 6. *[4] The following problem is NPC:*

$2DIRPATH$:
INPUT: A directed graph $G = (V, E)$ and four nodes $x, y, z, w \in V$
QUESTION: Are there two edge disjoint directed paths,
   one from x to y and the other from z to w in G ?

Theorem 7. *For any $\epsilon > 0$, extended UFP cannot be approximated better than $\Omega(m^{1-\epsilon})$.*

Proof. For a given instance A of $2DIRPATH$ with $|A|$ edges and a small constant ϵ, we construct an instance of extended UFP composed of l copies of A, $A^1, A^2, ..., A^l$ where $l = |A|^{\lceil \frac{1}{\epsilon} \rceil}$. The instance A^i is composed of edges of capacity 2^{l-i}. A special node y^0 is added to the graph. Two edges are added for each A^i, (y^{i-1}, x^i) of capacity $2^{l-i} - 1$ and (y^{i-1}, z^i) of capacity 2^{l-i}. All l requests share y^0 as a source node. The sink of request $1 \leq i \leq l$ is w^i. The demand of request i is 2^{l-i} and its profit is 1. The above structure is shown in the following figure for the hypothetical case where $l = 4$. Each diamond indicates a copy of A with x, y, z, w being its left, right, top and bottom corners respectively. The number inside each diamond indicates the capacity of A's edges in this copy.

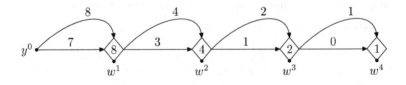

Fig. 1. The UFP instance for the case $l = 4$

We claim that for a given YES instance of $2DIRPATH$ the maximal profit gained from the extended UFP instance is l. We route request $1 \leq i \leq l$ through $[y^0, x^1, y^1, x^2, y^2, ..., y^{i-1}, z^i, w^i]$. Note that the path from x^j to y^j and from z^j to w^j is a path in A^j given by the YES instance.

For a NO instance, we claim that at most one request can be routed. That is because the path chosen for a request i ends at w^i. So, it must arrive from either z^i or x^i. The only edge entering x^i is of capacity $2^{l-i} - 1$ so z^i is the only option. The instance A^i is a NO instance of capacity 2^{l-i} through which a request of demand 2^{l-i} is routed form z^i to w^i. No other path can therefore be routed through A^i so requests $j > i$ are not routable. Since i is arbitrary, we conclude that at most one request can be routed through the extended UFP instance and its profit is 1.

The gap created is $l = |A|^{\frac{1}{\epsilon}}$ and the number of edges is $l \cdot (|A|+2) = O(l^{1+\epsilon})$. Hence, the gap is $\Omega(m^{\frac{1}{1+\epsilon}}) = \Omega(m^{1-\epsilon'})$ and since ϵ is arbitrary we complete the proof.

Theorem 8. *For any $\epsilon > 0$ extended UFP with any ratio $d_{max}/u_{min} \geq 2$ cannot be approximated better than $\Omega(m^{\frac{1}{2}-\epsilon}\sqrt{\lfloor \log(\frac{d_{max}}{u_{min}})\rfloor})$.*

Proof. Omitted.

6   Online Applications

Somewhat surprisingly, variants of the algorithms considered so far can be used in the online setting with slightly worse bounds. For simplicity, we present here an algorithm for the unweighted K-bounded UFP in which $r_j = d_j$ for every $j \in T$.

First note that for unweighted K-bounded UFP, both $EKROUTE$ and $BKROUTE$ can be used as online deterministic algorithms since sorting the requests becomes unnecessary. By splitting T into T_1 and T_2 as in $CKROUTE$ we can combine the two algorithms:

$ONLINECKROUTE(T)$:
   Choose one of the two routing methods below with equal probabilities:
   Route T_1 by using $BKROUTE$ and reject all requests in T_2
   Route T_2 by using $EKROUTE$ and reject all requests in T_1

Theorem 9. *Algorithm $ONLINECKROUTE$ is an $O(K \cdot D^{\frac{1}{K}})$ competitive online algorithm for unweighted K-bounded UFP.*

Proof. The expected value of the total accepted demand of the algorithm for any given input is the average between the total accepted demands given by the two routing methods. Since each method is $O(K \cdot D^{\frac{1}{K}})$ competitive on its part of the input, the theorem follows.

Theorem 10. *The competitive ratio of any deterministic on-line algorithm for the K-bounded UFP is at least $\Omega(K \cdot n^{\frac{1}{K}})$.*

Proof. Omitted.

7   Conclusion

Using combinatorial methods we showed algorithms for all three variants of the UFP problem. We improve previous results and provide the best approximations for UFP by using strongly polynomial algorithms. Due to their relative simplicity we believe that further analysis should lead to additional performance guarantees such as non linear bounds. Also, the algorithms might perform better over specific networks. It is interesting to note that no known lower bound exists for the half-disjoint case and we leave that as an open question.

References

[1] B. Awerbuch, Y. Azar, and S. Plotkin. Throughput-competitive online routing. In *34th IEEE Symposium on Foundations of Computer Science*, pages 32–40, 1993.

[2] A. Baveja and A. Srinivasan. Approximation algorithms for disjoint paths and related routing and packing problems. *To appear in Mathematics ofOperations Research*.

[3] A. Borodin and R. El-Yaniv. Online computation and competitive analysis (cambridge university press, 1998). *SIGACTN: SIGACT News (ACM Special Interest Group on Automata and Computability Theory)*, 29, 1998.

[4] S. Fortune, J. Hopcroft, and J. Wyllie. The directed homeomorphism problem. *Theoretical Computer Science*, 10:111–121, 1980.

[5] V. Guruswami, S. Khanna, R. Rajaraman, B. Shepherd, and M. Yannakakis. Near-optimal hardness results and approximation algorithms for edge-disjoint paths and related problems. *Proc. of STOC '99*, pages 19–28.

[6] R.M. Karp. *Reducibility among Combinatorial Problems, R.E. Miller and J.W. Thatcher (eds.), Complexity of Computer Computations*. Plenum Press, 1972.

[7] J. Kleinberg. *Approximation Algorithms for Disjoint Paths Problems*. PhD thesis, Massachusetts Institue of Technology, 1996.

[8] J. Kleinberg. Decision algorithms for unsplittable flow and the half-disjoint paths problem. In *Proceedings of the 30th Annual ACM Symposium on Theory of Computing (STOC '98)*, pages 530–539, New York, May 23–26 1998. ACM Press.

[9] J. Kleinberg and É. Tardos. Approximations for the disjoint paths problem in high-diameter planar networks. *Proc. of STOC '95*, pages 26–35.

[10] S. Kolliopoulos and C. Stein. Approximating disjoint-path problems using greedy algorithms and packing integer programs. In *IPCO: 6th Integer Programming and Combinatorial Optimization Conference*, 1998.

[11] P. Raghavan and C.D. Thompson. Provably good routing in graphs: Regular arrays. In *Proc. 17th ACM Symp. on Theory of Computing*, May 1985.

[12] N. Robertson and P. D. Seymour. An outline of a disjoint paths algorithm. In *Paths, Flows and VLSI Design, Algorithms and Combinatorics*, volume 9, pages 267–292, 1990.

[13] N. Robertson and P. D. Seymour. Graph minors. XIII. the disjoint paths problem. *JCTB: Journal of Combinatorial Theory, Series B*, 63, 1995.

[14] A. Srinivasan. Improved approximations for edge-disjoint paths, unsplittable flow, and related routing problems. In *Proc. 38th IEEE Symp. on Found. of Comp. Science*, pages 416–425.

A   Appendix

A.1   Proof of Theorem 3

Proof. The proofs of Theorem 1 and of Theorem 2 hold also for the extended case. The only part which has to be proved is Lemma 2. The following replaces the lemma:

Lemma 3. $E^k \subseteq E_{heavy}$

Proof. Let $e_j \in E^k$ be an edge with $j \in R^k$, a request corresponding to it. We claim that when the algorithm fails finding a path for j, $L_j(e_j) \geq \frac{1}{4}$. For the case $i' = 1$, the claim is obvious as before. For the case $i' > 1$, we know that $2^{i'-3}u_{min} < d_j \leq 2^{i'-2}u_{min}$. In case $u(e_j) > 2^{i'-1}u_{min}$, the only way to overflow it with demands of size at most $2^{i'-2}u_{min}$ is when the edge is loaded at least $u(e_j) - 2^{i'-2}u_{min} \geq u(e_j)/2$. Otherwise, $u(e_j) \leq 2^{i'-1}u_{min}$ and since j is routed through this edge in the optimal solution $d_j \leq u(e_j)$. Therefore, the edge cannot be empty. Since we only route requests from $T_{i'}$ the edge's load must be at least $2^{i'-3}u_{min} \geq u(e_j)/4$.

The number of iterations $ESPROUTE$ performs is at most l since we ignore empty T_i's. For T_1, the number of iterations of $Routine_2$ is the same as in $SPROUTE$. For a set T_i, $i > 1$, the number of iterations of $Routine_2$ is $\log(n \frac{r_{max}}{r_{min}} \frac{u_{max}}{u_{min}})$. As before, the preprocessing of $Routine_3$ reduces this number to $\log(nl^2 \frac{d_{max}}{d_{min}})$. Since the ratio $\frac{d_{max}}{d_{min}}$ is at most 2 in each T_i, we conclude that $ESPROUTE$ is strongly polynomial.

A.2   Proof of Theorem 4

Proof. The first thing to note is that the algorithms never overflow an edge. For the first algorithm, the demands are at most $\frac{u_{min}}{K}$ and the only way to exceed an edge capacity is to route request j through an edge e that holds at least $\lfloor \frac{K \cdot u(e)}{u_{min}} \rfloor$ requests. For such an edge, $L_{j-1}(e) \geq 1$ and $\mu^{L_{j-1}(e)} - 1 \geq \mu - 1 \geq D$. For the second algorithm, it is sufficient to show that in case $L_{j-1}(e) > 1 - \frac{1}{K}$ for some e then $\mu^{L_{j-1}(e)} - 1 \geq D$; that is true since $\mu^{L_{j-1}(e)} - 1 \geq ((2D)^{1+\frac{1}{K-1}})^{1-\frac{1}{K}} - 1 = 2D - 1 \geq D$. Therefore, the algorithms never overflow an edge.

Now we lower bound the total demand accepted by our algorithms. We denote by $\mathcal{Q}$ the set of requests in the optimal solution and by $\mathcal{P}$ the requests accepted by either of our algorithm. For $j \in \mathcal{Q}$ denote by Q_j the path chosen for it in the optimal solution and for $j \in \mathcal{P}$ let P_j be the path chosen for it by our algorithm.

We consider prefixes of the input so let $Q^k = Q \cap \{1, ..., k\}$ and $P^k = P \cap \{1, ..., k\}$ for $k = 1, ..., l$. We prove that

$$d(P^k) \geq \frac{\sum_e u(e)(\mu^{L_k(e)} - 1)}{6KD\mu^{\frac{1}{k}}}.$$

The proof is by induction on k and the induction base is trivial since the above expression is zero. Thus, it is sufficient to show that for an accepted request j

$$\frac{\sum_{e \in P_j} u(e)(\mu^{L_j(e)} - \mu^{L_{j-1}(e)})}{6KD\mu^{\frac{1}{k}}} \leq d_j.$$

Note that for any $e \in P_j$, $L_j(e) - L_{j-1}(e) \leq \frac{1}{K}$ for both algorithms. In addition, for both algorithms $L_j(e) - L_{j-1}(e) \leq 3\frac{d_j}{u(e)}$ where the factor 3 is only necessary for $EKROUTE$ where the virtual load is higher than the actual increase in relative load. The worst case is when $K = 2$, $u(e) = (1.5 - \epsilon)u_{min}$ and $d_j = (\frac{1}{3} + \epsilon)u_{min}$: the virtual load increases by $\frac{1}{2}$ whereas $\frac{d_j}{u(e)}$ is about $\frac{2}{9}$. Looking at the exponent,

$$\mu^{L_j(e)} - \mu^{L_{j-1}(e)} = \mu^{L_{j-1}(e)}(\mu^{L_j(e) - L_{j-1}(e)} - 1)$$
$$= \mu^{L_{j-1}(e)}((\mu^{\frac{1}{k}})^{K(L_j(e) - L_{j-1}(e))} - 1)$$
$$\leq \mu^{L_{j-1}(e)}\mu^{\frac{1}{k}}K(L_j(e) - L_{j-1}(e))$$
$$\leq \mu^{L_{j-1}(e)}\mu^{\frac{1}{k}}3K\frac{d_j}{u(e)}$$

where the first inequality is due to the simple relation $x^y - 1 \leq xy$ for $0 \leq y \leq 1, 0 \leq x$ and that for $e \in P_j$, $L_j(e) - L_{j-1}(e) \leq \frac{1}{K}$. Therefore,

$$\sum_{e \in P_j} u(e)(\mu^{L_j(e)} - \mu^{L_{j-1}(e)}) \leq \sum_{e \in P_j} \mu^{L_{j-1}(e)}\mu^{\frac{1}{k}}3Kd_j$$
$$= 3K\mu^{\frac{1}{k}}d_j \sum_{e \in P_j} \mu^{L_{j-1}(e)}$$
$$= 3K\mu^{\frac{1}{k}}d_j(\sum_{e \in P_j} (\mu^{L_{j-1}(e)} - 1) + |P_j|)$$
$$\leq 3K\mu^{\frac{1}{k}}(D + D)d_j$$
$$= 6KD\mu^{\frac{1}{k}}d_j$$

where the last inequality holds since the algorithm routes the request through P_j and the length of P_j is at most D.

The last step in the proof is to upper bound the total demand accepted by an optimal algorithm. Denote the set of requests rejected by our algorithm and accepted by the optimal one by $R^k = Q^k \setminus P^k$. For $j \in R^k$, we know that

$\sum_{e \in Q_j}(\mu^{L_{j-1}(e)} - 1) \geq D$ since the request is rejected by our algorithm. Hence,

$$D \cdot d(R^k) \leq \sum_{j \in R^k} \sum_{e \in Q_j} d_j(\mu^{L_{j-1}(e)} - 1)$$

$$\leq \sum_{j \in R^k} \sum_{e \in Q_j} d_j(\mu^{L_k(e)} - 1)$$

$$= \sum_{e} \sum_{j \in R^k | e \in Q_j} d_j(\mu^{L_k(e)} - 1)$$

$$= \sum_{e}(\mu^{L_k(e)} - 1) \sum_{j \in R^k | e \in Q_j} d_j$$

$$\leq \sum_{e}(\mu^{L_k(e)} - 1)u(e),$$

where the last inequality holds since the optimal algorithm cannot overflow an edge.

By combining the two inequalities shown above,

$$d(Q^k) \leq d(\mathcal{P}^k) + d(R^k) \leq d(\mathcal{P}^k) + d(\mathcal{P}^k)\frac{6KD}{D}\mu^{\frac{1}{k}} = (1 + 6K\mu^{\frac{1}{k}})d(\mathcal{P}^k)$$

The algorithm followed a non-increasing order of $\frac{r_j}{d_j}$ and by Lemma 1 we obtain the same inequality above for profits. So, the approximation ratio of the algorithm is

$$1 + 6K\mu^{\frac{1}{k}} = O(K \cdot \mu^{\frac{1}{k}})$$

which, by assigning the appropriate values of μ, yields the desired results.

Edge Covers of Setpairs and the Iterative Rounding Method

Joseph Cheriyan[1] and Santosh Vempala[2]

[1] Department of Combinatorics and Optimization, University of Waterloo,
Waterloo, ON, CANADA, N2L 3G1.
jcheriyan@math.uwaterloo.ca
http://www.math.uwaterloo.ca/~jcheriyan
[2] Department of Mathematics, M.I.T., Cambridge, MA 02139, USA.
vempala@math.mit.edu
http://www-math.mit.edu/~vempala

Abstract. Given a digraph $G = (V, E)$, we study a linear programming relaxation of the problem of finding a minimum-cost edge cover of pairs of sets of nodes (called *setpairs*), where each setpair has a nonnegative integer-valued demand. Our results are as follows: (1) An extreme point of the LP is characterized by a noncrossing family of tight setpairs, $\mathcal{L}$ (where $|\mathcal{L}| \leq |E|$). (2) In any extreme point x, there exists an edge e with $x_e \geq \Theta(1)/\sqrt{|\mathcal{L}|}$, and there is an example showing that this lower bound is best possible. (3) The iterative rounding method applies to the LP and gives an integer solution of cost $O(\sqrt{|\mathcal{L}|}) = O(\sqrt{|E|})$ times the LP's optimal value. The proofs rely on the fact that $\mathcal{L}$ can be represented by a special type of partially ordered set that we call *diamond-free*.

1 Introduction

Many NP-hard problems in network design including the Steiner tree problem and its generalizations are captured by the following formulation. We are given an (undirected) graph $G = (V, E)$ where each edge e has a nonnegative cost c_e, and each subset of nodes S has a nonnegative integer requirement $f(S)$. The problem is to find a minimum-cost subgraph H that satisfies all the requirements, i.e., H should have at least $f(S)$ edges in every cut $(S, V - S)$. This can be modelled as an integer program.

$$(SIP) \quad \text{minimize} \sum_e c_e x_e$$

$$\text{subject to} \sum_{e \in \delta(S)} x_e \geq f(S), \qquad \forall S \subset V$$

$$x_e \in \{0, 1\}, \qquad \forall e \in E.$$

Let (SLP) be the linear programming relaxation of (SIP). The requirement function $f(\cdot)$ should be such that (SIP) models some interesting problems in network design, (SLP) has a provably small integrality ratio, and (SLP) is solvable in

K. Aardal, B. Gerards (Eds.): IPCO 2001, LNCS 2081, pp. 30–44, 2001.
© Springer-Verlag Berlin Heidelberg 2001

polynomial time. Approximation algorithms based on (SIP) and (SLP) were designed and analyzed by Goemans and Williamson [4], Williamson et al [7], and Goemans et al [3]. Then Jain [2] gave a 2-approximation algorithm for the case of *weakly supermodular* requirement functions via a new technique called *iterative rounding*. A key discovery in [2] is that every non-zero extreme point x of (SLP) has $\max_{e \in E}\{x_e\} \geq \frac{1}{2}$. Subsequently, Melkonian and Tardos [5] studied the problem on directed graphs, and proved that if the requirement function is *crossing supermodular*, then every non-zero extreme point of their linear programming relaxation has an edge of value at least $\frac{1}{4}$.

Although (SIP) is quite general, there are several interesting problems in network design that elude this formulation, such as the problem of finding a minimum-cost k-node connected spanning subgraph. Frank and Jordan [1] gave a more general formulation where *pairs* of node sets have requirements (also, see Schrijver [6] for earlier related results). In this formulation, we are given a digraph $G = (V, E)$ and each edge e has a nonnegative cost c_e. A setpair is an ordered pair of nonempty node sets $W = (W_1, W_2)$, where $W_1 \subseteq V$ is called the *tail*, denoted $t(W)$, and $W_2 \subseteq V$ is called the *head*, denoted $h(W)$. Let $\mathcal{A}$ be the set of all setpairs. For a setpair W, $\delta(W)$ denotes the set of edges covering W, i.e., $\delta(W) = \{uv \in E | u \in t(W), v \in h(W)\}$. Each setpair W has a nonnegative, integer requirement $f(W)$. The problem is to find a minimum-cost subgraph that satisfies all the requirements. (Note that the requirement function $f(\cdot)$ of (SIP) is the special case where every setpair with positive requirement is a partition of V and has the form $(S, V - S)$ where $S \subset V$.)

$$(IP) \quad \text{minimize} \sum_e c_e x_e$$

$$\text{subject to} \sum_{e \in \delta(W)} x_e \geq f(W), \qquad \forall W \in \mathcal{A}$$

$$x_e \in \{0, 1\}, \qquad \forall e \in E.$$

Frank and Jordan used this formulation to derive min-max results for special cost functions under the assumption that the requirement function is *crossing bisupermodular* (defined in Section 2). They aso also showed that the linear programming relaxation is solvable in polynomial time, under this assumption. The problem of finding a minimum-cost k-node connected spanning subgraph of a digraph may be modeled by (IP) by taking the requirement function to be $f(W_1, W_2) = k - (|V| - |W_1 \cup W_2|)$, where W_1, W_2 are nonempty node subsets; this function is crossing bisupermodular.

We study the linear programming relaxation (LP) for arbitrary nonnegative cost functions. In Section 2 we show that for any extreme point of (LP), the space of incidence vectors of tight setpairs (setpairs whose requirement is satisfied exactly) is spanned by the incidence vectors of a *noncrossing* family of tight setpairs. A noncrossing family of setpairs is the analogue of a laminar family of sets. Recall that two sets are *laminar* if they are either disjoint or one is contained in the other. Two setpairs W, Y are *comparable* if either $t(W) \supseteq t(Y)$, $h(W) \subseteq h(Y)$, (denoted as $W \preceq Y$), or $t(W) \subseteq t(Y)$, $h(W) \supseteq h(Y)$, (denoted as $W \succeq Y$).

Setpairs W, Y are *noncrossing* if either they are comparable, or their heads are disjoint $(h(W) \cap h(Y) = \emptyset)$, or their tails are disjoint $(t(W) \cap t(Y) = \emptyset)$; otherwise W, Y *cross*. A family of setpairs $\mathcal{L} \subseteq \mathcal{A}$ is called *noncrossing* if every two setpairs in $\mathcal{L}$ are noncrossing.

In Section 3, we study noncrossing families of setpairs by representing them as partially ordered sets (posets). It turns out that the Hasse diagram of such a poset has a special property — any two chains (dipaths) of the poset have at most one subchain in common. We refer to such posets as *diamond-free* posets. Based on this, we prove the following result. (Note the contrast with (SLP), where the lower bound is $\Omega(1)$, see [2,5].)

Theorem 1. *For any digraph $G = (V, E)$, any nonzero extreme point x of (LP) satisfies*

$$\max_{e \in E}\{x_e\} \geq \frac{\Theta(1)}{\sqrt{|E|}}.$$

A direct application of iterative rounding then yields an $O(\sqrt{|E|})$ approximation algorithm for the setpairs formulation (IP). Iterative rounding is based on two properties: (1) the linear programming relaxation has an optimum extreme point with an edge of high value, say at least $\frac{1}{\phi(G)}$ where $\phi(G)$ is a nondecreasing function of the size of G, and (2) the linear program is self-reducible, i.e., on fixing any edge variable at 1, the residual linear program continues to have these properties. Under these conditions, iterative rounding gives a $\phi(G)$-approximation guarantee by finding an edge of highest value in an optimum extreme point, setting it to 1, and recursively rounding the residual linear program (for details, see [2]). In our case, the second property holds for (LP) provided the requirement function is crossing bisupermodular (see Section 2). Theorem 1 guarantees the first property. This raises the question of whether iterative rounding can achieve a better approximation guarantee for (IP). In Section 4, we show that the bound in Theorem 1 is the best possible, up to a constant factor.

Theorem 2. *Given any sufficiently large integer $|E|$, there exists a digraph $G = (V, E)$ such that (LP) has an extreme point x satisfying*

$$\max_{e \in E}\{x_e\} \leq \frac{\Theta(1)}{\sqrt{|E|}}.$$

In the rest of the paper, an edge means a directed edge of the input digraph G.

2   Characterizing Extreme Points via Noncrossing Families

For two crossing setpairs W, Y let $W \otimes Y$ denote the setpair $(t(W) \cup t(Y), h(W) \cap h(Y))$ and let $W \oplus Y$ denote the setpair $(t(W) \cap t(Y), h(W) \cup h(Y))$. Note that W (similarly, Y) is $\preceq W \oplus Y$ and is $\succeq W \otimes Y$. If both W and Y are partitions of V, so $W = (V - h(W), h(W))$, $Y = (V - h(Y), h(Y))$, then note that $W \otimes Y$ is the partition of V with head $h(W) \cap h(Y)$, and $W \oplus Y$ is the partition of V with head $h(W) \cup h(Y)$. Also, see Figure 1. A nonnegative integer-valued requirement function $f : \mathcal{A} \rightarrow \mathbb{Z}_+$ is called *crossing bisupermodular* if for every two setpairs W, Y,

$$f(W) + f(Y) \leq f(W \oplus Y) + f(W \otimes Y)$$

holds whenever $f(W) > 0$, $f(Y) > 0$, $h(W) \cap h(Y) \neq \emptyset$, and $t(W) \cap t(Y) \neq \emptyset$. If the reverse inequality holds for every two setpairs (without any conditions), then f is called *bisubmodular*. For any nonnegative vector $x : E \rightarrow \mathbb{R}_+$ on the edges, the function on setpairs $x(\delta(W)) = \sum\{x_e \mid e \in \delta(W)\}$, $W \subset \mathcal{A}$, is bisubmodular; this is discussed in [1, Claim 2.1].

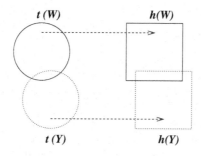

Fig. 1. Illustration of crossing setpairs. The dashed edges contribute to $x(\delta(W)) + x(\delta(Y))$ but not to $x(\delta(W \otimes Y)) + x(\delta(W \oplus Y))$.

For a feasible solution x of (LP), a setpair W is called *tight* (w.r.t. x) if $x(\delta(W)) = f(W)$. Let $vec(W)$ denote the zero-one incidence vector of $\delta(W)$.

Theorem 3. *Let x be an extreme point solution of (LP) such that $x_e < 1$ for each edge e, and let $F = \{e \in E \mid x_e > 0\}$ be the support of x. Then there exists a noncrossing family of tight setpairs $\mathcal{L}$ such that*

(i)   every setpair $W \in \mathcal{L}$ has $f(W) \geq 1$,
(ii) $|\mathcal{L}| = |F|$,
(iii) the vectors $vec(W)$, $W \in \mathcal{L}$ are linearly independent, and
(iv) x is the unique solution to $\{x(\delta(W)) = f(W), \forall W \in \mathcal{L}\}$.

The proof is based on the next two lemmas. The first of these lemmas "uncrosses" two tight setpairs that cross.

Lemma 4. *Let $x : E \rightarrow \mathbb{R}$ be a feasible solution of (LP). If two setpairs W, Y with $f(W) > 0, f(Y) > 0$ are tight and crossing, then the setpairs $W \otimes Y, W \oplus Y$ are tight. Moreover, if $x_e > 0$ for each edge e, then*

$$vec(W) + vec(Y) = vec(W \otimes Y) + vec(W \oplus Y).$$

Proof. The requirement function $f(\cdot)$ is crossing bisupermodular, and the "edge supply" function $x(\delta(\cdot))$ satisfies the bisubmodular inequality

$$x(\delta(W)) + x(\delta(Y)) \geq x(\delta(W \otimes Y)) + x(\delta(W \oplus Y)).$$

Therefore, we have

$$f(W \otimes Y) + f(W \oplus Y) \leq x(\delta(W \otimes Y)) + x(\delta(W \oplus Y)) \leq$$
$$x(\delta(W)) + x(\delta(Y)) = f(W) + f(Y) \leq f(W \otimes Y) + f(W \oplus Y).$$

Hence, all the inequalities hold as equations, and so $W \otimes Y, W \oplus Y$ are tight.

If the bisubmodular inequality holds with equality, then all edges contributing to the l.h.s. contribute also to the r.h.s., hence

$$x(\delta(t(W) - t(Y), h(W) - h(Y))) = 0, \text{ and } x(\delta(t(Y) - t(W), h(Y) - h(W))) = 0.$$

The second statement in the lemma follows since there are no edges in both $\delta(t(W) - t(Y), h(W) - h(Y))$ and $\delta(t(Y) - t(W), h(Y) - h(W))$. □

Lemma 5. *Let L and S be two crossing setpairs. Let $N = S \otimes L$ (or, let $N = S \oplus L$). If another setpair J crosses N, then either J crosses S or J crosses L.*

Proof. We prove the lemma for the case $N = S \otimes L$; the other case is similar. The proof is by contradiction. Suppose the lemma fails. Then there is a setpair $J \in \mathcal{L}$ such that J, N cross (so $t(J) \cap t(N) \neq \emptyset$ and $h(J) \cap h(N) \neq \emptyset$), but both J, L and J, S are noncrossing.

We have four main cases, depending on whether J, L are head disjoint, tail disjoint, $J \succeq L$ or $J \preceq L$.

(i) J, L are head disjoint: Then J, N are head disjoint (by $h(N) = h(S) \cap h(L)$) so J, N do not cross.

(ii) J, L are tail disjoint: We have three subcases, depending on the tails of J, S.

- $t(J)$ properly intersects $t(S)$:
  Then J, S are head disjoint (since J, S are noncrossing) so J, N are also head disjoint, and do not cross.
- $t(J) \subseteq t(S)$:
  Since J, S are noncrossing, either J, S are head disjoint, in which case J, N are head disjoint and do not cross, or $h(J) \supseteq h(S)$, in which case $h(J) \supseteq h(N)$, so $J \succeq N$ and J, N do not cross.
- $t(J) \supseteq t(S)$:
  This is not possible, since J, L are tail disjoint, and $t(S)$ intersects $t(L)$ (since L, S cross).

(iii) $J \succeq L$: Then $J \succeq N$ since $h(J) \supseteq h(L) \supseteq h(N)$ and $t(J) \subseteq t(L) \subseteq t(N)$.

(iv) $J \preceq L$: As in case(ii), we have three subcases, depending on the tails of J, S.

- $t(J)$ properly intersects $t(S)$:
  Similar to case(ii) above, first subcase.
- $t(J) \subseteq t(S)$:
  Similar to case(ii) above, second subcase.

- $t(J) \supseteq t(S)$:
  Since J, S do not cross, either J, S are head disjoint, in which case J, N are head disjoint and do not cross, or $h(J) \subseteq h(S)$, in which case $J \preceq N$ since $h(J) \subseteq h(S) \cap h(L) = h(N)$ and $t(J) \supseteq t(S) \cup t(L) = t(N)$ (note that $h(J) \subseteq h(L)$ and $t(J) \supseteq t(L)$).

This concludes the proof of the lemma.   □

Proof. (of Theorem 3) Our proof is inspired by Jain's proof of [2, Theorem 3.1]. Since x is an extreme point solution (basic solution) with $0 < x < 1$, there exists a set of $|F|$ tight setpairs such that the vectors $vec(W)$ corresponding to these setpairs W are linearly independent.

Let $\mathcal{L}$ be an (inclusionwise) maximal noncrossing family of tight setpairs. Let $span(\mathcal{L})$ denote the vector space spanned by the vectors $vec(W)$, $W \in \mathcal{L}$. We will show that $span(\mathcal{L})$ equals the vector space spanned by the vectors $vec(Y)$ where Y is any tight setpair. The theorem then follows by taking a basis for $span(\mathcal{L})$ from the set $\{vec(W) \mid W \in \mathcal{L}\}$.

Suppose there is a tight setpair S such that $vec(S) \notin span(\mathcal{L})$. Choose such an S that crosses the minimum number of setpairs in $\mathcal{L}$ (this is a key point). Next, choose any setpair $L \in \mathcal{L}$ such that S crosses L. By Lemma 4,

$$vec(S) = vec(S \otimes L) + vec(S \oplus L) - vec(L).$$

Hence, either $vec(S \otimes L) \notin span(\mathcal{L})$ or $vec(S \oplus L) \notin span(\mathcal{L})$. Suppose the first case holds. (The argument is similar for the other case, and is omitted.) Let $N = S \otimes L = (t(S) \cup t(L), h(S) \cap h(L))$. The next claim follows from Lemma 5.
Claim. Any setpair $J \in \mathcal{L}$ that crosses N also crosses S (note that J, L do not cross since both are in $\mathcal{L}$).

Clearly, L does *not* cross N (since $L \succeq N$), but L crosses S. This contradicts our choice of S (since N is a tight setpair that crosses fewer setpairs in $\mathcal{L}$ and $vec(N) \notin span(\mathcal{L})$).   □

3   An Edge of High Value in an Extreme Point

This section has the proof of Theorem 1. The theorem is proved by representing the noncrossing family $\mathcal{L}$ as a poset and examining the Hasse diagram.

3.1   Diamond-Free Posets

Recall that a poset $\mathcal{P}$ is a set of elements together with a binary relation ($\preceq$) that is reflexive, transitive and antisymmetric. Elements W, Z in $\mathcal{P}$ are called *comparable* if either $W \preceq Z$ or $Z \preceq W$, otherwise they are alled *incomparable*. The Hasse diagram of the poset, also denoted by $\mathcal{P}$, is a directed acyclic graph that has a node for each element in the poset, and for elements W, Z there is an arc (W, Z) if $W \preceq Z$ and there is no element Y such that $W \preceq Y \preceq Z$ (the Hasse diagram has no arcs that are implied by transitivity). Throughout this section, a *node* means a node of $\mathcal{P}$, not a node of G. An *arc* means an arc of $\mathcal{P}$, whereas an *edge* means a directed edge of G. A node Z is called a *predecessor* (or *successor*) of a node W if the arc (Z, W) (or (W, Z)) is present. A directed path in $\mathcal{P}$ is called a *chain*. An *antichain* of $\mathcal{P}$ is a set of nodes that are pairwise incomparable. If S is a chain or an antichain of $\mathcal{P}$, then $|S|$ denotes the number of nodes in S; the number of nodes in $\mathcal{P}$ is denoted by $|\mathcal{P}|$. For an arbitrary poset, define a *diamond* to be a set of four (distinct) elements a, b, c, d such that b, c are incomparable, $a \succeq b \succeq d$ and $a \succeq c \succeq d$. A poset is called *diamond-free* if it contains no diamond. In other words, any two chains of such a poset have at most one subchain in common.

Let $\mathcal{L}$ be a noncrossing family of setpairs. We define the poset $\mathcal{P}$ representing $\mathcal{L}$ as follows. The elements of $\mathcal{P}$ are the setpairs in $\mathcal{L}$ and the relation between elements is the same as the relation between setpairs (for two setpairs W and Y, if $t(W) \supseteq t(Y)$, $h(W) \subseteq h(Y)$, then $W \preceq Y$; if $t(W) \subseteq t(Y)$, $h(W) \supseteq h(Y)$ then $Y \preceq W$; otherwise they are incomparable). In the Hasse diagram, an arc (W, Y) indicates that $h(W) \subseteq h(Y)$ and $t(W) \supseteq t(Y)$.

Lemma 6. *Let $\mathcal{L}$ be a noncrossing family of setpairs, and let $\mathcal{P}$ be the poset representing $\mathcal{L}$. Then $\mathcal{P}$ is diamond-free.*

Proof. Suppose that $\mathcal{L}$ has four setpairs W, X, Y, Z such that X, Y are incomparable, $W \succeq X \succeq Z$ and $W \succeq Y \succeq Z$. Since X, Y are incomparable, either they are head disjoint, or tail disjoint. Moreover, $h(X) \supseteq h(Z)$ since $X \succeq Z$, and $h(Y) \supseteq h(Z)$ since $Y \succeq Z$. Then X, Y are not head disjoint, since both heads contain the head of Z, which is nonempty. Similarly, it can be seen that X, Y are not tail disjoint, since both tails contain the tail of W, which is nonempty. This contradiction proves that $\mathcal{P}$ contains no diamond. $\square$

We call a node W *unary* if the Hasse diagram has exactly one arc incoming to W and exactly one arc outgoing from W. Consider the maximum cardinality of an antichain in a diamond-free poset $\mathcal{P}$. This may be as small as one, since $\mathcal{P}$ may be a chain. The next result shows that this quantity cannot be so small if $\mathcal{P}$ has no unary nodes.

Proposition 7. *(1) If a diamond-free poset $\mathcal{P}$ has no unary nodes, then it has an antichain of cardinality at least $\sqrt{|\mathcal{P}|/2}$.*
(2) If $\mathcal{P}$ is a diamond-free poset such that neither the predecessor nor the successor of a unary node is another unary node, then it has an antichain of cardinality at least $\frac{1}{2}\sqrt{|\mathcal{P}|}$.

Proof. We prove part (1); the proof of part (2) is similar.

If $\mathcal{P}$ has an antichain of cardinality at least $\sqrt{|\mathcal{P}|/2}$, then we are done. Otherwise, by Dilworth's theorem (the minimum number of disjoint chains required to cover all the nodes of a poset equals the maximum cardinality of an antichain), $\mathcal{P}$ has a chain, call it C, with $|C| > |\mathcal{P}|/\sqrt{|\mathcal{P}|/2} = \sqrt{2|\mathcal{P}|}$. Let $C = W_1, W_2, \ldots, W_\ell$. Each of the internal nodes $W_2, \ldots, W_{\ell-1}$ is non-unary, so it has either two predecessors or two successors. Clearly, one of the two predecessors (or one of the two successors) is not in C. Let C_p be the set of nodes in $\mathcal{P} - C$ that are predecessors of nodes in C, and similarly let C_s be the set of nodes in $\mathcal{P} - C$ that are successors of nodes in C. Then either $|C_p| \geq (|C| - 2)/2$ or $|C_s| \geq (|C| - 2)/2$. Suppose that the first case holds (the argument is similar for the other case). Let us add W_1 (the first node of C) to C_p. Now, we claim that C_p is an antichain. Observe that part (1) follows from this claim, because $|C_p| \geq |C|/2 > \sqrt{|\mathcal{P}|/2}$.

To prove that C_p is an antichain, focus on any two (distinct) nodes $Y_i, Y_j \in C_p$. Let W_i and W_j be the nodes in C such that Y_i is the predecessor of W_i and Y_j is the predecessor of W_j. First, suppose that $W_i \neq W_j$, and (w.l.o.g.) assume that $W_i \preceq W_j$. We cannot have $Y_i \preceq Y_j$, otherwise, the nodes W_j, W_{j-1}, Y_j, Y_i will form a diamond, where W_{j-1} is the predecessor of W_j in C (note that the four nodes are distinct, and W_{j-1}, Y_j are incomparable, since both are predecessors of W_j). Also, we cannot have $Y_j \preceq Y_i$, otherwise, we have $Y_j \preceq Y_i \preceq W_i \preceq W_j$ and so the arc (Y_j, W_j) is implied by transitivity. Hence, Y_i, Y_j are incomparable, if $W_i \neq W_j$. If $W_i = W_j$, then Y_i, Y_j are incomparable (by transitivity).   □

3.2   A Proof of Theorem 1

Theorem 1. *For any digraph $G = (V, E)$, any nonzero extreme point x of (LP) satisfies $\max_{e \in E}\{x_e\} \geq \dfrac{\Theta(1)}{\sqrt{|E|}}$.*

The proof is by contradiction. Let x be an extreme point of (LP), and let $F = \{e \in E \mid x_e > 0\}$. For convenience, assume that no edges e with $x_e = 0$ are present. Also, assume that each edge e has $x_e < 1$, otherwise the proof is done.

Let $\mathcal{L}$ be a noncrossing family of tight setpairs defining x and satisfying the conditions in Theorem 3, and let $\mathcal{P}$ be the poset representing $\mathcal{L}$. Note that $|\mathcal{P}| = |\mathcal{L}| = |F|$. Let U be the set of unary nodes of $\mathcal{P}$, and call a maximal chain of unary nodes a U-*chain*. Let $\mathcal{P}'$ be the "reduced" poset formed by replacing each U-chain by a single unary node. Note that $\mathcal{P}'$ is diamond-free, since $\mathcal{P}$ is diamond-free.

Let C be a maximum-cardinality antichain of $\mathcal{P}$. By Proposition 7(2), $|C| \geq \frac{1}{2}\sqrt{|\mathcal{P}'|}$. We may assume that each unary node of C (if any) is a bottom node of a U-chain. By an *upper* U-chain we mean one that has all nodes $\succeq$ some node in C, and by a *lower* U-chain we mean one that has all nodes $\preceq$ some node in C. Let U_0 be the set of bottom nodes of all the upper U-chains together with the set of top nodes of all the lower U-chains. Let U_* be the set of nodes $W \in U - U_0$ in upper U-chains such that the predecessor Y of W has $f(W) = f(Y)$, together with the set of nodes $W \in U - U_0$ in lower U-chains such that the successor Z of W has $f(W) = f(Z)$. Let U_1 be the set of nodes $W \in U - U_0$ in upper U-chains such that the predecessor Y of W has $f(W) > f(Y)$, together with the set of nodes $W \in U - U_0$ in lower U-chains such that the successor Z of W has $f(W) < f(Z)$. Similarly, let U_2 be the set of nodes $W \in U - U_0$ in upper U-chains such that the predecessor Y of W has $f(W) < f(Y)$, together with the set of nodes $W \in U - U_0$ in lower U-chains such that the successor Z of W has $f(W) > f(Z)$.

Clearly,

$$U = U_0 \cup U_1 \cup U_2 \cup U_* \quad \text{and} \quad |\mathcal{P}| - |\mathcal{P}'| = |U_1| + |U_2| + |U_*|.$$

Claim. If α is a number such that $x_e < 1/\alpha$, $\forall e \in E$, then

$$|F| > \alpha \cdot \max\{|C|, |U_1|, |U_2|\} + |U_*|.$$

We defer the proof of the claim, and complete the proof of the theorem. Let $\alpha = 4\sqrt{|\mathcal{P}|}$. Suppose that $x_e < 1/\alpha$ for each edge $e \in F$. Then, by the claim,

$$
\begin{aligned}
|F| &> 4\sqrt{|\mathcal{P}|} \cdot \max\{|C|, |U_1|, |U_2|\} + |U_*| \\
&\geq 4\sqrt{|\mathcal{P}|} \cdot (\frac{1}{2}|C| + \frac{1}{4}|U_1| + \frac{1}{4}|U_2|) + |U_*| \\
&\geq 4\sqrt{|\mathcal{P}|} \cdot (\frac{1}{4}\sqrt{|\mathcal{P}'|} + \frac{1}{4}|U_1| + \frac{1}{4}|U_2|) + |U_*| \\
&\geq |\mathcal{P}'| + |U_1| + |U_2| + |U_*| \\
&\geq |\mathcal{P}|.
\end{aligned}
$$

This is a contradiction, since $|F| = |\mathcal{P}|$. Hence, there exists an edge e with $x_e \geq 1/\alpha = 1/(4\sqrt{|\mathcal{P}|})$. This proves the theorem.

Proof. (of the Claim) We need to prove the three inequalities separately. Consider the first inequality:

$$|F| > \alpha \cdot |C| + |U_*|.$$

Each setpair $W \in \mathcal{L}$ has $f(W) \geq 1$, so W is covered by $> \alpha$ edges (otherwise, $x(\delta(W)) < 1$). Hence, each node $W \in \mathcal{P}$ is covered by $> \alpha$ edges. We assign all of the edges covering a node $W \in C$ to W; note that no edge covers two distinct nodes of C. This assigns a total of $> \alpha|C|$ edges. Now, consider a node $W_i \in U_*$ that is in an upper U-chain $W_1, \ldots, W_\ell$, where $1 < i \leq \ell$. Since $f(W_{i-1}) = f(W)$ and $vec(W_{i-1}) \neq vec(W_i)$, there is an edge in $\delta(W_i) - \delta(W_{i-1})$. We assign this edge to W_i. Similarly, for a node $W_i \in U_*$ in a lower U-chain $W_1, \ldots, W_\ell$, where $1 \leq i < \ell$, we assign to W_i an edge in $\delta(W_i) - \delta(W_{i+1})$. It can be seen that no edge is assigned to two different nodes. Hence, the first inequality follows.

Consider the second inequality: $|F| > \alpha \cdot |U_1| + |U_*|$. Let $W_i \in U_1$ be any node in an upper U-chain $W_1, \ldots, W_\ell$, where $1 < i \leq \ell$. Since $f(W_i) \geq f(W_{i-1}) + 1$, there must be $> \alpha$ edges in $\delta(W_i) - \delta(W_{i-1})$. We assign all these edges to W_i. Similarly, for a node $W_i \in U_1$ that is in a lower U-chain $W_1, \ldots, W_\ell$, where $1 \leq i < \ell$, we assign $> \alpha$ edges in $\delta(W_{i+1}) - \delta(W_i)$ to W_i. Finally, for nodes $W_i \in U_*$, if W_i is in an upper U-chain $W_1, \ldots, W_\ell$, then we assign to W_i an edge in $\delta(W_i) - \delta(W_{i-1})$, and if W_i is in a lower U-chain $W_1, \ldots, W_\ell$, then we assign to W_i an edge in $\delta(W_{i+1}) - \delta(W_i)$. The second inequality follows, since no edge is assigned to two different nodes.

The proof of the third inequality is similar to the proof of the second inequality. $\qquad\square$

4   A Tight Example

In this section, we present an example of an extreme point x of (LP) such that $0 < x_e \leq \Theta(1)/\sqrt{|E|}$ for all edges $e \in E$. Thus the lower bound in Theorem 1 is tight (up to a constant factor), and hence, $O(\sqrt{|E|})$ is the best possible approximation guarantee for (IP) via iterative rounding. An extreme point x of (LP) is defined by a system of $|E|$ tight constraints, where each is of the form $x(\delta(W)) = f(W)$, for some setpair W (we assume $0 < x < 1$ so the constraints $x_e \geq 0$, $x_e \leq 1$ are redundant). Let $\mathcal{L}$ be the noncrossing family of tight setpairs defining x (see Theorem 3), and let $\mathcal{P}$ be the poset (and the Hasse diagram) representing $\mathcal{L}$. In this section, a node means a node of $\mathcal{P}$ (not a node of G). Each edge $e \in E$ corresponds to a path $p(e)$ in $\mathcal{P}$, where the nodes of $p(e)$ are the setpairs $W \in \mathcal{L}$ that are covered by e, that is, $p(e) = W_1, \ldots, W_\ell$, where $W_1 \preceq \ldots \preceq W_\ell$ and $e \in \delta(W_i)$ $(i = 1, \ldots, \ell)$. We refer to such paths $p(e)$ as e-paths.

Let $m = |E(G)|$. Our example is a poset $\mathcal{P}$ with m nodes and m e-paths (see Figure 2), so $|\mathcal{P}| = m$. Define the incidence matrix A to be an $m \times m$ matrix whose rows correspond to the nodes, and whose columns correspond to the e-paths, such that the entry for node W and e-path p, A_{Wp}, is 1 if W is in p and is 0 otherwise. We will prove that A has rank $m - 1$ and the system $Ax = \mathbf{1}$ has a solution where each entry of x is $\Theta(1)/\sqrt{m}$. (Note that x assigns a real number to each of the e-paths, and it corresponds to a solution of the LP.)

The poset $\mathcal{P}$ consists of several copies of the following $path\ structure\ Q$. Let t be a parameter (we will fix $t = \sqrt{m/12}$), and let there be $3t$ nodes $1, 2, \ldots, 3t$. Then Q consists of a path $[1, \ldots, 3t]$ on these nodes, together with $2t$ $local\ e$-$paths$, call them $p_1, \ldots, p_{2t}$, where each p_j is a subpath of the path $[1, \ldots, 3t]$. For odd j $(j = 1, 3, 5, \ldots, 2t - 1)$, p_j consists of the first j nodes (so $p_j = [1, \ldots, j]$), and for even j $(j = 2, 4, 6, \ldots, 2t)$, p_j consists of all the nodes, except the first $j - 2$ nodes (so $p_j = [j - 1, j, \ldots, 3t]$).

Call the nodes $1, 3, 5, \ldots, 2t - 1$ the $black$ nodes, the nodes $2, 4, 6, \ldots, 2t - 2$ the $white$ nodes, and the remaining nodes $2t, 2t + 1, \ldots, 3t$ the red nodes. Note that each black node is incident to $t + 1$ local e-paths, and each of the other nodes is incident to t local e-paths.

We take $4t$ copies of Q, and partition them into two sets, the top path-structures $T_1, \ldots, T_{2t}$, and the $bottom$ path-structures $B_1, \ldots, B_{2t}$. (We will also refer to these as top paths and bottom paths.) Finally, we add another $4t^2$ $nonlocal\ e$-$paths$ such that the following conditions hold:

- each node is incident to a total of $t + 1$ e-paths;
- each nonlocal e-path is incident to exactly two nodes, one in a top path T_i and one in a bottom path B_j; moreover, for every T_i and every B_j, there is exactly one nonlocal e-path incident to both T_i and B_j;
- each nonlocal e-path is incident to either two red nodes, or one red node and one white node;
- each top/bottom path T_i or B_j is incident to exactly two red-red nonlocal e-paths, where
  (i)   there is an e-path incident to the last node of B_i and the last node of

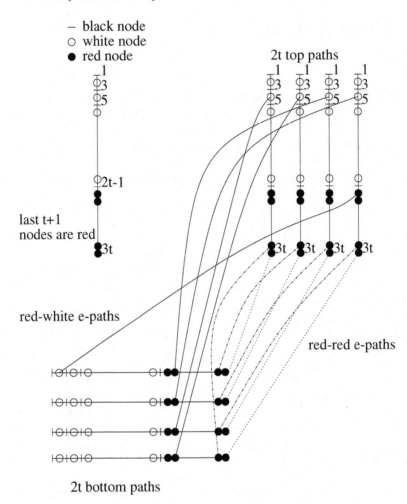

Fig. 2. Illustration of the poset $\mathcal{P}$.

T_i $(i = 1, \ldots, 2t)$, and

(ii)   there is an e-path incident to the 2nd last node of B_i and the 2nd last node of T_{i+1} $(i = 1, \ldots, 2t)$; the indexing is modulo $2t$, so $2t + 1$ means 1; note that there is cyclic shift by 1 in the index of the top versus bottom paths;

− the red-white nonlocal e-paths are fixed according to the first two conditions, and are as follows: for $\ell = 1, 2, \ldots, t - 1$, there is an e-path incident to the 2ℓth node of B_i and the $(2t - 1 + \ell)$th node of $T_{i+1+\ell}$ $(i = 1, \ldots, 2t)$, indexing modulo $2t$; note that there is a cyclic shift by $\ell + 1$ in the index of the top versus bottom paths; similarly, for $\ell = 1, 2, \ldots, t - 1$, there is an e-path incident to the 2ℓth node of T_i and the $(2t - 1 + \ell)$th node of $B_{i+1+\ell}$ $(i = 1, \ldots, 2t)$, indexing modulo $2t$.

Let $\mathbf{0}$ and $\mathbf{1}$ denote column vectors with all entries at 0 and 1, respectively, where the dimension of the vector will be clear from the context.

Proposition 8. *Let t be a positive integer, and let $m = 12t^2$. Let A be the $m \times m$ incidence matrix of the poset $\mathcal{P}$ and the e-paths (constructed above). Then $\operatorname{rank}(A) \geq m - 1$ and a solution to the system $Ax = \mathbf{1}$ is given by $x = \frac{1}{t+1} \cdot \mathbf{1}$.*

Proof. A column vector of dimension ℓ with all entries at 0 (or, 1) is denoted by 0_ℓ (or, 1_ℓ). Let e_i denote the ith column of the $s \times s$ identity matrix I_s, where s is a positive integer. Let f_i denote $\sum_{j=1}^{i} e_j$; so f_i is a column vector with a 1 in entries $1, \ldots, i$ and a 0 in entries $i+1, \ldots, s$.

Let the rows of A be ordered according to the nodes $1, \ldots, 3t$ of $T_1, \ldots, T_{2t}$, followed by the nodes $1, \ldots, 3t$ of $B_1, \ldots, B_{2t}$.

First, consider a bottom path-structure B_i; top path-structures T_i are handled similarly, and this is sketched later.

Let M denote the incidence matrix of B_i versus all the e-paths. Then M is $3t \times m$ matrix, where the rows $1, \ldots, 3t$ correspond to the nodes $1, \ldots, 3t$ of B_i, and the columns of M are ordered as follows:

- the $2t$ local e-paths of B_i, $p_1, p_2, \ldots, p_{2t}$,
- the $t - 1$ red-white e-paths whose red ends are in B_i (these are the e-paths incident to nodes $2t, 2t + 1, \ldots, 3t - 2$ of B_i),
- the two red-red e-paths incident to nodes $3t - 1$ and $3t$ of B_i,
- the remaining e-paths (among these are $t - 1$ red-white e-paths incident to B_i, such that the white end is one of the white nodes of B_i).

Let M^{beg} denote the submatrix of M formed by the first $2t$ columns, so M^{beg} is the incidence matrix of the nodes versus the local e-paths of B_i. Let M^{end} denote the submatrix of M formed by excluding the first $3t+1$ columns (keeping only the columns of the "remaining e-paths"). Then

$$
M = \left[\begin{array}{c|ccc|cc|c}
 & & 0_{2t-1} \ldots 0_{2t-1} & & 0_{2t-1} & 0_{2t-1} & \\
M^{beg} & & I_{t-1} & & 0_{t-1} & 0_{t-1} & M^{end} \\
 & & 0 \ldots 0 & & 1 & 0 & \\
 & & 0 \ldots 0 & & 0 & 1 &
\end{array} \right].
$$

Note that the rows and columns of M^{end} may be reordered such that the submatrix in the first $t - 1$ rows (make these the rows of the white nodes of B_i) and the first $t - 1$ columns (make these the columns of the red-white e-paths incident to the white nodes of B_i) is the identity matrix I_{t-1}, and every other entry of the matrix is zero. Then

$$
M^{beg} = [f_1, \ \mathbf{1}, \ f_3, \ \mathbf{1} - f_2, \ f_5, \ \mathbf{1} - f_4, \ \ldots, \ f_{2t-1}, \ \mathbf{1} - f_{2t-2}].
$$

Using elementary column operations, we can rewrite this matrix as

$$
[e_1, e_2, e_3, \ldots, e_{2t-1}, \mathbf{1} - f_{2t-1}] = \left[\begin{array}{cc} I_{2t-1} & 0_{2t-1} \\ 0_{t+1} \ldots 0_{t+1} & 1_{t+1} \end{array} \right].
$$

Then it is clear that the matrix $[M^{beg} \ M^{end}]$ may be rewritten using elementary column operations as

$$\begin{bmatrix} I_{2t-1} & 0_{2t-1} & \\ 0_{t+1}\ldots0_{t+1} & 1_{t+1} & \mathbf{0}\ldots\mathbf{0} \end{bmatrix}.$$

Going back to M, observe that it may be rewritten using elementary column operations as

$$\left[\begin{array}{cccc|cc c} I_{2t-1} & 0_{2t-1}\,0_{2t-1}\ldots0_{2t-1} & 0_{2t-1}\,0_{2t-1} & \\ 0_{t-1}\ldots0_{t-1}\,1_{t-1} & I_{t-1} & 0_{t-1}\ 0_{t-1} & \\ 0\ldots0 & 1 & 0\ldots0 & 1\quad\ 0 & \mathbf{0}\ldots\mathbf{0} \\ 0\ldots0 & 1 & 0\ldots0 & 0\quad\ 1 \end{array}\right],$$

or as

$$M^* \ = \ \left[\begin{array}{ccc|ccc c} I_{2t-1} & 0_{2t-1}\ldots0_{2t-1} & 0_{2t-1}\,0_{2t-1}\,0_{2t-1} & \\ 0_{t-1}\ldots0_{t-1} & I_{t-1} & 0_{t-1}\ 0_{t-1}\ 1_{t-1} & \\ 0\ldots0 & 0\ldots0 & 1\quad\ 0\quad\ 1 & \mathbf{0}\ldots\mathbf{0} \\ 0\ldots0 & 0\ldots0 & 0\quad\ 1\quad\ 1 \end{array}\right].$$

Now, focus on the matrix A (the incidence matrix of the nodes of $\mathcal{P}$ versus all the e-paths), and its column vectors. Consider the two red-red e-paths incident to B_i and their column vectors in A. Let r_{3t-1} and r_{3t} denote the two red-red e-paths incident to the red nodes $3t-1$ and $3t$ (of B_i), respectively, and let the column vectors in A of these red-red e-paths also be denoted by the same symbols. Note that r_{3t} has two nonzero entries, namely, a 1 for node $3t$ of B_i and a 1 for node $3t$ of T_i. Similarly, r_{3t-1} has two nonzero entries, namely, a 1 for node $3t-1$ of B_i and a 1 for node $3t-1$ of T_{i+1} (indexing modulo $2t$). Keep the column r_{3t-1}, but use elementary column operations to replace r_{3t} by $r'_{3t} = r_{3t} + r_{3t-1} + M^*_1 + \ldots + M^*_{2t} + M^*_{3t-2} - M^*_{3t+1}$, where M^*_j denotes the column vector of dimension m obtained from the jth column vector of M^* by padding with zeros (fixing entries of rows not in M^* at 0). Clearly, r'_{3t} has two nonzero entries, namely a 1 for node $3t$ of T_i and a 1 for node $3t-1$ of T_{i+1} (indexing modulo $2t$).

Let $M^{bot,i}$ be the $m \times m$ matrix obtained from M^* by replacing the $(3t-1)$th column by r_{3t-1}, deleting the $3t$th column, and padding with zeros (fixing at 0 all entries except those in M^* or in the $(3t-1)$th column).

The construction for a top path T_{i+1} ($i = 1, 2, \ldots, 2t$, indexing modulo $2t$) is similar, except for the handling of columns $3t-1$ and $3t$ of M^*. Let $M^{top,i+1}$ be the $m \times m$ matrix obtained from M^* by replacing the $3t$th column by r'_{3t}, deleting the $(3t-1)$th column, and padding with zeros.

Let A^* be the $m \times m$ matrix obtained from A by elementary column operations, where

$$A^* \ = \ \sum_{i=1}^{2t} M^{top,i} \ + \ \sum_{i=1}^{2t} M^{bot,i}.$$

Figure 3 illustrates the zero-nonzero pattern of A^*.

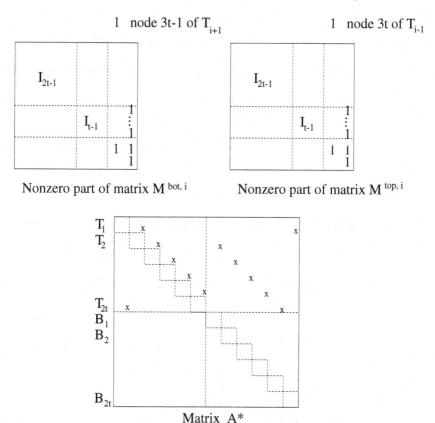

Fig. 3. Illustration of matrices $M^{bot,\,i}$ and $M^{top,\,i}$, and the nonzero pattern of matrix A^*.

With the exception of one entry, A^* is an upper triangular matrix with every diagonal entry at 1; the exceptional entry is in row $6t$ (node $3t$ of T_{2t}) and column $3t-1$ (2nd last column of T_1). Then, deleting row $6t$ and column $6t$ of A^* we get an upper triangular matrix with determinant 1. This proves that $\text{rank}(A) \geq m-1$.

By construction, each node of $\mathcal{P}$ is incident to exactly $t+1$ e-paths, hence fixing $x_e = 1/(t+1)$ for each e-path $p(e)$ gives a solution to the system $Ax = \mathbf{1}$. The proposition follows.   □

Proposition 9. *Let P be the polytope $\{x \in \mathbb{R}^m \mid \widetilde{A}x \leq \widetilde{b},\ 0 \leq x \leq 1\}$ and let F be the face $\{x \in P \mid Ax = b\}$, where $Ax \leq b$ is a subsystem of $\widetilde{A}x \leq \widetilde{b}$. If the matrix A has rank $m-1$ and there exists an $x \in F$ such that each entry of x is $\leq \alpha$, then P has an extreme point $\widetilde{x}$ such that each entry of $\widetilde{x}$ is $\leq 2\alpha$.*

Proof. F is a line-segment and so it has two extreme points, call them y and z. Note that y and z must be extreme points of P also. Hence, $x = a \cdot y + (1-a) \cdot z$, where $0 \leq a \leq 1$. Suppose $a \geq 1/2$ (the other case is similar). Then $y \leq 2(x - (1 - a) \cdot z) \leq 2x$, since $z \geq 0$, so each entry of y is $\leq 2\alpha$.   □

Theorem 2, which is the main result of this section, follows from Propositions 8 and 9.

Theorem 2. *Given any sufficiently large integer* $|E|$, *there exists a digraph* $G = (V, E)$ *such that (LP) has an extreme point* x *satisfying* $\max_{e \in E}\{x_e\} \leq \dfrac{\Theta(1)}{\sqrt{|E|}}$.

5   Conclusion

In conclusion, we mention that the results generalize from crossing bisupermodular functions to (nonnegative, integral) functions $f : \mathcal{A} \rightarrow \mathbb{Z}_+$ such that for every two setpairs W, Y with $f(W) > 0, f(Y) > 0$, the following holds:

$$f(W) + f(Y) \leq \max \left\{ \begin{array}{l} f(W \otimes Y) \; + \; f(W \oplus Y), \\ f(t(W), \; h(W) - h(Y)) \; + \; f(t(Y), \; h(Y) - h(W)), \\ f(t(W) - t(Y), \; h(W)) \; + \; f(t(Y) - t(W), \; h(Y)) \end{array} \right\}.$$

References

1. A. Frank and T. Jordan, "Minimal edge-coverings of pairs of sets," *J. Combinatorial Theory, Series B*, 65:73–110, 1995.
2. K. Jain, "A factor 2 approximation algorithm for the generalized Steiner network problem," *Proc. IEEE Foundations of Computer Science*, 1998.
3. M. Goemans, A. Goldberg, S. Plotkin, D. Shmoys, E. Tardos and D. Williamson, "Improved approximation algorithms for network design problems," in *Proc. ACM SIAM Symposium on Discrete Algorithms*, 1994, 223-232.
4. M. Goemans and D. Williamson, "A general approximation technique for constrained forest problems," *SIAM Journal on Computing*, 24:296-317, 1995.
5. V. Melkonian and E. Tardos, "Approximation algorithms for a directed network design problem," In the *Proceedings of the 7th International Integer Programming and Combinatorial Optimization Conference* (IPCO'99), Graz, Austria, 1999.
6. A. Schrijver, "Matroids and linking systems," *J. Combinatorial Theory, Series B*, 26:349–369, 1979.
7. D. Williamson, M. Goemans, M. Mihail, and V. Vazirani, "A primal-dual approximation algorithm for generalized Steiner network problems," *Combinatorica*, 15:435-454, 1995.

The Asymptotic Performance Ratio of an On-Line Algorithm for Uniform Parallel Machine Scheduling with Release Dates[*]

Cheng-Feng Mabel Chou[1], Maurice Queyranne[2], and David Simchi-Levi[3]

[1] Northwestern University, Evanston IL 60208, USA
[2] University of British Columbia, Vancouver B.C., Canada V6T 1Z2
[3] Massachusetts Institute of Technology, Cambridge MA 02139, USA

Abstract. Jobs arriving over time must be non-preemptively processed on one of m parallel machines, each of which running at its own speed, so as to minimize a weighted sum of the job completion times. In this on-line environment, the processing requirement and weight of a job are not known before the job arrives. The Weighted Shortest Processing Requirement (WSPR) on-line heuristic is a simple extension of the well known WSPT heuristic, which is optimal for the single machine problem without release dates. We prove that the WSPR heuristic is asymptotically optimal for all instances with bounded job processing requirements and weights. This implies that the WSPR algorithm generates a solution whose relative error approaches zero as the number of jobs increases. Our proof does not require any probabilistic assumption on the job parameters and relies extensively on properties of optimal solutions to a single machine relaxation of the problem.

1 Introduction

In the **uniform parallel machine minsum scheduling problem with release dates**, jobs arrive over time and must be allocated for processing to one of m given parallel machines. Machine M_i ($i = 1, \ldots, m$) has *speed* $s_i > 0$ and can process at most one job at a time. Let n denote the total number of jobs to be processed and let $N = \{1, 2, \ldots, n\}$. Job $j \in N$ has *processing requirement* $p_j \geq 0$, *weight* $w_j > 0$, and *release date* $r_j \geq 0$. The processing of job j cannot start before its release date r_j and cannot be interrupted once started on a machine. If job j starts processing at time S_j on machine M_i, then it is completed p_j/s_i time units later; that is, its completion time is $C_j = S_j + p_j/s_i$. In the single machine case, i.e., when $m = 1$, we may assume that $s_1 = 1$ and in this case the processing requirement of a job is also referred to as the job *processing time*.

---

[*] Research supported in part by ONR Contracts N00014-95-1-0232 and N00014-01-1-0146, NSF Contracts DDM-9322828 and DMI-9732795, and a research grant from the Natural Sciences and Research Council of Canada (NSERC).

We seek a feasible schedule of all n jobs, which minimizes the *minsum* objective $\sum_{j=1}^n w_j C_j$, the weighted sum of completion times. In standard scheduling notation, see, e.g., [5], this problem is denoted $Q|r_j|\sum w_j C_j$. Our main result concerns the case where the set of m parallel machines is held fixed, which is usually denoted as problem $Qm|r_j|\sum w_j C_j$.

In practice, the precise processing requirement, weight and release date (or arrival time) of a job may not be known before the job actually arrives for processing. Thus, we consider an *on-line* environment where these data p_j, w_j and r_j are not known before time r_j. This implies that scheduling decisions have to be made over time, using at any time only information about the jobs already released by that time; see, e.g., [10] for a survey of on-line scheduling.

In a competitive analysis, we compare the objective value, $Z^A(I)$, of the schedule obtained by applying a given (deterministic) on-line algorithm A to an instance I to the optimum (off-line) objective value $Z^*(I)$ of this instance. The *competitive ratio* of algorithm A, relative to a class $\mathcal{I}$ of instances (such that $Z^*(I) > 0$ for all $I \in \mathcal{I}$), is

$$c_{\mathcal{I}}(A) = \sup\left\{\frac{Z^A(I)}{Z^*(I)} : I \in \mathcal{I}\right\}.$$

One of the best known results [11] in minsum scheduling is that the single machine problem without release dates, $1||\sum w_j C_j$, is solved to optimality by the following **Weighted Shortest Processing Time** (WSPT) algorithm: process the jobs in nonincreasing order of their weight-to-processing time ratio w_j/p_j. Thus the competitive ratio $c_{\mathcal{I}}(WSPT) = 1$ for the class $\mathcal{I}$ of all instances of the single machine problem $1||\sum w_j C_j$. Unfortunately, this result does not extend to problems with release dates or with parallel machines; in fact the single machine problem with release dates and equal weights, $1|r_j|\sum C_j$, and the identical parallel machine problem, $P||\sum w_j C_j$ are already NP-hard [8]. Consequently, a great deal of work has been devoted to the development and analysis of heuristics, in particular, Linear Programming (LP) based heuristics, with attractive competitive ratios.

A departure from this line of research was presented in [6]. To present their results, define the *asymptotic performance ratio* $R_{\mathcal{I}}^\infty(A)$ of an algorithm A, relative to instance class $\mathcal{I}$, as

$$R_{\mathcal{I}}^\infty(A) \doteq \inf\left\{r \geq 1 \mid \exists n_0 \text{ such that } \frac{Z^A(I)}{Z^*(I)} \leq r,\right.$$

$$\left. \text{for all instances } I \in \mathcal{I} \text{ with } n \geq n_0\right\}.$$

Thus, the asymptotic performance ratio characterizes the maximum relative deviation from optimality for all "sufficiently large" instances in $\mathcal{I}$. When A is an on-line algorithm, and $Z^*(I)$ still denotes the off-line optimum objective value, we call $R_{\mathcal{I}}^\infty(A)$ the *asymptotic competitive ratio* of A relative to instance class $\mathcal{I}$.

[6] focused on the single machine total completion time problem with release dates, i.e., problem $1|r_j|\sum C_j$ and analyzed the effectiveness of a simple on-line

dispatch rule, referred to as the **Shortest Processing Time among Available** jobs (SPTA) heuristic. In this algorithm, at the completion time of any job, one considers all the jobs which have been released by that date but not yet processed, and select the job with the smallest processing time to be processed next. If no job is available the machine is idle until at least one job arrives. The results in [6] imply that the asymptotic competitive ratio of the SPTA heuristic is equal to *one* for all classes of instances with bounded processing times, i.e., instance classes $\mathcal{I}$ for which there exist constants $\bar{p} \geq \underline{p} \geq 0$ such that $\underline{p} < p_j < \bar{p}$ for all jobs j in every instance $I \in \mathcal{I}$.

It is natural to try and extend the WSPT and SPTA heuristics to problems with release dates and/or parallel machines. A simple extension to the problem $Q|r_j| \sum w_j C_j$ considered herein, with both uniform parallel machines and release dates, is the following *WSPR algorithm:* whenever a machine becomes idle, start processing on it an available job, if any, with largest w_j/p_j ratio; otherwise, wait until the next job release date. This is a very simple on-line algorithm which is fairly myopic and clearly suboptimal, for at least two reasons. First, it is "non-idling", that is, it keeps the machines busy so long as there is work available for processing; this may be suboptimal if a job k with large weight w_k (or short processing requirement p_k) is released shortly thereafter and is forced to wait because all machines are then busy. Second, for machines with different speeds, the WSPR algorithm arbitrarily assigns jobs to the idle machines, irrespective of their speeds; thus an important job may be assigned to a slow machine while a faster machine is currently idle or may become idle soon thereafter. Thus, it is easy to construct instance classes $\mathcal{I}$, for example with two jobs on a single machine, for which the WSPR heuristic performs very poorly; this implies that its competitive ratio $c_{\mathcal{I}}(WSPR)$ is unbounded.

In contrast, the main result of this paper is that the *asymptotic* competitive ratio $R_{\mathcal{I}}^{\infty}(WSPR)$ of the WSPR heuristic is equal to *one* for all classes $\mathcal{I}$ of instances with a fixed set of machines and with bounded job weights and processing requirements.

Formally, our main result is presented in the following theorem.

Theorem 1. *Consider any class $\mathcal{I}$ of instances of the uniform parallel machines problem $Qm|r_j| \sum w_j C_j$ with a fixed set of m machines, and with bounded weights and processing times, that is, for which there exist constants $\bar{w} \geq \underline{w} > 0$ and $\bar{p} \geq \underline{p} > 0$ such that*

$$\underline{w} \leq w_j \leq \bar{w} \quad and \quad \underline{p} \leq p_j \leq \bar{p} \quad for \ all \ jobs \ j \ in \ every \ instance \ I \in \mathcal{I}.$$

Then the asymptotic competitive ratio of the WSPR heuristic is $R_{\mathcal{I}}^{\infty}(WSPR) = 1$ for instance class $\mathcal{I}$.

To put our results in perspective, it is appropriate at this point to refer to the work of Uma and Wein [12] who perform extensive computational studies with various heuristics including WSPR as well as linear programming based approximation algorithms for the single machine problem $1|r_j| \sum w_j C_j$. While Uma and Wein note that it is trivial to see that the worst-case performance of the WSPR heuristic is unbounded, they find that, on most data sets they used, this heuristic is superior to all the LP relaxation based approaches. The results in the present paper provide a nice explanation of this striking behavior reported by Uma and Wein. Indeed, our results show that if the job parameters, i.e., weights and processing times, are bounded, then the WSPR algorithm generates a solution whose relative error decreases to zero as the number of jobs increases. Put differently, WSPR has an unbounded worst-case performance *only when* the job parameters are unbounded.

2   A Mean Busy Date Relaxation for Uniform Parallel Machines

Let $N = \{1, \ldots, n\}$ be a set of jobs to be processed, with a given vector $p = (p_1, \ldots, p_n)$ of job processing requirements. Given any (preemptive) schedule we associate with each job $j \in N$ its *processing speed function* σ_j, defined as follows: for every date t we let $\sigma_j(t)$ denote the speed at which job j is being processed at date t. For example, for uniform parallel machines, $\sigma_j(t) = s_{i(j,t)}$ is the speed of the machine $M_{i(j,t)}$ processing job j at date t, and $\sigma_j(t) = 0$ if job j is idle at that date. Thus, for a single machine with unit speed, $\sigma_j(t) = 1$ if the machine is processing job j at date t, and 0 otherwise. We consider schedules that are *complete* in the following sense. First we assume that $0 \leq \sigma_j(t) \leq \bar{s}$ for all j and t, where $\bar{s}$ is a given upper bound on the maximum speed at which a job may be processed. Next, we assume that all processing occurs during a finite time interval $[0, T]$, where T is a given upper bound on the latest job completion time in a schedule under consideration. For the single machine problem we may use $\bar{s} = 1$ and $T = \max_j r_j + \sum_j p_j$. For the uniform parallel machines problem, we may use $\bar{s} = \max_i s_i$ and $T = \max_j r_j + \sum_j p_j / \min_i s_i$. The assumption

$$\int_0^T \sigma_j(\tau) \, d\tau = p_j$$

then express the requirement that, in a complete schedule, each job is entirely processed during this time interval $[0, T]$. The preceding assumptions imply that all integrals below are well defined.

The *mean busy date* M_j of job j in a complete schedule is the average date at which job j is being processed, that is,

$$M_j = \frac{1}{p_j} \int_0^T \sigma_j(\tau) \, \tau \, d\tau \ .$$

We let $M = (M_1, \ldots, M_n)$ denote the mean busy date vector, or *MBD vector*, of the schedule. When the speed function σ_j is piecewise constant, we may express the mean busy date M_j as the weighted average of the midpoints of the time intervals during which job j is processed at constant speed, using as weights the fraction of its work requirement p_j processed in these intervals. Namely, if

$\sigma_j(t) = s_{j,k}$ for $a_k < t < b_k$, with $0 \leq a_1 < b_1 < \ldots < a_K < b_K \leq T$ and $\sum_{k=1}^{K} s_{j,k}(b_k - a_k) = p_j$ then

$$M_j = \sum_{k=1}^{K} \frac{s_{j,k}(b_k - a_k)}{p_j} \frac{a_k + b_k}{2}. \tag{1}$$

Thus, if job j is processed without preemption at speed $s_{j,1}$, then its completion time is $C_j - M_j \mid \frac{1}{2}p_j/s_{j,1}$. In any complete schedule, the completion time of every job j satisfies $C_j \geq M_j + \frac{1}{2}p_j/\bar{s}$, with equality if and only if job j is processed without preemption at maximum speed $\bar{s}$.

Let $w = (w_1, \ldots, w_n)$ denote the vector of given job weights. We use scalar product notation, and let $w^\top p = \sum_j w_j p_j$ and $w^\top C = \sum_j w_j C_j$, the latter denoting the minsum objective of instance I of the scheduling problem under consideration. We call $w^\top M = \sum_j w_j M_j$ the *mean busy date objective*, or *MBD objective*, and the problem

$$Z^{P-MBD}(I) = \min\left\{ w^\top M : M \text{ is the MBD vector of a feasible}\right.$$

$$\left.\text{preemptive schedule for instance } I \right\} \tag{2}$$

the *preemptive MBD problem*. Letting $Z^*(I)$ denote the optimum minsum objective value $w^\top C$ of a feasible nonpreemptive schedule, it follows from $w \geq 0$ and the preceding observations that $Z^*(I) \geq Z^{P-MBD}(I) + (1/2\bar{s})w^\top p$. Accordingly, we shall also refer to the preemptive MBD problem as the *preemptive MBD relaxation* of the original nonpreemptive minsum scheduling problem.

The preemptive MBD problem is well solved, see [3] and [4], for the case of a single machine with constant speed $s > 0$ and job release dates. To present the result, define the *LP schedule* as the following schedule: whenever a new job is released or a job is completed, we compare the weight to (original) processing time ratios w_j/p_j of all the available jobs: the one with the largest ratio is selected and starts (or resumes) processing immediately, even if this forces the preemption of a currently in-process job.

Theorem 2 ((Goemans)). *The LP schedule defines an optimal solution to the preemptive MBD problem* $1|r_j, pmtn| \sum w_j M_j$.

Let C^{LP} and M^{LP} denote the completion time vector, resp., the MBD vector, of the LP schedule, and let $Z^{MBD}(I) = Z^{P-MBD}(I) + (1/2\bar{s})w^\top p$. Theorem 2 implies that $Z^*(I) \geq Z^{MBD}(I) = w^\top M^{LP} + (1/2\bar{s})w^\top p$.

We will bound the maximum delay that certain amounts of "work" can incur in the WSPR schedule, relative to the LP schedule. For this, we now present a decomposition of the MBD objective $w^\top M$ using certain "nested" job subsets. Some of the results below were introduced in [3] in the context of single machine scheduling.

We consider a general scheduling environment and a *complete schedule*, as defined at the beginning of this Section. Assume, without loss of generality, that the jobs are indexed in a nonincreasing order of their ratios of weight to processing requirement:

$$w_1/p_1 \geq w_2/p_2 \geq \ldots \geq w_n/p_n \geq w_{n+1}/p_{n+1} = 0. \tag{3}$$

Accordingly, job k has *lower WSPR priority* than job j if and only if $k > j$. In case of ties in (3), we consider WSPR priorities and an LP schedule which are consistent with the WSPR schedule. For $h = 1, \ldots, n$, let $\Delta_h = w_h/p_h - w_{h+1}/p_{h+1}$, and let $[h] = \{1, 2, \ldots, h\}$ denote the set of the h jobs with highest priority. For any feasible (preemptive) schedule we have:

$$w^\top M = \sum_{j=1}^{n} \frac{w_j}{p_j} p_j M_j = \sum_{j=1}^{n} \left(\sum_{k=j}^{n} \Delta_k \right) p_j M_j = \sum_{h=1}^{n} \Delta_h \sum_{j \in [h]} p_j M_j .$$

For any subset $S \subseteq N = \{1, \ldots, n\}$, let $p(S) = \sum_{j \in S} p_j$ denote its total processing time, and let $\sigma_S = \sum_{j \in S} \sigma_j$ denote its processing speed function. Define its mean busy date $M_S = (1/p(S)) \int_0^T \sigma_S(\tau) \tau \, d\tau$. Note that, in a feasible schedule, $\int_0^T \sigma_S(\tau) \, d\tau = p(S)$ and $\sum_{j \in S} p_j M_j = \int_0^T \sigma_S(\tau) \tau \, d\tau = p(S) M_S$. Therefore, we obtain the *MBD objective decomposition*

$$w^\top M = \sum_{h=1}^{n} \Delta_h \, p([h]) \, M_{[h]} . \tag{4}$$

This decomposition allows us to concentrate on the mean busy dates $M_{[h]}$ of the job subsets $[h]$ ($h = 1, \ldots, n$).

For any date $t \leq T$, let $R_S(t) = \int_t^T \sigma_S(\tau) \, d\tau = p(S) - \int_0^t \sigma_S(\tau) \, d\tau$ denote the *unprocessed work* from set S at date t. (Note that this unprocessed work may include the processing time of jobs not yet released at date t.) Since the unprocessed work function $R_S(t)$ is nonincreasing with time t, we may define its (functional) inverse $\bar{R}_S$ as follows: for $0 \leq q \leq p(S)$ let $\bar{R}_S(q) = \inf\{t \geq 0 : R_S(t) \leq q\}$. Thus the *processing date* $\bar{R}_S(q)$ is the earliest date at which $p(S) - q$ units of work from set S have been processed. For any feasible schedule with a finite number of preemptions we have

$$\int_0^T R_S(t) \, dt = \int_0^{p(S)} \bar{R}_S(q) \, dq .$$

The mean busy date M_S can be expressed using the processing date function $\bar{R}_S$:

$$p(S)M_S = \int_0^T \sigma_S(t) t \, dt = \int_0^T \int_0^t \sigma_S(t) \, d\tau \, dt = \int_0^T \int_\tau^T \sigma_S(t) \, dt \, d\tau$$

$$= \int_0^T R_S(\tau) \, d\tau = \int_0^{p(S)} \bar{R}_S(q) \, dq .$$

Combining this with equation (4) allows us to express the MBD objective using the processing date function:

$$w^\top M = \sum_{h=1}^{n} \Delta_h \int_0^{p([h])} \bar{R}_{[h]}(q)\, dq \; . \tag{5}$$

We now present a mean busy date relaxation for uniform parallel machines and then use the above expression (5) to bound the difference between the min-sum objectives of the WSPR and LP schedules.

Assume that we have m parallel machines $M_1, \ldots, M_m$, where machine M_i has *speed* $s_i > 0$. Job j has processing requirement $p_j > 0$; if it is processed on machine M_i then its actual processing time is $p_{ij} = p_j/s_i$. We assume that the set of machines and their speeds are fixed and, without loss of generality, that the machines are indexed in nonincreasing order of their speeds, that is, $s_1 \geq s_2 \geq \ldots \geq s_m > 0$. We have job release dates $r_j \geq 0$ and weights $w_j \geq 0$, and we seek a nonpreemptive feasible schedule in which no job j is processed before its release date, and which minimizes the minsum objective $\sum_j w_j C_j$. Since the set of m parallel machines is fixed, this problem is usually denoted as $Qm|r_j| \sum w_j C_j$. First, we present a fairly natural single machine preemptive relaxation, using a machine with speed $s^{[m]} = \sum_{i=1}^{m} s_i$. We then compare the processing date functions for the high WSPR priority sets $[h]$ between the WSPR schedule on the parallel machines and LP schedule on the speed-$s^{[m]}$ machine. We show an $O(n)$ additive error bound for the WSPR heuristic, for every instance class with a fixed set of machines and bounded job processing times and weights. This implies the asymptotic optimality of the WSPR heuristic for such classes of instances.

Consider any feasible preemptive schedule on the parallel machines, with completion time vector C. Recall that $\sigma_j(t)$ denotes the speed at which job j is being processed at date t in the schedule, and that the mean busy date M_j of job j is $M_j = (1/p_j) \int_0^T \sigma_j(\tau) \tau\, d\tau$ (where T is an upper bound on the maximum job completion time in any schedule being considered).

To every instance I of the uniform parallel machines problem we associate an instance $I^{[m]}$ of the single machine preemptive problem with the same job set N and in which each job $j \in N$ now has processing time $p_j^{[m]} = p_j/s^{[m]}$. The job weights w_j and release dates r_j are unchanged. Thus we have replaced the m machines with speeds $s_1, \ldots, s_m$ with a single machine with speed $s^{[m]} = \sum_{i=1}^{m} s_i$. Consider any feasible preemptive schedule for this single machine problem and let $C^{[m]}$ denote its completion time vector. Let $I_j^{[m]}(t)$ denote the speed (either $s^{[m]}$ or 0) at which job j is being processed at date t. Thus the mean busy date $M_j^{[m]}$ of job j for this single machine problem is $M_j^{[m]} = (1/p_j) \int_0^T I_j^{[m]}(\tau)\, \tau\, d\tau$.

In the following Lemma, the resulting inequality $C_j^{[m]} \leq C_j$ on all job completion times extends earlier results of [1] for the case of identical parallel machines (whereby all $s_i = 1$), and of [9] for a broad class of shop scheduling problems (with precedence delays but without parallel machines). To our knowledge, the mean busy date result, $M_j^{[m]} = M_j$, which we use later on, is new.

Lemma 1 ((Preemptive Single Machine Relaxation Lemma)). *To every feasible (preemptive or nonpreemptive) schedule with a finite number of pre-emptions[1] and with mean busy date vector M and completion time vector C on the uniform parallel machines, we can associate a feasible preemptive schedule with mean busy date vector $M^{[m]}$ and completion time vector $C^{[m]}$ on the speed-$s^{[m]}$ machine, such that $M_j^{[m]} = M_j$ and $C_j^{[m]} \leq C_j$ for all jobs $j \in N$.*

Proof. Let S_j denote the start date of job j in the given parallel machines schedule. Partition the time interval $[\min_j S_j, \max_j C_j]$ into intervals $[a_{t-1}, a_t]$ $(t = 1, \ldots, \tau)$ such that exactly the same jobs are being processed by exactly the same machines throughout each interval. Thus $\{a_t : t = 0, \ldots, \tau\}$ is the set of all job start dates and completion times, and all dates at which some job is preempted. Partition each job j into τ pieces (j, t) with *work amount* $q_{jt} = s_{i(j,t)}(a_t - a_{t-1})$ if job j is being processed during interval $[a_{t-1}, a_t]$ on a machine $M_{i(j,t)}$, and zero otherwise. Since each job j is performed in the given schedule, its processing requirement is $p_j = \sum_{t=1}^{\tau} q_{jt}$. Since each machine processes at most one job during interval $[a_{t-1}, a_t]$, we have $\sum_{j \in N} q_{jt} \leq s^{[m]}(a_t - a_{t-1})$ for all t, with equality iff no machine is idle during interval $[a_{t-1}, a_t]$. Therefore the speed-$s^{[m]}$ machine has enough capacity to process all the work $\sum_{j \in N} q_{jt}$ during this interval. Construct a preemptive schedule on the speed-$s^{[m]}$ machine as follows. For each $t = 1, \ldots, \tau$, fix an arbitrary sequence $(j_1, t), \ldots, (j_{n(t)}, t)$ of the $n(t)$ pieces (j, t) with $q_{jt} > 0$. Starting at date a_{t-1} process half of each such piece (j, t) (i.e., for $\frac{1}{2} q_{jt}/s^{[m]}$ time units) in the given sequence. This processing is complete no later than date $\mu_t = \frac{1}{2}(a_{t-1} + a_t)$, the midpoint of the interval $[a_{t-1}, a_t]$. Then "mirror" this partial schedule about this midpoint μ_t by processing the other half of each piece in reverse sequence so as to complete this mirrored partial schedule precisely at date a_t. Since no job starts before its release date, all the processing requirement of every job is processed, and the speed-$s^{[m]}$ machine processes at most one job at a time, the resulting preemptive schedule is indeed feasible. Furthermore each job j completes at the latest at date $\max\{a_t : q_{jt} > 0\} = C_j$, so $C_j^{[m]} \leq C_j$. Finally, the "mirroring" applied in each interval $[a_{t-1}, a_t]$ ensures that, for all jobs $j \in N$

$$\int_{a_{t-1}}^{a_t} \sigma_j^{[m]}(\tau) \, \tau \, d\tau = \frac{q_{jt}}{s^{[m]}} \, s^{[m]} \, \mu_t = \frac{q_{jt}}{s_i} \, s_i \, \mu_t = \int_{a_{t-1}}^{a_t} \sigma_j(\tau) \, \tau \, d\tau$$

where s_i is the speed of the machine M_i on which job j is processed during interval $[a_{t-1}, a_t]$ in the given parallel machines schedule. Adding over all intervals implies $M_j^{[m]} = M_j$ for all jobs $j \in N$. The proof is complete.

Lemma 1 implies that the preemptive single machine problem, with a speed-$s^{[m]}$ machine, is a relaxation of the original uniform parallel machines problem, for any objective function (including the minsum objective with all $w_j \geq 0$) which is nondecreasing in the job completion times. For the minsum objective $\sum_j w_j C_j$, we may combine this result with Theorem 2 and obtain:

---

[1] The finiteness restriction may be removed by appropriate application of results from open shop theory, as indicated in [7], but this is beyond the scope of this paper.

Corollary 1. *Let $Z^*(I)$ denote the optimum objective value for instance I of the parallel machines problem $Q|r_j|\sum w_j C_j$. Let $M^{LP[m]}$ denote the mean busy date vector of the LP schedule for the corresponding instance $I^{[m]}$ of the single machine problem. Then*

$$Z^{MBD[m]}(I) \doteq w^\top M^{LP[m]} + \frac{1}{2\,s^{[m]}}\, w^\top p \le Z^*(I) . \tag{6}$$

Proof. Let $Z^{[m]}(I)$ denote the optimum value of the minsum objective $\sum_j w_j C^{[m]}_j$ among all feasible preemptive schedules for instance $I^{[m]}$. From Theorem 2 it follows that $w^\top M^{LP[m]} + \frac{1}{2\,s^{[m]}}\, w^\top p \le Z^{[m]}(I)$. From the inequalities $C^{[m]}_j \le C_j$ in Lemma 1 and $w \ge 0$, it follows that $Z^{[m]}(I) \le Z^*(I)$. This suffices to prove the corollary.

Remark 1. For the problem $P||\sum w_j C_j$ with identical parallel machines and all release dates $r_j = 0$, each $s_i = 1$. Therefore, for any nonpreemptive parallel machines schedule, the mean busy date M_j and completion time C_j of every job j satisfy $M^j = C^j - \frac{1}{2} p_j$. On the other hand $s^{[m]} = m$ and, since the LP schedule is nonpreemptive for identical release dates, $M^{LP[m]}_j = C^{LP[m]}_j - p_j/2m$. Applying the mean busy date relationships $M^{[m]}_j = M^*_j$ of Lemma 1 to the MBD vector $M^* = C^* - \frac{1}{2}p$ of an optimal parallel machine schedule, we obtain the slightly stronger bound:

$$Z^*(I) = w^\top C^* = w^\top M^* + \frac{1}{2}\,w^\top p = w^\top M^{[m]} + \frac{1}{2}\,w^\top p$$

$$\ge w^\top M^{LP[m]} + \frac{1}{2}\,w^\top p = w^\top C^{LP[m]} + \frac{1}{2}\left(1 - \frac{1}{m}\right)w^\top p . \tag{7}$$

Let $Z_n(I) = w^\top p$ denote the optimum value of the n-machine version of the problem, and $Z_1(I) = w^\top \left(M^{LP[m]} + \frac{1}{2m}w^\top p\right)$ the minsum objective value of the LP schedule for instance $I^{[m]}$ of the single speed-m machine version of the problem. Recall that, in the absence of release dates, $Z_1(I)$ is the optimum value of a feasible nonpreemptive schedule on a single machine operating at m times the speed of each given parallel machine. Inequality (7) may be written as

$$Z^*(I) - \frac{1}{2}Z_n(I) \ge \frac{1}{m}\left(Z_1(I) - \frac{1}{2}Z_n(I)\right)$$

which is precisely the lower bound obtained in [2] using algebraic and geometric arguments.

3   Asymptotic Optimality of the WSPR Rule for Uniform Parallel Machines

We now show the asymptotic optimality of the WSPR rule for uniform parallel machines. The simple version of the WSPR heuristic considered herein is defined as follows: whenever a machine becomes idle, start processing the available job, if any, with highest WSPR priority, i.e., job j such that $j < k$ according to (3); if no job is available, wait until the next job release date. Note that we allow the assignment of jobs to machines to be otherwise arbitrary. (We suspect that one can design versions of the uniform parallel machine WSPR heuristic which may be preferable according to some other performance measure, but this is not needed for the present asymptotic analysis.) As before, let C_j^{WSPR} (resp., M_j^{WSPR}) denote the completion time (resp., mean busy date) of job j in the WSPR schedule.[2] Recall that s_m is the speed of the slowest machine and, to simplify, let $p_{\max} = \max_{j \in N} p_j$.

Following [2], it is easy to obtain a job-by-job bound for the WSPR schedule in the absence of release dates:

Lemma 2 ((Job-by-Job Bound Without Release Dates)). *For the uniform parallel machines problem $Q||\sum w_j C_j$ without release dates, the completion time vectors of the WSPR and LP schedules satisfy*

$$C_j^{WSPR} \le C_j^{LP} + \left(\frac{1}{s_m} - \frac{1}{s^{[m]}}\right) p_{\max} \quad \text{for all } j \in N. \tag{8}$$

Proof. Assuming the jobs are ranked according to WSPR order (3), the completion time of job j in the LP schedule is $C_j^{LP} = p([j])/s^{[m]}$. In the WSPR schedule, job j starts at the earliest completion time of a job in $[j-1]$, that is, no later than $p([j-1])/s^{[m]}$, and completes at most p_j/s_m time units later. Therefore $C_j^{WSPR} \le C_j^{LP} + (1/s_m - 1/s^{[m]}) p_j$. This implies (8). $\square$

We now turn to the case with release dates $r_j \ge 0$. Let $N(i)$ denote the set of jobs processed on machine M_i in the WSPR schedule. Since $M_j^{WSPR} = C_j^{WSPR} - \frac{1}{2} w_j p_j / s_i$ for all $j \in N(i)$, we have

$$Z^{WSPR} = w^\top M^{WSPR} + \frac{1}{2} \sum_{i=1}^{m} \sum_{j \in N(i)} w_j \frac{p_j}{s_i} \le w^\top M^{WSPR} + \frac{1}{2 s_m} w^\top p. \tag{9}$$

Combining inequalities (6) and (9) with the decomposition (5) of the MBD objective, we only need to compare the processing date functions of the speed-$s^{[m]}$ machine LP schedule and of the parallel machines WSPR schedule. The next Lemma shows that, for any instance with a fixed set of machines, no amount of work from any set $[h]$ can, in the parallel machines WSPR schedule, be delayed, relative to the single machine LP schedule, by more than a constant multiple of $p_{\max}$ time units.

---

[2] To properly speak of "*the* WSPR schedule" we would need to define a rule for assigning jobs to machines in case several machines are available when a job starts processing. For example, we may assign the highest priority available job to a fastest available machine. In fact, our analysis applies to *any* nonpreemptive feasible schedule which is consistent with the stated WSPR priority rule, irrespective of the details of such machine assignments.

Lemma 3 ((Uniform Parallel Machines Work Delay Lemma)).   *Assume the jobs are ranked according to the WSPR order (3). Consider the WSPR schedule on uniform parallel machines, and the speed-$s^{[m]}$ machine LP schedule defined above. Then, for all $h \le n$,*

$$\bar{R}_{[h]}^{WSPR}(q) \le \bar{R}_{[h]}^{LP}(q) + \left(\frac{1}{s_1} + \frac{m-1}{s_m} + \frac{s^{[m]}}{(s_m)^2} \right) p_{\max} \quad \text{for all } 0 < q \le p([h]) \, .$$

$$(10)$$

Proof. We fix $h \in \{1, \dots, n\}$ and we define

$$\alpha = m - 1 + \frac{s^{[m]}}{s_m} \, .$$

We start by considering the LP schedule on the speed-$s^{[m]}$ machine. Let $[a_k, b_k]$ (where $k = 1, \dots, K$) denote the disjoint time intervals during which set $[h]$ is being processed continuously in the LP schedule. Thus $0 \le a_1$ and $b_{k-1} < a_k$ for $k = 2, \dots, K$. The unprocessed work function $R_{[h]}^{LP}$ starts with $R_{[h]}^{LP}(t) = p([h])$ for $0 \le t \le a_1$; decreases at rate $s^{[m]}$ in the intervals $[a_k, b_k]$ while remaining constant outside these intervals; and it ends with $R_{[h]}^{LP}(t) = 0$ for $b_K \le t \le T$. Let $J(k) = \{j \in [h] : a_k < C_j^{LP} \le b_k\}$ denote the set of jobs in $[h]$ that are processed during time interval $[a_k, b_k]$ in the LP schedule. Note that $a_k = \min_{j \in J(k)} r_j$ and $b_k = a_k + p(J(k))/s^{[m]}$. Furthermore, $Q_k = \sum_{\ell > k} p(J(\ell))$ is the total work from set $[h]$ released *after* date b_k, where $Q_K = 0$ and $Q_0 = p([h])$. For all $k = 1, \dots, K$ we have $Q_k = Q_{k-1} - p(J(k))$. In the interval $[Q_k, Q_{k-1})$ the processing date function $\bar{R}_{[h]}^{LP}$ decreases at rate $1/s^{[m]}$ from $\bar{R}_{[h]}^{LP}(Q_k) = b_k$. Thus $\bar{R}_{[h]}^{LP}(q) = a_k + (Q_{k-1} - q)/s^{[m]}$ for all $Q_k \le q < Q_{k-1}$.

Now consider the WSPR schedule on the uniform parallel machines and fix an interval $[a_k, b_k)$. We claim that, for every $k = 1, \dots, K$ and every date $a_k \le t < b_k$ the unprocessed work

$$R_{[h]}^{WSPR}(t) \le R_{[h]}^{LP}(t) + \alpha \, p_{\max} \, .$$

$$(11)$$

By contradiction, assume that (11) is violated at date $t \in [a_k, b_k)$. Let $\hat{t} = \inf\{t : (11) \text{ is violated}\}$. Since the functions $R_{[h]}^{WSPR}$ and $R_{[h]}^{LP}(t)$ are continuous, $R_{[h]}^{WSPR}(\hat{t}) \ge R_{[h]}^{LP}(\hat{t}) + \alpha \, p_{\max}$, and the difference $R_{[h]}^{WSPR}(t) - R_{[h]}^{LP}(t)$ is strictly increasing immediately to the right of $\hat{t}$. But since $R_{[h]}^{LP}$ is constant outside the intervals $[a_k, b_k]$ and $R_{[h]}^{WSPR}$ then we must have $a_k \le \hat{t} < b_k$ for some $k \in \{1, \dots, K\}$. This implies that at least one machine M_i is not processing a job in $[h]$ immediately after date $\hat{t}$. If at least one machine is idle just after date $\hat{t}$ then let $\theta = \hat{t}$; otherwise, let $\theta \le \hat{t}$ be the latest start date of a job not in $[h]$ and in process just after date $\hat{t}$. Since no job in $[h]$ was available for processing at date θ, then all work released no later than θ must either have been completed, or be started on a machine $M_u \neq M_i$. Note that at least $p([h]) - R_{[h]}^{LP}(\theta)$ units of work have been released by date θ. On the other hand, a total of at most

$(m-1)p_{\max}$ units of work can be started on machines $\mathsf{M}_u \neq \mathsf{M}_i$ just after date $\hat{t}$. Therefore

$$p([h]) - R_{[h]}^{LP}(\theta) \le p([h]) - R_{[h]}^{WSPR}(\theta) + (m-1)p_{\max}$$

If $\theta < \hat{t}$ then the job $j \notin [h]$ started at date θ has processing requirement $p_j \le p_{\max}$ and is processed at least at the slowest machine speed s_m. Since this job is still in process at date $\hat{t}$, we must have $\hat{t} < \theta + p_{\max}/s_m$. The unprocessed work function $R_{[h]}^{LP}$ decreases by at most $s^{[m]}(\hat{t} - \theta)$ between dates θ and $\hat{t}$, whereas $R_{[h]}^{WSPR}$ is nonincreasing. Therefore

$$
\begin{aligned}
R_{[h]}^{LP}(\hat{t}) &\ge R_{[h]}^{LP}(\theta) - s^{[m]}(\hat{t} - \theta) \\
&> R_{[h]}^{LP}(\theta) - s^{[m]}\frac{p_{\max}}{s_m} \\
&\ge R_{[h]}^{WSPR}(\theta) - (m-1)p_{\max} - s^{[m]}\frac{p_{\max}}{s_m} \\
&\ge R_{[h]}^{WSPR}(\hat{t}) - \alpha\, p_{\max} \\
&\ge R_{[h]}^{LP}(\hat{t}) \,,
\end{aligned}
$$

a contradiction. Thus claim (11) is proved.

Claim (11) implies that the processing date functions satisfy

$$\bar{R}_{[h]}^{WSPR}(q) \le \bar{R}_{[h]}(q) + \frac{\alpha\, p_{\max}}{s^{[m]}} \quad \text{whenever } Q_k + \alpha\, p_{\max} < q < Q_{k-1}\,. \quad (12)$$

Let $\hat{q} = \min\{Q_k + \alpha\, p_{\max},\ Q_{k-1}\}$ and consider the last $\hat{q}$ units of work released from set $J(k)$. If $\hat{q} < Q_{k-1}$ then claim (11) implies that $\bar{R}_{[h]}^{WSPR}(\hat{q}) \le b_k$, that is, the first $p([h]) - \hat{q}$ units of work are completed by date b_k. If some of the remaining $\hat{q}$ units of work is being processed at a date $\tilde{t} > b_k$ then, since all work from $J(k)$ has been released by date b_k, there will be no date τ at which no work from $J(k)$ is in process until all this work from $J(k)$ is completed, that is, until date $\bar{R}_{[h]}^{WSPR}(Q_k)$. Furthermore, this work is processed at least at the minimum speed $s_m > 0$, so $\bar{R}_{[h]}^{WSPR}(Q_k) \le \tilde{t} + \hat{q}/s_m$. Note also that, unless $\bar{R}_{[h]}^{WSPR}(Q_k) = b_k$, a machine becomes available for processing these $\hat{q}$ units of work between dates b_k and $b_k + p_{\max}/s_1$, where s_1 is the fastest speed of a machine. Thus, for $Q_k \le q \le Q_k + \hat{q}$ we have

$$\bar{R}_{[h]}^{LP}(q) = b_k - (q - Q_k)/s^{[m]} \quad \text{and} \quad \bar{R}_{[h]}^{WSPR}(q) \le b_k + \frac{p_{\max}}{s_1} + \frac{\alpha\, p_{\max} - q}{s_m}\,.$$
$$(13)$$

Inequalities (12) and (13) imply (10) and the proof is complete.

Integrating inequality (10) from 0 to $p([h])$ implies

$$\int_0^{p([h])} \left(\bar{R}_{[h]}^{WSPR}(q) - \bar{R}_{[h]}^{LP}(q) \right) dq \le \left(\frac{1}{s_1} + \frac{m-1}{s_m} + \frac{s^{[m]}}{(s_m)^2} \right) p_{\max}\, p([h])\,.$$
$$(14)$$

The next theorem combines inequality (14) with the cost decomposition (5) and inequalities (6) and (9), to derive a $O(n)$ bound on the difference between the minsum objective values of the parallel machines WSPR schedule and the single machine LP schedule, for all instances with bounded weights and processing requirements.

Theorem 3. *Consider any instance of the uniform parallel machine problem* $Q|r_j|\sum w_j C_j$ *such that* $0 \leq w_j \leq \bar{w}$ *and* $0 < \underline{p} \leq p_j \leq \bar{p}$ *for all jobs* $j \in N$. *Then*

$$Z^{WSPR}(I) \leq Z^{MBD[m]}(I) + \beta\,\bar{w}\,\bar{p}\,n$$

$$\text{where} \quad \beta = \frac{\bar{p}}{\underline{p}}\left(\frac{1}{s_1} + \frac{m-1}{s_m} + \frac{s^{[m]}}{(s_m)^2}\right) + \frac{1}{2}\left(\frac{1}{s_m} - \frac{1}{s^{[m]}}\right). \quad (15)$$

Proof. Using (5), inequality (14), all $\Delta_h \geq 0$, and the given bounds on all w_j and p_j, we have

$$Z^{WSPR}(I) - Z^{MBD[m]}(I) \leq$$
$$\leq w^\top M^{WSPR} + \frac{1}{2s_m} w^\top p - \left(w^\top M^{LP[m]} + \frac{1}{2s^{[m]}} w^\top p\right)$$
$$= \sum_{h=1}^n \Delta_h \int_0^{p([h])} \left(\bar{R}^{WSPR}_{[h]}(q) - \bar{R}^{LP}_{[h]}(q)\right) dq + \left(\frac{1}{2s_m} - \frac{1}{2s^{[m]}}\right) w^\top p$$
$$\leq \sum_{h=1}^n \Delta_h \left(\frac{1}{s_1} + \frac{m-1}{s_m} + \frac{s^{[m]}}{(s_m)^2}\right) p_{\max} p([h]) + \frac{1}{2}\left(\frac{1}{s_m} - \frac{1}{s^{[m]}}\right) n\,\bar{w}\,\bar{p}$$
$$\leq \frac{w_1}{p_1}\left(\frac{1}{s_1} + \frac{m-1}{s_m} + \frac{s^{[m]}}{(s_m)^2}\right) n\,\bar{p}^2 + \frac{1}{2}\left(\frac{1}{s_m} - \frac{1}{s^{[m]}}\right) n\,\bar{w}\,\bar{p}$$
$$\leq \bar{w}\,\bar{p}\left(\frac{\bar{p}}{\underline{p}}\left(\frac{1}{s_1} + \frac{m-1}{s_m} + \frac{s^{[m]}}{(s_m)^2}\right) + \frac{1}{2}\left(\frac{1}{s_m} - \frac{1}{s^{[m]}}\right)\right) n.$$

This proves Theorem 3.

Now we are ready to prove Theorem 1.

Proof (of Theorem 1). For every instance $I \in \mathcal{I}$, let $Z^*(I)$ (resp., $Z^{MBD}(I)$; resp., $Z^{WSPR}(I)$) denote the minsum objective of an optimal non-preemptive schedule (resp., the LP schedule; resp., a WSPR schedule). Theorem 2 and Theorem 3 imply

$$Z^{MBD[m]}(I) \leq Z^*(I) \leq Z^{WSPR}(I) \leq Z^{MBD[m]}(I) + \beta\,\bar{w}\,\bar{p}\,n,$$

where β is as defined in (15). Note that $Z^{MBD[m]}(I) \geq \frac{\underline{w}}{s^{[m]}} \frac{n(n+1)}{2} \underline{p}$. Therefore

$$\frac{Z^{WSPR}(I)}{Z^*(I)} \leq 1 + \frac{2s^{[m]}}{n+1} \frac{\bar{w}\,\bar{p}}{\underline{w}\,\underline{p}}.$$

Thus, for every $r > 1$, there exists n_0 such that for all instances $I \in \mathcal{I}$ with $n \geq n_0$ we have $Z^{WSPR}(I)/Z^*(I) \leq r$. The proof is complete.

Zero processing time jobs: Assume now that we have a set Z, disjoint from N, of *zero jobs* j with $p_j = 0$. The total number of jobs is now $n' = n + |Z|$. Note that, for all $j \in Z$, $C_j^{LP} = r_j$ since every job $j \in Z$ is immediately inserted into the LP schedule at date r_j. On the other hand, $C_j^{WSPR} < r_j + p_{max}/s_1$, where $p_{max} = \max_{j \in N} p_j \leq \bar{p}$ denotes the longest processing requirement, and s_1 is the fastest machine speed; indeed, in the WSPR schedule every job $j \in Z$ is processed either at date r_j or else at the earliest completion of a job in N in process at date r_j. Therefore, with β as defined in Theorem 1 and assuming $w_j \leq \bar{w}$ for all $j \in Z$, we have

$$\sum_{j \in N \cup Z} w_j C_j^{WSPR} - \sum_{j \in N \cup Z} w_j C_j^{LP} \leq \beta \, \bar{w} \, \bar{p} \, n + w(Z) \, \frac{\bar{p}}{s_1} \leq \bar{w} \, \bar{p} \, \beta \, n'$$

since $\beta \geq 1/s_1$. So the $O(n')$ bound in Theorem 1 extends to the case of zero processing time jobs if one defines $\underline{p} = \min\{p_j : p_j > 0\}$.

For Theorem 3 to extend to this case as well, it suffices that the number n of nonzero jobs grow faster than the square root $\sqrt{n'}$ of the total number of jobs. Indeed in such a case a lower bound on the MBD objective value $Z^{MBD}(I)$, which is quadratic in n, grows faster than linearly in n'. In this respect, one may recall the class of "bad instances" presented in [4] for the single machine problem, which we rescale here by dividing processing times by n' and multiplying weights by n', so $\bar{p} = \bar{w} = 1$. For these instances the objective value $Z^{WSPR}(I)$ approaches $e \approx 2.718$ whereas the MBD lower bound $Z^{MBD}(I)$ approaches $e - 1$. Thus one cannot use this MBD lower bound to establish the asymptotic optimality of the WSPR schedule in this case. This is due to the fact that, for these instances, the number of nonzero jobs is in fact constant (equal to one), and the optimum objective value does not grow at a faster rate than the additive error bound.

References

1. Chekuri, C., Motwani, R., Natarajan, B., Stein, C.: (1997). Approximation Techniques for Average Completion Time Scheduling. Proceedings of the Eight Annual ACM-SIAM Symposium on Discrete Algorithms (1997) 609–618
2. Eastman, W. L., Even, S., Isaacs, I. M.: Bounds for the Optimal Scheduling of n Jobs on m Processors. Management Science **11** (1964) 268–279
3. Goemans, M. X.: Improved Approximation Algorithms for Scheduling with Release Dates. Proceedings of the 8th ACM-SIAM Symposium on Discrete Algorithms (1997) 591-598
4. Goemans, M. X., Queyranne, M., Schulz, A. S., Skutella, M., Wang, Y.: Single Machine Scheduling with Release Dates. Report 654, Fachbereich Mathematik (1999), Technische Universität Berlin, Germany. Available at URL:
   `http://www.math. tu-berlin.de/coga/publications/techreports/1999/`
   `Report-654-1999.html`
5. Graham, R.L., Lawler, E. L., Lenstra, J. K., Rinnooy Kan, A. H. G.: Optimization and approximation in deterministic sequencing and scheduling: a survey. Annals of Discrete Mathematics **5** (1979) 287–326

6. Kaminsky, P., Simchi-Levi, D.: Probabilistic Analysis of an On-Line Algorithm for the Single Machine Completion Time Problem With Release Dates. Under review (1997)

7. Lawler, E. L., Lenstra, J. K., Rinnooy Kan, A. H. G., Shmoys, D. B.: Sequencing and Scheduling: Algorithms and Complexity. In: S. C. Graves, A. H. G. Rinnooy Kan and P. H. Zipkin (eds.), Logistics of Production and Inventory, Handbooks in Operations Research and Management Science 4 (1993), North–Holland, Amsterdam.

8. Lenstra, J. K., Rinnooy Kan, A. H. G., Brucker, P.: Complexity of Machine Scheduling Problems. Annals of Discrete Math 1 (1977) 343–362

9. Queyranne, M., Sviridenko, M.: Approximation algorithms for shop scheduling problems with minsum objective. Faculty of Commerce, University of British Columbia (1999)

10. Sgall, J.: On-line scheduling — a survey. In: A. Fiat and G.J. Woeginger (eds.), Online Algorithms: The State of the Art, Lecture Notes in Computer Science 1442 (1998) 196–231, Springer, Berlin.

11. Smith, W.: Various optimizers for single-stage production. Naval Res. Logist. Quart. 3 (1956) 59–66

12. Uma, R. N., Wein, J.: On the Relationship between Combinatorial and LP-Based Approaches to NP-hard Scheduling Problems. In: R. E. Bixby, E. A. Boyd and R. Z. Rios-Mercado (eds.), Integer Programming and Combinatorial Optimization. Proceedings of the Sixth International IPCO Conference, Lecture Notes in Computer Science 1412 (1998) 394–408, Springer, Berlin.

Approximate k-MSTs and k-Steiner Trees via the Primal-Dual Method and Lagrangean Relaxation

Fabián A. Chudak[1], Tim Roughgarden[2], and David P. Williamson[3]

[1] Tellabs Research Center, 1 Kendall Square, Building 100, Suite 202,
Cambridge, MA, 02139, USA
Fabian.Chudak@tellabs.com

[2] Cornell University, Department of Computer Science,
Upson Hall, Ithaca, NY, 14853, USA
timr@cs.cornell.edu

[3] IBM Almaden Research Center, 650 Harry Rd. K53/B1, San Jose, CA, 95120, USA
dpw@almaden.ibm.com
www.almaden.ibm.com/cs/people/dpw

Abstract. We consider the problem of computing the minimum-cost tree spanning at least k vertices in an undirected graph. Garg [10] gave two approximation algorithms for this problem. We show that Garg's algorithms can be explained simply with ideas introduced by Jain and Vazirani for the metric uncapacitated facility location and k-median problems [15], in particular via a Lagrangean relaxation technique together with the primal-dual method for approximation algorithms. We also derive a constant-factor approximation algorithm for the k-Steiner tree problem using these ideas, and point out the common features of these problems that allow them to be solved with similar techniques.

1 Introduction

Given an undirected graph $G = (V, E)$ with non-negative costs c_e for the edges $e \in E$ and an integer k, the k-MST problem is that of finding the minimum-cost tree in G that spans at least k vertices. The *rooted* version of the problem has a root vertex r as part of its input, and the tree output must contain r. For technical reasons, we will consider the rooted version of the problem. The more natural unrooted version reduces easily to the rooted one, by trying all n possible roots and returning the cheapest of the n solutions obtained.

The k-MST problem is known to be NP-hard [9]; hence, researchers have attempted to find approximation algorithms for the problem. An α-approximation algorithm for a minimization problem runs in polynomial time and produces a solution of cost no more than α times that of an optimal solution. The value α is the *performance guarantee* or *approximation ratio* of the algorithm. The first non-trivial approximation algorithm for the k-MST problem was given by Ravi et al. [17], who achieved an approximation ratio of $O(\sqrt{k})$. This ratio

K. Aardal, B. Gerards (Eds.): IPCO 2001, LNCS 2081, pp. 60–70, 2001.

was subsequently improved to $O(\log^2 k)$ by Awerbuch et al. [4] and $O(\log k)$ by Rajagopalan and Vazirani [16] before a constant-factor approximation algorithm was discovered by Blum et al. [7]. Garg [10] improved upon the constant, giving a simple 5-approximation algorithm and a somewhat more involved 3-approximation algorithm for the problem. Using Garg's algorithm as a black box, Arya and Ramesh [3] gave a 2.5-approximation algorithm for the unrooted version of the problem, and Arora and Karakostos [2] gave a $(2 + \epsilon)$-approximation algorithm for any fixed $\epsilon > 0$ for the rooted version. Finally, Garg [11] has recently announced that a slight modification of his 3-approximation algorithm gives a performance guarantee of 2 for the unrooted version of the problem.

In addition to the practical motivations given in [4,17], the k-MST problem has been well-studied in recent years due to its applications in the context of other approximation algorithms, such as the k-TSP problem (the problem of finding the shortest tour visiting at least k vertices) [2,10] and the minimum latency problem (the problem of finding a tour of n vertices minimizing the average distance from the starting vertex to any other vertex along the tour) [1, 6,12].

This paper is an attempt to simplify Garg's two approximation algorithms for the k-MST problem. In particular, Jain and Vazirani [15] recently discovered a new approach to the primal-dual method for approximation algorithms, and demonstrated its applicability with constant-factor approximation algorithms for the metric uncapacitated facility location and k-median problems. One novel aspect of their approach is the use of their facility location heuristic as a subroutine in their k-median approximation algorithm, the latter based on the technique of Lagrangean relaxation. This idea cleverly exploits the similarity of the integer programming formulations of the two problems. We show that Garg's algorithms can be regarded as another application of this approach, that is, as a Lagrangean relaxation algorithm employing a primal-dual approximation algorithm for a closely related problem as a subroutine. We also give a constant-factor approximation algorithm for the k-Steiner tree problem, via a similar analysis. We believe that these results will give a clearer and deeper understanding of Garg's algorithms, while simultaneously demonstrating that the techniques of Jain and Vazirani should find application beyond the two problems for which they were originally conceived.

This paper is structured as follows. In Section 2, we give linear programming relaxations for the k-MST problem and the closely related prize-collecting Steiner tree problem. In Section 3 we describe and analyze Garg's 5-approximation algorithm for the k-MST problem. In Section 4 we discuss extensions to the k-Steiner tree problem and outline improvements to the basic 5-approximation algorithm. We conclude in Section 5 with a discussion of the applicability of Jain and Vazirani's technique.

2   Two Related LP Relaxations

The rooted k-MST problem can be formulated as the following integer program

$$\text{Min} \sum_{e \in E} c_e x_e$$

subject to:

(kMST)
$$\sum_{e \in \delta(S)} x_e + \sum_{T:T \supseteq S} z_T \geq 1 \qquad \forall S \subseteq V \setminus \{r\} \qquad (1)$$

$$\sum_{S:S \subseteq V \setminus \{r\}} |S| z_S \leq n - k \qquad (2)$$

$$x_e \in \{0,1\} \qquad \forall e \in E$$

$$z_S \in \{0,1\} \qquad \forall S \subseteq V \setminus \{r\}$$

where $\delta(S)$ is the set of edges with exactly one endpoint in S. The variable $x_e = 1$ indicates that the edge e is included in the solution, and the variable $z_S = 1$ indicates that the vertices in the set S are not spanned by the tree. Thus the constraints (1) enforce that for each $S \subseteq V \setminus \{r\}$ either some edge e is selected from the set $\delta(S)$ or that the set S is contained in some set T of unspanned vertices. Collectively, these constraints ensure that all vertices not in any set S with $z_S = 1$ will be connected to the root vertex r. The constraint (2) enforces that at most $n - k$ vertices are not spanned. We can relax this integer program to a linear program by replacing the integrality constraints with nonnegativity constraints.

This formulation is not the most natural one, but we chose it to highlight the connection of the k-MST problem with another problem, the prize-collecting Steiner tree problem. In the prize-collecting Steiner tree problem, we are given a undirected graph $G = (V, E)$ with non-negative costs c_e on edges $e \in E$, a specified root vertex r, and non-negative penalties π_i on the vertices $i \in V$. The goal is to choose a set $S \subseteq V \setminus \{r\}$ and a tree $F \subseteq E$ spanning the vertices of $V \setminus S$ so as to minimize the cost of F plus the penalties of the vertices in S. An integer programming formulation of this problem is

$$\text{Min} \sum_{e \in E} c_e x_e + \sum_{S \subseteq V \setminus \{r\}} \pi(S) z_S$$

subject to:

(PCST)
$$\sum_{e \in \delta(S)} x_e + \sum_{T:T \supseteq S} z_T \geq 1 \qquad \forall S \subseteq V \setminus \{r\}$$

$$x_e \in \{0,1\} \qquad \forall e \in E$$

$$z_S \in \{0,1\} \qquad \forall S \subseteq V \setminus \{r\},$$

where $\pi(S) = \sum_{i \in S} \pi_i$. The interpretation of the variables and the constraints is as above, and again we can relax the integer program to a linear program by replacing the integrality constraints with nonnegativity constraints.

The existing constant-factor approximation algorithms for the k-MST problem [2,7,10] all use as a subroutine a primal-dual 2-approximation algorithm for the prize-collecting Steiner tree problem due to Goemans and Williamson [13,14] (which we will refer to on occasion as "the prize-collecting algorithm"). The integer programming formulations for the two problems are remarkably similar, and recent work on the k-median problem by Jain and Vazirani [15] gives a methodology for exploiting such similarities. Jain and Vazirani present an approximation algorithm for the k-median problem that applies Lagrangean relaxation to a complicating constraint in a formulation of the problem (namely, that at most k facilities can be chosen). Once relaxed, the problem is an uncapacitated facility location problem for which the Lagrangean variable is the cost of opening a facility. By adjusting this cost and applying an approximation algorithm for the uncapacitated facility location problem, they are able to extract a solution for the k-median problem.

One can show that the same dynamic is at work in Garg's algorithms. In particular, if we apply Lagrangean relaxation to the complicating constraint $\sum_{S:S \subseteq V \setminus \{r\}} |S| z_S \leq n - k$ in the relaxation of $(kMST)$, we obtain the following for fixed Lagrangean variable $\lambda \geq 0$:

$$\text{Min} \sum_{e \in E} c_e x_e + \lambda \left(\sum_{S \subseteq V \setminus \{r\}} |S| z_S - (n - k) \right)$$

subject to:

$$(LRk) \qquad \sum_{e \in \delta(S)} x_e + \sum_{T:T \supseteq S} z_T \geq 1 \qquad \forall S \subseteq V \setminus \{r\}$$

$$x_e \geq 0 \qquad \forall e \in E$$

$$z_S \geq 0 \qquad \forall S \subseteq V \setminus \{r\}.$$

For fixed λ, this is nearly identical to $(PCST)$ with $\pi_i = \lambda$ for all i, the only difference being the constant term of $-(n-k)\lambda$ in the objective function. Observe that any solution feasible for $(kMST)$ is also feasible for (LRk) with no greater cost, and so the value of (LRk) is a lower bound on the cost of an optimal k-MST.

Before discussing how Garg's algorithms work, we need to say a little more about the prize-collecting algorithm. The algorithm constructs a primal-feasible solution (F, A), where F is a tree including the root r, and A is the set of vertices not spanned by F. The algorithm also constructs a feasible solution y for the dual of $(PCST)$, which is

$$\text{Max} \sum_{S \subseteq V \setminus \{r\}} y_S$$

subject to:

$$(PCST - D) \qquad \sum_{S:e \in \delta(S)} y_S \leq c_e \qquad \forall e \in E$$

$$\sum_{T:T\subseteq S} y_T \le \pi(S) \qquad \forall S \subseteq V \setminus \{r\}$$

$$y_S \ge 0 \qquad \forall S \subseteq V \setminus \{r\}.$$

Then the following is true:

Theorem 1 (Goemans and Williamson [13]). *The primal solution* (F, A) *and the dual solution* y *produced by the prize-collecting algorithm satisfy*

$$\sum_{e \in F} c_e + \left(2 - \frac{1}{n-1}\right) \pi(A) \le \left(2 - \frac{1}{n-1}\right) \sum_{S \subseteq V \setminus \{r\}} y_S.$$

Note that, by weak duality and the feasibility of y, $\sum_{S \subseteq V \setminus \{r\}} y_S$ is a lower bound for the cost of any solution to the prize-collecting Steiner tree problem.

Suppose we set $\pi_i = \lambda \ge 0$ for all $i \in V$ and run the prize-collecting algorithm. The theorem statement implies that we obtain (F, A) and y such that

$$\sum_{e \in F} c_e + 2|A|\lambda \le 2 \sum_{S \subseteq V \setminus \{r\}} y_S. \tag{3}$$

We wish to reinterpret the tree F as a feasible solution for the k-MST instance, and extract a lower bound on the cost of an optimal k-MST from y. Toward this end, we consider the dual of the (LRk) LP, as follows (recall that λ is a fixed constant):

$$\text{Max} \sum_{S \subseteq V \setminus \{r\}} y_S - (n-k)\lambda$$

subject to:

$(LRk - D)$
$$\sum_{S:e\in\delta(S)} y_S \le c_e \qquad \forall e \in E$$

$$\sum_{T:T\subseteq S} y_T \le |S|\lambda \qquad \forall S \subseteq V \setminus \{r\}$$

$$y_S \ge 0 \qquad \forall S \subseteq V \setminus \{r\}.$$

The dual solution y created by the prize-collecting algorithm is feasible for $(LRk - D)$ when all prizes π_i are set to λ. Furthermore, its value will be no greater than the cost of an optimal k-MST. After subtracting $2(n-k)\lambda$ from both sides of (3), by weak duality we obtain the following:

$$\sum_{e \in F} c_e + 2\lambda(|A| - (n-k)) \le 2\left(\sum_{S \subseteq V \setminus \{r\}} y_S - (n-k)\lambda\right) \tag{4}$$

$$\le 2 \cdot OPT_k, \tag{5}$$

where OPT_k is the optimal solution to the k-MST problem. In the lucky event that $|A| = n - k$, F is a feasible solution having cost no more than twice optimal.

Otherwise, our solution will either not be feasible (if $|A| > n - k$) or the relations (4) and (5) will not give a useful upper bound on the cost of the solution (if $|A| < n - k$). However, in the next section we combine these ideas with a Lagrangean relaxation approach to derive an algorithm that always produces a near-optimal feasible solution (though with a somewhat inferior performance guarantee).

3 Garg's 5 Approximation Algorithm

We begin with three assumptions, each without loss of generality. First, by standard techniques [17], one can show that it is no loss of generality to assume that the edge costs satisfy the triangle inequality. Second, we assume that the distance between any vertex v and the root vertex r is at most OPT_k; this is accomplished by "guessing" the distance D of the farthest vertex from r in the optimal solution (there are but $n - 1$ "guesses" to enumerate) and deleting all nodes of distance more than D from r. Note that $D \leq OPT_k$. The cheapest feasible solution of these $n - 1$ subproblems is the final output of the algorithm. Third, we assume that $OPT_k \geq c_{\min}$, where $c_{\min}$ denotes the smallest non-zero edge cost. If this is not true, then $OPT_k = 0$ and the optimal solution is a connected component containing r of at least k nodes in the subgraph of zero-cost edges. We can easily check whether such a solution exists before we run Garg's algorithm.

Garg's algorithm is essentially a sequence of calls to the prize-collecting algorithm, each with a different value for the Lagrangean variable λ. We will use the fact that for λ sufficiently small (e.g., $\lambda = 0$), the prize-collecting algorithm will return $(\emptyset, V \setminus \{r\})$ as a solution (that is, the degenerate solution of the empty tree trivially spanning r) and for λ sufficiently large (e.g., $\lambda = \sum_{e \in E} c_e$) the prize-collecting algorithm will return a tree spanning all n vertices. Also, if any call to the prize-collecting algorithm returns a tree T spanning precisely k vertices, then by the analysis of the previous section, T is within a factor 2 of optimal, and the k-MST algorithm can halt with T as its output. Thus, by a straightforward bisection search procedure consisting of polynomially many subroutine calls to the prize-collecting algorithm, Garg's algorithm either finds a tree spanning precisely k vertices (via a lucky choice of λ) or two values $\lambda_1 < \lambda_2$ such that the following two conditions hold:

(i) $\lambda_2 - \lambda_1 \leq \frac{c_{\min}}{2n(2n+1)}$, where (as above) $c_{\min}$ denotes the smallest non-zero edge cost and

(ii) for $i = 1, 2$, running the prize-collecting algorithm with λ set to λ_i yields a primal solution (F_i, A_i) spanning k_i vertices and a dual solution $y^{(i)}$, with $k_1 < k < k_2$.

Henceforth we assume the algorithm failed to find a value of λ resulting in a tree spanning exactly k vertices. The final step of the algorithm combines the two primal solutions, (F_1, A_1) and (F_2, A_2), into a single tree spanning precisely k vertices. For the analysis, the two dual solutions will also be combined.

From Theorem 1, we have the following inequalities:

$$\sum_{e \in F_1} c_e \le \left(2 - \frac{1}{n}\right)\left(\sum_{S \subseteq V \setminus \{r\}} y_S^{(1)} - |A_1|\lambda_1\right) \tag{6}$$

$$\sum_{e \in F_2} c_e \le \left(2 - \frac{1}{n}\right)\left(\sum_{S \subseteq V \setminus \{r\}} y_S^{(2)} - |A_2|\lambda_2\right) \tag{7}$$

We would like to take a convex combination of these two inequalities so as to get a bound on the cost of F_1 and F_2 in terms of OPT_k. Let $\alpha_1, \alpha_2 \ge 0$ satisfy $\alpha_1|A_1| + \alpha_2|A_2| = n - k$ and $\alpha_1 + \alpha_2 = 1$, and for all $S \subseteq V \setminus \{r\}$, let $y_S = \alpha_1 y_S^{(1)} + \alpha_2 y_S^{(2)}$. Note that

$$\alpha_1 = \frac{n - k - |A_2|}{|A_1| - |A_2|} \quad \text{and} \quad \alpha_2 = \frac{|A_1| - (n - k)}{|A_1| - |A_2|}.$$

Lemma 1.

$$\alpha_1 \sum_{e \in F_1} c_e + \alpha_2 \sum_{e \in F_2} c_e < 2OPT_k.$$

Proof. From inequality (6) we have

$$\sum_{e \in F_1} c_e \le \left(2 - \frac{1}{n}\right)\left(\sum_{S \subseteq V \setminus \{r\}} y_S^{(1)} - |A_1|(\lambda_1 + \lambda_2 - \lambda_2)\right)$$

$$\le \left(2 - \frac{1}{n}\right)\left(\sum_{S \subseteq V \setminus \{r\}} y_S^{(1)} - |A_1|\lambda_2\right) + \left(2 - \frac{1}{n}\right)\frac{c_{\min}|A_1|}{2n(2n+1)}$$

$$< \left(2 - \frac{1}{n}\right)\left(\sum_{S \subseteq V \setminus \{r\}} y_S^{(1)} - |A_1|\lambda_2\right) + \frac{c_{\min}}{2n+1}.$$

By a convex combination of this inequality and inequality (7), it follows that

$$\alpha_1 \sum_{e \in F_1} c_e + \alpha_2 \sum_{e \in F_2} c_e < \left(2 - \frac{1}{n}\right)\left(\sum_{S \subseteq V \setminus \{r\}} y_S - \lambda_2(\alpha_1|A_1| + \alpha_2|A_2|)\right) + \frac{\alpha_1 c_{\min}}{2n+1}$$

$$= \left(2 - \frac{1}{n}\right)\left(\sum_{S \subseteq V \setminus \{r\}} y_S - \lambda_2(n - k)\right) + \frac{\alpha_1 c_{\min}}{2n+1} \tag{8}$$

$$\le \left(2 - \frac{1}{n}\right)OPT_k + \frac{\alpha_1 c_{\min}}{2n+1} \tag{9}$$

$$\le \left(2 - \frac{1}{n}\right)OPT_k + \frac{1}{2n+1}OPT_k \tag{10}$$

$$\le 2OPT_k.$$

Equality (8) follows by our choice of α_1, α_2. Inequality (9) follows from the feasibility of y for $(LRk - D)$ with the Lagrangean variable set to λ_2 (since the feasible region is convex and $\lambda_2 > \lambda_1$). Inequality (10) follows since $\alpha_1 \leq 1$ and $OPT_k \geq c_{\min}$. $\square$

Garg considers two different solutions to obtain a 5-approximation algorithm. First, if $\alpha_2 \geq \frac{1}{2}$, then F_2 is already a good solution; since $|A_2| < n - k$, it spans more than k vertices, and

$$\sum_{e \in F_2} c_e \leq 2\alpha_2 \sum_{e \in F_2} c_e \leq 4 \cdot OPT_k$$

by Lemma 1. Now suppose $\alpha_2 < \frac{1}{2}$. In this case the tree F_1 is supplemented by vertices from F_2. Let $\ell \geq k_2 - k_1$ be the number of vertices spanned by F_2 but not F_1. Doubling the edges of the tree F_2, shortcutting any resulting Euler tour down to a simple tour of the ℓ vertices spanned solely by F_2, and choosing the cheapest path of $k - k_1$ consecutive vertices from this tour, we obtain a tree (in fact, a path) on $k - k_1$ vertices of cost at most

$$2\frac{k - k_1}{k_2 - k_1} \sum_{e \in F_2} c_e.$$

This set of vertices can be connected to F_1 by adding an edge from the root to the set, which will have cost no more than OPT_k (due to the second assumption made at the beginning of this section). Since

$$\frac{k - k_1}{k_2 - k_1} = \frac{n - k_1 - (n - k)}{n - k_1 - (n - k_2)} = \frac{|A_1| - (n - k)}{|A_1| - |A_2|} = \alpha_2,$$

the total cost of this solution is

$$\sum_{e \in F_1} c_e + 2\alpha_2 \sum_{e \in F_2} c_e + OPT_k \leq 2 \left(\alpha_1 \sum_{e \in F_1} c_e + \alpha_2 \sum_{e \in F_2} c_e \right) + OPT_k$$
$$\leq 4OPT_k + OPT_k,$$

since $\alpha_2 < \frac{1}{2}$ implies $\alpha_1 > \frac{1}{2}$, and by Lemma 1.

4   Extensions

The k-Steiner tree problem is defined as follows: given an undirected graph $G = (V, E)$ with non-negative costs c_e for the edges $e \in E$, a set $R \subseteq V$ of *required vertices* (also called *terminals*), and an integer k, find the minimum-cost tree in G that spans at least k of the required vertices. Of course, the problem is only feasible when $k \leq |R|$. The k-Steiner tree problem includes the classical Steiner tree problem (set $k = |R|$) and is thus both NP-hard and MAX SNP-hard [5]. The problem was studied by Ravi et al. [17], who gave a simple reduction showing that an α-approximation algorithm for the k-MST problem yields a 2α-approximation algorithm for the k-Steiner tree problem. Thus, the result of the previous section implies the existence of a 10-approximation algorithm for the problem. However, we can show that a modification of Garg's 5-approximation algorithm achieves a performance guarantee of 5 for this problem as well. Consider the following LP relaxation for the k-Steiner tree problem

$$\text{Min} \sum_{e \in E} c_e x_e$$

subject to:

(kST)
$$\sum_{e \in \delta(S)} x_e + \sum_{T:T \supseteq S} z_T \geq 1 \qquad \forall S \subseteq V \setminus \{r\}$$

$$\sum_{S:S \subseteq V \setminus \{r\}} |S \cap R| z_S \leq |R| - k$$

$$x_e \geq 0 \qquad \qquad \forall e \in E$$

$$z_S \geq 0 \qquad \qquad \forall S \subseteq V \setminus \{r\}.$$

We modify Garg's algorithm at the point where the prize-collecting algorithm is called as a subroutine with a fixed value of λ inside the main Lagrangean relaxation loop. To reflect that we are only interested in how many required vertices are spanned by a solution, we assign required vertices a penalty of λ and Steiner (non-required) vertices a penalty of 0. In the notation of the linear program $(PCST)$, we put $\pi_i = \lambda$ for $i \in R$ and $\pi_i = 0$ for $i \notin R$. An analog of Lemma 1 can then be shown, leading as in Section 3 to a 5-approximation algorithm for the k-Steiner tree problem.

We now discuss improving the approximation ratio of the two algorithms. Using ideas of Arora and Karakostos [2], the k-MST and k-Steiner tree algorithms can be refined to achieve performance guarantees of $(4 + \epsilon)$, for an arbitrarily small constant ϵ. Roughly speaking, their idea is as follows. Garg's algorithm essentially "guesses" one vertex that appears in the optimal solution, namely the root r. Instead, one can "guess" $O(\frac{1}{\epsilon})$ vertices and edges in the optimal solution (for fixed ϵ, there are but polynomially many guesses to enumerate) such that any other vertex in the optimal solution has distance at most $O(\epsilon OPT)$ from the guessed subgraph H. After H is guessed, all vertices of distance more than $O(\epsilon OPT)$ from H can be deleted. It is not difficult to modify the prize-collecting algorithm to handle the additional guessed vertices. Finally, when creating a feasible solution from two subsolutions as at the end of Section 3, the final edge connecting the two subtrees costs no more than ϵOPT, leading to a final upper bound of $(4 + \epsilon)OPT$. The reader is referred to [2] for the details of this refinement.

In addition to the 5-approximation algorithm discussed in Section 3, Garg [10] gave a more sophisticated 3-approximation algorithm for the k-MST problem. Unfortunately, the analysis seems to require a careful discussion of the inner workings of the prize-collecting algorithm, a task we will not undertake here. (Similarly, improving Jain and Vazirani's 6-approximation algorithm for the k-

median problem to a 4-approximation algorithm required a detailed analysis of the primal-dual facility location subroutine; see the paper of Charikar and Guha [8].) However, we believe that Garg's 3-approximation algorithm can also be recast in the language of Jain and Vazirani and of this paper, and that it will extend to a 3-approximation algorithm for the k-Steiner tree problem as well. The same ideas that led from a 5-approximation algorithm to one with performance guarantee $(4+\epsilon)$ should then yield $(2+\epsilon)$-approximation algorithms for the k-MST problem (as in [2]) and the k-Steiner tree problem.

5 Conclusion

We have shown that the techniques of Jain and Vazirani [15], invented for a constant-factor approximation algorithm for the k-median problem, also give constant-factor approximation algorithms for the k-MST problem (essentially reinventing older algorithms of Garg [10]) and the k-Steiner tree problem. A natural direction for future research is the investigation of the applicability and limitations of this Lagrangean relaxation approach. The three problems solved in this framework so far share several characteristics. First, each problem admits an LP relaxation with an obvious "complicating" constraint. Moreover, once the complicating constraint is lifted into the objective function, the new linear program corresponds to the relaxation of a problem known to be well-approximable (in our cases by a primal-dual approximation algorithm). Lastly, and perhaps most importantly, the subroutine for the relaxed problem produces a pair of primal and dual solutions such that the portion of the primal cost corresponding to the constraint of the original problem (e.g., the $\sum_S \pi(S) z_S$ term in the prize-collecting Steiner tree objective function) is bounded above by the value of the dual. Note that this is a stronger condition than merely ensuring that the primal solution has cost no more than some constant times the dual solution value. For example, in Theorem 1, the total primal cost is upper-bounded by twice the value of the dual solution, $2 \sum_S y_S$, and in addition the second term of the primal cost is bounded above by $\sum_S y_S$. (Note such a statement does not hold in general for the first primal cost term of Theorem 1.) This last property seems necessary for extracting lower bounds for the problem of interest (via the dual LP) from the dual solutions returned by the subroutine, and may turn out to be the primary factor limiting the applicability of the Lagrangean relaxation approach. It would be of great interest to find further problems that can be approximately solved in this framework, and to devise more general variants of the framework that apply to a broader class of problems.

Acknowledgements. The first author was supported by an IBM Postdoctoral Research Fellowship. The second author is supported by ONR grant N00014-98-1-0589, an NSF Fellowship, and a Cornell University Fellowship. His research was carried out while visiting IBM.

References

1. S. Arora and G. Karakostas. Approximation schemes for minimum latency problems. In *Proceedings of the 31st Annual ACM Symposium on the Theory of Computing*, pages 688–693, 1999.
2. S. Arora and G. Karakostas. A $2 + \epsilon$ approximation algorithm for the k-MST problem. In *Proceedings of the 11th Annual ACM-SIAM Symposium on Discrete Algorithms*, pages 754–759, 2000.
3. S. Arya and H. Ramesh. A 2.5-factor approximation algorithm for the k-MST problem. *Information Processing Letters*, 65:117–118, 1998.
4. B. Awerbuch, Y. Azar, A. Blum, and S. Vempala. Improved approximation guarantees for minimum-weight k-trees and prize-collecting salesmen. *SIAM Journal on Computing*, 28(1):254–262, 1999.
5. M. Bern and P. Plassmann. The Steiner problem with edge lengths 1 and 2. *Information Processing Letters*, 32:171–176, 1989.
6. A. Blum, P. Chalasani, D. Coppersmith, W. Pulleyblank, P. Raghavan, and M. Sudan. The minimum latency problem. In *Proceedings of the 26th Annual ACM Symposium on the Theory of Computing*, pages 163–171, 1994.
7. A. Blum, R. Ravi, and S. Vempala. A constant-factor approximation algorithm for the k-MST problem. *Journal of Computer and System Sciences*, 58(1):101–108, 1999.
8. M. Charikar and S. Guha. Improved combinatorial algorithms for the facility location and k-median problems. In *Proceedings of the 40th Annual Symposium on Foundations of Computer Science*, pages 378–388, 1999.
9. M. Fischetti, H. Hamacher, K. Jørnsten, and F. Maffioli. Weighted k-cardinality trees: Complexity and polyhedral structure. *Networks*, 24:11–21, 1994.
10. N. Garg. A 3-approximation for the minimum tree spanning k vertices. In *Proceedings of the 37th Annual Symposium on Foundations of Computer Science*, pages 302–309, 1996.
11. N. Garg. Personal communication, 1999.
12. M. Goemans and J. Kleinberg. An improved approximation ratio for the minimum latency problem. *Mathematical Programming*, 82:111–124, 1998.
13. M. X. Goemans and D. P. Williamson. A general approximation technique for constrained forest problems. *SIAM Journal on Computing*, 24:296–317, 1995.
14. M. X. Goemans and D. P. Williamson. The primal-dual method for approximation algorithms and its application to network design problems. In D. S. Hochbaum, editor, *Approximation Algorithms for NP-Hard Problems*, chapter 4, pages 144–191. PWS Publishing Company, 1997.
15. K. Jain and V. V. Vazirani. Primal-dual approximation algorithms for metric facility location and k-median problems. In *Proceedings of the 40th Annual Symposium on Foundations of Computer Science*, pages 2–13, 1999.
16. S. Rajagopalan and Vijay V. Vazirani. Logarithmic approximation of minimum weight k trees. Unpublished manuscript, 1995.
17. R. Ravi, R. Sundaram, M. V. Marathe, D. J. Rosenkrantz, and S. S. Ravi. Spanning trees short or small. *SIAM Journal on Discrete Mathematics*, 9:178–200, 1996.

On the Rank of Mixed 0,1 Polyhedra*

Gérard Cornuéjols and Yanjun Li

Graduate School of Industrial Administration
Carnegie Mellon University, Pittsburgh, USA
{gc0v, yanjun}@andrew.cmu.edu

Abstract. Eisenbrand and Schulz showed recently (IPCO 99) that the maximum Chvátal rank of a polytope in the $[0,1]^n$ cube is bounded above by $O(n^2 log n)$ and bounded below by $(1 + \epsilon)n$ for some $\epsilon > 0$. It is well known that Chvátal's cuts are equivalent to Gomory's fractional cuts, which are themselves dominated by Gomory's mixed integer cuts. What do these upper and lower bounds become when the rank is defined relative to Gomory's mixed integer cuts? An upper bound of n follows from existing results in the literature. In this note, we show that the lower bound is also equal to n. We relate this result to bounds on the disjunctive rank and on the Lovász-Schrijver rank of polytopes in the $[0,1]^n$ cube. The result still holds for mixed 0,1 polyhedra with n binary variables.

1 Introduction

Consider a mixed integer program $P_I \equiv \{(x,y) \in Z^n_+ \times R^p_+ | Ax + Gy \leq b\}$ where A and G are given rational matrices (dimensions $m \times n$ and $m \times p$ respectively) and b is a given rational column vector (dimension m). Let $P \equiv \{(x,y) \in R^{n+p}_+ | Ax + Gy \leq b\}$ be its standard linear relaxation. In [13], Gomory introduced a family of valid inequalities for P_I, called *mixed integer inequalities*, that can be used to strengthen P. These inequalities are obtained from P by considering an equivalent equality form. Let $P' = \{(x,y,s) \in R^{n+p+m}_+ | Ax + Gy + s = b\}$ and $P'_I = \{(x,y,s) \in Z^n_+ \times R^{p+m}_+ | Ax + (G,I)\binom{y}{s} = b\}$. Introduce $z = \binom{y}{s}$. For any $u \in R^m$, let $\bar{a} = uA$, $\bar{g} = u(G, I)$ and $\bar{b} = ub$. Let $\bar{a}_i = \lfloor \bar{a}_i \rfloor + f_i$ and $\bar{b} = \lfloor \bar{b} \rfloor + f_0$. Gomory showed that the following inequality is valid for P'_I:

$$\sum_{(i:f_i \leq f_0)} f_i x_i + \frac{f_0}{1-f_0} \sum_{(i:f_i > f_0)} (1-f_i)x_i + \sum_{(j:\bar{g}_j \geq 0)} \bar{g}_j z_j - \frac{f_0}{1-f_0} \sum_{(j:\bar{g}_j < 0)} \bar{g}_j z_j \geq f_0.$$

Plugging $s = b - Ax - Gy$ into it, we get a valid inequality for P_I. Any such inequality $\alpha_u x + \gamma_u y \leq \beta_u$ is called a *mixed integer inequality*. The convex set P^1 defined as the intersection of all mixed integer inequalities is called the *mixed integer closure* of P. In fact P^1 is a polyhedron (see below). By recursively taking the mixed integer closure of P^{k-1}, for integers $k \geq 2$, we obtain the polyhedron

* Supported by NSF grant DMI-9802773 and ONR grant N00014-97-1-0196.

K. Aardal, B. Gerards (Eds.): IPCO 2001, LNCS 2081, pp. 71–77, 2001.

P^k. Clearly $P_I \subseteq P^k \subseteq P^{k-1} \ldots \subseteq P^1 \subseteq P$. For mixed 0,1 programs, there is always a finite k such that $P_I = P^k$ (see below). The smallest such k is called the *mixed integer rank* of P. In this note, we show the following.

Theorem 1. *The maximum mixed integer rank of P, taken over all mixed 0,1 programs P_I with n binary variables, is equal to n.*

In particular, the maximum mixed integer rank for a pure integer program in the $[0,1]^n$ cube is equal to n. This is in contrast to the maximum Chvátal rank which was shown by Eisenbrand and Schulz to lie in the interval $(1 + \epsilon)n$ to $O(n^2 log n)$ for some $\epsilon > 0$.

To prove Theorem 1, we use the equivalence between Gomory's mixed integer inequalities and Balas's disjunctive inequalities shown by Nemhauser and Wolsey [17]. Given the polyhedron $P \equiv \{(x,y) \in R_+^{n+p} | Ax + Gy \le b\}$, an inequality is called a *disjunctive inequality* if it is valid for $\text{Conv}((P \cap \{x| \pi x \le \pi_0\}) \cup (P \cap \{x| \pi x \ge \pi_0 + 1\}))$ for some $(\pi, \pi_0) \in Z^{n+1}$, where $\text{Conv}(A \cup B)$ denotes the convex hull of the union of two sets A and B. Disjunctive inequalities were introduced by Balas [1,2]. They were also studied by Cook, Kannan and Schrijver [9] under the name of *split cuts*. Many of the classical cutting planes can be interpreted as disjunctive inequalities. For instance, in the pure case, Chvátal's cuts [5] are disjunctive inequalities where at least one of the two polyhedra $P \cap \{x| \pi x \le \pi_0\}$ or $P \cap \{x| \pi x \ge \pi_0 + 1\}$ is empty. (Indeed, if say $P \cap \{x| \pi x \ge \pi_0 + 1\}$ is empty, then $\pi x < \pi_0 + 1$ is valid for P, which implies that the disjunctive inequality $\pi x \le \pi_0$ is a Chvátal cut and, conversely, any Chvátal cut can be obtained this way). As another example, it is well known that the lift-and-project cuts [3] are disjunctive inequalities obtained from the disjunction $x_j \le 0$ or $x_j \ge 1$, i.e. they are valid inequalities for $\text{Conv}((P \cap \{x| x_j \le 0\}) \cup (P \cap \{x| x_j \ge 1\}))$.

The convex set defined as the intersection of all disjunctive inequalities is called the *disjunctive closure* of P. Cook, Kannan and Schrijver [9] showed that the disjunctive closure of P is a polyhedron. Nemhauser and Wolsey [17] showed that the disjunctive closure of P is identical to the mixed integer closure P^1 defined above. The proof of this equivalence can also be found in [10]. Therefore P^1 is a polyhedron. Consider P^k defined above. If there exists an integer k such that $P_I = P^k$, the smallest such k was defined above as the mixed integer rank of P and we also refer to it as the *disjunctive rank* of P. In general, mixed integer programs do not have a finite disjunctive rank, as shown by Cook, Kannan and Schrijver [9] using a simple example with two integer variables and one continuous variable. Mixed 0,1 programs have the property that the disjunction $x_j \le 0$ or $x_j \ge 1$ is facial, i.e. both $P \cap \{x| x_j \le 0\}$ and $P \cap \{x| x_j \ge 1\}$ define faces of P. If follows from a result of Balas [2] on facial disjunctive programs that the disjunctive rank of a mixed 0,1 program is at most n. This shows the upper bound in Theorem 1. New results in this direction were obtained recently by Balas and Perregaard [4]. In this note, we exhibit a lower bound of n, thus completing the proof of Theorem 1.

We show that the disjunctive rank of the following well-known polytope studied by Chvátal, Cook, and Hartmann [6] is exactly n:

$$P_n = \{x \in [0,1]^n |\ \sum_{j \in J} x_j + \sum_{j \notin J}(1 - x_j) \geq \frac{1}{2}, \text{ for all } J \subseteq \{1, 2, \cdots, n\}\}, \quad (1)$$

just as its Chvátal rank is.

2 Proof

In this section, we prove that the disjunctive rank of P_n in (1) is exactly n.

Definition 1. *Let P be a polyhedron in R^n and $\alpha x \leq \beta$ a disjunctive inequality of P from the disjunction $\pi x \leq \pi_0$ or $\pi x \geq \pi_0 + 1$, where $(\pi, \pi_0) \in Z^{n+1}$. If at least one of the two polyhedra $P \cap \{x|\ \pi x \leq \pi_0\}$ or $P \cap \{x|\ \pi x \geq \pi_0 + 1\}$ is empty, the inequality is said to be of* type 1, *and if both are nonempty, the inequality is said to be of* type 2.

As pointed out in Section 1, Chvátal cuts are type 1 inequalities. Conversely, any type 1 inequality $\alpha x \leq \beta$ is dominated by a Chvátal cut $\pi x \leq \pi_0$ where $(\pi, \pi_0) \in Z^{n+1}$ ($\pi x \leq \pi_0$ dominates $\alpha x \leq \beta$ means that $P \cap \{x|\ \pi x \leq \pi_0\} \subseteq P \cap \{x|\ \alpha x \leq \beta\}$).

In the n-dimensional $0, 1$-cube, let S_n be the set of all vectors that have $n - 1$ elements with value $1/2$ and one element with value either 0 or 1. Let Q_n denote the convex hull of S_n.

Lemma 1. *None of the disjunctive inequalities of Q_n can remove the point $c^n = (1/2, 1/2, \cdots, 1/2) \in R^n$ for $n \geq 2$.*

Proof. In order to show that the type 1 inequalities of Q_n cannot eliminate the center point c^n, it suffices to prove that the Chvátal cuts of Q_n cannot remove c^n. This was proved by Chvátal, Cook and Hartmann [6]. We give the proof for completeness. Consider an arbitrary inequality $\pi x < \pi_0 + 1$ valid for Q_n, where $(\pi, \pi_0) \in Z^{n+1}$. We want to show $\pi c^n \leq \pi_0$. Since $c^n \in Q_n$, $\pi c^n < \pi_0 + 1$, so the result holds if πc^n is an integer. Assume now that πc^n is not an integer. Consider an index i such that $\pi_i \neq 0$. Let u be the vertex of Q_n which is different from c^n only in the ith component where $u_i = 0$, and v the vertex of Q_n which is only different from c^n in the ith component where $v_i = 1$. Then $\max(\pi u, \pi v) \geq \pi c^n + 1/2$, because π_i is integer and $c^n_i = 1/2$. Since $u, v \in Q_n$, $\max(\pi u, \pi v) < \pi_0 + 1$. Thus we know $\pi c^n < \pi_0 + 1/2$. Therefore, $\pi c^n \leq \pi_0$ follows from the fact that $2\pi c^n$ is an integer.

Now consider a type 2 inequality of Q_n, say $\alpha x \leq \beta$. Let u and v be two extreme points of Q_n that respectively minimize and maximize πx over Q_n. Since Q_n is center symmetric relative to the point c^n, we can choose u and v symmetric relative to c^n. Since $\alpha x \leq \beta$ is an inequality of type 2, $u \in Q_n \cap \{x|\ \pi x \leq \pi_0\}$

and $v \in Q_n \cap \{x| \pi x \geq \pi_0 + 1\}$. Therefore neither u nor v is removed by the inequality $\alpha x \leq \beta$. This implies that the middle point c^n of the segment uv cannot be removed by a type 2 inequality. $\qquad\square$

Lemma 2. *Let $P \subseteq [0,1]^n$ and $F \subseteq [0,1]^{n-1}$ be two polytopes. For $1 \leq i \leq n$, assume that the face $T = P \cap \{x \in R^n| x_i = 0\}$ (or $U = P \cap \{x \in R^n| x_i = 1\}$) of P is a polytope identical to F. Then, for any integer $k \geq 1$, the polytope F^k is identical to the face $P^k \cap \{x \in R^n| x_i = 0\}$ (or $P^k \cap \{x \in R^n| x_i = 1\}$) of P^k.*

Proof. It suffices to prove the lemma for $k = 1$.

Assume that the polytope U, viewed as a subset of $[0,1]^{n-1}$, is identical to $F \subseteq [0,1]^{n-1}$.

We first show that F^1 is contained in the face $P^1 \cap \{x \in R^n| x_i = 1\}$ of P^1. Let $\alpha x \leq \beta$ be a disjunctive inequality of P from the disjunction $\pi x \leq \pi_0$ or $\pi x \geq \pi_0 + 1$, where $(\pi, \pi_0) \in Z^{n+1}$. If the hyperplane $\pi x = \pi_0$ is parallel to the hyperplane $x_i = 1$, the inequality $\alpha x \leq \beta$ removes no points in U. Now consider the case where the hyperplane $\pi x = \pi_0$ intersects the hyperplane $x_i = 1$. Let π' be the restriction of π to the components distinct from i and let $\pi'_0 = \pi_0 - \pi_i$. Similarly, let α' (respectively x') be the restriction of α (respectively x) to the components distinct from i and let $\beta' = \beta - \alpha_i$. The inequality $\alpha'x' \leq \beta'$ is valid for $\text{Conv}((F \cap \{x' \in R^{n-1}| \pi'x' \leq \pi'_0\}) \cup (F \cap \{x' \in R^{n-1}| \pi'x' \geq \pi'_0 + 1\}))$. It follows that $\alpha'x' \leq \beta'$ is a disjunctive inequality of F (it may not be of the same type as $\alpha x \leq \beta$ for P).

Conversely, we show that the face $P^1 \cap \{x \in R^n| x_i = 1\}$ of P^1 is contained in F^1. Let $\alpha'x' \leq \beta'$ be a disjunctive inequality of F from the disjunction $\pi'x' \leq \pi'_0$ or $\pi'x' \geq \pi'_0 + 1$, where $(\pi', \pi'_0) \in Z^n$. Define $\pi_j = \pi'_j$ $(j \neq i)$ and $\pi_i = 0$. Let K_0 and K_1 be the extreme points of the polytopes $P \cap \{x \in R^n| \pi x \leq \pi_0\}$ and $P \cap \{x \in R^n| \pi x \geq \pi_0 + 1\}$ respectively and let L be the points of $K_0 \cup K_1$ that satisfy $x_i < 1$. Clearly L is finite. In order to get a disjunctive inequality $\alpha x \leq \beta$ of P, we let $\alpha_j = \alpha'_j$ $(j \neq i)$, $\beta = \beta' + \alpha_i$ and we choose α_i sufficiently large so that all points of L satisfy $\alpha x \leq \beta$. Then $\alpha x \leq \beta$ is a disjunctive inequality of P from the disjunction $\pi x \leq \pi_0$ or $\pi x \geq \pi_0 + 1$.

Similarly the result can be proved for T as well. $\qquad\square$

Theorem 2. *The disjunctive rank of P_n is exactly n.*

Proof. It was shown in the previous section that the disjunctive rank of P_n is at most n. In order to prove the theorem, it remains to show that the disjunctive rank of P_n is at least n. We obtain this result by showing that the $(n-1)^{st}$ disjunctive closure of P_n is nonempty, i.e. $P_n^{n-1} \neq \emptyset$. This implies that at least one more step of the disjunctive procedure is needed since $(P_n)_I = \emptyset$. To show that $P_n^{n-1} \neq \emptyset$, we prove that $c^n = (1/2, 1/2, \cdots, 1/2) \in R^n$ belongs to P_n^{n-1}. The proof is by induction on n. Since $P_2 = Q_2$, Lemma 1 shows that the result holds for $n = 2$. Now assume that $c^n \in P_n^{n-1}$ for some $n \geq 2$. We will prove that $c^{n+1} \in P_{n+1}^n$.

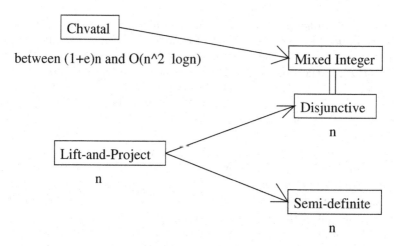

Fig. 1. Maximum rank of polytopes in the $[0,1]^n$ cube

We first show that $Q_{n+1} \subseteq P_{n+1}^{n-1}$. Let $T_i = P_{n+1} \cap \{x \in R^{n+1} |\ x_i = 0\}$ and $U_i = P_{n+1} \cap \{x \in R^{n+1} |\ x_i = 1\}$ for $1 \leq i \leq n+1$. One can verify directly from the definition of P_n in (1) that the polytopes T_i and U_i are n-dimensional polytopes identical to the polytope P_n in $[0,1]^n$. Let t_i (u_i respectively) be the point in T_i (U_i respectively) with n elements with value $1/2$ and its ith element with value 0 (1 respectively). We claim that $t_i \in P_{n+1}^{n-1}$ and $u_i \in P_{n+1}^{n-1}$. We know $c^n \in P_n^{n-1}$ by the induction hypothesis. Because T_i and U_i are polytopes identical to P_n, it follows by Lemma 2 that $t_i \in P_{n+1}^{n-1} \cap \{x \in R^{n+1} |\ x_i = 0\}$ and $u_i \in P_{n+1}^{n-1} \cap \{x \in R^{n+1} |\ x_i = 1\}$. This proves the claim. Hence $S_{n+1} \subseteq P_{n+1}^{n-1}$. Since P_{n+1}^{n-1} is a convex set, this implies $Q_{n+1} \subseteq P_{n+1}^{n-1}$.

By Lemma 1, c^{n+1} cannot be removed by any disjunctive inequality of Q_{n+1}, i.e. $c^{n+1} \in Q_{n+1}^1$. Since $Q_{n+1}^1 \subseteq (P_{n+1}^{n-1})^1 = P_{n+1}^n$, it follows that $c^{n+1} \in P_{n+1}^n$.

$\square$

3   Concluding Remarks

In this note, we considered Gomory's mixed integer procedure and Balas's disjunctive procedure applied to polytopes P in the n-dimensional 0, 1-cube (it is known that the two procedures are equivalent [17]). We showed that, in the worst case, the mixed integer rank (or, equivalently, the disjunctive rank) of P is equal to n. Lovász and Schrijver [15] introduced a different procedure, based on a semi-definite relaxation of P_I for strengthening a polytope P in the n-dimensional 0, 1-cube. Recently, Cook and Dash [8] and Goemans and Tuncel [12] established that the semi-definite rank of polytopes in the n-dimensional 0, 1-cube is equal to n, in the worst case, by showing that the semi-definite rank of P_n (the same example as in Theorem 2 above) is equal to n. Neither of the two results implies the other since the disjunctive and semi-definite closures are incomparable (neither contains the other in general). Interestingly, both the semi-definite and disjunctive closures are contained in the lift-and-project closure as introduced by Balas, Ceria and Cornuéjols [3]. Since the lift-and-project rank is at most n [3] and the semi-definite and disjunctive ranks of P_n equal n, it follows that, in the worst case, all three procedures have rank n. We summarize this in Figure 1 where $A \to B$ means that the corresponding elementary closures satisfy $P_A \supseteq P_B$ and the inclusion is strict for some instances, and A not related to B in the figure means that for some instances $P_A \not\subseteq P_B$ and for other instances $P_B \not\subseteq P_A$.

Cook and Dash [8] also considered the intersection of the Chvátal closure and the semi-definite closure. They showed that, even for this Chvátal + semi-definite closure, it is still the case that the rank of P_n equals n. In a similar way, we can define the disjunctive + semi-definite closure of a mixed 0,1 program P_I as the intersection of the disjunctive closure and of the semi-definite closure of P. Using the approach of Cook and Dash and Theorem 2 above, it is easy to show that the disjunctive + semi-definite rank of P_n is equal to n.

Theorem 3. *The disjunctive + semi-definite rank of P_n is exactly n.*

References

1. Balas, E.: Disjunctive programming: cutting planes from logical conditions. In: Mangasarian, O., et al. (eds.): Nonlinear Programming, Vol. 2. Academic Press, New York (1975) 279–312
2. Balas, E.: Disjunctive programming. Annals of Discrete Mathematics **5** (1979) 3–51
3. Balas, E., Ceria, S., Cornuéjols, G.: A lift-and-project cutting plane algorithm for mixed 0-1 programs. Mathematical Programming **58** (1993) 295–324
4. Balas, E., Perregaard, M.: A Precise correspondence between lift-and-project cuts, simple disjunctive cuts and mixed integer Gomory cuts for 0-1 programming. Management Science Research Report MSRR-631, Carnegie Mellon University (2000)
5. Chvátal, V.: Edmonds polytopes and a hierarchy of combinatorial optimization. Discrete Mathematics **4** (1973) 305–337
6. Chvátal, V., Cook, W., Hartmann, M.: On cutting-plane proofs in combinatorial optimization. Linear Algebra and its Applications **114/115** (1989) 455–499
7. Cook, W., Cunningham, W., Pullyblank, W., and Schrijver, A.: Combinatorial Optimization. John Wiley, New York (1998).
8. Cook, W., Dash, S.: On the matrix-cut rank of polyhedra. preprint, Department of Computational and Applied Mathematics, Rice University, Houston, Texas (1999), to appear in Mathematics of Operations Research.
9. Cook, W., Kannan, R., Schrijver, A.: Chvátal closures for mixed integer programming problems. Mathematical Programming **47** (1990) 155–174
10. Cornuéjols, G., Li, Y.: Elementary closures for integer programs. Operations Research Letters **28** (2001) 1–8
11. Eisenbrand, F., Schulz, A.: Bounds on the Chvátal rank of polytopes in the 0/1-cube. In: Cornuéjols, G., et al, (eds.): Integer Programming and Combinatorial Optimization. Lecture Notes in Computer Science, **1610** (1999) 137–150
12. Goemans, M., Tuncel, L.: When does the positive semidefiniteness constraint help in lifting procedures. preprint, Department of Combinatorics and Optimization, University of Waterloo, Ontario, Canada (2000), to appear in Mathematics of Operations Research.

13. Gomory, R.: An algorithm for the mixed integer problem. Technical Report RM-2597, The RAND Corporation (1960)
14. Li, Y.: Bounds on the ranks of polytopes in the 0/1-cube. research report, Carnegie Mellon University (2000), submitted.
15. Lovász, L., Schrijver, A.: Cones of matrices and set-functions and 0-1 optimization. SIAM Journal of Optimization **1** (1991) 166–190
16. Nemhauser, G., Wolsey, L.: Integer and Combinatorial Optimization. John Wiley and Sons, New York (1988)
17. Nemhauser, G., Wolsey, L.: A recursive procedure to generate all cuts for 0-1 mixed integer programs. Mathematical Programming **46** (1990) 379–390

Fast 2-Variable Integer Programming

Friedrich Eisenbrand[1] and Günter Rote[2]

[1] Max-Planck-Institut für Informatik, Stuhlsatzenhausweg 85, 66123 Saarbrücken,
Germany, eisen@mpi-sb.mpg.de
[2] Institut für Informatik, Freie Universität Berlin, Takustraße 9, 14195 Berlin,
Germany, rote@inf.fu-berlin.de

Abstract. We show that a 2-variable integer program defined by m
constraints involving coefficients with at most s bits can be solved with
$O(m + s \log m)$ arithmetic operations or with $O(m + \log m \log s)M(s)$ bit
operations, where $M(s)$ is the time needed for s-bit integer multiplication.

1 Introduction

Integer linear programming is related to convex geometry as well as to algorithmic number theory, in particular to the algorithmic geometry of numbers. It
is well known that some basic number theoretic problems, such as the greatest
common divisor or best approximations of rational numbers can be formulated
as integer linear programs in two variables. Thus it is not surprising that current polynomial methods for *integer programming in fixed dimension* [7,12] use
lattice reduction methods, related to the reduction which is part of the classical
Euclidean algorithm for integers, or the computation of the *continued fraction
expansion* of a rational number. Therefore, integer programming in fixed dimension has a strong flavor of algorithmic number theory, and the running times of
the algorithms also depend on the binary encoding length of the input.

In this paper, we want to study this relation more carefully for the case of
2-*dimensional integer programming*. The classical Euclidean algorithm for computing the greatest common divisor (GCD) of two s-bit integers requires $\Theta(s)$
arithmetic operations and $\Theta(s^2)$ bit operations in the worst case. For example,
when it is applied to two consecutive Fibonacci numbers, it generates all the predecessors in the Fibonacci sequence (see e.g. [10]). Schönhage's algorithm [17]
improves this complexity to $O(M(s) \log s)$ bit operations, where $M(s)$ is the bit
complexity of s-bit integer multiplication. Thus the greatest common divisor of
two integers can be computed with a close to linear number of *bit operations*, if
one uses the fastest methods for integer multiplication [19]. The speedup technique by Schönhage has not yet been incorporated into current methods for two
variable integer programming. The best known algorithms for the integer programming problem in two dimensions [4,22,6] use $\Theta(s)$ arithmetic operations
and $\Omega(s^2)$ bit operations when the number of constraints is fixed. This number
of steps is required because these algorithms construct the complete sequence of
convergents of certain rational numbers that are computed from the input.

K. Aardal, B. Gerards (Eds.): IPCO 2001, LNCS 2081, pp. 78–89, 2001.
© Springer-Verlag Berlin Heidelberg 2001

Our goal is to show that integer programming in two variables is not harder than greatest common divisor computation. We achieve this goal for the case that the number of constraints is fixed. As one allows an arbitrary number of constraints, the nature of the problem also becomes combinatorial. For this general case we present an algorithm which requires $O(\log m)$ gcd-like computations, where m is the number of constraints. This improves on the best previously known algorithms.

Previous work. The 2-variable integer programming problem was extensively studied by various authors. The polynomiality of the problem was settled by Hirschberg and Wong [5] and Kannan [9] for special cases and by Scarf [15,16] for the general case before Lenstra [12] established the polynomiality of integer programming in any fixed dimension. Since then, several algorithms for the 2-variable problem have been suggested. We summarize them in the following table, for problems with m constraints involving (integer) numbers with at most s bits.

Method for integer programming	arithmetic complexity	bit complexity
Feit [4]	$O(m \ \log m + ms)$	$O(m \ \log m + ms)M(s)$
Zamanskij and Cherkasskij [22]	$O(m \ \log m + ms)$	$O(m \ \log m + ms)M(s)$
Kanamaru et al. [6]	$O(m \ \log m + s)$	$O(m \ \log m + s)M(s)$
Clarkson [2] (randomized)[1]	$O(m + s^2 \ \log m)$	$O(m + \log m \ s^2)M(s)$
This paper (Theorem 4)	$O(m + \log m \ s)$	$O(m + \log m \ \log s)M(s)$
Checking a point for feasibility	$\Theta(m)$	$\Theta(m)M(s)$
Shortest vector [18,21], GCD [17]	$O(s)$	$O(\log s)M(s)$

Thus our algorithm is better in the arithmetic model if the number of constraints is large, whereas the bit complexity of our algorithm is superior to the previous methods in all cases.

For comparison, we have also given the complexity of a few basic operations. The greatest common divisor of two integers a and b can be calculated by the special integer programming problem $\min\{ ax_1 + bx_2 \mid ax_1 + bx_2 \geq 1, \ x_1, x_2 \in \mathbb{Z} \}$ in two variables with one constraint. Also, checking whether a given point is feasible should be no more difficult than finding the optimum. So the sum of the last two lines of the table is the goal that one might aim for (short of trying to improve the complexity of integer multiplication or GCD itself). The complexity of our algorithm has still an extra $\log m$ factor in connection with the terms involving s, compared to the "target" of $O(m + s)$ and $O(m + \log s)M(s)$, respectively. However, we identify a nontrivial class of polygons, called *lower polygons*, for which we achieve this complexity bound.

---

[1] Clarkson claimed a complexity of $O(m + \log m \ s)$, because he mistakenly relied on algorithms from the literature to optimize small integer programs, whereas they only solve the integer programming *feasibility* problem.

Outline of the parametric lattice width method. The key concept of Lenstra's polynomial algorithm for integer programming in fixed dimension [12] is the lattice width of a convex body. Let $K \subseteq \mathbb{R}^d$ be a convex body and $\Lambda \subseteq \mathbb{R}^d$ be a lattice. The *width* of K along a direction $c \in \mathbb{R}^d$ is the quantity $w_c(K) = \max\{c^\mathrm{T}x \mid x \in K\} - \min\{c^\mathrm{T}x \mid x \in K\}$. The *lattice width* of K, $w_\Lambda(K)$, is the minimum of its widths along nonzero vectors c of the dual lattice Λ^* (see Section 2.1 for definitions related to lattices). For the standard integer lattice $\Lambda = \mathbb{Z}^d$, we denote $w_{\mathbb{Z}^d}(K)$ by $w(K)$ and call this number the *width* of K.[2] Thus if a convex body K has lattice width ℓ, then its lattice points can be covered by $\lfloor \ell + 1 \rfloor$ parallel lattice hyperplanes. If K does not include any lattice points, then K must be "flat" along a nonzero vector in the dual lattice. This fact is known as Khinchin's *flatness theorem* (see [8]).

Theorem 1 (Flatness theorem). *There exists a constant f_d depending only on the dimension d, such that each convex body $K \subseteq \mathbb{R}^d$ containing no lattice points of Λ has lattice width at most f_d.*

Lenstra [12] applies this fact to the *integer programming feasibility* problem as follows: Compute the lattice width ℓ of the given d-dimensional polyhedron P. If $\ell > f_d$, then one is certain that P contains integer points. Otherwise all lattice points in P are covered by at most $f_d + 1$ parallel hyperplanes. Each of the intersections of these hyperplanes with P corresponds to a $(d-1)$-dimensional feasibility problem. These are solved recursively. The actual integer programming *optimization* problem is reduced to the feasibility problem via binary search. This brings an additional factor of s into the running time.

Our approach avoids this binary search by letting the objective function slide into the polyhedron, until the lattice width of the truncated polyhedron equals f_d. The approach can roughly be described for $d = 2$ as follows. For solving the integer program

$$\max\{c^\mathrm{T}x \mid (x_1, x_2)^\mathrm{T} \in P \cap \mathbb{Z}^2\} \tag{1}$$

over a polygon P, we determine the smallest $\ell \in \mathbb{Z}$ such that the width of the truncated polyhedron $P \cap (c^\mathrm{T}x \geq \ell)$ is at most f_2. The optimum of (1) can then be found among the optima of the constant number of 1-dimensional integer programs formed by hyperplanes of the corresponding flat direction. We shall describe this parametric approach more precisely in Section 3. The core of the algorithm, which allows us to solve the parametric problem in essentially one single shortest vector computation, is presented in Section 4.3 (Proposition 4).

In the remaining part of the paper, we restrict our attention to the 2-di-men-sion-al case of integer programming. We believe that the flatness constant of Theorem 1 in two dimensions is $f_2 = 1 + \sqrt{4/3}$, but we have not found any result of this sort in the literature.

---

[2] This differs from the usual geometric notion of width, which is the minimum of w_c over all unit vectors c.

Complexity models. We analyze our algorithms both in the *arithmetic complexity* model and in the *bit complexity* model. The arithmetic model treats arithmetic operations $+, -, *$ and $/$ as unit-cost operations. This is the most common model in the analysis of algorithms and it is appropriate when the numbers are not too large and fit into one machine word. In the bit complexity model, every single bit-operation is counted. Addition and subtraction of s-bit integers takes $O(s)$ time. The current state of the art method for multiplication [19] shows that the bit complexity $M(s)$ of multiplication and division is $O(s \log s \log \log s)$, see [1, p. 279]. The difference between the two models can be seen in the case of the gcd-computation, which is an inherent ingredient of integer programming in small dimension: The best algorithm takes $\Theta(s)$ arithmetic operations, which amounts to $O(M(s)s)$ bit operations. However, a gcd can be computed in $O(M(s) \log s)$ bit operations [17]. The bit complexity model permits a more refined analysis of the asymptotic behavior of such algorithms.

2   Preliminaries

The symbols $\mathbb{N}$ and $\mathbb{N}_+$ denote the nonnegative and positive integers respectively. The *size* of an integer a is the length of its binary encoding. The size of a vector, a matrix, or a linear inequality is defined as the size of the largest entry or coefficient occurring in it. The *standard triangle* T^0 is the triangle with vertices $(0,0)$, $(1,0)$ and $(0,1)$.

The general 2-variable integer programming problem is as follows: given an integral matrix $A \in \mathbb{Z}^{m \times 2}$ and integral vectors $b \in \mathbb{Z}^m$ and $c \in \mathbb{Z}^2$, determine $\max\{ c^T x \mid x \in P(A,b) \cap \mathbb{Z}^2 \}$, where $P = P(A,b) = \{ x \in \mathbb{R}^2 \mid Ax \le b \}$ is the polyhedron defined by A and b.

We can assume without loss of generality that P is bounded (see e.g. [20, p. 237]). We can also restrict ourselves to problems where c is the vector $(0,1)^T$, by means of an appropriate unimodular transformation. These operations (as well as all the other reductions and transformations that will be applied in this paper) increase the size of the involved numbers by at most a constant factor. Therefore we define the 2-variable integer programming problem as follows.

Problem 1 (2IP). Given an integral matrix $A \in \mathbb{Z}^{m \times 2}$ and an integral vector $b \in \mathbb{Z}^m$ defining a polygon $P(A,b)$, determine $\max\{ x_2 \mid x \in P(A,b) \cap \mathbb{Z}^2 \}$.

2.1   The GCD, Best Approximations, and Lattices

The Euclidean algorithm for computing the *greatest common divisor* $\gcd(a_0, a_1)$ of two integers $a_0, a_1 > 0$ computes the remainder sequence $a_0, a_1, \ldots, a_k \in \mathbb{N}_+$, where a_i, $i \ge 2$ is given by $a_{i-2} = a_{i-1}q_{i-1} + a_i$, $q_i \in \mathbb{N}$, $0 < a_i < a_{i-1}$, and a_k divides a_{k-1} exactly. Then $a_k = \gcd(a_0, a_1)$. The *extended Euclidean algorithm* keeps track of the unimodular matrices $M^{(j)} = \prod_{i=1}^{j} \left(\begin{smallmatrix} q_i & 1 \\ 1 & 0 \end{smallmatrix} \right)$, $0 \le j \le k - 1$. One has $\left(\begin{smallmatrix} a_0 \\ a_1 \end{smallmatrix} \right) = M^{(j)} \left(\begin{smallmatrix} a_j \\ a_{j+1} \end{smallmatrix} \right)$. The representation $\gcd(a_0, a_1) = ua_0 + va_1$ with two integers u, v with $|u| \le a_1$ and $|v| \le a_0$ can be computed with the extended

Euclidean algorithm with $O(s)$ arithmetic operations or with $O(M(s) \log s)$ bit operations [17]. More generally, given two integers $a_0, a_1 > 0$ and some integer K with $a_0 > K > \gcd(a_0, a_1)$, one can compute the elements a_i and a_{i+1} of the remainder sequence $a_0, a_1, \ldots, a_k$ such that $a_i \geq L > a_{i+1}$, together with the matrix $M^{(i)}$ with $O(M(s) \log s)$ bit operations using the so-called *half-gcd* approach [1, p. 308].

The fractions $M_{1,1}^{(i)}/M_{2,1}^{(i)}$ are called the *convergents* of $\alpha = a_0/a_1$. A fraction x/y, $y \geq 1$ is called a *best approximation* to α, if one has $|y\alpha - x| < |y'\alpha - x|$ for all other fractions x'/y', $0 < y' \leq y$. A best approximation to α is a convergent of α.

A 2-*dimensional (rational) lattice* Λ is a set of the form $\Lambda(A) = \{ Ax \mid x \in \mathbb{Z}^2 \}$, where $A \in \mathbb{Q}^{2\times 2}$ is a nonsingular rational matrix. The matrix A is called a *basis* of Λ. One has $\Lambda(A) = \Lambda(B)$ for $B \in \mathbb{Q}^{2\times 2}$ if and only if $B = AU$ with some *unimodular matrix* U, i.e., $U \in \mathbb{Z}^{2\times 2}$ and $\det(U) = \pm 1$. Every lattice $\Lambda(A)$ has a unique basis of the form $\left(\begin{smallmatrix} a & b \\ 0 & c \end{smallmatrix} \right) \in \mathbb{Q}^{2\times 2}$, where $c > 0$ and $a > b \geq 0$, called the *Hermite normal form, HNF* of Λ. The Hermite normal form can be computed with an extended-gcd computation and a constant number of arithmetic operations. The *dual lattice* of $\Lambda(A)$ is the lattice $\Lambda^*(A) = \{ x \in \mathbb{R}^2 \mid x^{\mathrm{T}} v \in \mathbb{Z}, \ \forall v \in \Lambda(A) \}$. It is generated by $(A^{-1})^{\mathrm{T}}$. A *shortest vector* of Λ (w.r.t. some given norm) is a nonzero vector of Λ with minimal norm. A shortest vector of a 2-dimensional lattice $\Lambda(A)$, where A has size at most s, can be computed with $O(s)$ arithmetic operations [11]. Asymptotically fast algorithms with $O(M(s) \log s)$ bit operations have been developed by Schönhage [18] and Yap [21], see also Eisenbrand [3] for an easier approach.

2.2   Homothetic Approximation

We say that a body P *homothetically approximates* another body Q with *homothety ratio* $\lambda \geq 1$, if $P + t_1 \subseteq Q \subseteq \lambda P + t_2$ for some translation $x \mapsto x + t_1$ and some homothety (scaling and translation) $x \mapsto \lambda x + t_2$.

This concept is important for two reasons: (i) The lattice width of Q is determined by the lattice width of P up to a multiplicative error of at most λ, i.e., $w_\Lambda(P) \leq w_\Lambda(Q) \leq \lambda w_\Lambda(P)$. (ii) A general convex body Q can be approximated by a simpler body P; for example, any plane convex body can be approximated by a triangle with homothety ratio 2.

In this way, one can compute an approximation to the width of a triangle. Let $a\colon x \mapsto Bx + t$ be some affine transformation. Clearly the width $w(K)$ of a convex body K is equal to the lattice width $w_{\Lambda(B)}$ of the transformed body $a(K)$. Thus we get the following lemma.

Lemma 1. *Let $T \subseteq \mathbb{R}^2$ be a triangle which is mapped to the standard triangle T^0 by the affine transformation $x \mapsto Bx + t$. Let $\bar{v}$ be a shortest vector of $\Lambda^*(B)$ with respect to the ℓ_2-norm. Then*

$$(1 - \sqrt{1/2}) \, \|\bar{v}\|_2 \leq w(T) \leq 1/\sqrt{2} \, \|\bar{v}\|_2.$$

Moreover, the integral vector $v := B^T \bar{v}$ *is a good substitute for the minimum-width direction:*

$$w(T) \leq w_v(T) \leq (\sqrt{2} + 1)\, w(T) \qquad \qquad \Box$$

With linear programming, one can easily find a good approximating triangle $T \subseteq P$ for a given polygon $P = P(A, b)$. For example, we can take the longest horizontal segment ef contained in P. It is characterized by having two parallel supporting lines through e and f which enclose P, and it can be computed as a linear programming problem in three variables in $O(m)$ steps, by Megiddo's algorithm [13]. (Actually, one can readily adapt Megiddo's simple algorithm for *two-variable* linear programming to this problem.) Together with a point $g \in P$ which is farthest away from the line through e and f (this point can be found by another linear programming problem), we obtain a triangle $T = efg$ which is a homothetic approximation of P with homothety ratio 3. Together with the previous lemma, and the known algorithms for computing shortest lattice vectors, we get the following lemma.

Lemma 2. *Let $P \subseteq \mathbb{R}^2$ be a polygon defined by m constraints each of size s. Then one can compute with $O(m + s)$ arithmetic operations or with $O(m + \log s)M(s)$ bit operations an integral direction $v \in \mathbb{Z}^2$ with*

$$w_v(P) = \Theta(w(P)).$$

$\hspace{1cm}\Box$

2.3 "Checking the Width"

We will several times use the following basic subroutine, which we call *checking the width*.

For a given polygon P, we first compute its approximate lattice width by Lemma 2, together with a direction v. This gives us an interval $[K, \alpha K]$ for the lattice width of P. If $K \geq f_2 + 1$, then we say that P is *thick*, and we know that P contains an integral point. This is one possible outcome of the algorithm. Otherwise, P is *thin* and we solve 2IP for P as follows. Enumerate the at most $\alpha(f_2 + 1) = O(1)$ lattice lines through P, solving a one-dimensional integer program for each of them. We may find that P is *empty*, that is, it contains no integral point, or otherwise find the optimum point in P. These are the other two possible results of the algorithm, and this will always mean that no further processing of P is required. It is easy to see that the following bounds hold.

Lemma 3. *Checking the width takes $O(m + s)$ arithmetic operations or $O(m + \log s)M(s)$ bit operations.*

$\hspace{1cm}\Box$

3   The Approximate Parametric Lattice Width (APLW) Problem

As in the case of Lenstra's algorithm, the lattice width is approximated via homothetic approximations and a shortest vector computation. This brings in some error which complicates the parametric lattice width method described in the introduction. The following problem, called *approximate parametric lattice width problem*, APLW for short, is an attempt to live up to the involved approximation error. If P is a polygon, we denote by P_ℓ the *truncated polygon* $P_\ell = P \cap (x_2 \geq \ell)$.

Problem 2 (APLW). This problem is parameterized by an approximation ratio $\gamma \geq 1$. The input is a number $K \in \mathbb{N}$ and a polygon $P \subseteq \mathbb{R}^2$ with $w(P) \geq K$. The task is to find some $\ell \in \mathbb{Z}$ such that the width of the truncated polygon P_ℓ satisfies

$$K \leq w(P_\ell) \leq \gamma\, K.$$

Integer programming can be reduced to the APLW problem:

Proposition 1. *Suppose that the APLW problem with any fixed approximation ratio γ can be solved in $A(m, s)$ bit operations or in $\widetilde{A}(m, s)$ arithmetic operations for a polygon P, described by m constraints of size at most s. Then 2IP can be solved in $T(m, s)$ bit operations or in $\widetilde{T}(m, s)$ arithmetic operations, with*

$$T(m, s) = O(A(m, s) + (m + \log s)\, M(s)),$$
$$\widetilde{T}(m, s) = O(\widetilde{A}(m, s) + m + s).$$

Proof. First we check the width of P. If P is thin we are done with the claimed time bounds (see Lemma 3). Otherwise solve APLW for $K = f_2 + 1$, yielding an integer $\ell \in \mathbb{Z}$ such that $f_2 + 1 \leq w(P_\ell) \leq \gamma\,(f_2 + 1)$. Then the polytope $P_\ell = P \cap (x_2 \geq \ell)$ must contain an integer point. Therefore the optimum of 2IP over P is the optimum of 2IP over P_ℓ. Compute an integral direction v with $w_v(P_\ell) = O(1)$ by Lemma 2. As in the case of checking the width, the optimum can then be found among the corresponding constant number of univariate integer programs. $\square$

4   Solving the Integer Program

An *upper polygon* P has a horizontal line segment ef as an edge and a pair of parallel lines through the points e and f enclosing P, and it lies above ef. A *lower polygon* is defined analogously, see Fig. 1. We now describe efficient algorithms for APLW for upper triangles and lower polygons. This enables us to solve 2IP for polygons with a fixed number of constraints. Polygons described by a fixed number of constraints are the base case of our prune-and-search algorithm for general 2IP.

4.1   Solving APLW for Upper Triangles

Let T be an upper triangle. By translating T, we can assume that the top vertex is at the origin, and hence T is described by inequalities of the form $Ax \leq 0$, $x_2 \geq L$, where $L < 0$. All the truncated triangles $T_\ell = T \cap (x_2 \geq \ell)$

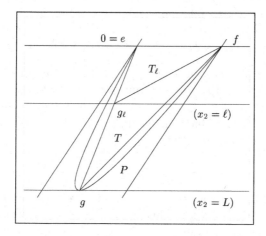

Fig. 1. The approximation of P_ℓ

for $0 > \ell \geq L$ are scaled copies of T, and the width of T_ℓ scales accordingly: $w(T_\ell) = |\ell|\, w(T_{-1})$. Therefore we simply have to compute an approximation to the width of T by Lemma 1 and choose the scaling factor $|\ell|$ accordingly so that $K \leq w(T_\ell) \leq \gamma K$ holds. Hence we have the following fact.

Proposition 2. *APLW can be solved with $O(s)$ arithmetic operations or with $O(M(s) \log s)$ bit operations for an upper triangle which is described by constraints of size at most s.*

4.2 Solving APLW for Lower Polygons

Since the width is invariant under translation, we can assume that the left vertex e of the upper edge ef is at the origin. We want to find an $\ell \in \mathbb{Z}$ with $K \leq w(P_\ell) \leq \gamma K$ for some constant $\gamma \geq 1$.

Let $g = (g_1, g_2) \in P$ be a vertex of P with smallest second component $g_2 = L$. Let g_ℓ be the point of intersection between the line segment eg and the line $x_2 = \ell$, for $0 > \ell \geq L$ and denote the triangle $0fg_\ell$ by T_ℓ (see Fig. 1).

Lemma 4. *The triangle T_ℓ is a homothetic approximation to the truncated lower polygon P_ℓ with homothety ratio 2.*   □

Thus we can solve APLW for P by finding the largest $\ell \in \mathbb{Z}$, $0 \geq \ell \geq L$ such that $w(T_\ell) \geq K$. For any $2{\times}2$ matrix A and a number p, we use the notation

$$A_p := \begin{pmatrix} 1 & 0 \\ 0 & p \end{pmatrix} A$$

for the matrix whose second row is multiplied by p. If the matrix B maps the triangle T_{-1} to the standard triangle T^0, then $B_{1/|\ell|}$ maps T_ℓ to T^0. By Lemma 4 and Lemma 1 we have the relation

$$(1 - \sqrt{1/2}) \, \|v\|_2 \leq w(P_\ell) \leq \sqrt{2} \, \|v\|_2,$$

where v is a shortest vector of $\Lambda^*(B_{1/|\ell|})$ w.r.t. the ℓ_2-norm.

We can thus solve APLW by determining the smallest $p \in \mathbb{N}$ such that the length of a shortest vector v of $\Lambda^*(B_{1/p})$ w.r.t. the ℓ_2-norm satisfies the relation $\|v\|_2 \geq (1 - \sqrt{1/2})^{-1} K$. Since one has $\|v\|_\infty \leq \|v\|_2 \leq \sqrt{2}\|v\|_\infty$ we can as well search for the smallest p such that

$$\|v\|_\infty \geq (1 - \sqrt{1/2})^{-1} K,$$

where v is a shortest vector of $\Lambda^*(B_{1/p})$ w.r.t. the ℓ_∞-norm. Observe that one has $\Lambda^*(B_{1/p}) = \Lambda(((B^{-1})^{\mathrm{T}})_p)$. This shows that we can translate APLW into the following problem which we call the *parametric shortest vector problem*. The parameter is the factor p with which the second coordinate of the lattice is multiplied and we want to find the largest value of p such that the norm of the shortest vector does not exceed a given bound.

Problem 3 (PSV). Given a nonsingular matrix $A \in \mathbb{Q}^{2\times 2}$ and a constant $K \in \mathbb{N}_+$, find the largest $p \in \mathbb{N}_+$ such that $\|v\|_\infty \leq K$, where v is a shortest vector of $\Lambda(A_p)$ w.r.t. the ℓ_∞-norm.

The following statement is proved in [3]. It shows that a shortest vector of a 2-dimensional integral lattice can be found among the best approximations a rational number computed from the Hermite normal form of the lattice.

Proposition 3. *Let $\Lambda \subseteq \mathbb{Z}^2$ be given by its Hermite normal form $\left(\begin{smallmatrix} a & b \\ 0 & c \end{smallmatrix}\right) \in \mathbb{Z}^{2\times 2}$. A shortest vector of Λ with respect to the ℓ_∞-norm is either $\left(\begin{smallmatrix} a \\ 0 \end{smallmatrix}\right)$ or $\left(\begin{smallmatrix} b \\ c \end{smallmatrix}\right)$, or it is of the form $\left(\begin{smallmatrix} -xa+yb \\ yc \end{smallmatrix}\right)$, $x \in \mathbb{N}$, $y \in \mathbb{N}_+$ where the fraction x/y is a best approximation to the number b/a.*

By the relation between best approximations and the remainder sequence of the Euclidean algorithm we can now efficiently solve PSV.

Proposition 4. *PSV can be solved with $O(s)$ arithmetic operations or with $O(M(s) \log s)$ bit operations for matrices A and integers K of size s.*

Proof. Assume without loss of generality that A is an integral matrix. Let $\left(\begin{smallmatrix} a & b \\ 0 & c \end{smallmatrix}\right)$ be the HNF of A, the HNF of $\Lambda(A_p)$ is then the matrix $\left(\begin{smallmatrix} a & b \\ 0 & pc \end{smallmatrix}\right)$. Either $\left(\begin{smallmatrix} a \\ 0 \end{smallmatrix}\right)$ or $\left(\begin{smallmatrix} b \\ pc \end{smallmatrix}\right)$ is a shortest vector (these cases can be treated easily), or there exists a shortest vector $\left(\begin{smallmatrix} -xa+yb \\ pyc \end{smallmatrix}\right)$ such that x/y is a best approximation of b/a, thus a convergent of b/a.

Since we want to maximize p we have to find the convergent x/y of b/a with minimal $y \geq 1$ which satisfies $|-xa + yb| \leq K$. This convergent yields the candidate with which we can achieve a best scaling factor p. The best scaling factor is then simply $p = \lfloor K/(yc) \rfloor$.

The required convergent x/y can be determined by exploiting the relation between convergents and the Euclidean algorithm. Let $a_0, a_1, \dots, a_k$ be the remainder sequence of $b = a_0$ and $a = a_1$. Multiplying the equation $\binom{a_0}{a_1} = M^{(i)} \binom{a_i}{a_{i+1}}$ by the inverse of the unimodular matrix $M^{(i)}$ gives the following equation for the i-th convergent $x/y = M_{1,1}^{(i)}/M_{2,1}^{(i)}$:

$$\pm(-xa + yb) = \pm(-M_{1,1}^{(i)}a_1 + M_{2,1}^{(i)}a_0) = a_{i+1}$$

Since subsequent convergents have strictly increasing denominators, we are looking for the first index i with $a_{i+1} \le K$. This index is determined by the conditions $a_i > K$ and $a_{i+1} \le K$. As mentioned in Section 2.1, the corresponding convergent $x/y = M_{1,1}^{(i)}/M_{2,1}^{(i)}$ can be computed with $O(M(s)\log(s))$ bit operations, or with $O(s)$ arithmetic operations.   □

Theorem 2. *APLW for lower polygons defined by m constraints of size at most s can be solved with $O(m + s)$ arithmetic operations or with $O(m + \log s)M(s)$ bit operations.*

Proof. After one has found the point g which minimizes $\{x_2 \mid x \in P\}$ with Megiddo's algorithm for linear programming [13,14] one has to solve PSV for the matrix B which maps T_{-1} to the standard triangle. The time bound thus follows from Proposition 4.   □

4.3 An Efficient Algorithm for 2IP with a Fixed Number of Constraints

Theorem 3. *A 2IP problem defined by a constant number of constraints of size at most s can be solved with $O(s)$ arithmetic operations or with $O(M(s)\log s)$ bit operations.*

Proof. We compute the underlying polygon, triangulate it, and cut each triangle into an upper and a lower triangle. We get a constant number of 2IP's on upper and lower triangles, defined by inequalities of size $O(s)$. The complexity follows thus from Proposition 1, Proposition 2, and Theorem 2.   □

4.4 Solving 2IP for Upper Polygons

It follows from Sect. 4.2 that 2IP over lower polygons can be solved with $O(m+s)$ basic operations or with $O((m + \log s)M(s))$ bit operations. Any polygon can be dissected into an upper and a lower part (by solving a linear programming problem); thus we are left with solving 2IP for upper polygons. Unfortunately, we cannot solve APLW for upper polygons directly. Instead, we use Megiddo's prune-and-search technique [13] to reduce the polygon to a polygon with a constant number of sides, to which the method from Theorem 3 in the previous section is then applied. This procedure works for general polygons and not only for upper polygons.

Our algorithm changes the polygon P by discarding constraints and by introducing bounds $l \leq x_2 \leq u$ such that the solution to 2IP remains invariant. Initially we check the width of P. If P is thin, we are done. Otherwise, we start with $l = -\infty$ and $u = \infty$. We gradually narrow down the interval $[l, u]$ and at the same time remove constraints that can be ignored without changing the optimum.

One iteration proceeds as follows: Ignoring the constraints of the form $l \leq x_2 \leq u$, the m constraints defining P can be classified into *left* and *right* constraints, depending on whether the feasible side lies to their right or left, respectively. We arbitrarily partition all left constraints into pairs and compute the intersection points of their corresponding lines, and similarly for the right constraints. We get roughly $m/2$ intersection points, and we compute the median μ of their x_2-coordinates. Now we check the width of P_μ. If P_μ is thick, we replace l by μ. In addition, for each of the $m/4$ intersection points *below* μ, there is one constraint among the two constraints defining it which is redundant in the new range $\mu \leq x_2 \leq u$. Removing these constraints reduces the number of constraints by a factor of $\frac{3}{4}$.

If P_μ is thin and contains integer points, we are done. If P_μ is empty, we replace u by μ. We remove constraints as above, but we now consider the intersection points *above* μ.

In this way, after $O(\log m)$ iterations, we have either solved the problem, or we have reduced it to a polygon with at most four constraints, which can be solved by Theorem 3.

Each iteration involves one operation of checking the width, plus $O(m)$ arithmetic operations for computing intersections and their median. Since the number m of constraints is geometrically decreasing in successive steps, we get a total of $O(m + \log m \; s)$ arithmetic operations and $O(m + \log m \log s)M(s)$ bit operations. The additional complexity for dealing with the final quadrilateral is dominated by this. This gives rise to our main result.

Theorem 4. *The two-variable integer programming problem with m constraints of size at most s can be solved in $O(m + \log m \; s)$ arithmetic operations or in $O(m + \log m \log s)M(s)$ bit operations.*

References

1. A. V. Aho, J. E. Hopcroft, and J. D. Ullman. *The Design and Analysis of Computer Algorithms.* Addison-Wesley, Reading, 1974.
2. K. L. Clarkson. Las Vegas algorithms for linear and integer programming when the dimension is small. *Journal of the Association for Computing Machinery,* 42:488–499, 1995.
3. F. Eisenbrand. Short vectors of planar lattices via continued fractions. *Information Processing Letters,* 2001. to appear. http://www.mpi-sb.mpg.de/~eisen/report_lattice.ps.gz
4. S. D. Feit. A fast algorithm for the two-variable integer programming problem. *Journal of the Association for Computing Machinery,* 31(1):99–113, 1984.

5. D. S. Hirschberg and C. K. Wong. A polynomial algorithm for the knapsack problem in two variables. *Journal of the Association for Computing Machinery*, 23(1):147–154, 1976.
6. N. Kanamaru, T. Nishizeki, and T. Asano. Efficient enumeration of grid points in a convex polygon and its application to integer programming. *Int. J. Comput. Geom. Appl.*, 4(1):69–85, 1994.
7. R. Kannan. Minkowski's convex body theorem and integer programming. *Mathematics of Operations Research*, 12(3):415–440, 1987.
8. R. Kannan and L. Lovász. Covering minima and lattice point free convex bodies. *Annals of Mathematics*, 128:577–602, 1988.
9. Ravindran Kannan. A polynomial algorithm for the two-variable integer programming problem. *Journal of the Association for Computing Machinery*, 27(1):118–122, 1980.
10. D. E. Knuth. *The Art of Computer Programming*, volume 2. Addison-Wesley, 1969.
11. J. C. Lagarias. Worst-case complexity bounds for algorithms in the theory of integral quadratic forms. *Journal of Algorithms*, 1:142–186, 1980.
12. H. W. Lenstra. Integer programming with a fixed number of variables. *Mathematics of Operations Research*, 8(4):538–548, 1983.
13. N. Megiddo. Linear time algorithms for linear programming in R^3 and related problems. *SIAM Journal on Computing*, 12:759–776, 1983.
14. N. Megiddo. Linear programming in linear time when the dimension is fixed. *Journal of the Association for Computing Machinery*, 31:114–127, 1984.
15. H. E. Scarf. Production sets with indivisibilities. Part I: generalities. *Econometrica*, 49:1–32, 1981.
16. H. E. Scarf. Production sets with indivisibilities. Part II: The case of two activities. *Econometrica*, 49:395–423, 1981.
17. A. Schönhage. Schnelle Berechnung von Kettenbruchentwicklungen. (Fast computation of continued fraction expansions). *Acta Informatica*, 1:139–144, 1971.
18. A. Schönhage. Fast reduction and composition of binary quadratic forms. In *International Symposium on Symbolic and Algebraic Computation, ISSAC 91*, pages 128–133. ACM Press, 1991.
19. A. Schönhage and V. Strassen. Schnelle Multiplikation großer Zahlen (Fast multiplication of large numbers). *Computing*, 7:281–292, 1971.
20. A. Schrijver. *Theory of Linear and Integer Programming*. John Wiley, 1986.
21. C. K. Yap. Fast unimodular reduction: Planar integer lattices. In *Proceedings of the 33rd Annual Symposium on Foundations of Computer Science*, pages 437–446, Pittsburgh, 1992.
22. L. Ya. Zamanskij and V. D. Cherkasskij. A formula for determining the number of integral points on a straight line and its application. *Ehkonomika i Matematicheskie Metody*, 20:1132–1138, 1984.

Approximating k-Spanner Problems for $k > 2$

Michael Elkin and David Peleg[*]

Department of Applied Mathematics and Computer Science, The Weizmann Institute of Science, Rehovot, 76100 Israel. {elkin,peleg}@wisdom.weizmann.ac.il.

Abstract. Given a graph $G = (V, E)$, a subgraph $G' = (V, H)$, $H \subseteq E$ is a *k-spanner* (respectively, *k-DSS*) of G if for any pair of vertices $u, w \in V$ it satisfies $d_H(u, w) \leq k \cdot d_G(u, w)$ (resp., $d_H(u, w) \leq k$). The *basic k-spanner (resp., k-DSS)* problem is to find a k-spanner (resp., k-DSS) of a given graph G with the smallest possible number of edges.

This paper considers approximation algorithms for these and some related problems for $k > 2$. Both problems are known to be $\Omega(2^{\log^{1-\epsilon} n})$-inapproximable [11,13]. The basic k-spanner problem over undirected graphs with $k > 2$ has been given a sublinear ratio approximation algorithm (with ratio roughly $O(n^{\frac{2}{k+1}})$), but no such algorithms were known for other variants of the problem, including the directed and the client-server variants, as well as for the k-DSS problem. We present the first approximation algorithms for these problems with sublinear approximation ratio.

The second contribution of this paper is in characterizing some wide families of graphs on which the problems do admit a logarithmic and a polylogarithmic approximation ratios. These families are characterized as containing graphs that have optimal or "near-optimal" spanners with certain desirable properties, such as being a tree, having low arboricity or having low girth. All our results generalize to the directed and the client-server variants of the problems. As a simple corollary, we present an algorithm that given a graph G builds a subgraph with $\tilde{O}(n)$ edges and stretch bounded by the *tree-stretch* of G, namely the minimum maximal stretch of a spanning tree for G.

The analysis of our algorithms involves the novel notion of *edge-dominating systems* developed in the paper. The technique introduced in the paper enables us to reduce the studied algorithmic questions of approximability of the k-spanner and k-DSS problems to purely graph-theoretical questions concerning the existence of certain combinatorial objects in families of graphs.

1 Introduction

1.1 Motivation

Spanners for general graphs were introduced in [23,21], and were shown to be the underlying graph structure in a number of constructions in distributed systems

---

[*] Supported in part by grants from the Israel Science Foundation and from the Israel Ministry of Science and Art.

and communication networks. Spanners of Euclidean graphs have turned out to be relevant also in the area of computational geometry (cf. [4,8,2]).

Given a graph $G = (V, E)$ and a subgraph $G' = (V, E')$, $E' \subseteq E$, let

$$stretch(G') = \max \left\{ \frac{d_{G'}(u, w)}{d_G(u, w)} \mid u, w \in V \right\},$$

where $d_H(u, w)$ denotes the distance between u and w in H. The subgraph G' is a k-*spanner* of G if $stretch(G') \leq k$. The combinatorial problem of finding the sparsest k-spanner of a given graph G (in terms of number of edges) is called the (basic) k-*spanner problem*. This problem is NP-hard, which makes it interesting to study its approximability properties.

For $k = 2$, the approximation ratio on arbitrary n-vertex graphs is known to be $\Theta(\log n)$ [18,17]. For $k > 2$, however, the situation is not as good. To begin with, it is known that under the assumption that $NP \not\subseteq DTIME(n^{\text{polylog } n})$, the approximation ratio of any polynomial time algorithm for the k-spanner problem, for any constant $k > 2$ and for any $0 < \epsilon < 1$, is $\Omega(2^{\log^{1-\epsilon} n})$ [11]. On the positive side, it is known that every n-vertex graph has a (polynomial-time constructible) k-spanner of size $O(n^{1+\frac{2}{k+1}})$ for odd k and $O(n^{1+\frac{2}{k+2}})$ for even k [21,16]. Since any spanner must contain at least $n - 1$ edges, the algorithm for constructing such spanners can be thought of as a "universal" approximation algorithm for the problem, with ratio $O(n^{\frac{2}{k+1}})$ for odd k and $O(n^{\frac{2}{k+2}})$ for even k. This performance is poor for small constant values of k, e.g., for $k = 3$ it yields an $O(n^{1/2})$-approximation algorithm.

The situation is even worse for some related edge deletion problems. For instance, the *directed k-spanner* problem (namely, the problem on directed graphs) has no known sublinear approximation ratio. In particular, it is known that for certain n-vertex directed graphs, *any* k-spanner requires $\Omega(n^2)$ edges, hence the problem does not enjoy a "universal" sublinear ratio approximation algorithm in the above sense.

The same applies to the k-*diameter spanning subgraph* (or k-*DSS*) problem, defined as follows. Given a graph $G = (V, E)$, a subgraph $G' = (V, H)$, $H \subseteq E$ is a k-*DSS* of G if the distance in H between any two vertices $u, w \in V$ satisfies $d_H(u, w) \leq k$. The k-DSS problem calls for finding the sparsest k-DSS of a given graph G (in terms of number of edges). Again, the results of [21,10] indicate that no "universal" sublinear ratio approximation algorithm for the (directed or undirected) k-DSS problem exists.

A third example involves the *client-server (CS) k-spanner* problem. This is a generalization of the basic k-spanner problem in which every edge may be either a client, a server, or both a client and a server. The goal is to k-span all the client edges by server edges, using a minimal number of server edges. In the *all-client (AC)* variant of the CS k-spanner problem, all the edges are clients, and in the *all-server (AS)* variant, all the edges are servers. All of these problems are known to be $\Omega(2^{\log^{1-\epsilon} n})$-inapproximable for $k > 2$ and any $0 < \epsilon < 1$ [10]. Moreover, with the exception of the AS k-spanner problem, none of these problems currently enjoys a sublinear ratio approximation algorithm.

1.2   Our Results

The current paper is concerned with two main directions. The first involves obtaining sublinear ratio approximation algorithms for a number of the edge deletion problems discussed above. In particular, the paper presents $\tilde{O}(n^{2/3})$-approximation algorithms for the directed 3-spanner, the (directed and undirected) CS 3-spanner and the (directed and undirected) 3-DSS problems. (We denote $\tilde{O}(f(n)) = O(f(n)\text{polylog }(n))$.) In fact, our approximation algorithm usually provides a better ratio, and, in particular, we show that for graphs with $O(n^{1+\beta})$ edges, the algorithm has an $\tilde{O}(n^{\frac{\beta+1}{3}})$-approximation ratio.

The second direction aims at developing a better understanding for the causes of the apparent difficulty of the k-spanner problem, by identifying specific parameters which affect its approximability. Specifically, our approach to this problem is based on examining various restricted graph classes with special properties, and attempting to approximate the problem on these classes.

The families of graphs we consider are characterized as containing graphs that have optimal or "near-optimal" spanners with certain properties. Intuitively, we prove that if the given input graph G has a "near-optimal" spanner H of some convenient structure, then it is possible to find a "relatively good" spanner for G (namely, one which is close to H in sparsity).

As a first and most extreme example, we consider the family of graphs that enjoy a *tree k-spanner*, namely, a k-spanner in the form of a tree. Let $S_{TREE}(G)$ denote the minimum k such that G has a tree k-spanner. Finding a tree attaining $S_{TREE}(G)$ is known to be NP-hard even restricted to planar graphs [15]. The problem of computing $S_{TREE}(G)$ for a given graph G is known to be $(1+\sqrt{5})/2$-inapproximable [22], and no nontrivial approximation algorithm for the problem is known. Denote by $SP_k(TREE)$ the family of graphs that admit a tree k-spanner.

The k-spanner problem restricted to the class $SP_k(TREE)$ was shown in [6] to be polynomially solvable for $k = 2$ and NP-complete for $k \geq 4$. In Section 5.2 we present an algorithm providing an $O(\min\{k^2 \log n, \frac{\log^3 n}{(\log \log n)^2}\})$-approximation ratio for the problem on $SP_k(TREE)$ for arbitrary (not necessarily constant) values of k. In particular, for any graph G this algorithm builds a subgraph H of G with $\tilde{O}(n)$ edges and $stretch(H) \leq S_{TREE}(G)$.

We next turn to wider classes of graphs, enjoying a spanner with low arboricity or low girth. For any graph G, a k-spanner H is called a *near-optimal k-spanner* of G if any other k-spanner H' of G satisfies $\frac{|H|}{|H'|} = O(\text{polylog } n)$. Denote by $SP_k(PL)$ the family of graphs that admit a planar k-spanner. Denote by $SP_k(BA)$ (respectively, $SP_k(\log A)$) the family of graphs that admit a near-optimal k-spanner with arboricity (or genus) bounded by a constant (respectively, by polylog n). For any fixed integer g denote by $SP_k(GIRTH(g))$ the family of graphs that admit a near-optimal k-spanner with girth no smaller than g.

In Section 5.1 we present an algorithm providing an $O(\log n \cdot a(H))$-approximation ratio for the 3-spanner problem on general graphs, where $a(G')$ is the arboricity of the graph G' and H is the optimal 3-spanner of the input graph. It follows that the problem admits a logarithmic approximation ratio when restricted to graphs in the class $SP_3(BA)$ (and in particular $SP_3(PL)$), and a polylogarithmic ratio when restricted to graphs in $SP_3(\log A)$. In the full paper these results are extended to the directed and CS variants of the problem.

All the above results can be easily adapted to the k-DSS problem as well. In particular, define the graph family $DSS_k(TREE)$ (respectively, $DSS_k(PL)$, $DSS_k(BA)$, $DSS_k(\log A)$ or $DSS_k(GIRTH(g))$) as the set of graphs that admit a k-DSS in the form of a tree (resp., which is planar, with arboricity bounded by a constant, with arboricity bounded by polylog n, or with girth at least g). In the full paper we present an $O(\min\{k^2 \log n, \frac{\log^3 n}{(\log \log n)^2}\})$-approximation algorithm for the k-DSS problem on $DSS_k(TREE)$, and an $O(\log n \cdot a(H))$-approximation algorithm for the 3-DSS problem on general graphs, where H is the optimal 3-DSS of the input graph, yielding a logarithmic approximation over the class $DSS_3(BA)$ and a polylogarithmic ratio over $DSS_3(\log A)$.

For problems whose definition involves a free parameter (like the parameter k in all the above problems), it is instructive to study their approximability as a function of this parameter. In particular, a set of problems $\{\Pi_p\}$ indexed by a parameter p is said to enjoy the *ratio degradation property* with respect to the parameter if the approximability of the problem Π_p decreases exponentially fast with the inverse of the parameter p. The result of [21] shows that the basic k-spanner problem enjoys the ratio degradation property with respect to the stretch requirement k, whereas the results of [10,13] show that the CS, the AC and the directed k-spanner problems, as well as the k-DSS problem, do not enjoy it (with respect to k).

We analyze the behavior of the 3-spanner and the 3-DSS problems over the graph classes $SP_3(GIRTH(g))$ and $DSS_3(GIRTH(g))$. Formally, let 3-spanner(g) (resp., 3-DSS(g)) be the 3-spanner (resp., 3-DSS) problem restricted to the family $SP_3(GIRTH(g))$ (resp., $DSS_3(GIRTH(g))$). We show that the problem families $\{3\text{-spanner}(g)\}_{g=1}^{\infty}$ and $\{3\text{-DSS}(g)\}_{g=1}^{\infty}$ enjoy the ratio degradation property with respect to this parameter. All the results mentioned above generalize to the directed and the client-server variants of the problems.

Section 6 concerns bicriteria approximation algorithms. It presents a bicriteria $(O(n^{1/2}), +2)$-approximation algorithm for the AC k-spanner and k-DSS problems. In other words, the algorithm produces an AC $(k+2)$-spanner (respectively, $(k+2)$-DSS subgraph) which is greater by a factor of at most of $O(n^{1/2})$ than an optimal AC k-spanner (k-DSS subgraph). We also present a bicriteria $(O(n^{1/d}), 2d-1)$-approximation algorithm for the AC k-spanner problem and k-DSS problem for any positive integer d. These algorithms can be interpreted also as $O(n^{1/2})$-approximation algorithms for the k-DSS problem restricted to the graphs of diameter at most $k-2$ and as $O(n^{1/d})$-approximation algorithms for the k-DSS problem restricted to the graphs of diameter at most $\frac{k}{2d-1}$ for any positive integer d. We also prove an analogous statement for the AC k-spanner problem.

Finally, in the full paper we consider the k-spanner problem on the class $SP_k(DEG(\Delta))$ of graphs that admit a k-spanner with maximum degree bounded by Δ, and provide an algorithm with $O(\Delta^{k-2} \cdot \log n)$ approximation ratio, for $k > 2$. Note, however, that the k-spanner problem enjoys a trivial $O(\Delta^k)$-approximation algorithm, hence the new algorithm is advantageous only when $\Delta = \Omega(\sqrt{\log n})$.

1.3 Our Techniques

Generally speaking, our algorithms generalize the logarithmic approximation algorithm of [18] for the 2-spanner problem. That algorithm is based on decomposing the problem into a finite number of appropriate subproblems and iteratively solving them, where every such subproblem involves performing some *density* computations over a small neighborhood in the graph.

However, as discussed earlier, the k-spanner problem for $k > 2$ is significantly more difficult than the 2-spanner case. Hence a generalization of the algorithm for the 2-spanner problem to the k-spanner one requires the introduction of novel algorithmic and analytic techniques.

The technique introduced here for handling these problems involves a new graph construct called *edge-dominating system*, based on a special type of graph decomposition. Using these systems, we define an algorithmic procedure (for density computation) which is applied to each component of this decomposition, and gives rise to an approximation algorithm for the k-spanner problem. We demonstrate that the approximation ratio of our algorithm equals the *sparsity* of the edge-dominating system used by the algorithm, up to a factor logarithmic in n.

Our approach thus reduces the algorithmic question of the approximability of the k-spanner problem to the pure graph-theoretic question of existence of combinatorial objects with desirable properties (namely, edge-dominating systems with bounded sparsity). In particular, we show that a proof of existence of an edge-dominating system for the 3-spanner problem of sparsity bounded by some power $0 < \delta < 1$ of the arboricity of some near-optimal 3-spanner leads to an approximation algorithm for the 3-spanner problem within an approximation ratio of $\tilde{O}(n^{\frac{3\delta}{2(2+\delta)}})$ (the current state of the art is $O(n^{1/2})$ by the universal construction of [14,9]; note that the state of the art of the δ-edge-dominating systems is $\delta = 1$ which unfortunately yields a similar $\tilde{O}(n^{1/2})$-approximation ratio). We also show that the lower bound of [11] on the approximability of the 3-spanner problem yields a lower bound on the sparsity of the possible edge-dominating systems.

Consequently, we present some constructions of the edge-dominating systems. To illustrate the concept, we start by presenting a construction of constant sparsity for the 2-spanner problem on general graphs. This yields an $O(\log n)$-approximation algorithm for the 2-spanner problem (which is, in fact, simply a more general presentation of the algorithm of [18] and its analysis). We proceed with presenting a construction of constant sparsity (i.e., not depending on n, but depending on k) for the k-spanner problem on $SP_k(TREE)$. This construction yields an $O(k^2 \log n)$-approximation algorithm for the k-spanner problem on $SP_k(TREE)$. Finally, we present a construction for general graphs for $k = 3$ (for the 3-spanner problem), but the sparsity of this construction is linear in the arboricity of a near-optimal spanner of the input graph. This construction yields logarithmic and polylogarithmic approximation ratio algorithms for the 3-spanner problem on $SP_3(BA)$ (in particular, on $SP_3(PL)$) and on $SP_3(\log A)$, respectively, and $n^{O(1/g)}$-approximation ratio on $SP_3(GIRTH(g))$.

We beleive that our techniques may enable in the future to get an improvement for the basic 3-spanner too, and to get an $o(n^{2/3})$-approximation ratio for the aforementioned problems. Another challenging direction is to generalize our algorithms in such a way that they would provide to a sublinear approximation ratio for these problems for $k > 3$.

2   Density Computation

Throughout, we denote the number of vertices in the graph G by n. The *girth* of graph G, denoted $g(G)$, is the length of the shortest cycle in the graph. For a set of nodes W let $E(W)$ be the set of edges with both endpoints in W. The *arboricity* of graph G is defined as $a(G) = \max\left\{\frac{|E(W)|}{|W|-1}\right\}$. We will use Scheinerman's theorem, stating that $a(G) = O(\sqrt{\mu(G)})$ [24], where $\mu(G)$ denotes the *genus* of graph G.

The Nash-Williams theorem (cf. [3]) states that the edge set of a graph G with arboricity t can be decomposed into t edge-disjoint forests, and furthermore, that this decomposition can be computed polynomially. The notion of *graph density*, denoted by $\rho(G)$, is very similar to arboricity, except that it has $|W|$ instead of $|W| - 1$ in the denominator. Note that the relation between the two notions is

$$\rho(G) \leq a(G) \leq 2 \cdot \rho(G), \tag{1}$$

Definition 1. *For every graph G and subgraph H for G, the vertex $v \in V$ is said to (H,k)-dominate the edge $e = (u, z)$ if H contains a path $P = (u_1, \ldots, u_r, z)$ of $r + 1 \leq k$ edges connecting u and z and going through v, or formally, s.t. $(u, u_1), (u_1, u_2), ..., (u_{r-1}, u_r), (u_r, z) \in H$ and $v \in \{u, u_1, u_2, .., u_r, z\}$.*

The algorithm that we present in Section 4 constructs the spanner subgraph denoted H. The algorithm proceeds in steps, in each step increasing the initially empty set H. Through the algorithm H^u denotes the set of edges that are still uncovered.

For any vertex v and any two subsets $H, H^u \subseteq E$ of the edge set E, define $COV_k(H, v, H^u)$ as the subset of H^u edges that are (H, k)-dominated by v.

Define the ψ_k-*density* of a subset of edges H and a vertex v with respect to an edge set H^u as $\psi_k(H, v, H^u) \triangleq \frac{|COV_k(H,v,H^u)|}{|H|}$. Intuitively, a high-density edge subset H is a good candidate for participating in the spanner, as it covers many edges at a low cost. For a graph $G = (V, E)$, a subset of vertices $W \subseteq V$

and a vertex $v \in W$, define a *breadth-first search (BFS) tree* rooted at v for the set W to be a tree T rooted at v, whose vertex set coincides with W and such that for any node $u \in W$, $d_T(v, u) = d_G(v, u)$. For a vertex set U and a vertex v, let $T(U, v)$ denote the set of all possible non-empty BFS trees rooted at v and contained in $E(U)$, and let $\hat{T}(U, v)$ denote the set of all possible non-empty trees rooted at v and contained in $E(U)$. Now define the ψ_k^l-density (respectively, $\hat{\psi}_k^l$-density) of a node v with respect to an edge set H^u to be the maximum density achievable by a BFS tree (respectively, arbitrary tree) rooted at v and spanning some part of v's l-neighborhood, i.e.,

$$\psi_k^l(v, H^u) \triangleq \max_{T \in T(\Gamma_l(v), v)} \{\psi_k(T, v, H^u)\} \ , \tag{2}$$

$$\hat{\psi}_k^l(v, H^u) \triangleq \max_{\hat{T} \in \hat{T}(\Gamma_l(v), v)} \left\{\psi_k(\hat{T}, v, H^u)\right\} \ . \tag{3}$$

The following lemma shows that in order to approximate $\hat{\psi}_k^l(v, H^u)$ it suffices to compute the value of $\psi_k^l(v, H^u)$. (All proofs are omitted.)

Lemma 1. *For any integer $k > 1$, vertex $v \in V$ and subset of edges $H^u \subseteq E$,*
$$\psi_k^{\lceil k/2 \rceil}(v, H^u) \le \hat{\psi}_k^{k-1}(v, H^u) \le (k-1)\psi_k^{\lceil k/2 \rceil}(v, H^u) \ .$$

We denote by T_v the tree that maximizes the value of $\psi_k(\hat{T}, v, H^u)$ and by $COV_k(v, H^u)$ the set $COV_k(T_v, v, H^u)$. Specifically, our algorithm needs to compute the value of $\psi_k^l(v, H^u)$, where $l = \lceil k/2 \rceil$. Observe that an l-deep tree T automatically $2l$-spans all the edges between its vertices.

Furthermore, we next show that the density $\psi_k^l(v, H^u)$ can be approximated with a ratio linear in l in polynomial time on *any graph*.

Hereafter we fix $l = \lceil k/2 \rceil$ and denote the function ψ_k^l by ψ. Also we denote by small letters the sizes of sets denoted by capital letters. We denote $\hat{\Gamma}_l(v) = \{u \mid d_G(u, v) \le l\}$ and $\Gamma_l(v) = \{u \mid d_G(u, v) = l\}$.

Consider the subgraph $\tilde{G} = (\tilde{V}, \tilde{E})$, where $\tilde{V} = \hat{\Gamma}_l(v)$ and

$$\tilde{E} = \begin{cases} E(\hat{\Gamma}_l(v)) \cap H^u, & k \text{ is even,} \\ E(\hat{\Gamma}_l(v)) \cap (H^u \setminus E(\Gamma_l(v))), & k \text{ odd.} \end{cases} \tag{4}$$

The following close relation holds between $\rho(\tilde{G})$ and $\psi(v, H^u)$.

Lemma 2. $\rho(\tilde{G}) \le \psi(v, H^u) \le 2 \cdot \rho(\tilde{G})$.

Corollary 1. $\rho(\tilde{G}) \le \hat{\psi}(v, H^u) \le 2l \cdot \rho(\tilde{G})$.

Lemma 3. *Given a graph $G = (V, E)$, a vertex v in G and a subset of edges $H^u \subseteq E$, the value of $\psi(v, H^u)$ can be approximated with ratio $2l$ in polynomial time.*

3   Constructions for Dominating Systems

3.1   Dominating Systems

At this point we introduce the notion of *(H,k)-dominating systems* which will be used in the analysis.

Definition 2. *For a graph G and a spanner H of G, an H-system for G is a pair (D, S) such that $D \subseteq V$ and $S = [Sp(v)]_{v \in D}$, where $Sp(v)$ is an edge set contained in H for every $v \in D$.*

The H-system (D, S) is called an (H,k)-dominating system for the graph $G = (V, E)$ if for every edge $e \in E$ there exists a vertex $v \in D$ such that v $(Sp(v), k)$-dominates e.

Our construction for k-spanners makes use of a specific type of (H,k)-dominating system, in which the set $Sp(v)$ is a subtree of the BFS tree rooted at v, of depth at most $k - 1$. To define it formally, we need the following terminology.

For a non-root vertex v in a tree T, let $p_T(v)$ denote its parent node in T and $Sub_T^s(v)$ denote the edge set of the s-deep subtree rooted by v. Also let r_T denote the root of the tree T and L_T denote its set of leaves.

In order to build a good spanner, we need a "sparse" (H,k)-dominating system. The sparsity of an (H,k)-dominating system (D, S) is defined as

$$Sparsity_H(D, S) \overset{\Delta}{=} (\textstyle\sum_{v \in D} sp(v))/|H|).$$

3.2 $(H, 2)$-Dominating Systems

To illustrate the concept of dominating systems we present a simple construction of an $(H, 2)$-dominating system with constant sparsity for arbitrary graphs. This construction can be used to simplify the analysis of the logarithmic ratio approximation algorithm for the 2-spanner problem due to [18].

Construction A
1. $D \leftarrow V$.
2. For every vertex v in D set $Sp(v) = \{(u, v) \in E\}$.
3. $S \leftarrow \{Sp(v)\}_{v \in D}$.

Lemma 4. *For any 2-spanner H of a graph G, the H-system (D, S) constructed by Construction A is an $(H, 2)$-dominating system with $Sparsity_H(D, S) = 2$.*

3.3 (H, k)-Dominating Systems for $SP_k(TREE)$

Now we show that if a graph G admits a tree-spanner H then it has a sparse (H,k)-dominating system, i.e., with $Sparsity_H(D, S) = O(k)$.

Assume that G has a tree spanner H. Consider the following construction of an H-system for G.

Construction B
1. $D \leftarrow V \setminus L_H$.

2. For every vertex v in D, set $Sp(v)$ to be the depth $(k-1)$ subtree of H rooted at v plus

the edge from v to its parent (unless v is the root), i.e.,

$$Sp(v) = \{(p_H(v), v)\} \cup Sub_H^{k-1}(v) \text{ for } v \neq r_H \text{ and } Sp(v) = Sub_H^{k-1}(v) \text{ for}$$

$v = r_H$.

3. Define S to be the set $\{Sp(v)\}_{v \in D}$.

Lemma 5. *For any tree-k-spanner H of a graph G, the H-system (D, S) constructed by Construction B is an (H, k)-dominating system with $Sparsity_H(D, S)$ at most k.*

3.4 $(H, 3)$-Dominating Systems for $SP_3(BA)$

We now generalize the above construction of the (H, k)-dominating system from tree spanners to spanners with bounded arboricity, albeit only for $k = 3$.

By the Nash-Williams theorem, every graph H with arboricity $a = a(H)$ has a decomposition in to a edge-disjoint forests $F_1, .., F_a$, where for every $i = 1, .., a$, the forest F_i is a collection of vertex-disjoint trees, i.e., $F_i = \{T_i^1, .., T_i^{j_i}\}$. Since the forests are edge-disjoint, every edge participates in exactly one tree. Since the trees in each forest are vertex-disjoint, each vertex may participate in at most one tree per forest. Hence each vertex participates in at most a trees.

Consider the following construction. For every vertex v, denote by $T_i(v)$ the tree in the forest F_i in which v participates. An edge $e = (w, u) \in F_j$ is said to *cross* the node v in F_i, for $i \neq j$, if u or w is a neighbor of v in tree $T_i(v)$, but e does not belong to F_i.

Essentially, this construction is based on repeating Construction B for every tree T_i^j individually, and adding the crossing edges. A formal description follows.

Construction C

1. Set $D = V$.

2. For every vertex v set $Sp(v) \leftarrow \bigcup_i(\{(p_{T_i(v)}(v), v)\} \cup Sub_{k-1}^{T_i(v)}(v))$.

(If v is a root of T_i^j then the edge $\{(p_{T_i^j}(v), v)\}$ is not included.)

3. For every vertex v, add to $Sp(v)$ also every edge that crosses v in F_i for some $i = 1, .., a$.

4. Define S to be the set $\{Sp(v)\}_{v \in D}$.

Remark 1: Note that the edges that do not cross to a different tree are taken as well for the children of v but not for its parent. For the parent, we are not interested in taking all its neighboring edges in its own tree, but only the edges that cross between different trees.

Remark 2: For a neighbor w of v the edge (w, z) is said to cross to a different tree if the edge (v, w) is located on the different tree than (w, z). This is well-defined, since each edge belongs to exactly one tree.

Lemma 6. *For any graph G with a 3-spanner H, the H-system (D, S) constructed by Construction C is an $(H, 3)$-dominating system with $Sparsity_H(D, S)$ at most $O(a(H))$.*

4 Spanner Construction Algorithm

This section presents an algorithm for building a k-spanner for a given graph $G = (V, E)$. The algorithm is a modification of the 2-spanner approximation algorithm devised in [18]. Its main part is a loop, repeated while there is at least one vertex with positive ψ-density.

Inside the loop, we pick a vertex v that maximizes the ψ-density, and compute for v the corresponding tree T_v that attains it. The algorithm is described formally next.

Algorithm *Sparse_Spanner*
Input: graph $G = (V, E)$, integer $k \geq 2$.
Output: subset $H \subseteq E$.
1. $H^u \leftarrow E$; $H^c \leftarrow \phi$; $H \leftarrow \phi$; $l \leftarrow \lceil k/2 \rceil$
2. While $\exists v$ s.t. $\hat{\psi}(v, H^u) > 0$ do :

   1. Choose a vertex v that maximizes $\hat{\psi}(v, H^u)$.
   2. Approximate $\hat{\psi}(v, H^u)$ with ratio $O(l)$ for this vertex;
      let T_v be the corresponding densest tree of $\hat{\Gamma}_l(v)$.
   3. $H^c \leftarrow H^c \cup COV_k(T_v, v, H^u)$; $H \leftarrow H \cup T_v$; $H^u \leftarrow H^u \setminus H^c$

3. Return(H)

It can be easily seen that if the $\hat{\psi}(v, H^u)$-density is zero for every vertex v in the graph, then H^u is empty. Thus H^c contains all the graph edges. Since an edge is inserted into H^c only when it is 3-spanned by some path contained in of the spanner H, and no edges are removed from H, it follows that the algorithm leaves the loop and returns H only when H is a 3-spanner for G.

The termination of the algorithm follows from the fact that in each iteration, at least one edge is removed from H^u, hence the algorithm terminates after at most $|E|$ iterations.

In what follows, we analyze the algorithm as if it computes the density $\hat{\psi}(v, H^u)$ exactly, rather than approximates it. Later we show that approximating the density by a factor of ρ instead of computing it exactly decreases the approximation ratio of the algorithm by the same factor ρ only.

5 Analysis of the Spanner Construction Algorithm

5.1 Analysis of the 3-Spanner Case

In this section we prove that Algorithm Sparse_Spanner provides an approximation ratio of $\tilde{O}(a(H^*))$ for the basic 3-spanner problem, where H^* is an optimal 3-spanner for the given input graph G.

For every $j \geq 1$, let H_j be the spanner at the beginning of the jth iteration, let H_j^c and H_j^u be the corresponding sets of covered and uncovered edges, let v_j be the vertex chosen in the jth iteration, and let $T_j = T_{v_j}$ be the corresponding tree of neighbors selected, i.e., the $\hat{\psi}$-densest partial spanning tree of $\Gamma_2(v_j)$.

Observe that since H^u decreases at every step, the value of $\hat{\psi}(v, H^u)$ is monotonically decreasing as well. Let us partition the iterations of the algorithm into phases as follows. Let $r = \frac{|E|}{|V|}$ and $f = \lceil \log r \rceil$. The first phase includes all iterations during which each vertex v_j chosen by the algorithm satisfies $\hat{\psi}(v_j, H^u) \geq \frac{r}{2}$. For $2 \leq i \leq f$, let the ith phase consist of the iterations during which the chosen v_j satisfies $\frac{r}{2^{i-1}} > \hat{\psi}(v_j, H^u) \geq \frac{r}{2^i}$.

Following [12], denote by P_i the set of indices of iterations in the ith phase. Let $H[i]$ and $H^c[i]$ denote the sets of new edges added to H and H^c, respectively, during the ith phase, and let $H^u[i]$ denote the set of edges left in H^u at the end of the ith phase. Also denote the spanner at the end of the ith phase by $\hat{H}[i] = \bigcup_{j \leq i} H[j]$.

Observe that for every $v \in V$,

$$\hat{\psi}(v, H^u[i]) < \frac{r}{2^i}, \tag{5}$$

because before the first iteration j of the $(i+1)$st phase, $H^u_j = H^u[i]$ and so the vertex v_j chosen at this iteration satisfies $\hat{\psi}(v_j, H[i]) < \frac{r}{2^{i+1-1}} = \frac{r}{2^i}$, and this v_j maximizes $\hat{\psi}$.

Let X_i be the set of vertices chosen during the ith phase. For $v \in X_i$, denote by $P_i(v)$ the subset of P_i containing only those ith phase iterations at which v was chosen, i.e., $P_i(v) = \{j \in P_i \mid v_j = v\}$. Let $H[i, v] = \bigcup_{j \in P_i(v)} T_j$, $H^c[i, v] = \bigcup_{j \in P_i(v)} COV_3(T_j, v, H^u_j)$.

Since for any integer $2 \leq i \leq f$, vertex $v \in V$ and $j \in P_i(v)$ the set $COV_3(T_j, v, H^u_j)$ includes only edges from H^u and only edges from some set $COV_3(T_j, v, H^u_j)$ are put into H^c and removed from H^u, and since edges taken to T_j are inserted into H we have that $h[i, v] \leq \sum_{j \in P_i(v)} |T_j|$, $h^c[i, v] = \sum_{j \in P_i(v)} cov_3(T_j, v, H^u_j)$. Therefore we can state the following lemma.

Lemma 7. For every $v \in X_i$, $h^c[i] \geq h[i] \cdot \frac{r}{2^i}$.

Now let $\bar{H}$ be some 3-spanner for G and $\bar{H}_i$ be (a possibly Steiner) 3-spanner for $H^u[i]$ of size no greater than $\bar{H}$ and satisfying $a(\bar{H}_i) \leq a(\bar{H})$. Specifically, $\bar{H}_i$ is allowed to use all E edges and not only those of $H^u[i]$. This implies the existence of such a spanner, since in particular $\bar{H}$ itself spans $H^u[i]$.

Lemma 8. $h/\bar{h} = O(a(\bar{H}) \cdot \log r)$.

Finally, we observe that exactly as in [18,12] it suffices to approximate the densities instead of computing them precisely. This is implied by the following lemma.

Lemma 9. Approximating the densities by a factor $\alpha > 1$ instead of computing them exactly increases the approximation ratio of Algorithm Sparse_Spanner by the same factor.

Denote by $\mathcal{H}(G)$ the set of all 3-spanners of the graph G. We have proved the following theorem.

Theorem 1. *For any graph G, running Algorithm Sparse_Spanner with $k = 3$ yields a 3-spanner H' such that $|H'| = O(\min_{H \in \mathcal{H}(G)} \{|H| \cdot a(H) \cdot \log r\})$.*

We next generalize Theorem 1 to establish a general connection between the approximability of the 3-spanner problem and the sparsity of possible edge-dominating systems.

Theorem 2. *Consider a family $\mathcal{F}$ of graphs which admit $(H, 3)$-dominating systems (D, S) with $Sparsity_H(D, S) = O(a(H)^\delta)$, for some $0 < \delta < 1$. Then the 3-spanner problem restricted to the family of $\mathcal{F}$-spanned graphs admits an $\tilde{O}(n^{\frac{3\delta}{2(2+\delta)}})$-approximation ratio.*

We now prove a lower bound on the sparsity of the possible edge-dominating systems.

Theorem 3. *If for every graph G and its optimal 3-spanner H there exists an $(H, 3)$-dominating system (D, S) such that $Sparsity_H(D, S) = O(2^{\log^{1-\epsilon} a(H)})$ for some $0 < \epsilon < 1$, then $NP \subseteq DTIME(n^{\text{polylog } n})$.*

Denote by $\mathcal{H}_1(G)$ the subfamily of $\mathcal{H}(G)$ of 3-spanners whose size is close to the size of an optimal 3-spanner up to a constant factor and by $\mathcal{H}_2(G)$ the subfamily of $\mathcal{H}(G)$ of spanners whose size is close to the size of an optimal 3-spanner up to a polylogarithmic factor in n. In particular, we get

Corollary 2. *Sparse_Spanner with $k = 3$ is an $O(\log r)$-approximation algorithm for the 3-spanner problem on the class $SP_3(BA)$, and an $O(\text{polylog } n)$-approximation algorithm on the class $SP_3(\log A)$.*

Denote the girth of a graph G by $g(G)$. Recall that $SP_3(GIRTH(g))$ is the family of graphs admitting a near-optimal 3-spanner with girth no smaller than g.

Lemma 10. *For any graph G with $g(G) \geq g$ and m edges, $a(G) \leq m/n^{g-2/g}$.*

Theorem 4. *Algorithm Sparse_Spanner with $k = 3$ provides an $\tilde{O}(n^{\frac{4g-4}{(g-2)g}})$-approximation for the 3-spanner problem on $SP_3(GIRTH(g))$.*

Note that the exponent tends to zero as g tends to infinity. Denote by 3-$spanner(g)$ the 3-spanner problem on $SP_3(GIRTH(g))$. Then

Corollary 3. *The set $\{3\text{-}spanner(g)\}_{g=1}^{\infty}$ problem enjoys the ratio degradation property in g.*

5.2   Analysis of the k-Spanner Case

In this section we show that executing Algorithm $Sparse\_Spanner$ with integer parameter $k \geq 2$ provides a logarithmic approximation ratio for the k-spanner problem on $SP_k(TREE)$. The proof of this fact involves the dominating systems constructed in Section 3.3.

The analysis is analogous to that of the previous section. The notions of $H_c[i, v]$ and $h^c[i, v]$ are changed to $H^c[i, v] = \bigcup_{j \in P_i(v)} COV_k(T_j, v, H_j^u)$, $h^c[i, v] = \sum_{j \in P_i(v)} cov_k(T_j, v, H_j^u)$.

Lemma 7 holds as is. Let $\bar{H}$ be some tree k-spanner for G and $\bar{H}_i$ be (a possibly Steiner) tree k-spanner for $H^u[i]$.

Lemma 11. $h^u[i]/\bar{h} \leq k \cdot \frac{r}{2^{i-1}}$.

Next, we show that for every $1 \leq i \leq f$, $h[i]/\bar{h} < O(k)$, which allows us to conclude that $h/\bar{h} = O(k \log r)$, hence we have

Theorem 5. *For any $G \in SP_k(TREE)$, Algorithm Sparse_Spanner finds a k-spanner of $O(nk^2 \log r)$ edges.*

Remark: One factor of k follows from Lemma 5 and the other because the value of maximal density is not computed exactly but only approximated to a factor of $O(l) = O(k)$.

Corollary 4. *The k-spanner problem is $O(k^2 \log n)$-approximable for any k over $SP_k(TREE)$.*

Corollary 5. *The k-spanner problem is $O(\frac{\log^3 n}{(\log \log n)^2})$-approximable for any k over $SP_k(TREE)$.*

Recall that $S_{TREE}(G)$ denotes the minimum stretch of any spanning tree T for the graph G. As mentioned earlier, the problem of finding such a tree is known to be $(1 + \sqrt{5})/2$-inapproximable [22].

Theorem 6. *There is a polynomial time algorithm A that given a graph G constructs a $S_{TREE}(G)$-spanner H with $|H| = \tilde{O}(n)$.*

6   Bicriteria Approximations

In this section we provide some bicriteria upper bounds on the CS k-spanner problem and on the k-DSS problem. We say that an algorithm is a *bicriteria $(a, +b)$-approximation algorithm* for the CS k-spanner (respectively, k-DSS) problem if given an instance of the problem it returns a solution for the CS $(k+b)$-spanner (resp., $(k+b)$-DSS) problem of size that is no more than a times bigger than the optimal solution for the CS k-spanner (resp., k-DSS) problem. Analogously a *bicriteria (a, b)-approximation algorithm* for the CS k-spanner and k-DSS problems is defined as one supplying, for a given instance of the CS k-spanner (resp., k-DSS) problem, a solution whose size is greater than the optimal CS $(k \cdot b)$-spanner (resp., $(k \cdot b)$-DSS) by a factor of at most a.

Theorem 7. *The AC k-spanner problem admits an $(O(n^{1/2}), +2)$-bicriteria approximation algorithm.*

For any instance $(G = (V, E), \mathcal{C}, \mathcal{S})$ of the CS k-spanner problem, define $Diam(\mathcal{C}, \mathcal{S}) = \max_{(u,w) \in \mathcal{C}} dist_{\mathcal{S}}(u, w)$. The proof of Theorem 7 yields the following.

Theorem 8. *The AC k-spanner problem restricted to the case of $Diam(E, \mathcal{S}) \leq k - 2$ admits an $O(n^{1/2})$-approximation algorithm.*

As already mentioned, the k-DSS problem is reducible to the AC k-spanner problem. Subsequently, we have:

Theorem 9. *1. The k-DSS problem admits an $(O(n^{1/2}), +2)$-bicriteria approximation algorithm.*
2. The k-DSS problem restricted to instances with $Diam(G) \leq k - 2$ admits an $O(n^{1/2})$-approximation algorithm.
3. For any integers $l > 0$ and $k \geq 2l - 1$, the AC k-spanner and k-DSS problems admit $(O(n^{1/l}), 2l - 1)$-bicriteria approximation algorithms.
4. For any integers $l > 0$ and $k \geq 2l - 1$, the AC k-spanner problem restricted to instances with $Diam(E, \mathcal{S}) \leq \frac{k}{2l-1}$ and the k-DSS problem restricted to instances with $Diam(G) \leq \frac{k}{2l-1}$ admit an $O(n^{1/l})$-approximation ratio algorithms.

References

1. Baruch Awerbuch, Alan Baratz, and David Peleg. Efficient broadcast and lightweight spanners. Unpublished manuscript, November 1991.
2. I. Althöfer, G. Das, D. Dobkin, and D. Joseph, Generating sparse spanners for weighted graphs, *Proc. 2nd Scnadinavian Workshop on Algorithm Theory*, Lect. Notes in Comput. Sci., Vol. 447, pp. 26-37, Springer-Verlag, New York/Berlin, 1990.
3. B. Bollobas, *Extremal Graph Theory*, Academic Press, New York, 1978.
4. L.P. Chew, There is a planar graph almost as good as the complete graph, *Proc. ACM Symp. on Computational Geometry*, 1986, pp. 169-177
5. L. Cai, NP-completeness of minimum spanner problems. *Discrete Applied Math.*, 48:187-194, 1994
6. L. Cai and D.G. Corneil, Tree Spanners, *SIAM J. on Discrete Mathematics* **8**, (1995), 359-387.
7. B. Chandra, G. Das, G. Narasimhan, J. Soares, New Sparseness Results on Graph Spanners, *Proc. 8th ACM Symp. on Computational Geometry*, pp. 192-201, 1992.
8. D.P. Dobkin, S.J. Friedman and K.J. Supowit, Delaunay graphs are almost as good as complete graphs, *Proc. 31st IEEE Symp. on Foundations of Computer Science*, 1987, pp. 20-26.
9. D. Dor, S. Halperin, U. Zwick, All pairs almost shortest paths, *Proc. 37th IEEE Symp. on Foundations of Computer Science*, 1997, pp. 452-461.
10. M.-L. Elkin and D. Peleg, The Hardness of Approximating Spanner Problems, *Proc. 17th Symp. on Theoretical Aspects of Computer Science*, Lille, France, Feb. 2000, 370-381.

11. M.-L. Elkin and D. Peleg, Strong Inapproximability of the Basic k-Spanner Problem, *Proc. 27th International Colloquim on Automata, Languages and Programming*, Geneva, Switzerland, July 2000. See also Technical Report MCS99-23, the Weizmann Institute of Science, 1999.
12. M.-L. Elkin and D. Peleg, The Client-Server 2-Spanner Problem and Applications to Network Design, Technical Report MCS99-24, the Weizmann Institute of Science, 1999.
13. M.-L. Elkin, Additive Spanners and Diameter Problem, manuscript, 2000.
14. M. -L. Elkin and D. Peleg, Spanners with Stretch $1 + \epsilon$ and Almost Linear Size, manuscript, 2000.
15. S. P. Fekete, J. Kremer, Tree Spanners in Planar Graphs, Angewandte Mathematik und Informatik Universitat zu Koln, Report No. 97.296
16. S. Halperin, U. Zwick, Private communication, 1996.
17. G. Kortsarz, On the Hardness of Approximating Spanners, *Proc. 1st Int. Workshop on Approximation Algorithms for Combinatorial Optimization Problems*, Lect. Notes in Comput. Sci., Vol. 1444, pp. 135-146, Springer-Verlag, New York/Berlin, 1998.
18. G. Kortsarz and D. Peleg, Generating Sparse 2-Spanners. *J. Algorithms*, **17** (1994) 222-236.
19. E.L. Lawler, *Combinatorial Optimization: Networks and Matroids*, Holt, Rinehart, Winston, New York, 1976.
20. D. Peleg, *Distributed Computing: A Locality-Sensitive Approach*, SIAM, Philadelphia, PA, 2000.
21. D. Peleg and A. Schäffer, Graph Spanners, *J. Graph Theory* **13** (1989), 99-116.
22. D. Peleg and E. Reshef, A Variant of the Arrow Distributed Directory with Low Average Complexity, *Proc. 26th Int. Colloq. on Automata, Languages & Prog.*, Prague, Czech Republic, July 1999, 615–624.
23. D. Peleg and J.D. Ullman, An optimal synchronizer for the hypercube, *SIAM J. Computing* **18** (1989), pp. 740-747.
24. E. Scheinerman, The maximal integer number of graphs with given genus, *J. Graph Theory* **11** (1987), no. 3, 441-446.

A Matroid Generalization of the Stable Matching Polytope

Tamás Fleiner*

Alfréd Rényi Institute of Mathematics,
Hungarian Academy of Sciences,
POB 127, H-1364 Budapest, Hungary
fleiner@renyi.hu

Abstract. By solving a constrained matroid intersection problem, we give a matroid generalization of the stable marriage theorem of Gale and Shapley. We describe the related matroid-kernel polytope, an extension of the stable matching polytope. Linear conditions of the characterizations resemble to the ones that describe the matroid intersection polytope.

1 Introduction

Gale and Shapley published their ground-breaking stable marriage theorem in 1962 [9], a result that proves the existence of a bipartite matching satisfying a certain stability constraint. The theorem can be naturally formulated in terms of marriages: if each of n men and n women has a linear preference order on the members of the opposite sex then there exists a stable marriage scheme, that is a set of n disjoint man-woman pairs in such a way that no man and woman mutually prefer each other to their partners. This matching model of Gale and Shapley turned out to be extremely applicable in Game Theory and Mathematical Economics, but a non-negligible part of the vast literature on stable matchings is concerned with its combinatorial links. Such result is the one of Vande Vate [13] (later generalized by Rothblum [12]) that characterizes the stable matching polytope (that is the convex hull of the incidence vectors of stable matchings) in terms of linear constraints.

Recently, Fleiner has developed a fixed point theorem based approach to the theory of bipartite stable matchings [3,4,5] that provided a common framework for several seemingly distant topics in Combinatorics. He proved a matroid generalization of the stable marriage theorem of Gale and Shapley, a result that was prophesied by Roth (cf. [11]).

In Section 2, we describe and prove this particular result on matroid kernels. In Section 3, using the theory of lattice polyhedra and of blocking polyhedra, we characterize certain matroid-kernel related polyhedra. We indicate how our

* Research was done as part of the author's PhD studies at the Centrum voor Wiskunde en Informatica (CWI), POB 94079, NL-1090GB, Amsterdam and it was supported by the Netherlands Organization for Scientific Research (NWO) and by OTKA T 029772.

K. Aardal, B. Gerards (Eds.): IPCO 2001, LNCS 2081, pp. 105–114, 2001.
© Springer-Verlag Berlin Heidelberg 2001

description relates to Rothblum's and conclude with a speculation on a possible generalization of our result that would also contain Edmonds' characterization of the matroid intersection polytope as a special case.

2   Matroid-Kernels

A triple $\mathcal{M} = (E, \mathcal{I}, <)$ is an *ordered matroid*, if $(E, \mathcal{I})$ is a matroid on groundset E with family $\mathcal{I}$ of independent sets and $<$ is a linear order on E. Let $\mathcal{M} = (E, \mathcal{I}, <)$ be an ordered matroid. A subset E' of E *dominates* element e of E if $e \in E'$ or there is an independent subset C_e of E' in such a way that $\{e\} \cup C_e \notin \mathcal{I}$ (that is, C_e spans e) and $c < e$ for all elements c of C_e. The set of elements dominated by E' is denoted by $D_{\mathcal{M}}(E')$.

Let $\mathcal{M}_A = (E, \mathcal{I}_A, <_A)$ and $\mathcal{M}_B = (E, \mathcal{I}_B, <_B)$ be two ordered matroids on the same ground set E. We say that subset K of E is an $\mathcal{M}_A \mathcal{M}_B$-*kernel* if K is a common independent set of the underlying matroids, and any element e of E is dominated by $\mathcal{M}_A$ or by $\mathcal{M}_B$. That is, if $K \in \mathcal{I}_A \cap \mathcal{I}_B$ and

$$D_{\mathcal{M}_A}(K) \cup D_{\mathcal{M}_B}(K) = E \ . \tag{1}$$

For example, assume that we have a finite simple bipartite graph $G = (A \cup B, E)$ with colour classes A and B. Fix linear orders $<_v$ on the star of v for each vertex v of G. Let $(E, \mathcal{I}_A)$ be the partition matroid on E defined by the A-stars, and $(E, \mathcal{I}_B)$ be the partition matroid for the B-stars of G. Define linear orders $<_A$ and $<_B$ on E in such a way that

$$\begin{aligned} e <_A f \text{ whenever } e <_a f \text{ for some } a \in A, \text{ and} \\ e <_B f \text{ whenever } e <_b f \text{ for some } b \in B. \end{aligned} \tag{2}$$

Such linear orders exist. Define ordered matroids $\mathcal{M}_A := (E, \mathcal{I}_A, <_A)$ and $\mathcal{M}_B := (E, \mathcal{I}_B, <_B)$. With these definitions, it is easy to check that an $\mathcal{M}_A \mathcal{M}_B$-kernel is exactly a matching K of G with the property that for any edge $e \notin K$ of G there is an edge k of K such that $k <_v e$ for some vertex v of G. In other words, K is a stable matching.

For any ordered matroid $\mathcal{M} = (E, \mathcal{I}, <)$, there is an $\mathcal{M}\mathcal{M}$-kernel, it is unique and it can be constructed with the greedy algorithm of Edmonds the following way. Let $E' \subseteq E = \{e_1, e_2, \dots, e_n\}$ with $e_1 < e_2 <_A \dots < e_n$, $K_0^{\mathcal{M}}(E') := K_0(E') := \emptyset$. Define, for $1 \leq i \leq n$, subset $K_i^{\mathcal{M}}(E') := K_i(E')$ of E' by

$$K_i(E') = \begin{cases} K_{i-1}(E') & \text{if } e_i \notin E' \text{ or} \\ & \text{if there is a subset } C \text{ of } K_{i-1}(E') \\ & \text{such that } \{e_i\} \cup C \in \mathcal{C} \\ K_{i-1}(E') \cup \{e_i\} & \text{else.} \end{cases} \tag{3}$$

It is clear from the algorithm that $|K_n^{\mathcal{M}}(E')| = \text{rank}_{\mathcal{M}}(E')$ and that

$$E' \subseteq D_{\mathcal{M}}(K_n(E')) \ . \tag{4}$$

As by definition $K_n(E)$ is independent, (4) implies that $K_n(E)$ is an $\mathcal{M}\mathcal{M}$-kernel. Hence the notion of $\mathcal{M}_A\mathcal{M}_B$-kernel can be regarded as a generalization of minimum cost matroid basis.

The following fact is a straightforward consequence of a well-known result of Edmonds [2].

Theorem 1. If $\mathcal{M} = (E, \mathcal{I}, <)$ is an ordered matroid then

$$K_n(E') = \{e \in F' \cdot e \notin D_\mathcal{M}(E' \setminus \{e\}\} \tag{5}$$

for any subset E' of E. □

For a setfunction $\mathcal{F} : 2^E \to 2^E$ let us define setfunction $\overline{\mathcal{F}} : 2^E \to 2^E$ by

$$\overline{\mathcal{F}}(A) := A \setminus \mathcal{F}(A) . \tag{6}$$

It follows from (5) that $\overline{K_n}$ is a monotone function, that is

$$E' \subseteq E'' \subseteq E \qquad \Rightarrow \qquad \overline{K_n}(E') \subseteq \overline{K_n}(E'') . \tag{7}$$

Theorem 2. If $\mathcal{M}_A = (E, \mathcal{I}_A, <_A)$ and $\mathcal{M}_B = (E, \mathcal{I}_B, <_B)$ are ordered matroids then there exists an $\mathcal{M}_A\mathcal{M}_B$-kernel K.

Proof. For subset E' of E let $f(E') := E \backslash \overline{K_n^{\mathcal{M}_B}}(E \backslash \overline{K_n^{\mathcal{M}_A}}(E')) = K_n^{\mathcal{M}_B}(g(E')) \cup (E \setminus g(E'))$, where $g(E') := (E \setminus E') \cup K_n^{\mathcal{M}_A}(E')$. It is straightforward to check that $K_n^{\mathcal{M}_A}(E')$ is an $\mathcal{M}_A\mathcal{M}_B$-kernel whenever $f(E') = E'$. It follows from (7) that $E' \subseteq E'' \subseteq E$ implies $f(E') \subseteq f(E'')$. Hence $f(\emptyset) \subseteq f(f(\emptyset)) \subseteq \dots$, so $f^{k+1}(\emptyset) = f^k(\emptyset)$ for some $k \le n$. Then $f^k(\emptyset)$ is a fixed point of f. □

Note that the iteration of f in the proof generalizes the proposal algorithm of Gale and Shapley. Let us denote the set of $\mathcal{M}_A\mathcal{M}_B$-kernels by $\mathcal{K}_{\mathcal{M}_A\mathcal{M}_B}$. For subsets K and L of E define

$$K \vee L := K_n^{\mathcal{M}_A}(K \cup L) \text{ and}$$
$$K \wedge L := K_n^{\mathcal{M}_B}(K \cup L) .$$

Theorem 3. If $K, L \in \mathcal{K}_{\mathcal{M}_A\mathcal{M}_B}$ then $K \vee L, K \wedge L \in \mathcal{K}_{\mathcal{M}_A\mathcal{M}_B}$ and K and L span the same set in $\mathcal{M}_A$. Moreover, $(\mathcal{K}_{\mathcal{M}_A\mathcal{M}_B}, \vee, \wedge)$ is a lattice.

Proof. Let $N := (K \cup L) \cup D_{\mathcal{M}_B}(K) \cup D_{\mathcal{M}_B}(L)$. By (5),

$$N \subseteq K_n^{\mathcal{M}_B}(D_{\mathcal{M}_B}(K) \cap D_\mathcal{M}(L)) , \tag{8}$$

hence N is independent. From (1) and (5) it follows that

$$K \vee L = K_n^{\mathcal{M}_A}(D_{\mathcal{M}_A}(K) \cup D_{\mathcal{M}_A}(L)) \subseteq N . \tag{9}$$

On the other hand, we have

$$
\begin{aligned}
|K| = \mathrm{rank}_{\mathcal{M}_A}(K) = \mathrm{rank}_{\mathcal{M}_A}(D_{\mathcal{M}_A}(K)) &\leq \mathrm{rank}_{\mathcal{M}_A}(D_{\mathcal{M}_A}(K) \cup D_{\mathcal{M}_A}(L)) = \\
= |K_n^{\mathcal{M}_A}(D_{\mathcal{M}_A}(K) \cup D_{\mathcal{M}_A}(L))| \leq |N| &\leq |K_n^{\mathcal{M}_B}(D_{\mathcal{M}_B}(K) \cap D_{\mathcal{M}}(L))| \leq \\
&\leq \mathrm{rank}_{\mathcal{M}_B}(D_{\mathcal{M}_B}(K)) = \mathrm{rank}_{\mathcal{M}_B}(K) = |K| \ ,
\end{aligned}
\tag{10}
$$

thus there must be equality throughout. It follows that in (8) and (9) we have equalities as well, so $N = K \vee L$. We also see that $D_{\mathcal{M}_A}(K) \cup D_{\mathcal{M}_A}(L) \subseteq D_{\mathcal{M}_A}(N)$ and $D_{\mathcal{M}_B}(K) \cap D_{\mathcal{M}_B}(L) \subseteq D_{\mathcal{M}_B}(N)$, that is $D_{\mathcal{M}_A}(N) \cup D_{\mathcal{M}_B}(N) = E$. This means that $N = K \vee L$ is an $\mathcal{M}_A \mathcal{M}_B$-kernel, as claimed. It can be proved with the interchange of the role of $\mathcal{M}_A$ and $\mathcal{M}_B$ that $K \wedge L$ is an $\mathcal{M}_A \mathcal{M}_B$-kernel.

It also follows from (10) that $|K| = |N| = |K \vee L| = \mathrm{rank}_{\mathcal{M}_A}(K \cup L)$. This means that $\mathcal{M}_A \mathcal{M}_B$-kernels span the same set in $\mathcal{M}_A$. In particular, $\mathcal{M}_A \mathcal{M}_B$-kernels have the same size.

To prove the lattice property, introduce partial orders $<_A$ and $<_B$ on the set of $\mathcal{M}_A \mathcal{M}_B$-kernels by saying that $K \leq_A L$ if $D_{\mathcal{M}_A}(K) \subseteq D_{\mathcal{M}_A}(K)$ and $K \leq_B L$ if $D_{\mathcal{M}_B}(K) \subseteq D_{\mathcal{M}_B}(K)$. For these partial orders, any two $\mathcal{M}_A \mathcal{M}_B$-kernels K and L have a least upper bound. This is because we have seen above that

$$
K \vee L = K_n^{\mathcal{M}_A \mathcal{M}_B}(K \cup L) = K_n^{\mathcal{M}_A \mathcal{M}_B}(D_{\mathcal{M}_A \mathcal{M}_B}(K) \cup D_{\mathcal{M}_A \mathcal{M}_B}(L))
$$

and $D_{\mathcal{M}_A \mathcal{M}_B}(K) \cup D_{\mathcal{M}_A \mathcal{M}_B}(L) \subseteq D_{\mathcal{M}_A \mathcal{M}_B}(K \vee L)$, so $K \vee L$ is the least $<_A$-upper bound of K and L. Similarly, the least $<_B$-upper bound of K and L is $K \wedge L$.

What left is to show that $<_A = <_B^{-1}$. So assume that $K <_A L$ for $\mathcal{M}_A \mathcal{M}_B$-kernels K and L. This means that $D_{\mathcal{M}_A}(K) \subseteq D_{\mathcal{M}_A}(L)$, so $L = K_n^{\mathcal{M}_A}(K \cup L)$. As K is a $\mathcal{M}_A \mathcal{M}_B$-kernel, each element of $L \setminus K$ must be dominated by K, thus $K = K_n^{\mathcal{M}_B}(K \cup L)$. Hence $D_{\mathcal{M}_B}(K) \supseteq D_{\mathcal{M}_B}(L)$, that is, $L <_B K$. □

From Theorem 3, it follows that $\mathcal{M}_A \mathcal{M}_B$-kernels have the same size, that we shall denote by k.

Let $\leq$ be the partial order on $\mathcal{K}_{\mathcal{M}_A \mathcal{M}_B}$ that corresponds to the lattice in Theorem 3. We need two important properties of this lattice.

Theorem 4. *For any $\mathcal{M}_A \mathcal{M}_B$-kernels K, L,*

$$
\chi^K + \chi^L = \chi^{K \vee L} + \chi^{K \wedge L} \ .
\tag{11}
$$

Moreover, if $K \leq L \leq M$ for $\mathcal{M}_A \mathcal{M}_B$-kernels K, L and M then $K \cap M \subseteq L$.

Proof. Observe that

$$
\mathbf{1} \cdot (\chi^K + \chi^L) = 2k = \mathbf{1} \cdot (\chi^{K \vee L} + \chi^{K \wedge L}) \ ,
$$

so for the modular property (11) it is enough to prove that $\chi^{K \vee L} + \chi^{K \wedge L} \leq \chi^K + \chi^L$. As these vectors are nonnegative integral and $\chi^{K \vee L} + \chi^{K \wedge L} \leq 2\chi^{K \cup L}$, we only have to show that if $e \in K \triangle L$ then $e \notin (K \vee L) \cap (K \wedge L)$. If, say $e \notin L$, then without limiting of generality we may assume that $e \in D_{\mathcal{M}_A}(L)$, that is, by the definition of dominance, there is a subset C_e of L that spans e and $c <_{\mathcal{M}_A} e$ for all elements c of C_e. But then $e \notin K_n^{\mathcal{M}_A}(K \cup L) = K \vee L$ follows. This shows (11).

For the second part, let e be an element of $K \cap M$. As $K \leq L \leq M$, we have $e \in K = K_n^{\mathcal{M}_B}(K \cup L)$ and $e \in M = K_n^{\mathcal{M}_A}(L \cup M)$. But L can dominate e only if e belongs to L. $\square$

3   Matroid-Kernel Polyhedra

For ordered matroids $\mathcal{M}_A$ and $\mathcal{M}_B$ on the same ground-set E, let us denote by

$$\mathcal{B}_{\mathcal{M}_A\mathcal{M}_B} := \{B \subseteq E : B \cap K \neq \emptyset \text{ for any } K \in \mathcal{K}_{\mathcal{M}_A\mathcal{M}_B}\} \text{ and}$$
$$\mathcal{A}_{\mathcal{M}_A\mathcal{M}_B} := \{A \subseteq E : |A \cap K| \leq 1 \text{ for any member } K \text{ of } \mathcal{K}_{\mathcal{M}_A\mathcal{M}_B}\}$$

the *blocker* and the *antiblocker* of $\mathcal{K}_{\mathcal{M}_A\mathcal{M}_B}$, respectively. Further, for $X := E \setminus \bigcup \mathcal{K}_{\mathcal{M}_A\mathcal{M}_B}$, define

$$\mathcal{P}_{\mathcal{K}_{\mathcal{M}_A\mathcal{M}_B}} := \text{conv}\{\chi^K : K \in \mathcal{K}_{\mathcal{M}_A\mathcal{M}_B}\} ,$$
$$\mathcal{C}_{\mathcal{K}_{\mathcal{M}_A\mathcal{M}_B}} := \text{cone}\{\chi^K : K \in \mathcal{K}_{\mathcal{M}_A\mathcal{M}_B}\} = \{\lambda \cdot x : \lambda \in \mathbb{R}_+, x \in \mathcal{P}_{\mathcal{K}_{\mathcal{M}_A\mathcal{M}_B}}\} ,$$
$$\mathcal{P}_{\mathcal{K}_{\mathcal{M}_A\mathcal{M}_B}}^{\uparrow} := \mathcal{P}_{\mathcal{K}_{\mathcal{M}_A\mathcal{M}_B}} + \mathbb{R}_+^E = \{x + y : x \in \mathcal{P}_{\mathcal{K}_{\mathcal{M}_A\mathcal{M}_B}}, y \geq 0\} ,$$
$$\mathcal{P}_{\mathcal{K}_{\mathcal{M}_A\mathcal{M}_B}}^{\downarrow} := (\mathcal{P}_{\mathcal{K}_{\mathcal{M}_A\mathcal{M}_B}} - \mathbb{R}_+^E) \cap \mathbb{R}_+^E =$$
$$= \{x - y : x \in \mathcal{P}_{\mathcal{K}_{\mathcal{M}_A\mathcal{M}_B}}, y \geq 0\} \cap \mathbb{R}_+^E ,$$
$$\mathcal{P}_{\mathcal{B}_{\mathcal{M}_A\mathcal{M}_B}}^{\uparrow} := \{\chi^B : B \in \mathcal{B}_{\mathcal{M}_A\mathcal{M}_B}\}^{\uparrow} =$$
$$= \{x + y : x \in \text{conv}\{\chi^B : B \in \mathcal{B}_{\mathcal{M}_A\mathcal{M}_B}\}, y \geq 0\} , \text{ and}$$
$$\mathcal{P}_{\mathcal{A}_{\mathcal{M}_A\mathcal{M}_B}}^{\downarrow} := \{\chi^A : A \in \mathcal{A}_{\mathcal{M}_A\mathcal{M}_B}\}^{\downarrow} + C_M =$$
$$= C_X + \{x - y : x \in \text{conv}\{\chi^A : A \in \mathcal{A}_{\mathcal{M}_A\mathcal{M}_B}\}, y \geq 0\} \cap \mathbb{R}_+^E$$

the $\mathcal{M}_A\mathcal{M}_B$-*kernel polytope*, the $\mathcal{M}_A\mathcal{M}_B$-*kernel cone*, the *dominant* of the $\mathcal{M}_A\mathcal{M}_B$-kernel polytope and the *submissive* of the kernel polytope, the $\mathcal{M}_A\mathcal{M}_B$-*blocker polyhedron* and the $\mathcal{M}_A\mathcal{M}_B$-*antiblocker polyhedron*, respectively, where $C_X := \{x \in \mathbb{R}^E : x \geq 0, x_e = 0 \text{ for } e \notin X\}$. We are going to characterize these polyhedra in terms of linear constraints. For $\mathcal{P}_{\mathcal{B}_{\mathcal{M}_A\mathcal{M}_B}}^{\uparrow}$ and $\mathcal{P}_{\mathcal{A}_{\mathcal{M}_A\mathcal{M}_B}}^{\downarrow}$ we apply the theory of lattice polyhedra, and then the theory of blocking and antiblocking polyhedra gives descriptions for the rest.

To state the Hoffman-Schwartz theorem, a basic result on lattice polyhedra, we need to formulate some assumptions. Fix a ground-set E and a family $\mathcal{K}$ of subsets of E. A partial order $\prec$ on $\mathcal{K}$ is called *consecutive* if $K \cap M \subseteq L$ holds for any members K, L, M of $\mathcal{K}$ with $K \preceq L \preceq M$. Family $\mathcal{K}$ is *clutter* if there is a lattice on $\mathcal{K}$ with lattice operations $\wedge, \vee$ such that the underlying partial order is consecutive and $\chi^K + \chi^L = \chi^{K \wedge L} + \chi^{K \vee L}$ holds for any members K, L of $\mathcal{K}$.

Theorem 5 (Hoffman-Schwartz [10]). *Let $(\mathcal{K}, \prec, \wedge, \vee)$ be a consecutive lattice on ground-set E, $\mathcal{K}$ be a clutter for this lattice and $d : E \to \mathbb{N} \cup \{\infty\}$ be and arbitrary function. If $r : \mathcal{K} \to \mathbb{N}$ is submodular then system*

$$\{x \in \mathbb{R}^E : \mathbf{0} \le x \le d, \ x(K) \le r(K) \text{ for any } K \in \mathcal{K}\} \tag{12}$$

is TDI.

If and $r : \mathcal{K} \to \mathbb{N}$ is supermodular then system

$$\{x \in \mathbb{R}^E : \mathbf{0} \le x \le d, \ x(K) \ge r(K) \text{ for any } K \in \mathcal{K}\} \tag{13}$$

is TDI. □

(Here, $r : \mathcal{K} \to \mathbb{N}$ is *submodular* if $r(K) + r(L) \ge r(K \wedge L) + r(K \vee L)$ holds for any $K, L \in \mathcal{K}$; r is *supermodular* if the reverse inequality is true.)

Theorem 4 can be translated into clutter-language as follows.

Corollary 1. *If $\mathcal{M}_A, \mathcal{M}_B$ are ordered matroids on the same ground-set E then family $\mathcal{K}_{\mathcal{M}_A \mathcal{M}_B}$ of $\mathcal{M}_A \mathcal{M}_B$-kernels is a clutter.* □

Hence Theorem 5 is relevant in our setting. Applying the Hoffman-Schwartz theorem on $\mathcal{K}_{\mathcal{M}_A \mathcal{M}_B}$, we get the following characterizations.

Theorem 6. *If $\mathcal{M}_A$ and $\mathcal{M}_B$ are ordered matroids on ground-set E then*

$$\mathcal{P}^\uparrow_{\mathcal{B}_{\mathcal{M}_A \mathcal{M}_B}} = \{x \in \mathbb{R}^X : x \ge \mathbf{0} \text{ and } x(K) \ge 1 \text{ for any } K \in \mathcal{K}_{\mathcal{M}_A \mathcal{M}_B}\} \ , \tag{14}$$

$$\mathcal{P}^\downarrow_{\mathcal{A}_{\mathcal{M}_A \mathcal{M}_B}} = \{x \in \mathbb{R}^X : x \ge \mathbf{0} \text{ and } x(K) \le 1 \text{ for any } K \in \mathcal{K}_{\mathcal{M}_A \mathcal{M}_B}\} \ . \tag{15}$$

Proof. Obviously, the polyhedra on the left hand side of (14,15) are the integer hulls of the polyhedra described by right hand sides.

By Observation 1, $\mathcal{K}_{\mathcal{M}_A \mathcal{M}_B}$ is a clutter. Let $d(v) := \infty$ (for (14)), $d(v) := \mathbf{1}$ (for (15)) and $r(K) := 1$ for all $v \in X$ and $K \in \mathcal{K}_{\mathcal{M}_A \mathcal{M}_B}$. Clearly, r is sub- and supermodular. By Theorem 5, linear systems in (14,15) are TDI, hence polyhedra on the right hand sides of (14, 15) are integer. □

We introduce some basic notions from the theory of blocking and antiblocking polyhedra to be able to describe the other kernel-related polyhedra.

Polyhedron $P \subseteq \mathbb{R}^d_+$ is a *blocking type polyhedron* if $P = P + \mathbb{R}^d_+$, and it is an *antiblocking type polyhedron* if $P = \mathbb{R}^d_+ \cap (P - \mathbb{R}^d_+)$. Any finite subset H of $\mathbb{R}^d_+$ defines a blocking and an antiblocking polyhedron by

$$H^\uparrow := \mathrm{conv}(H) + \mathbb{R}^d_+ \qquad \text{and}$$

$$H^\downarrow := \mathbb{R}^d_+ \cap (\mathrm{conv}(H) - \mathbb{R}^d_+) \ ,$$

respectively. For a polyhedron P

$$B(P) := \{x \in \mathbb{R}^d_+ : x^T y \ge 1 \text{ for all } y \in P\} \text{ and}$$

$$A(P) := \{x \in \mathbb{R}^d_+ : x^T y \le 1 \text{ for all } y \in P\}$$

are the *blocking* and *antiblocking polyhedron* of P, respectively. As suggested by the name, if P is a polyhedron then both $A(P)$ and $B(P)$ are polyhedra.

Theorem 7 (Fulkerson [6,7,8]). *If P is a blocking type polyhedron then $B(P)$ is a blocking type polyhedron and $P = B(B(P))$. If P is an antiblocking type polyhedron then $A(P)$ is an antiblocking type polyhedron and $P = A(A(P))$. Furthermore,*

$$B(\{x_1, x_2, \dots, x_n\}^\uparrow) = \{y \in \mathbb{R}_+^d : \quad y^T x_i \geq 1 \text{ for } i \in [n]\} \quad (16)$$

$$A(\{x_1, x_2, \dots, x_n\}^\downarrow) + C_M = \{y \in \mathbb{R}_+^d : \quad y^T x_i \leq 1 \text{ for } i \in [n] \text{ and}$$
$$y(m) - 0 \text{ for } m \in M\} \quad (17)$$

for any $n \in \mathbb{N}$ and elements x_i ($i \in [n]$) of $\mathbb{R}_+^d$ where for subset M of $[d]$ cone $C_M := \{x \in \mathbb{R}^d : x \geq 0 \text{ and } x(m) = 0 \text{ for } m \in [d] \setminus M\}$ is the projection of the positive orthant to $\mathbb{R}^d$. □

With these tools, we can justify that the following descriptions correspond to our kernel polyhedra.

Theorem 8. *If $\mathcal{M}_A$ and $\mathcal{M}_B$ are ordered matroids on ground-set E then*

$$\mathcal{P}^\uparrow_{\mathcal{K}_{\mathcal{M}_A \mathcal{M}_B}} = \{x \in \mathbb{R}^E : x \geq 0, \ x(B) \geq 1 \ \text{for } B \in \mathcal{B}_{\mathcal{M}_A \mathcal{M}_B}\}, \quad (18)$$

$$\mathcal{P}^\downarrow_{\mathcal{K}_{\mathcal{M}_A \mathcal{M}_B}} = \{x \in \mathbb{R}^E : x \geq 0 \text{ and } x(A) \leq 1 \text{ for any } A \in \mathcal{K}_{\mathcal{M}_A \mathcal{M}_B}, \text{ and}$$
$$x(X) = 0\}, \quad (19)$$

$$\mathcal{P}_{\mathcal{K}_{\mathcal{M}_A \mathcal{M}_B}} = \{x \in \mathbb{R}^E : x \geq 0, \ \mathbf{1}^T x \leq k, \ x(B) \geq 1 \ \text{for } B \in \mathcal{B}_{\mathcal{M}_A \mathcal{M}_B}\}, \quad (20)$$

$$\mathcal{P}_{\mathcal{K}_{\mathcal{M}_A \mathcal{M}_B}} = \{x \in \mathbb{R}^E : x \geq 0, \ \mathbf{1}^T x \geq k, \ x(A) \leq 1 \ \text{for } A \in \mathcal{A}_{\mathcal{M}_A \mathcal{M}_B}, \text{ and}$$
$$x(X) = 0\}, \quad (21)$$

$$\mathcal{C}_{\mathcal{K}_{\mathcal{M}_A \mathcal{M}_B}} = \{x \in \mathbb{R}^E : x \geq 0, \ k \cdot x(B) \geq \mathbf{1}^T x \text{ for } B \in \mathcal{B}_{\mathcal{M}_A \mathcal{M}_B}\}, \text{ and} \quad (22)$$

$$\mathcal{C}_{\mathcal{K}_{\mathcal{M}_A \mathcal{M}_B}} = \{x \in \mathbb{R}^E : x \geq 0, \ k \cdot x(A) \leq \mathbf{1}^T x \text{ for } A \in \mathcal{A}_{\mathcal{M}_A \mathcal{M}_B}, \text{ and}$$
$$x(X) = 0\}, \quad (23)$$

where k is the common size of $\mathcal{M}_A \mathcal{M}_B$-kernels and X is the set of those elements of E that can not be contained in a $\mathcal{M}_A \mathcal{M}_B$-kernel.

Proof. By (14) and (16), $\mathcal{P}^\uparrow_{\mathcal{B}_{\mathcal{M}_A \mathcal{M}_B}} = B(\mathcal{P}^\uparrow_{\mathcal{K}_{\mathcal{M}_A \mathcal{M}_B}})$. From Theorem 7, we get that $\mathcal{P}^\uparrow_{\mathcal{K}_{\mathcal{M}_A \mathcal{M}_B}} = B(\mathcal{P}^\uparrow_{\mathcal{B}_{\mathcal{M}_A \mathcal{M}_B}})$, and (18) follows from (16). Similarly, $\mathcal{P}^\downarrow_{\mathcal{A}_{\mathcal{M}_A \mathcal{M}_B}} = A(\mathcal{P}^\downarrow_{\mathcal{K}_{\mathcal{M}_A \mathcal{M}_B}})$ from (15) and (17). Theorem 7 gives that $\mathcal{P}^\downarrow_{\mathcal{K}_{\mathcal{M}_A \mathcal{M}_B}} = A(\mathcal{P}^\downarrow_{\mathcal{A}_{\mathcal{M}_A \mathcal{M}_B}})$, so (19) follows from (17). As each $\mathcal{M}_A \mathcal{M}_B$-kernel has the same size k, (20) follows directly from (18), and (21) from (19).

Clearly, both cones C and C' described on the right hand sides of (22) and (23) contain $\mathcal{C}_{\mathcal{K}_{\mathcal{M}_A \mathcal{M}_B}}$. Let $x \geq \mathbf{0}$ be a vector outside $\mathcal{C}_{\mathcal{K}_{\mathcal{M}_A \mathcal{M}_B}}$, and $\lambda = \frac{k}{\mathbf{1}^T x}$. Then $\mathbf{1}^T(\lambda \cdot x) = k$ and $\lambda \cdot x \notin \mathcal{P}^\uparrow_{\mathcal{K}_{\mathcal{M}_A \mathcal{M}_B}} \cup \mathcal{P}^\downarrow_{\mathcal{K}_{\mathcal{M}_A \mathcal{M}_B}}$, hence there is a member B of $\mathcal{B}_{\mathcal{M}_A \mathcal{M}_B}$ such that $\lambda \cdot x(B) < 1$ and if all $x(X) = 0$ holds then there is a member A of $\mathcal{A}_{\mathcal{M}_A \mathcal{M}_B}$ such that $\lambda \cdot x(A) > 1$. This means that $k \cdot x(B) < \frac{k}{\lambda} = \mathbf{1}^T x$ and $k \cdot x(A) > \frac{k}{\lambda} = \mathbf{1}^T x$. Thus $x \notin C$ and $x \notin C'$, justifying (22) and (23). □

Note that we gave two different descriptions for both $\mathcal{P}_{\mathcal{K}_{\mathcal{M}_A\mathcal{M}_B}}$ and $\mathcal{C}_{\mathcal{K}_{\mathcal{M}_A\mathcal{M}_B}}$. Apart from the nonnegativity conditions, there is no constraint that appears in both of the descriptions. In particular, this means that $\mathcal{C}_{\mathcal{K}_{\mathcal{M}_A\mathcal{M}_B}}$ can neither be full dimensional nor 1-codimensional. It is also interesting to observe that the linear description by Vande Vate [13] and Rothblum [12] of the convex hull of bipartite stable matchings is related to (20).

Theorem 9 (Rothblum [12], see also Vande Vate [13]). *Let* $G = (V, E)$ *be a finite bipartite graph and for each* $v \in V$ *let* $\prec_v$ *be a linear order on* $D(v)$, *the edges incident to* v. *Define* $\phi(e) := \{f \in E : f \preceq_u e \text{ or } f \preceq_v e\}$ *for edge* $e = uv \in E$. *Then*

$$\text{conv}\{\chi^M : M \subseteq E \text{ is a stable matching of } G\} = \tag{24}$$
$$\{x \in \mathbb{R}^E : 0 \leq x, x(D(v)) \leq 1 \text{ for } v \in V, x(\phi(e)) \geq 1 \text{ for } e \in E\} \ .$$

□

In (24), conditions $x(D(v)) \leq 1$ are special cases of conditions of type $x(A) \leq 1$ of (21) and together with the nonnegativity of x, these are responsible for that any solution x is a convex combination of bipartite matchings. Constraints $x(\phi(e)) \geq 1$ are special cases of $x(B) \geq 1$ type constraints in (20). In fact, as $\mathbf{1}^T x \geq k$ for any $x \in \mathcal{P}^\uparrow_{\mathcal{K}_{\mathcal{M}_A\mathcal{M}_B}}$ and $\mathbf{1}^T x \leq k$ for any $x \in \mathcal{P}^\downarrow_{\mathcal{K}_{\mathcal{M}_A\mathcal{M}_B}}$, we have that $\mathcal{P}_{\mathcal{K}_{\mathcal{M}_A\mathcal{M}_B}} = \mathcal{P}^\uparrow_{\mathcal{K}_{\mathcal{M}_A\mathcal{M}_B}} \cap \mathcal{P}^\downarrow_{\mathcal{K}_{\mathcal{M}_A\mathcal{M}_B}}$. It follows that

$$\mathcal{P}_{\mathcal{K}_{\mathcal{M}_A\mathcal{M}_B}} = \{x \in \mathbb{R}^E : x \geq 0,\ x(B) \geq 1 \text{ for } B \in \mathcal{B}_{\mathcal{M}_A\mathcal{M}_B},$$
$$x(A) \leq 1 \text{ for } A \in \mathcal{A}_{\mathcal{M}_A\mathcal{M}_B}\} \ , \tag{25}$$

because for a vector x of the right hand side, define x' by zeroing the coordinates of x that correspond to elements $e \in X$ and add $x(X)$ to some other coordinate of x. It is easy to check that $x' \in \mathcal{P}^\uparrow_{\mathcal{K}_{\mathcal{M}_A\mathcal{M}_B}} \cap \mathcal{P}^\downarrow_{\mathcal{K}_{\mathcal{M}_A\mathcal{M}_B}}$, hence $\mathbf{1}^T x' = k$. But then $x \in \mathcal{P}^\uparrow_{\mathcal{K}_{\mathcal{M}_A\mathcal{M}_B}}$ can only hold if $x = x'$, that is, condition $x(X) = 0$ automatically holds in (25). Note that characterization (25) resembles very much to (24).

Another interesting question whether linear descriptions (18-23) are good characterizations, that is, whether the separation problem over those polyhedra can be solved efficiently. The answer is yes, and a possible way is explained in [5].

The notion of matroid-kernel can be generalized as follows. Let $\mathcal{M}_A$ and $\mathcal{M}_B$ be two matroids on the same ground-set E, and let $w_A, w_B : E \to \mathbb{N}$ be arbitrary (weight)functions. A subset K is called a *generalized* $\mathcal{M}_A\mathcal{M}_B$-*kernel* if K is a common independent set of $\mathcal{M}_A$ and $\mathcal{M}_B$ and for any element e of $E \setminus K$ there is a subset C_e of K such that

$$\{e\} \cup C_e \in \mathcal{C}_A \text{ and } w_A(c) \leq_A w_A(e) \text{ for any } c \in C_e \text{ or}$$
$$\{e\} \cup C_e \in \mathcal{C}_B \text{ and } w_B(c) \leq_B w_B(e) \text{ for any } c \in C_e. \tag{26}$$

Clearly, if functions w_A and w_B are injective then they induce linear orders $<_A$ and $<_B$ on E (by $e <_A f$ if $w_A(e) < w_A(f)$ and $e <_B f$ if $w_B(e) < w_B(f)$). A generalized $\mathcal{M}_A\mathcal{M}_B$ kernel turns out to be a $\mathcal{M}_A\mathcal{M}_B$-kernel for these orders. If functions w_A, w_B are not injective, then still we can find orders $<_A$ and $<_B$ with the property that $e <_A f$ implies $w_A(e) \leq w_A(f)$ and $e <_B f$ implies $w_B(e) \leq w_B(f)$. An $\mathcal{M}_A\mathcal{M}_B$-kernel for these orders is a generalized $\mathcal{M}_A\mathcal{M}_B$-kernel for weights w_A and w_B. The problem is that these linear orders are not unique in general, and depending on our choice the set of $\mathcal{M}_A\mathcal{M}_B$-kernels can be different. So if we would like to characterize generalized $\mathcal{M}_A\mathcal{M}_B$-kernels related polyhedra, we have to take this into account. Here, we only point out that for two (in a sense extreme) cases this characterization is known. Theorem 8 does it for injective functions w_A, w_B, and the following linear description of the matroid intersection polytope by Edmonds [1] does the job for constant functions w_A, w_B.

Theorem 10 (Edmonds [1]). *If $\mathcal{M}_A = (E, \mathcal{I}_A)$ and $\mathcal{M}_B = (E, \mathcal{I}_B)$ are matroids then*

$$\mathrm{conv}\{\chi^I : I \in \mathcal{I}_A \cap \mathcal{I}_B\} =$$
$$\{x : \mathbf{0} \leq x \in \mathbb{R}^E, x(F) \leq \min\{r_A(F), r_B(F)\} \text{ for any } F \subseteq E\} ,$$

where r_A and r_B are the rank functions of $\mathcal{M}_A$ and $\mathcal{M}_B$, respectively. □

It is a most challenging problem to find a characterization of the generalized matroid-kernel polytope. Such a common generalization of Theorem 8 and Theorem 10 could reveal interesting connections between the theory of lattice polyhedra, matroids, and bipartite stable matchings.

References

1. Jack Edmonds. Submodular functions, matroids, and certain polyhedra. In *Combinatorial Structures and their Applications (Proc. Calgary Internat. Conf., Calgary, Alta., 1969)*, pages 69–87. Gordon and Breach, New York, 1970.
2. Jack Edmonds. Matroids and the greedy algorithm. *Math. Programming*, 1:127–136, 1971.
3. Tamás Fleiner. A fixed-point approach to stable matchings and some applications. Submitted to *Mathematics of Operartions Research*, 2000.
4. Tamás Fleiner. A fixed-point approach to stable matchings and some applications. Egres Technical Report TR-2001-01, http://www.cs.elte.hu/egres/, 2000 March.
5. Tamás Fleiner. Stable and srossing structures, 2000. PhD dissertation, www.renyi.hu/~fleiner.
6. D. R. Fulkerson. Blocking polyhedra. In *Graph Theory and its Applications (Proc. Advanced Sem., Math. Research Center, Univ. of Wisconsin, Madison, Wis., 1969)*, pages 93–112. Academic Press, New York, 1970.
7. D. R. Fulkerson. Blocking and anti-blocking pairs of polyhedra. *Math. Programming*, 1:168–194, 1971.
8. D. R. Fulkerson. Anti-blocking polyhedra. *J. Combinatorial Theory Ser. B*, 12:50–71, 1972.

9. D. Gale and L.S. Shapley. College admissions and stability of marriage. *Amer. Math. Monthly*, 69(1):9–15, 1962.
10. A. J. Hoffman and D. E. Schwartz. On lattice polyhedra. In *Combinatorics (Proc. Fifth Hungarian Colloq., Keszthely, 1976), Vol. I*, pages 593–598. North-Holland, Amsterdam, 1978.
11. Alvin E. Roth. Conflict and coincidence of interest in job matching: some new results and open questions. *Math. Oper. Res.*, 10(3):379–389, 1985.
12. Uriel G. Rothblum. Characterization of stable matchings as extreme points of a polytope. *Math. Programming*, 54(1, Ser. A):57–67, 1992.
13. John H. Vande Vate. Linear programming brings marital bliss. *Oper. Res. Lett.*, 8(3):147–153, 1989.

A 2-Approximation for Minimum Cost $\{0, 1, 2\}$ Vertex Connectivity

Lisa Fleischer

Graduate School of Industrial Administration
Carnegie Mellon University
5000 Forbes Ave,
Pittsburgh, PA 15217, USA
lkf@andrew.cmu.edu

Abstract. In survivable network design, each pair (i, j) of vertices is assigned a level of importance r_{ij}. The vertex connectivity problem is to design a minimum cost network such that between each pair of vertices with importance level r, there are r *vertex* disjoint paths. There is no approximation algorithm known for this general problem. In this paper, we give a 2-approximation for the problem when $r \in \{0, 1, 2\}^{V \times V}$, improving on a previous known 3-approximation. This matches the best known approximation for the easier problem that requires that the paths be only edge-disjoint.

Our algorithm extends an iterative rounding algorithm that gives a 2-approximation for the edge-connectivity problem, for arbitrary connectivity requirements r. (K. Jain, A factor 2 approximation for the generalized Steiner network problem.) This algorithm relies on well-known uncrossing lemma for tight edge cutsets. Our extension uses a new type of uncrossing lemma for tight cutsets that may include vertices as well as edges.

For $r \in \{1, k\}^{V \times V}$, $k \geq 3$, we show that a) uncrossing tight cutsets is not possible, and b) any analysis for iterative rounding that depends directly on the largest fractional value in the linear programming solution cannot provide approximation guarantees better than the maximum connectivity requirement.

1 Introduction

Let $G = (V, E)$ be an undirected graph on vertex set V and edge set E. Given $X \subset V$, define $\delta(X)$ as the set of edges with exactly one endpoint in X and $E(X)$ as the set of edges with both endpoints in X. Define $G - X$ as that graph obtained from G by removing all vertices in X and all edges in $\delta(X) \cup E(X)$. Given $F \subset E$, define $G - F$ to be that graph obtained from G by removing all edges in F. G is called *k-vertex connected* if $|V| > k$ and for every $X \subset V$ with $|X| < k$, $G - X$ is connected. G is called *k-edge connected* if $|V| > k$ and for every $F \subset E$ with $|F| < k$, $G - F$ is connected. A k-vertex connected graph is k-edge connected, but the converse does not typically hold. Given vertex connectivity requirements r_{ij} between any pair of vertices (i, j), G satisfies the connectivity

K. Aardal, B. Gerards (Eds.): IPCO 2001, LNCS 2081, pp. 115–129, 2001.
© Springer-Verlag Berlin Heidelberg 2001

requirements if for every subset $X \subseteq V - \{i, j\}$ with $|X| < r_{ij}$, i and j are in the same connected component of $G - X$. If the requirements are for edge connectivity instead, then G satisfies the connectivity requirements if for every subset $F \subseteq E$ with $|F| < r_{ij}$, i and j are in the same connected component of $G - F$.

Let c be a cost vector on the edges of G. The problem of finding the minimum cost subgraph of G so that G satisfies edge connectivity requirements $r \in \mathbf{Z}^{V \times V}$ is called the *minimum cost edge connectivity problem (MCEC)*. The equivalent problem for vertex connectivity is the *minimum cost vertex connectivity problem (MCVC)*. Both of these problems are Max-SNP hard since the Steiner tree problem is a special case of each.

In [10], Jain describes the first constant factor approximation for MCEC. He obtains this approximation by iteratively solving linear programs with the property that at least one variable in each program has solution value of at least $1/2$. His main contribution is to prove this property holds for the iterative problems generated by his algorithm. The previous best approximation algorithm, by Goemans et al., gives a $O(\log k)$ approximation [5]. This is a primal-dual based approximation algorithm that does not rely on solving linear programs.

No nontrivial approximation algorithm is known for MCVC. For the problem where r is restricted to $\{0, 1, 2\}^{V \times V}$, Ravi and Williamson [16] describe a primal-dual 3-approximation algorithm. We call this problem $\{0, 1, 2\}$-MCVC. This problem arises in the design of survivable communications networks [8,15].

In this paper we describe a 2-approximation algorithm for $\{0, 1, 2\}$-MCVC that iteratively rounds appropriately defined linear programs. This approximation guarantee now matches the best approximation guarantee for the corresponding edge-connectivity problem [10].

Our approximation algorithm extends the algorithm of Jain [10]. Jain considers basic solutions to a linear programming relaxation of an integer programming formulation of the problem. A *basic* solution is any solution corresponding to a vertex of the polytope defined by the inequalities describing the linear program. Any basic solution is uniquely defined by a set of $|E|$ inequalities that are satisfied at equality. Any inequality that is satisfied at equality is called *tight*. Jain shows that for any basic feasible solution to the LP, there exists a set of inequalities that define the solution that correspond to laminar subsets of vertices. Two sets S and T are called *laminar* if at least one of $S \cap T$, $S \backslash T$, $T \backslash S$, and $V \backslash (S \cup T)$ is empty. Otherwise S and T cross. A set of sets is laminar if every pair in the set is laminar. Using laminarity, Jain develops a charging scheme to bound from below the value of the highest solution coordinate.

For $\{0, 1, 2\}$ MCVC, we define an appropriate linear program whose set of integer solutions correspond to solutions to $\{0, 1, 2\}$ MCVC. We prove the existence of a laminar set of inequalities defining any basic feasible solution to this LP. The proof relies on a new uncrossing lemma. We can then use a very similar charging scheme to obtain the same lower bound on the value of the highest coordinate in the solution vector.

A natural question is if this can be extended to yield a constant factor approximation for $\{0, 1, \ldots, k\}$-vertex connectivity. We show the same proof technique will not work by exhibiting an infinite family of examples where the only sets of inequalities defining a basic solution to the appropriate linear program are highly non-laminar. In this example, the largest fraction after one rounding is $\frac{1}{k}$. This indicates that an approximation argument based simply on the size largest fraction in an iterative rounding scheme will not yield better than a k-approximation.

For the more specialized connectivity problems of constructing a minimum cost uniformly k-connected graph, the best known approximation guarantee is roughly factor k [3,13].[1] If c is a metric, then there are constant factor approximations for uniform k-connectivity [12,13]. Last IPCO, Melkonian and Tardos [14] extend Jain's iterative rounding analysis to obtain a 4-approximation for uniform k-edge connectivity on directed graphs. They also describe a different approach that yields a 2-approximation. There are numerous approximation results for other special cases of vertex and edge connectivity. For surveys, see [6,11].

We discuss our linear program formulation and the recursive algorithm in Section 2. We prove the existence of a variable with value at least $1/2$ in Section 3. In Section 4, we describe an infinite family of examples that show that these proof techniques do not extend when $r \in \{1, k\}^{V \times V}$.

2   A 2-Approximation

Our 2-approximation relies on formulating the problem as an integer program, solving the LP relaxation of the integer program, and showing that there is at least one edge with fractional value greater than or equal to $1/2$. If we include this edge in our final solution, its contribution to the cost of our solution is no more than twice its contribution to the linear program solution, the latter being a lower bound on the cost of an optimal integer solution. We then show that the remaining problem is of the same general form as our original problem, and that we can use recursion to obtain a complete integer solution that has cost at most twice the optimal solution to the original linear program. Thus it is a 2-approximation to our problem.

This general outline was suggested by Jain in [10]. He uses the following linear programming relaxation of MCEC. There is a variable $x(e)$ for every edge $e \in E$, and an inequality for every subset of vertices that requires that the number of edges leaving the set be at least the maximum connectivity requirement over all pairs of vertices that have exactly one member of the pair in the set. Let $f(S)$ be defined to take this value for subset S. Let $\delta(S)$ denote the set of edges with exactly one endpoint in S. The linear program is:

---

[1] In [16], a primal-dual algorithm is proposed, but the analysis has recently discovered to be flawed, and the algorithm does not provide the claimed $O(\log k)$-factor guarantee [17].

$$\begin{array}{ll} \min & cx \\ \text{s.t. } \sum_{e \in \delta(S)} x(e) \geq f(S), & \forall S \subset V \\ 0 \leq x(e) \leq 1, & \forall e \in E \end{array} \qquad \text{(MCEC)}$$

The key to the argument in [10] is establishing that any basic solution to (**MCEC**) contains at least one edge with fractional value at least one half. This is done by assuming that the ground set corresponds to the support of x. Then, if $x < 1$, he shows that any basic solution is defined by a set of tight inequalities that correspond to a set of laminar subsets of V. This is proven by demonstrating that f is weakly supermodular and using this to employ a well-known uncrossing lemma for tight edge cutsets. This is used in an innovative charging scheme that shows that laminarity implies the existence of an edge with sufficiently high value in the solution to (**MCEC**). The other part of the argument involves establishing that this technique may be invoked recursively. This is done using a general description of f as being *weakly supermodular*. A set function f is *weakly supermodular* if $f(S) + f(T) \leq \max\{f(S \cup T) + f(S \cap T), f(S \backslash T) + f(T \backslash S)\}$.

To extend these arguments to the vertex connectivity problem we 1) introduce an appropriate linear program relaxation of the MCVC 2) prove a new uncrossing lemma for $\{0, 1, 2\}$ MCVC 3) establish that this lemma may be invoked recursively, by extending the notion of weak supermodularity. The examples in Section 4 indicate that step 2) and step 3) do not hold for more general connectivity requirements. One problem is that the cutsets corresponding to inequalities of the linear program for MCVC consist of edges *and vertices*. The inclusion of vertices in cutsets makes uncrossing nontrivial, and when connectivity requirements are higher than 2, it is no longer possible.

We use a linear program description of the problem that contains a variable $x(e)$ for each edge that indicates whether or not the edge is selected in the final network. The formulation contains an exponential number of constraints. However, as long as we can find a constraint of the LP that is violated by a given vector $x \in \{0, 1\}^E$ in polynomial time, we can find an optimal, basic solution to the linear program in polynomial time [7]. We describe such a subroutine in Section 5.

Our results extend to the case where we are allowed to select an edge multiple times. In our case, this would be at most twice. In fact, the problem appears to be only harder when we are restricted to selecting an edge at most once.

We give the linear programming formulation below. The constraints are based on a theorem of Menger. (For multiple proofs and references, see [4]):

Theorem 1 (Menger). *Let $G = (V, E)$ be a graph, and $s, t \in V$ such that $(s, t) \notin E$. Then, the minimum number of vertices separating s from t in G is equal to the maximum number of vertex disjoint paths from s to t in G.*

For any subset $S \subset V$, and disjoint set of vertices $A \subset V \backslash S$, we express the connectivity required between S and $V \backslash (A \cup S)$ as $f(S, A)$. Similarly, we represent the x-value of edges with one endpoint in S and the other in $V \backslash (S \cup A)$ as $x(S, A)$. That is, $x(S, A) := \sum_{i \in S, j \in V \backslash (S \cup A)} x((i, j))$. Here, A is the subset of vertices in the cutset separating S from the rest of the graph. Since there may be at most one path from S to $V\backslash(S\cup A)$ through each vertex of A, the number of edges from S to $V\backslash(A\cup S)$ must therefore be at least $f(S,A)-|A|$. This yields the following formulation.

$$
\begin{aligned}
\min \quad & cx \\
\text{s.t.} \quad & x(S,A) \ge f(S,A)-|A|, \quad \forall\, S, A \subset V, S\cap A = \emptyset \qquad \text{(MCVC)} \\
& 0 \le x(e) \le 1, \qquad\qquad\qquad \forall\, e \in E
\end{aligned}
$$

The following lemma is a simple consequence of Menger's Theorem.

Lemma 1. *The set of integral solutions to the above LP equals the set of solutions to the corresponding vertex connectivity problem.* □

The following definitions generalize the one-set function notions of submodularity, supermodularity, and weak supermodularity. A two-set function f defined on the set of pairs of disjoint subsets of V that satisfies

$$
\begin{aligned}
f(S,A)+f(T,B) \ge \\
\max\{\; & f(S\cup T,(A\backslash T)\cup(B\backslash S))+f(S\cap T,(A\cap T)\cup(B\cap S)), \\
& f(S\backslash(T\cup B),(A\backslash T)\cup(B\cap S))+f(T\backslash(S\cup A),(B\backslash S)\cup(A\cap T))\}
\end{aligned}
$$

is called *two-submodular*. This is a different concept from bisubmodularity introduced in [1,2]. For the case when $A = B = \emptyset$, this reduces to submodularity for symmetric one-set functions.

If $-f$ is two-submodular, then f is *two-supermodular*. This definition is equivalent to replacing $\ge$ with $\le$ and max with min in the above definition. A two-set function f is *very weakly two-supermodular* if whenever $f(S,A) > 0$ and $f(T,B) > 0$ then

$$
\begin{aligned}
f(S,A)+f(T,B) \le \\
\max\{\; & f(S\cup T,(A\backslash T)\cup(B\backslash S))+f(S\cap T,(A\cap T)\cup(B\cap S)), && (1) \\
& f(S\backslash(T\cup B),(A\backslash T)\cup(B\cap S))+f(T\backslash(S\cup A),(B\backslash S)\cup(A\cap T)), && (2) \\
& \max\{\; f(S\cup T,(A\backslash T)\cup(B\backslash S)), f(S\cap T,(A\cap T)\cup(B\cap S)), \\
& \quad f(S\backslash(T\cup B),(A\backslash T)\cup(B\cap S)), f(T\backslash(S\cup A),(B\backslash S)\cup(A\cap T))\; \}, && (3) \\
\mathbf{OR}\ & \text{for some permutation of } S \text{ and } V\backslash(S\cup A),\ T \text{ and } V\backslash(T\cup B), \\
& f(S\cap T,(A\cap T)\cup(B\cap S))+f(T\backslash(S\cup A),(B\backslash S)\cup(A\cap T))\; \}
\end{aligned}
$$

While it may seem that (3) is included in (1) and (2), if f is allowed to take on negative values, this may not be the case.

Let $\delta_F(S,A)$ denote the set of edges in F that have exactly one endpoint in each of S and $V\backslash(S\cup A)$.

Lemma 2. *Both $x(S,A)$ and $|\delta_F(S,A)|$ are two-submodular.* □

Define f_k by $f_k(S,A) := \max\{r_{ij} | i \in S, j \in V\backslash(S\cup A)\}$, where $r_{ij} \in \{0, 1, \dots, k\}$ for all $i, j \in V$. Define g_k by $g_k(S,A) = f_k(S,A) - |A|$. In Section 3, we prove Lemmas 3 and 4 and Theorem 2.

Lemma 3. *The two-set function g_2 is very weakly two-supermodular.*

Lemma 4. *For any edge set F on V, $g_2(S, A) - |\delta_F(S, A)|$ is very weakly two-supermodular.*

Theorem 2. *For the function $f(S, A) := f_2(S, A) - |\delta_F(S, A)|$, any basic solution to* (**MCVC**) *has at least one component with value at least $\frac{1}{2}$.*

We now describe an algorithm that yields a two approximation to (**MCVC**). This algorithm mirrors the algorithm in [10] for MCEC.

Let x^* be an optimal basic solution to (**MCVC**). Let $E_{\frac{1}{2}+}$ be the set of edges which have x^*-value $\geq \frac{1}{2}$. Fix all values of edges in $E_{\frac{1}{2}+}$ to 1. Let $E_{res} = E - E_{\frac{1}{2}+}$, and consider the resulting *residual LP*:

$$
\begin{aligned}
\min \quad & cx \\
\text{s.t.} \quad & x(S, A) \geq g_2(S, A) - |\delta_E(S, A) \cap E_{\frac{1}{2}+}|, \quad \forall\, S, A \subset V, S \cap A = \emptyset \\
& 0 \leq x(e) \leq 1, \qquad\qquad\qquad\qquad\qquad\quad \forall e \in E_{res}
\end{aligned}
$$
$$\text{(\textbf{MCVC}}_2)$$

Let z^*_{res} be the optimal value of this LP and z^* be the optimal value of (**MCVC**). The following theorem follows from similar arguments presented by Jain for edge-connectivity [10].

Theorem 3. *If E_{res} is an integral solution to* (**MCVC**$_2$) *with value at most $2z^*_{res}$, then $E_{res} \cup E_{\frac{1}{2}+}$ is an integral solution to* (**MCVC**) *with value at most $2z^*$.*

The 2-approximation algorithm: 1) Find an optimal basic solution x^* to (**MCVC**), 2) Include all edges e with $x^*(e) \geq 1/2$, in the final solution. 3) Delete all edges that were included in 2), and solve the residual problem on E_{res}.

3 Uncrossing Lemmas and Proof of Theorem

In this section, we prove an uncrossing lemma (Lemma 5) that we then use to establish laminarity of a set of spanning tight subsets (Corollary 2). The proof of this lemma relies on Lemmas 3 and 4. Laminarity of the tight subsets determining a basic solution in turn implies the main theorem.

Given (S, A), there is a pair $i \in S$, $j \in V \backslash (S \cup A)$ that determines $f_2(S, A)$. Let $i(S, A)$ denote one such i and $j(S, A)$ denote the corresponding j.

Proof of Lemma 3. For $r \in \{0, 1, 2\}^{V \times V}$, the only values of $|A|$ that yield nontrivial inequalities for (**MCVC**) are $|A| = 0$ or 1. Since $f_2(S, A) = f_2(V \backslash (S \cup A), A)$, it suffices to show that weak two-supermodularity holds for S and T satisfying $S \cap B = T \cap A = \emptyset$. Hence $A\backslash T = A$ and $B\backslash S = B$, and it suffices to show that

$$g_2(S, A) + g_2(T, B) \leq \max\{\; g_2(S \cup T, A \cup B) + g_2(S \cap T, \emptyset), \tag{4}$$
$$g_2(S\backslash T, A) + g_2(T\backslash S, B), \tag{5}$$
$$g_2(S \cap T, \emptyset), \tag{6}$$
$$\textbf{OR} \text{ by perhaps swapping } (S, A) \text{ for } (T, B),$$
$$g_2(S \cap T, \emptyset) + g_2(T \backslash S, B) \; \}. \tag{7}$$

If $|A| = |B| = 0$, then the weak supermodularity of the one-set function f' defined by $f'(S) := \max\{r_{ij} | i \in S, j \in V\backslash S\}$ used in [5,10] implies that either (4) or (5) hold.

If $|A| = |B| = 1$, then $f_2(S, A) = f_2(T, B) = 2$. In this case, either
1) $i(S, A) \in S\backslash T$ and $i(T, B) \in T\backslash S$, or
2) $\{i(S, A), i(T, B)\} \cap (S \cap T)$ is nonempty.
In the first case, (5) holds. In the second case, (6) holds.

If $|A| = 1$, $B = \emptyset$, (the case $|B| = 1$, $A = \emptyset$ may be treated symmetrically) then T may have connectivity requirement 1 or 2. If $i(S, A) \in S\backslash T$ and $j(S, A) \in T\backslash S$ then $f_2(T, \emptyset) = 2$ and (5) holds. If $i(S, A) \in S\backslash T$ and there is no corresponding $j(S, A)$ in $T\backslash S$ then $f_2(T, \emptyset) = 1$. In this case, if $i(T, B) \in T\backslash S$, then (5) holds. Otherwise, (4) holds.

The remaining case has $i(S, A) \in S \cap T$. If $j(S, A) \in V\backslash(S \cup T)$, then $f_2(T, \emptyset) = 2$ as well, and (4) holds. Otherwise, $j(S, A) \in T\backslash S$. We consider the possible values of $f_2(T, \emptyset)$. If $f_2(T, \emptyset) = 1$, then (6) holds. Otherwise, $f_2(T, \emptyset) = 2$. If $j(T, B) \in S\backslash T$, then (5) holds. If $j(T, B) \in V\backslash(S \cup A \cup T)$, then (4) holds. Otherwise, $A = \{j(T, B)\}$. In this case only, none of (4)-(6) hold, and (7) holds.
□

The proof of Lemma 3 demonstrates why and when we require (7) in the description of g_2. We summarize this in the following corollary so that we may easily refer to it.

Corollary 1. *If $T \cap A = \emptyset = S \cap B$ and $g_2(S, A) + g_2(T, B)$ is strictly greater than the maximum of (4)-(6), then $|A| + |B| = 1$, and assuming $|A| = 1$ (the other case is symmetric), then $A = \{j(T, B)\}$, $i(S, A) \in S \cap T$, $j(S, A) \in T\backslash S$, and $f_2(T, B) = f_2(T, \emptyset) = 2$.* □

The example in Section 4 shows that the corresponding g_k for $r \in \{1, k\}^{V \times V}$ is in general not very weakly two-supermodular for any $k \geq 3$. For $k = 3$, this example also demonstrates that the very weak two-supermodularity inequalities do not hold in this case even for the simplification $A \cap T = B \cap S = \emptyset$ used in the above proof.

Proof of Lemma 4. Since the proof of the lemma is independent of choice of F, and the context is clear, we use δ for δ_F. Suppose $g_2(S, A) > 0$ and $g_2(T, B) > 0$. If $g_2(S, A) + g_2(T, B)$ satisfies any of (4)-(6), then by the two-submodularity of $|\delta|$, we have $(g_2 - |\delta|)(S, A) + (g_2 - |\delta|)(T, B)$ satisfies the same inequality.

If $g_2(S, A) + g_2(T, B)$ does not satisfy (4)-(6), then it satisfies (7), for some permutation of S, $V \setminus (S \cup A)$, T, and $V \setminus (T \cup B)$. If $|\delta(S, A)| + |\delta(T, B)| \geq |\delta(S \cap T, \emptyset)| + |\delta(T \setminus S, B)|$, then $g_2 - |\delta|$ also satisfies (7), and we are done. Otherwise, there is an edge from $S \cap T$ to $T \setminus S$ in F, since this is the only type of edge that contributes more to the right hand side of (7) than the left. Thus $|\delta(S, A)| \geq 1$. Since $g_2(S, A) + g_2(T, B)$ does not satisfy (4)-6), by Corollary 1, we must have that $|A| + |B| = 1$. By swapping S, A for T, B we may assume $|A| - 1$. Then $g_2(S, A) = 1$, and thus $g_2(S, A) - |\delta(S, A)| \leq 0$. Hence, none of (4)-(7) need apply to (S, A) to establish weak two-supermodularity of $g_2 - |\delta|$.   □

Let x be a basic solution to (**MCVC₂**) with the property that $x(e) < 1$ for all $e \in E_{res}$, and let E_x be the set of edges with nonzero x-value, and let F be the set of edges already included in the final solution. A pair (S, A) is *tight* if it satisfies

$$x(S, A) \geq f_2(S, A) - |A| - |\delta_F(S, A)| \tag{8}$$

at equality. Given x, define $\chi_x(S, A)$ to be the characteristic vector of the support of $x(S, A)$. If $x(S, A) = 0$, we say that (S, A) is *empty*. If $x(S, A) > 0$, then (S, A) is *non-empty*. If (S, A) is empty, then $\chi_x(S, A) = \mathbf{0}$.

Lemma 5 (Uncrossing Lemma). *If (S, A) and (T, B) are tight and non-empty, then for the appropriate permutation of S and $V \setminus (S \cup A)$ and T and $V \setminus (T \cup B)$ so that $A \cap T = B \cap S = \emptyset$, one of the following holds.*

   i. $(S \cap T, \emptyset)$ is tight, $(S \cup T, A \cup B)$ is either empty or tight, and
     $\chi_x(S, A) + \chi_x(T, B) = \chi_x(S \cap T, \emptyset) + \chi_x(S \cup T, A \cup B)$,
   ii. $(S - T, A)$ and $(T - S, B)$ are tight and
     $\chi_x(S, A) + \chi_x(T, B) = \chi_x(S - T, A) + \chi_x(T - S, B)$,
   iii. After perhaps swapping (S, A) for (T, B), then $B = \emptyset$, and $(S \cap T, \emptyset)$ and $(T - S, \emptyset)$ are tight, and
     $2\chi_x(S, A) + \chi_x(T, \emptyset) = \chi_x(S \cap T, \emptyset) + \chi_x(T - S, \emptyset)$.

Proof. For simplicity of notation, let $g' = g_2 - |\delta_F|$. Since x is a solution to (**MCVC₂**), $g' - x \leq 0$. Since g' is very weakly two-supermodular by Lemma 4, if (S, A) and (T, B) are both nonempty, then for appropriate permutations of S and $V \setminus (S \cup A)$, T and $V \setminus (T \cup B)$, we have that $g'(S, A) + g'(T, B)$ must satisfy one of (4)-(7) with g_2 replaced by g'. If it satisfies any of (4)-(6), then since x is two-submodular, $(g' - x)(S, A) + (g' - x)(T, B)$ satisfies the same inequality. Thus if (S, A) and (T, B) are tight, then the left hand side of the corresponding inequality in (4)-(6) equals 0. Since $g' - x \leq 0$, this implies that each part of the corresponding right hand side equals 0. Thus, if (4) is satisfied, then i. holds with $(S \cup T, A \cup B)$ tight; if (5) is satisfied then, ii. holds; and if (6) is satisfied, then i. holds with $(S \cup T, A \cup B)$ empty.

If $g'(S, A) + g'(T, B)$ does not satisfy any of (4)-(6), then neither does g_2 and by Corollary 1), by perhaps swapping (S, A) for (T, B), we have that $|A| = 1$, $B = \emptyset$, $g_2(S, A) = 1$, $g_2(T, \emptyset) = 2$, $i(S, A) \in S \cap T$ and $j(S, A) \in T \setminus S$. Thus, $g_2(S \cap T, \emptyset) = g_2(T \setminus S, \emptyset) = 2$. Since (S, A) and $(T, \emptyset)$ are tight, then in order to satisfy the inequalities in (**MCVC**) for $(S \cap T, \emptyset)$ and $(T\backslash S, \emptyset)$, we must have that all the edges crossing (S, A) must leave $S \cap T$ and enter $T\backslash S$. Then the weights of edges leaving T are evenly split among edges from $S \cap T$ to $S\backslash T$ and edges from $T\backslash S$ to $V\backslash(S \cup T)$. This means that $x(S \cap T, \emptyset) = 2$ and $x(T\backslash S, \emptyset) = 2$, so that $(S \cap T, \emptyset)$ and $(T - S, \emptyset)$ are tight and $2\chi_x(S, A) + \chi_x(T, \emptyset) = \chi_x(S \cap T, \emptyset) + \chi_x(T - S, \emptyset)$. Thus iii. holds. $\quad\square$

Let $\mathcal{T}$ be the set of tight set pairs for x. Set pairs (S, A) and (T, B) are called *pair-laminar* if T and S are laminar and if T or $T \cup B$ cross S or $S \cup A$ then $A = B$. Otherwise, they are said to cross. A subset $\mathcal{L} \subset \mathcal{T}$ is called pair-laminar if all the pairs of set pairs in $\mathcal{L}$ are pair-laminar. Before establishing that $\mathcal{T}$ is spanned by a collection of pair-laminar set pairs, we need the following technical lemma.

Lemma 6. *Suppose $A \cap T = B \cap S = \emptyset$, and (S, A) crosses (T, B). If (T', B') crosses at least one of $(S \cap T, \emptyset)$, $(S \cup T, A \cup B)$, $(S - T, A)$, $(T - S, B)$, and it does not cross (T, B), then it crosses (S, A).*

Proof. We use the following easy to see fact:

$$X \text{ crosses } Y \cap Z, \ Y \cup Z, \ Y - Z, \text{ or } Z - Y \text{ but not } Z \Rightarrow X \text{ crosses } Y. \quad (9)$$

If (T', B') crosses one of the four set pairs in the lemma, then either T' crosses one of the first sets in each set pair; or $T' \cup B'$ crosses one of the first sets or T' or $T' \cup B'$ cross one of the four unions of first and second sets, and B' does not equal the corresponding second set. We consider each case in turn, progressively assuming that the previous cases do not occur.

If T' crosses $S \cap T$, $S - T$, $T - S$, or $S \cup T$, then by setting $X = T'$, $Y = S$, and $Z = T$, (9) implies it crosses S.

Otherwise, if $T' \cup B'$ crosses $S \cap T$, $S - T$, $T - S$, or $S \cup T$, then by setting $X = T' \cup B'$, $Y = S$, and $Z = T$, (9) implies it crosses S. We need to now establish that if B' does not equal the corresponding second set, then $B' \neq A$. Note that if the second set contains more than one element, then the set pair is empty, so it is not included in a collection of tight set pairs as described in Lemma 5. Suppose $B' = A$. Then the only cases of interest are $T' \cup B'$ crosses $T - S$ and $B' \neq B$, or $T' \cup B'$ crosses $S \cap T$ since in the other non-empty set pair cases A is the second set. If $T' \cup B'$ crosses $S \cap T$ or $T - S$, since by assumption, $B' \cap T = A \cap T = \emptyset$, then either $T' \subset T - S$ or $T' \subset S \cap T$. But then $T' \cup B'$ crosses T and $B' \neq B$, which contradicts (T', B') and (T, B) pair-laminar.

Otherwise, if T' crosses one of the four unions of set pairs in the lemma, then setting $X = T'$, $Y = S \cup A$, $Z = T \cup B$, implies that either T' crosses $S \cup A$ or $T \cup B$. If T' crosses $T \cup B$ then $B' = B \neq \emptyset$, so $T \subset T'$, and the only possibility is that T' crosses $(S \cup A) - T$ with $B' = B \neq A \neq \emptyset$. If T' does not cross $S \cup A$, then $T' \subseteq S \cup A$. Since $B' \neq B \nsubseteq S \cup A$ and T' does not cross S, it must be that $T' \subseteq S$. But then $T' \cup B'$ crosses S, and thus (T', B') crosses (S, A). A symmetric argument for the case of T' crossing $S \cup A$ with $B' = A$ yields a contradiction to (T', B') and (T, B) being pair laminar.

Finally, if none of the above cases hold, and $T' \cup B'$ crosses one of the four unions of set pairs, then $T' \cup B'$ crosses either $T \cup B$ or $S \cup A$. If the former holds, then $B' = B \neq \emptyset$, and $T \cap T' = \emptyset$. Thus, the only possibility of the four are that $T' \cup B'$ crosses $(S \cup A) - T$ and $B' = B \neq A \neq \emptyset$. Since T' does not cross $S \cup A$, and $B \notin S \cup A$, we have that $T' \subseteq S \cup A$. But then $T' \cup B'$ crosses S and thus (T', B') crosses (S, A).   □

Corollary 2. *For any maximal, pair-laminar family $\mathcal{L}$ of tight set pairs, the following holds:* $\mathrm{Span}(\mathcal{L}) = \mathrm{Span}(\mathcal{T})$.

Proof. If $\mathrm{Span}(\mathcal{L}) \neq \mathrm{Span}(\mathcal{T})$, then $\mathrm{Span}(\mathcal{L}) \subset \mathrm{Span}(\mathcal{T})$, and there exists a pair $(S, A) \in \mathcal{T}$, with $(S, A) \notin \mathcal{L}$, such that (S, A) crosses a minimum number of set pairs in $\mathcal{L}$. Let (T, B) be one of those pairs. Then by Lemma 5, we can rewrite $\chi_x(S, A)$ as a linear combination of characteristic vectors of pair-laminar tight set pairs. Note that the new set pairs do not cross (T, B). Since $(S, A) \notin \mathrm{Span}(\mathcal{L})$, at least one of these new set pairs is also not in $\mathrm{Span}(\mathcal{L})$. By Lemma 6, any set pair $L \in \mathrm{Span}(\mathcal{L})$ crossing any of the new sets must also have crossed (S, A). Since the new sets do not cross (T, B), they have strictly fewer crossings with sets in $\mathcal{L}$ than S does, contradicting the choice of S.   □

Corollary 3. *There exists a collection $\mathcal{B}$ of pair-laminar tight set pairs satisfying*

1. $|\mathcal{B}| = |E_x|$,
2. *the vectors $\chi_x(S, A)$ for $(S, A) \in \mathcal{B}$ are linearly independent,*
3. $(g_2 - \delta_F)(S, A) \geq 1$ *for all $(S, A) \in \mathcal{B}$.*   □

We define containment on set pairs by $(S, A) \subseteq (T, B)$ if $S \subseteq T$ and $A \subseteq T \cup B$. It is easy to see that the containment relation is transitive, reflexive, and anti-symmetric. Thus it defines a partially ordered set (poset).

Lemma 7. *If $\mathcal{B}$ is a collection of pair-laminar set pairs, then the poset defined by the containment relation on the set pairs in $\mathcal{B}$ is described by a unique forest.*

The following theorem implies Theorem 2:

Theorem 4. *There is a tight set pair (S, A) with $(g_2 - \delta_F)(S, A) > 0$ and at most 2 edges in $\delta_{E_{res}}(S, A)$. Hence, at least one of these edges has x-value at least $\frac{1}{2}$.*

Proof. Using the following concept of incidence, along with Lemma 7, the proof is very similar to the proof of the corresponding statement for MCEC in [10]. Each edge $e = (i, j)$ in E has two endpoints, i_j and j_i. An endpoint i_j of an edge (i, j) is *incident* to node (S, A) if (S, A) is the lowest node in the tree among all nodes with either $i \in S$ or $\{i, j\} \in S \cup A$. A vertex i may be the endpoint of several edges; and each such endpoint may be incident to a different node of the forest. An edge *crosses* a node (S, A) if exactly one of its endpoints is incident to any node in the subtree rooted at (S, A). This assignment ensures that an edge (i, j) crosses a node (S, A) if and only if $i \in S$ and $j \in V \setminus (S \cup A)$.   □

4   Examples and Counterexamples

In [10], Jain gives an example to show that the analysis of this algorithm is tight for the edge connectivity problem with connectivity requirements in $\{0,1\}$. Since in this case the edge and vertex connectivity problems are the same, the same example shows that the analysis is also tight for the vertex connectivity problem.

A natural question is: Can we extend the arguments given here to give a constant factor approximation for vertex connectivity problems with higher connectivity requirements? We answer this question negatively for general $r \in \{1,k\}^{V \times V}$ by describing basic solutions to an infinite family of instances of ($\mathbf{MCVC_2}$) for which 1) the tight set pairs spanning the basis are highly non-laminar, and 2) the largest fraction is bounded above by $\frac{1}{\max r_{ij}}$.

Specifically, we construct a family of vertex connectivity instances with $r_{ij} = \min\{r_i, r_j\}$, and $r_i \in \{1,k\}$ for all $i \in V$. This family has the property that after solving the initial LP and fixing all edges e with $x_e = 1$, the residual LP has a basic solution with largest x-value equal to $\frac{1}{k}$.

We depict the family of instances in Figure 1: For each k construct a graph on $2k$ vertices. The first k vertices $V = \{v_1, \ldots, v_k\}$ have demand k, the second k vertices $U = \{u_1, \ldots, u_k\}$ have demand 1. The edge set consists of a clique of 0-cost edges on V, and a complete bipartite graph between V and U of cost 1 edges. The optimal LP solution will choose every edge in the clique at value 1 and every edge in the bipartite graph at value $1/k$. After fixing all edges with $x_e = 1$, the remaining optimal LP solution will still have every edge in the bipartite graph at value $1/k$. It is not hard to establish that this is an optimal solution. For instance, consider the solution to the dual linear program that sets $y_{S,A} = 1$ for $S = \{u_i\}$ for $1 \leq i \leq k$ and $A = \emptyset$, and 0 otherwise. This is feasible, and has value equal to the feasible primal solution. Hence both are optimal.

We now establish that this is a vertex of the polytope described by ($\mathbf{MCVC_2}$) with all cost 0 edges included in $E_{\frac{1}{2}+}$. We do this by describing a set of k^2 tight inequalities (note that k^2 is the number of fractional edges and hence variables in the remaining problem), constructing a matrix of the support of these inequalities, constructing a second matrix and arguing that the two matrices are inverses of each other, hence each are linearly independent. Since the solution is then the intersection of k^2 linearly independent halfspaces in $\mathbf{R}^{k^2}$, it is a vertex of the polytope.

The set of k^2 tight inequalities is divided into k blocks of k inequalities. Block 0 includes the k inequalities with $S = \{u_i\}$, $1 \leq i \leq k$, and $A = \emptyset$. Aside from these inequalities, the point is highly degenerate. The remaining $k - 1$ blocks of inequalities are described as follows. In block $i \in \{1, \ldots, k-1\}$, we have $v_{i+1} \in S$, $A^i = \{v_j | j \neq i, i+1\}$, and $v_i \in V \backslash (S \cup A^i)$. Thus there is exactly one cost-0 edge crossing this cut (edge (v_i, v_{i+1})), and the $(g_2 - \delta_F)$-value of the inequality is 1. Denote the set S for inequality $q \in \{1, \ldots, k\}$ in this block by S_q^i. We set $S_q^i := \{v_{i+1}, u_1, u_2, \ldots, u_{k-q+1}\}$. See Figure 1. Then $x(S_q^i, A^i)$ is

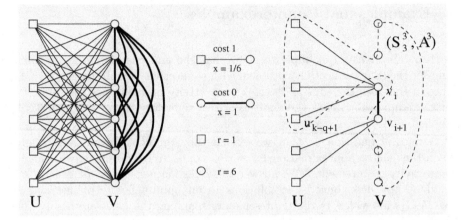

Fig. 1. On the left, a basic solution to (**MCVC₂**) after fixing the cost 0 edges to 1. The largest fraction in the solution is $\frac{1}{k}$, here $k = 6$. On the right, an example of a set (S_q^i, A^i) for $i = 3$, $q = 3$ with $f_2(S_q^i, A^i) = k$. All the edges crossing the cut are included in the figure. They have total value 2. Together with the 4 vertices in the cutset, this satisfies (8) at equality. The collection of cuts $\{(S_q^i, A^i)\}_{1 \le i \le k-1, \; 1 \le q \le k}$ are highly crossing.

determined by the $q - 1$ edges from U to v_{i+1} and the $k - q + 1$ edges from U to v_i, for a total value of $(q - 1 + k - q + 1)\frac{1}{k} = 1$. Thus, these cuts are tight.

Let matrix C be the support matrix of edges in each cutset above, with the rows of C corresponding to cutsets and the columns corresponding to edges. The rows of C are ordered first according to block, and then within each block, according to q. The first row block in C corresponds to the inequalities with $S = \{u_i\}$, i.e. it is the block 0 of the tight inequalities. The columns of C are ordered according to incidence to U, and then to V. All the edges incident to u_i are in the ith block. Within a block, the jth edge is the edge incident to v_j. See Figure 2.

Let matrix B be a $k^2 \times k^2$ matrix with its columns and rows ordered into blocks of k. The first block of columns (called column block 0) has a pattern that is slightly different from the rest. See Figure 3. The first column is 0 everywhere except in the last entry in the first row block, which is 1. The second thru k^{th} columns have the following pattern: the first row block consists of $k - 1$ entries of value $\frac{-1}{k}$ followed by a single entry of $\frac{k-1}{k}$. Then the qth column has the qth row block filled with $\frac{1}{k}$. All other entries are 0.

For the pattern of the ith column block, $i = 1, \ldots, k - 1$, see Figure 4. The first column of the first row block has $i - 1$ entries of $\frac{-k+i}{k}$ followed by $k - i$ entries of $\frac{i}{k}$ followed by a single entry of $\frac{-k+i}{k}$. This column vector is denoted X_i. The first column of the last row block is the vector Z_i containing i entries of $\frac{k-i}{k}$ followed by $k - i$ entries of $\frac{-i}{k}$. The qth column of the $k - q^{th}$ row block

	u_1	u_2	$\cdots$	u_{k-1}	u_k
block 0	1 1 $\cdots$ 1 1				
		1 1 $\cdots$ 1			
			$\ddots$		
				1 1 $\cdots$ 1	
					1 1 $\cdots$ 1
block 1	1 0 $\cdots$ 0 0	1 0 $\cdots$ 0	$\cdots$	1 0 0	1 0 $\cdots$ 0
	1 0 $\cdots$ 0 0	1 0 $\cdots$ 0	$\cdots$	1 0 $\cdots$ 0	0 1 $\cdots$ 0
	1 0 $\cdots$ 0 0	1 0 $\cdots$ 0	$\cdots$	0 1 $\cdots$ 0	0 1 $\cdots$ 0
	$\vdots$	$\vdots$	$\vdots$		
	1 0 $\cdots$ 0 0	0 1 $\cdots$ 0	$\cdots$	0 1 $\cdots$ 0	0 1 $\cdots$ 0
block 2	0 1 $\cdots$ 0 0	0 1 $\cdots$ 0	$\cdots$	0 1 $\cdots$ 0	0 1 $\cdots$ 0
	$\vdots$	$\vdots$	$\ddots$		
	$\vdots$	$\vdots$	$\ddots$		
block k-1	$\vdots$	$\vdots$	$\vdots$		
	0 0 $\cdots$ 1 0	0 0 $\cdots$ 1	$\cdots$	0 0 $\cdots$ 1	0 0 $\cdots$ 1

Fig. 2. Incidence Matrix C of k^2 Tight Set Pairs

column	1	2	3	$\cdots$	k
row	0	$\frac{-1}{k}$	$\frac{-1}{k}$	$\cdots$	$\frac{-1}{k}$
block 1	$\vdots$	$\vdots$	$\vdots$	$\ddots$	$\vdots$
	0	$\frac{-1}{k}$	$\frac{-1}{k}$	$\cdots$	$\frac{-1}{k}$
	1	$\frac{k-1}{k}$	$\frac{k-1}{k}$	$\cdots$	$\frac{k-1}{k}$
	0	$\frac{1}{k}$	0	$\cdots$	0
block 2	$\vdots$	$\vdots$	$\vdots$	$\ddots$	$\vdots$
	0	$\frac{1}{k}$	0	$\cdots$	0
	0	0	$\frac{1}{k}$	$\cdots$	0
block 3	$\vdots$	$\vdots$	$\vdots$	$\ddots$	$\vdots$
	0	0	$\frac{1}{k}$	$\cdots$	0
	$\vdots$	$\vdots$	$\vdots$	$\cdots$	$\vdots$
	0	0	0	$\cdots$	$\frac{1}{k}$
block k	$\vdots$	$\vdots$	$\vdots$	$\ddots$	$\vdots$
	0	0	0	$\cdots$	$\frac{1}{k}$

Fig. 3. Column Block 0 of Matrix B

	X_i	Y_i	Z_i
row 1	$\frac{-k+i}{k}$	$\frac{-k+i}{k}$	$\frac{k-i}{k}$
$\vdots$	$\vdots$	$\vdots$	$\vdots$
row i-1	$\frac{-k+i}{k}$	$\frac{-k+i}{k}$	$\frac{k-i}{k}$
row i	$\frac{i}{k}$	$\frac{-k+i}{k}$	$\frac{k-i}{k}$
row i+1	$\frac{i}{k}$	$\frac{i}{k}$	$\frac{-i}{k}$
$\vdots$	$\vdots$	$\vdots$	$\vdots$
row k-1	$\frac{i}{k}$	$\frac{i}{k}$	$\frac{-i}{k}$
row k	$\frac{-k+i}{k}$	$\frac{i}{k}$	$\frac{-i}{k}$

Col. Block i of B

$$
\begin{array}{ccccc}
X_i & 0 & \cdots & 0 & Z_i \\
0 & 0 & \cdots & Z_i & Y_i \\
0 & 0 & \cdots & Y_i & 0 \\
\vdots & \vdots & \ddots & \vdots & \vdots \\
0 & Z_i & \cdots & 0 & 0 \\
Z_i & Y_i & \cdots & 0 & 0
\end{array}
$$

Fig. 4. On the left, the pattern of column block i of matrix B. On the right, the composition of the vectors X_i, Y_i and Z_i that describe block i.

is Z_i for $1 \leq q \leq k$, and the qth column of the $k - q + 1^{st}$ row block is the vector Y_i with i entries of $\frac{-k+i}{k}$ followed by $k - i$ entries of $\frac{i}{k}$.

The following lemma follows by inspection of B and C.

Lemma 8. *For any k, matrices B and C are inverses.*

5   Algorithmic Details

To solve the LP in polynomial time, we need a separation algorithm for the connectivity constraints. We interpret x-values as capacities and transform the graph induced by the current fractional solution and the fixed edges into a directed graph by replacing every edge by oppositely oriented edges with the same capacity as the original undirected edge. We then perform a standard procedure of splitting vertices to model the fact that at most one path can pass through any vertex. Then, in the resulting graph, the maximum flow value between i and j is vertex connectivity between i and j. If this is less than r_{ij}, the minimum cut reveals a violated inequality.

Thus we have a polynomial time separation algorithm for (**MCVC$_2$**). Using ellipsoid algorithm, we can solve the LP in polynomial time. Once we have a solution, it may not be a basic solution. However, it can be transformed to a vertex solution in polynomial time, as described in [7,10].

Acknowledgment. I thank Joseph Cheriyan for his suggestion and encouragement to work on minimum cost vertex connectivity problems and for interesting conversations. I thank Kirsten Wickelgren for her enthusiasm in learning about combinatorial optimization, which led me to revisit this problem. I am also grateful to the Fields Institute, Toronto, Ontario and Bell Labs, Lucent Technologies, Murray Hill, New Jersey for hosting me during some parts of this work. Additional support was provided by NSF through grants EIA-9973858 and CCR-9985458.

References

1. A. Bouchet. Greedy algorithm and symmetric matroids. *Math. Programming*, 38:147–159, 1987.
2. R. Chandrasekaran and S. N. Kabadi. Pseudomatroids. *Discrete Math.*, 71:205–217, 1988.
3. J. Cheriyan, T. Jordán, and Z. Nutov. Approximating k-outconnected subgraph problems. In *Approximation algorithms for combinatorial optimization (Aarlborg)*, number 1444 in Lecture Notes in Comput. Sci., pages 77–88. Springer, Berlin, 1998.
4. R. Diestel. *Graph Theory*. Number 173 in Graduate Texts in Mathematics. Springer-Verlag, New York, 2nd edition, 2000.
5. M. X. Goemans, A. V. Goldberg, S. Plotkin, D. Shmoys, É. Tardos, and D. P. Williamson. Improved approximation algorithms for network design problems. In *Proc. 5th Annual ACM-SIAM Symp. on Discrete Algorithms*, pages 223–232, 1994.
6. M. X. Goemans and D. P. Williamson. The primal-dual method for approximation algorithms and its application to network design problems. In Hochbaum [9], pages 144–191.
7. M. Grötschel, L. Lovász, and A. Schrijver. *Geometric Algorithms and Combinatorial Optimization*. Springer-Verlag, 1988.
8. M. Grötschel, C. L. Monma, and M. Stoer. Computational results with a cutting plane algorithm for designing communication networks with low-connectivity constraints. *Operations Research*, 40(2):309–330, March-April 1992.
9. D. S. Hochbaum, editor. *Approximation Algorithms for NP-Hard Problems*. PWS Publishing Company, Boston, 1997.
10. K. Jain. A factor 2 approximation algorithm for the generalized Steiner network problem. In *39th Annual IEEE Symposium on Foundations of Computer Science*, 1998.
11. S. Khuller. Approximation algorithms for finding highly connected subgraphs. In Hochbaum [9], pages 236–265.
12. S. Khuller and B. Raghavachari. Improved approximation algorithms for uniform connectivity problems. *J. Algorithms*, 1996.
13. G. Kortsarz and Z. Nutov. Approximating node connectivity problems via set covers. In *Approximation Algorithms for Combinatorial Optimization (Proc. of APPROX 2000)*, number 1913 in Lecture Notes in Comp. Sci., pages 194–205. Springer-Verlag, 2000.
14. V. Melkonian and E. Tardos. Approximation algorithms for a directed network design problem. In *7th International Integer Programming and Combinatorial Optimization Conference*, pages 345–360, 1999.
15. C. L. Monma and D. F. Shallcross. Methods for designing communications networks with certain two-connected survivability constraints. *Operations Research*, 1989.
16. R. Ravi and D. P. Williamson. An approximation algorithm for minimum-cost vertex-connectivity problems. *Algorithmica*, 18(1):21–43, 1997.
17. R. Ravi and D. P. Williamson, November 2000. Personal communication.

Combined Connectivity Augmentation and Orientation Problems

András Frank[1,2]* and Tamás Király[1]**

[1] Department of Operations Research, Eötvös University, Kecskeméti u. 10–12,
Budapest, Hungary, H-1053
[2] Traffic Lab Ericsson Hungary, Laborc u. 1, Budapest, Hungary, H-1037

Abstract. Two important branches of graph connectivity problems are connectivity augmentation, which consists of augmenting a graph by adding new edges so as to meet a specified target connectivity, and connectivity orientation, where the goal is to find an orientation of an undirected or mixed graph that satisfies some specified edge-connection property. In the present work an attempt is made to link the above two branches, by considering degree-specified and minimum cardinality augmentation of graphs so that the resulting graph has an orientation satisfying a prescribed edge-connection requirement, such as (k, l)-edge-connectivity. Our proof technique involves a combination of the super-modular polyhedral methods used in connectivity orientation, and the splitting off operation, which is a standard tool in solving augmentation problems.

1 Introduction

In a connectivity augmentation problem the goal is to augment a graph or digraph by adding a cardinality- or degree-constrained new graph so as to meet a specified target connectivity. Initial deep results of the area are due to Lovász [6] and to Watanabe and Nakamura [10] on augmenting a graph to make it k-edge-connected. Since then, augmentation results for many different connectivity properties of graphs and digraphs have been proved, employing various versions of the splitting off technique, which was originally introduced by Lovász [6] and subsequently developed by Mader [7] and others.

In a connectivity orientation problem one is interested in the existence of an orientation of an undirected graph that satisfies some specified edge-connection properties. For example, classical results of Nash-Williams [8] and of Tutte [9] characterize graphs having k-edge-connected and rooted k-edge-connected orientations. For a common generalization of their results, call a digraph $D = (V, A)$

---

* Supported by the Hungarian National Foundation for Scientific Research, OTKA T029772. Part of research was done while this author was visiting the Institute for Discrete Mathematics, University of Bonn, July 2000.
** Supported by the Hungarian National Foundation for Scientific Research, OTKA T029772.

K. Aardal, B. Gerards (Eds.): IPCO 2001, LNCS 2081, pp. 130–144, 2001.

(k, l)-*edge-connected* for non-negative integers $k \geq l$ if there is a node $s \in V$ such that there are k edge-disjoint paths from s to any other node, and there are l edge-disjoint paths to s from any other node. Then (k, k)-edge-connectivity is equivalent to k-edge-connectivity, and $(k, 0)$-edge-connectivity is equivalent to rooted k-edge-connectivity from some node s. Good characterizations of undirected and mixed graphs having a (k, l)-edge-connected orientation were given in [1] and [3], with the help of submodular flows and related polyhedral methods.

In this paper an attempt is made to link these two branches of connectivity problems by studying combined augmentation and orientation problems. For example we characterize undirected and mixed graphs that can be augmented by an appropriate degree-specified undirected graph so as to have a (k, l)-edge-connected orientation. Another new result concerns the minimum number of new edges whose addition to an initial undirected graph results in a graph admitting a (k, l)-edge-connected orientation. Our proof methods for these characterizations combine the splitting off technique used in connectivity augmentation with extensions of the supermodular polyhedral techniques used in [3] to solve connectivity orientation problems. Since these methods are constructive from an algorithmic point of view, the proofs presented here give rise to polynomial algorithms for finding a feasible augmentation.

The results are presented in the customary framework for connectivity orientations. We consider graphs with no loops, but possibly with multiple edges. Given a graph $G = (V, E)$ and a set function $h : 2^V \rightarrow \mathbb{Z}$, an orientation $\vec{G}$ of G is said to *cover* h if $\varrho_{\vec{G}}(X) \geq h(X)$ for every set $X \subseteq V$, where $\varrho_{\vec{G}}(X)$ denotes the number of edges of the digraph $\vec{G}$ entering the set X. Throughout the paper we assume that $h(\emptyset) = h(V) = 0$. The h-orientation problem is to find an orientation of G that covers h. For general h this includes **NP**-complete problems, so special classes of set functions must be considered. A set function h is called *crossing G-supermodular* with respect to a given graph $G = (V, E)$ if

$$h(X) + h(Y) \leq h(X \cap Y) + h(X \cup Y) + d_G(X, Y) \qquad (1)$$

for every crossing pair (X, Y) (where the sets $X, Y \subseteq V$ are *crossing* if none of $X - Y$, $Y - X$, $X \cap Y$ and $V - (X \cup Y)$ are empty), and $d_G(X, Y)$ is the number of edges in E connecting $X - Y$ and $Y - X$. As in [3], we restrict our attention to crossing G-supermodular set functions. The *augmentation problem* corresponding to h-orientation is the following: given a graph G, find a graph G' (either with specified degrees, or with minimum number of edges), so that $G + G'$ has an orientation covering h.

It was shown in [1] that for a graph G and a non-negative crossing G-supermodular set function h the h-orientation problem can be solved in polynomial time. In Sect. 3 we solve the corresponding degree-specified and minimum cardinality augmentation problem, as well as minimum cost augmentation for node-induced cost functions.

These results are used in Sect. 4 to augment a graph to obtain one admitting a (k, l)-edge-connected orientation, and we show that in this special case the characterizations can be further simplified. The theorems obtained can also be interpreted independently of orientations. A graph G is called (k, l)-*tree-connected* if any graph obtained by deleting l edges from G contains k edge-disjoint spanning trees. It is known that if $k \geq l$, then (k, l)-tree-connected graphs are exactly those that have a (k, l)-edge-connected orientation; thus we can solve the (k, l)-tree-connectivity augmentation problem.

In [3], submodular flows were used to solve the h-orientation problem when h is a crossing G–supermodular set function that can have negative values; this implies for example that we can find a (k, l)-edge-connected orientation of a mixed graph M. In Sect. 5 we generalize this result by considering the h-orientation problem for positively crossing G–supermodular functions, and by solving the corresponding degree-specified augmentation problem. The proof exploits the TDI-ness of a system closely related to the intersection of two base polyhedra.

2   Preliminaries

A *family of sets* is a collection of subsets of the ground set V, with possible repetition. If every member of a family $\mathcal{F}$ is replaced by its complement, the resulting family is denoted by $\overline{\mathcal{F}}$. For an element $v \in V$, $d_{\mathcal{F}}(v)$ denotes the number of members of $\mathcal{F}$ containing v. A *composition* of a set $X \subseteq V$ is a family $\mathcal{F}$ for which $d_{\mathcal{F}} - d_{\{X\}}$ is constant. A composition of V is called a *regular family*. The *covering number* of a family $\mathcal{F}$ is $\min_{v \in V} d_{\mathcal{F}}(v)$.

For a function $x : V \to \mathbb{R}$ and a set $Z \subseteq V$, and analogously for a set function $p : 2^V \to \mathbb{Z} \cup \{-\infty\}$ and a family $\mathcal{F}$, we use the notations $x(Z) = \sum_{v \in Z} x(v)$ and $p(\mathcal{F}) = \sum_{X \in \mathcal{F}} p(X)$. The *upper truncation* of p is

$$p^{\wedge}(Z) := \max \{p(\mathcal{F}) \mid \mathcal{F} \text{ is a partition of } Z\} . \tag{2}$$

If p is intersecting supermodular, then $p^{\wedge}$ is fully supermodular. If p is crossing supermodular, then so is $p^{\wedge}$. To the set function p we associate the polyhedra

$$C(p) := \{x : V \to \mathbb{R} \mid x(Z) \geq p(Z) \; \forall Z \subseteq V\} , \tag{3}$$

$$B(p) := \{x : V \to \mathbb{R} \mid x(V) = p(V); \; x(Z) \geq p(Z) \; \forall Z \subseteq V\} . \tag{4}$$

Clearly, $C(p) = C(p^{\wedge})$. A polyhedron is a *contra-polymatroid* if it equals $C(p)$ for some monotone increasing fully supermodular function p; it is a *base polyhedron* if it can be represented as $B(p)$ for some fully supermodular function p. The following theorem of Fujishige [5] deals with base polyhedra given by crossing supermodular set functions.

Theorem 1 (Fujishige [5]). *Let* $p : 2^V \to \mathbb{Z} \cup \{-\infty\}$ *be a crossing supermodular function. Then* $B(p)$ *is non-empty if and only if*

$$\sum_{i=1}^{t} p(X_i) \leq p(V) ,$$

$$\sum_{i=1}^{t} p(\overline{X_i}) \leq (t-1)p(V)$$

both hold for every partition $\{X_1, \ldots, X_t\}$. *Furthermore, if* $B(p)$ *is non-empty, then it is a base polyhedron.*   □

Let $G = (V, E)$ be a graph. For a set $X \subseteq V$, $i_G(X)$ denotes the number of edges $uv \in E$ with $u, v \in X$. An important property of i_G is that if a set function h is crossing G–supermodular, then $h + i_G$ is crossing supermodular. For a family $\mathcal{F}$ of sets we define

$$e_G(\mathcal{F}) := \max \left\{ \varrho_{\vec{G}}(\mathcal{F}) \mid \vec{G} \text{ is an orientation of } G \right\} .$$

Note that $e_G(\mathcal{F})$ can be easily computed since we can orient the edges independently. For partitions it equals the number of cross-edges; more generally, if $\mathcal{F}$ is a regular family with covering number α, then

$$e_G(\mathcal{F}) = \alpha|E| - \sum_{X \in \mathcal{F}} i_G(X) . \tag{5}$$

A family $\mathcal{F}$ is *cross-free* if it has no crossing members. Simple examples are partitions and co-partitions; in fact, it is easy to show that these are the only minimal regular cross-free families:

Proposition 1. *Every regular cross-free family decomposes into partitions and co-partitions.*   □

3 Non-negative Crossing G–Supermodular Set Functions

The first result is a theorem on the degree-specified augmentation problem. The characterizations given are good in the sense that they provide an easily verifiable certificate if the augmentation is impossible. Moreover, the proof is constructive and gives rise to a polynomial algorithm, since it refers to polyhedral and splitting off problems that can be solved in polynomial time.

Theorem 2. *Let* $G = (V, E)$ *be a graph,* $h : 2^V \to \mathbb{Z}_+$ *a non-negative crossing G–supermodular set function on* V, *and* $m : V \to \mathbb{Z}_+$ *a degree specification with* $m(V)$ *even. There exists an undirected graph* $G' = (V, E')$ *such that* $G + G'$ *has an orientation covering* h *and* $d_{G'}(v) = m(v)$ *for all* $v \in V$, *if and only if the following hold for every partition* $\mathcal{F}$:

$$\frac{m(V)}{2} \geq h(\mathcal{F}) - e_G(\mathcal{F}) , \tag{6}$$

$$\min_{X \in \mathcal{F}} m(\overline{X}) \geq h(\mathcal{F}) - e_G(\mathcal{F}) , \tag{7}$$

$$\frac{m(V)}{2} \geq h(\overline{\mathcal{F}}) - e_G(\overline{\mathcal{F}}) , \tag{8}$$

$$\min_{X \in \mathcal{F}} m(\overline{X}) \geq h(\overline{\mathcal{F}}) - e_G(\overline{\mathcal{F}}) . \tag{9}$$

Proof. To see the necessity of these conditions, observe that $m(V)/2$ is the number of new edges, while $h(\mathcal{F}) - e_G(\mathcal{F})$ measures the deficiency of a partition $\mathcal{F}$, hence (6) simply requires that the deficiency of a partition should not exceed the number of new edges. The necessity of (7) is also straightforward since each new cross-edge must have an endnode in $\overline{X}$, so the number of new cross-edges, which should be at least the deficiency of $\mathcal{F}$, is at most $m(\overline{X})$. The necessity of (8) and (9) can be seen analogously.

To prove sufficiency, add a new node z to the set of nodes, and for every $v \in V$ add $m(v)$ parallel edges between v and z; the resulting graph is denoted by $G_0 = (V_0, E_0)$. Define the following extension of the set function h:

$$h_0(z) = h_0(V) := \frac{m(V)}{2} ,$$
$$h_0(X + z) = h_0(X) := h(X) \qquad \text{if } \emptyset \neq X \subset V .$$

The proof consists of finding an orientation of G_0 that covers h_0, and then splitting off the directed edges at z so that the resulting digraph on the ground set V covers h. To find an orientation covering h_0, we resort to a lemma that is a standard tool for orientation problems:

Lemma 1. *For a given vector* $x : V_0 \to \mathbb{Z}_+$*, there is an orientation* $\vec{G}_0$ *of* G_0 *such that* $\varrho_{\vec{G}_0}(v) = x(v)$ *for every* $v \in V_0$*, if and only if* $x(V_0) = |E_0|$ *and* $x(Z) \geq i_{G_0}(Z)$ *for every* $Z \subseteq V_0$.

Proof. The necessity is obvious. We prove the sufficiency by induction on the number of edges. Call a set Z *tight* if $x(Z) = i_{G_0}(Z)$. Let $uv \in E_0$ be an arbitrary edge. If there are no tight $\overline{u}v$-sets and $x(v) > 0$, then we can remove the edge uv, decrease $x(v)$ by one, find a feasible orientation of the resulting graph by induction, and add the directed edge uv. If $x(v) = 0$, then $x(u) > 0$ and there is no tight $\overline{v}u$-set X for otherwise $X + v$ would violate the condition. So we can assume that $x(u), x(v) > 0$, there is a tight $v\overline{u}$-set X, and similarly that there is a tight $u\overline{v}$-set Y. But then $d_{G_0}(X, Y) > 0$, thus $i_{G_0}(X) + i_{G_0}(Y) \leq i_{G_0}(X \cap Y) + i_{G_0}(X \cup Y) - d_{G_0}(X, Y)$ implies that either $x(X \cap Y) < i_{G_0}(X \cap Y)$ or $x(X \cup Y) < i_{G_0}(X \cup Y)$, contradicting the conditions.   □

Lemma 1 and the non-negativity of h imply that if we can find a vector $x : V_0 \to \mathbb{Z}_+$ that satisfies $x(V_0) = |E_0|$ and $x(X) \geq h_0(X) + i_{G_0}(X)$ for every $X \subseteq V_0$, then there is an orientation $\vec{G}_0$ of G_0 such that $\varrho_{\vec{G}_0}(v) = x(v)$ for every $v \in V_0$, and $\vec{G}_0$ covers h_0 since $\varrho_{\vec{G}_0}(X) = x(X) - i_{G_0}(X) \geq h_0(X)$. Such a vector x is called *feasible*. By the definition of h_0, $x(z)$ must be equal to $m(V)/2$; let $x' : V \to \mathbb{Z}_+$ denote the projection of x to V. Let

$$p_m(X) := h(X) + i_G(X) + \max\left\{0, m(X) - \frac{m(V)}{2}\right\} \qquad (X \subseteq V) .$$

Then it easily follows from the definition of h_0 that the vector x is feasible if and only if x' is an element of the polyhedron $B(p_m)$ as defined in (4).

Claim. The set function p_m is crossing supermodular.

Proof. The G–supermodularity of h implies that $h+i_G$ is crossing supermodular. Let $m^*(X) := \max\{0, m(X) - m(V)/2\}$; we show that this set function is fully supermodular. Indeed, if $m^*(Y) = 0$, then $m^*(X) + m^*(Y) = m^*(X) \leq m^*(X \cup Y) = m^*(X \cap Y) + m^*(X \cup Y)$. If $m^*(X), m^*(Y) > 0$, then $m^*(X) + m^*(Y) = m(X \cap Y) + m(X \cup Y) - m(V) \leq m^*(X \cap Y) + m^*(X \cup Y)$. The sum of a crossing supermodular and a fully supermodular function is crossing supermodular.   □

Claim. Suppose that (6)–(9) are true. Then $B(p_m)$ is non-empty.

Proof. By Theorem 1 it suffices to show that $p_m(\mathcal{F}) \leq |E| + m(V)/2$ and $p_m(\overline{\mathcal{F}}) \leq (t-1)(|E| + m(V)/2)$ for every partition $\mathcal{F}$ with t members. Observe that a partition has at most one member X with $m(X) > m(V)/2$. If there is no such member then (6) and the identity (5) imply that $p_m(\mathcal{F}) \leq |E| + m(V)/2$; if there is one such member, then (7) and (5) imply the same. Similarly, a co-partition has at most one member X with $m(X) < m(V)/2$, so (8) or (9) (depending on the existence of such a member) and (5) for the co-partition $\overline{\mathcal{F}}$ imply $p_m(\overline{\mathcal{F}}) \leq (t-1)(|E| + m(V)/2)$.   □

By Theorem 1, $B(p_m)$ is a base polyhedron, therefore it has an integral point x'; as we have seen, this and Lemma 1 implies that G_0 has an orientation $\vec{G}_0 = (V_0, \vec{E}_0)$ covering h_0.

Let $m_i(v)$ be the multiplicity of the edge zv in $\vec{G}_0$, and $m_o(v)$ the multiplicity of the edge vz in $\vec{G}_0$; let $\vec{G}$ denote the edges of $\vec{G}_0$ not incident with z. Then $m_i(X) \geq h(X) - \varrho_{\vec{G}}(X)$ and $m_o(V - X) \geq h(X) - \varrho_{\vec{G}}(X)$ for every $X \subseteq V$, since $\vec{G}_0$ covers h_0. By the crossing G–supermodularity of h, the set function $p(X) := h(X) - \varrho_{\vec{G}}(X)$ is crossing supermodular. Thus we can use the following result in [2], which generalizes Mader's directed splitting theorem:

Lemma 2 ([2]). *Let p be a positively crossing supermodular set function on V. Let m_i, m_o be non-negative integer-valued functions on V for which $m_i(V) = m_o(V)$. There exists a digraph $D = (V, A)$ such that $\varrho_D(v) = m_i(v)$, $\rho_D(V - v) = m_o(v)$ for every $v \in V$, and $\varrho_D(X) \geq p(X)$ for every $X \subseteq V$, if and only if*

$$m_i(X) \geq p(X) \text{ for every } X \subseteq V$$

and

$$m_o(V - X) \geq p(X) \text{ for every } X \subseteq V .$$

    □

To complete the proof of Theorem 2, observe that if G' is the underlying undirected graph of the digraph D given by Lemma 2, then G' satisfies the degree specification, and $G + G'$ has a feasible orientation, namely $\vec{G} + D$.   □

This theorem can be used to derive the following min-max theorem for minimum cardinality augmentation:

Theorem 3. *Let $G = (V, E)$ be a graph, and $h : 2^V \to \mathbb{Z}_+$ a non-negative crossing G-supermodular set function. There is an undirected graph $G' = (V, E')$ with γ edges such that $G + G'$ has an orientation covering h if and only if*

$$\gamma \geq h(\mathcal{F}) - e_G(\mathcal{F}) \tag{10}$$

holds for every partition and co-partition $\mathcal{F}$, and

$$2\gamma \geq h(\mathcal{F}) - e_G(\mathcal{F}) \tag{11}$$

holds for every cross-free regular family $\mathcal{F}$ that decomposes into a partition of some $X \subseteq V$ and a co-partition of $\overline{X}$.

Proof. Again, $h(\mathcal{F}) - e_G(\mathcal{F})$ measures the deficiency of the family $\mathcal{F}$, so the necessity follows easily by observing that an oriented new edge can cover at most one member of a (sub)partition or a (sub)-copartition.

Sufficiency can be proved by showing that if (10) and (11) hold, then there exists a vector $m : V \to \mathbb{Z}_+$ with $m(V) = 2\gamma$ satisfying (6)–(9); thus by Theorem 2 we can find a feasible augmentation with degree-specification m. The essential result in the proof is that the polyhedron

$$C := \{m : V \to \mathbb{Z}_+ \mid m \text{ satisfies (6)–(9)}\}$$

is a contra-polymatroid. Define the set functions

$$p_1(X) := h(X) + i_G(X) \,,$$

$$p_2(X) := h(\overline{X}) + i_G(\overline{X}) - |E| \,.$$

By the crossing G-supermodularity of h, the set functions p_1 and p_2 are crossing supermodular, therefore the set functions $p_1^\wedge$ and $p_2^\wedge$ (as defined in (2)) are also crossing supermodular. By the identity (5), a non-negative vector m satisfies (6)–(9) if and only if the following hold:

$$m(V) \geq 2 \left(\max_{X \subset V} p_1^\wedge(X) + p_2(X) \right) \,, \tag{12}$$

$$m(X) \geq p_1^\wedge(X) + p_2(X) \text{ for every } X \subset V \,, \tag{13}$$

$$m(V) \geq 2 \left(\max_{X \subset V} p_1(X) + p_2^\wedge(X) \right) \,, \tag{14}$$

$$m(X) \geq p_1(X) + p_2^\wedge(X) \text{ for every } X \subset V \,. \tag{15}$$

For a set $X \subset V$, define

$$p(X) := \max \{p_1^\wedge(X) + p_2(X), \ p_1(X) + p_2^\wedge(X), \ 0\} \,, \tag{16}$$

and let

$$p(V) := 2 \max_{X \subset V} p(X) \,. \tag{17}$$

Then the polyhedron C can be characterized as

$$C = \{m : V \to \mathbb{Z} \mid m(X) \geq p(X) \; \forall X \subseteq V\} \; .$$

To prove that C is a contra-polymatroid, we will show that the set function $p^\wedge$ is fully supermodular. First we establish some other properties of $p^\wedge$:

Proposition 2. *For every proper subset X of V, the value of $p^\wedge(X)$ is given by*

$$p^\wedge(X) = \max_{Y \subseteq X} \left(p_1^\wedge(Y) + p_2^\wedge(Y) \right) \; .$$

Proof. By definition $p^\wedge$ is less or equal to the maximum on the right side. For the other direction, suppose indirectly that there exists an $Y \subseteq X$ and partitions $\mathcal{F}_1$ and $\mathcal{F}_2$ of Y such that

$$p^\wedge(X) < p_1(\mathcal{F}_1) + p_2(\mathcal{F}_2) \; .$$

Repeat the following step as many times as possible:

- If $X \in \mathcal{F}_1$ and $Y \in \mathcal{F}_2$ are crossing, then replace X in $\mathcal{F}_1$ by $X - Y$, and replace Y in $\mathcal{F}_2$ by $Y - X$.

Observe that the resulting families are partitions of some proper subset of Y, so the procedure terminates after a finite number of steps. Furthermore, X and $\overline{Y}$ are crossing, so $h(X) + h(\overline{Y}) \leq h(X \cap \overline{Y}) + h(X \cup \overline{Y}) + d_G(X, \overline{Y})$; this implies that $p_1(X) + p_2(Y) \leq p_1(X - Y) + p_2(Y - X)$. Let $\mathcal{F}_1'$ and $\mathcal{F}_2'$ denote the families obtained at the end of the procedure; then $\mathcal{F}_1'$ and $\mathcal{F}_2'$ are partitions of some $Y' \subseteq Y$, and $p^\wedge(X) < p_1(\mathcal{F}_1') + p_2(\mathcal{F}_2')$. Moreover, $\mathcal{F}_1' + \mathcal{F}_2'$ is cross-free, which means that there is a partition $Y_1', \ldots, Y_t'$ of Y', such that for every i either $\mathcal{F}_1'$ contains Y_i' and $\mathcal{F}_2'$ contains a partition of Y_i', or vice versa. But then $p_1(\mathcal{F}_1') + p_2(\mathcal{F}_2') \leq p^\wedge(Y') \leq p^\wedge(X)$, a contradiction.   □

Proposition 3. *The set function p satisfies*

$$p(X) + p(Y) \leq p^\wedge(X \cap Y) + p^\wedge(X \cup Y) \tag{18}$$

for every pair (X, Y).

Proof. The inequality is obvious if one of $p(X)$ and $p(Y)$ is zero, or X and Y are not intersecting. If $X \cup Y = V$, then $p(X) + p(Y) \leq 2\max\{p(X), p(Y)\} \leq p(V) = p(X \cup Y) \leq p^\wedge(X \cap Y) + p^\wedge(X \cup Y)$.

By Proposition 2 it suffices to prove that if $p(X), p(Y) > 0$ and X and Y are crossing, then

$$p(X) + p(Y) \leq (p_1^\wedge + p_2^\wedge)(X \cap Y) + (p_1^\wedge + p_2^\wedge)(X \cup Y) \; .$$

Using the definition of p and the supermodularity of $p_1^\wedge$ and $p_2^\wedge$,

$$p(X) + p(Y) \leq p_1^\wedge(X) + p_2^\wedge(X) + p_1^\wedge(Y) + p_2^\wedge(Y)$$
$$\leq p_1^\wedge(X \cap Y) + p_2^\wedge(X \cap Y) + p_1^\wedge(X \cup Y) + p_2^\wedge(X \cup Y) \; .$$

□

This property is sufficient for the supermodularity of $p^\wedge$, as the following lemma states:

Lemma 3. *If a set function p (with $p(\emptyset) = 0$) satisfies (18) for every pair (X, Y), then $p^\wedge$ is fully supermodular.*

Proof. For a set $X \subseteq V$, let $\mathcal{F}_X$ denote a partition of X for which $p^\wedge(X) = p(\mathcal{F}_X)$. Let $X, Y \subseteq V$ be an arbitrary pair. Starting from the family $\mathcal{F} = \mathcal{F}_X + \mathcal{F}_Y$, repeat the following operation as many times as possible:

- If there is an intersecting pair X' and Y' in the family, remove both of them, and add the sets of $\mathcal{F}_{X' \cap Y'}$ and of $\mathcal{F}_{X' \cup Y'}$.

The operation doesn't change $d_\mathcal{F}$, and doesn't decrease $p(\mathcal{F})$, since p has the property (18). Since the operation either increases the cardinality of the family, or increases $\sum_{X \in \mathcal{F}} |X|^2$ without changing the cardinality, after a finite number of steps we get a laminar family $\mathcal{F}'$ for which $p(\mathcal{F}') \geq p(\mathcal{F})$. Such a family decomposes into a partition of $X \cap Y$ and a partition of $X \cup Y$, hence $p^\wedge(X) + p^\wedge(Y) \leq p^\wedge(X \cap Y) + p^\wedge(X \cup Y)$. □

Lemma 3 and Proposition 3 implies that $p^\wedge$ is fully supermodular, and it is obviously monotone increasing, hence the polyhedron C is a contra-polymatroid defined by $p^\wedge$. It is known that in this case the minimum cardinality of an integral element of the contra-polymatroid C is $p^\wedge(V)$. Thus, for a fix γ, there exists an element m of C with $m(V) = 2\gamma$ if and only if $p^\wedge(V) \leq 2\gamma$. This exactly gives conditions (10) and (11) of the theorem. □

Remark 1. The following example shows that (10) is not sufficient in Theorem 3. Let $V = \{v_1, v_2, v_3, v_4\}$, $E = \{v_1v_2, v_1v_3, v_1v_4\}$. Let $h = 1$ on the sets $\{v_2\}, \{v_3\}, \{v_4\}$ and on their complement; $h = 0$ on all other sets. We need at least 2 new edges for a feasible orientation, but (10) gives only $\gamma \geq 1$.

Remark 2. A cost function $c : E \to \mathbb{R}$ is called *node induced* if $c(uv) = c'(u) + c'(v)$ where $c' : V \to \mathbb{R}$ is a linear cost function on the nodes. To solve the minimum cost augmentation for node induced cost functions, one can find a minimum cost element m of the contra-polymatroid C according to the cost function c', using the greedy algorithm. Then this m can be used as a degree specification to find a minimum cost augmentation.

For general edge costs the problem is **NP**-complete: let G be the empty graph, and let $c(e) = 1$ on the edges of a fix graph G^*, $c(e) = 2$ on the other edges. Let $h(X) = 1$ if $X \neq \emptyset, V$; thus h is crossing supermodular. Now the minimum cost of the augmentation is $|V|$ if and only if G^* contains a Hamiltonian cycle.

4 (k, l)-Edge-Connected Orientations

In the introduction we defined (k, l)-edge-connectivity for non-negative integers $k \geq l$, and mentioned that the (k, l)-edge-connectivity orientation problem is a common generalization of k-edge-connectivity orientation (with $l = k$) and rooted k-edge-connectivity orientation (with $l = 0$). Recently, it was shown in [4] that the case $l = k - 1$ has an important role in orientations with parity constraints. As for the corresponding augmentation problems, the degree-specified and minimum cardinality augmentation of a graph to have a k-edge-connected orientation is already solved, but the minimum cost augmentation is **NP**-complete even for $k = 1$. Conversely, for rooted k-edge-connected orientations, the minimum cost augmentation is easily solvable by matroid techniques, but the degree constrained augmentation was hitherto unsolved.

To show how the results of the previous section can be used to solve degree-specified and minimum cardinality augmentation of a graph so that the new graph has a (k, l)-edge-connected orientation, fix a node $s \in V$, and introduce the following family of set functions:

$$h_{kl}(X) := \begin{cases} k & \text{if } s \notin X , \\ l & \text{if } s \in X . \end{cases} \tag{19}$$

Menger's Theorem implies that an orientation is (k, l)-edge-connected from root s if and only if it covers h_{kl}. The set function h_{kl} is crossing G–supermodular for any G. Note that if a digraph is (k, l)-edge-connected from root s, and for some $s' \in V - s$ we reverse the orientation of the edges of $k - l$ edge-disjoint paths from s to s', then we get a digraph that is (k, l)-edge-connected from root s'. Thus the root can be selected arbitrarily in orientation problems.

Theorem 4. *Let $G = (V, E)$ be a graph, and $m : V \to \mathbb{Z}_+$ a degree specification with $m(V)$ even. There exists an undirected graph $G' = (V, E')$ such that $G + G'$ has a (k, l)-edge-connected orientation and $d_{G'}(v) = m(v)$ for all $v \in V$, if and only if the following hold for every partition $\mathcal{F} = \{X_1, \ldots, X_t\}$ of V:*

$$\frac{m(V)}{2} \geq (t - 1)k + l - e_G(\mathcal{F}) , \tag{20}$$

$$\min_i m(\overline{X_i}) \geq (t - 1)k + l - e_G(\mathcal{F}) . \tag{21}$$

Proof. The necessity can be shown as in Theorem 2. As for the sufficiency, we can fix a node $s \in V$ and use Theorem 2 with h_{kl}. In this case the inequalities (8) and (9) in Theorem 2 are consequences of (6) and (7), since $h_{kl}(\mathcal{F}) \geq h_{kl}(\overline{\mathcal{F}})$ and $e_G(\mathcal{F}) = e_G(\overline{\mathcal{F}})$ for every partition $\mathcal{F}$. □

Theorem 5. *Let $G = (V, E)$ be a graph. There is a graph G' with γ edges such that $G + G'$ has a (k, l)-edge-connected orientation, if and only if the following two conditions are met:*

1. $\gamma \geq (t - 1)k + l - e_G(\mathcal{F})$ for every partition $\mathcal{F}$ with t members.

2. $2\gamma \geq t_1 k + t_2 l - e_G(\mathcal{F})$ *for every* $\mathcal{F} = \mathcal{F}_1 + \mathcal{F}_2$ *where* $\mathcal{F}_1$ *is a partition of some* X *with* t_1 *members,* $\mathcal{F}_2$ *is a co-partition of* $\overline{X}$ *with* t_2 *members, and every member of* $\mathcal{F}_2$ *is the complement of the union of some members of* $\mathcal{F}_1$.

Proof. As in the proof of Theorem 4, we demand that $G + G'$ should have an orientation covering h_{kl}. Going back to the proof of Theorem 3, the set function p defined in (16) can be defined in this case as

$$p(X) := \begin{cases} \max\{p_1^\wedge(X) + p_2(X), \ 0\} & \text{if } X \subset V , \\ 2\max_{Y \subset V} (p_1^\wedge(Y) + p_2(Y)) & \text{if } X = V . \end{cases} \tag{22}$$

As it was proved in Theorem 3, a feasible augmentation with γ edges exists if and only if $p^\wedge(V) \leq 2\gamma$; by the above characterization of p, this is equivalent to the conditions of the theorem. □

There are other equivalent characterizations of graphs that have a (k,l)-edge-connected orientation. For given non-negative integers k and l, a graph $G = (V, E)$ is called (k,l)-*tree-connected* if any graph obtained by deleting l edges from G contains k edge-disjoint spanning trees; it is called (k,l)-*partition-connected* if $e_G(\mathcal{F}) \geq k(t-1)+l$ for every partition $\mathcal{F}$ with t members. Tutte [9] proved that a graph is $(k,0)$-tree-connected if and only if it is $(k,0)$-partition-connected. This immediately implies that a graph is (k,l)-tree-connected if and only if it is (k,l)-partition-connected.

Simple calculation shows that for $k \leq l$, a graph G is (k,l)-tree-connected if and only if it is $(k+l)$-edge-connected; hence the (k,l)-tree-connectivity augmentation problem is interesting only for $k \geq l$.

Proposition 4. *For* $k \geq l$, *a graph* $G = (V, E)$ *is* (k,l)-*tree-connected if and only if it has a* (k,l)-*edge-connected orientation.*

Proof. It follows from the orientation theorem in [1] that for $k \geq l$, a graph has a (k,l)-edge-connected orientation if and only if it is (k,l)-partition-connected. □

Thus Theorems 4 and 5 solve the degree-specified and minimum cardinality (k,l)-tree-connectivity augmentation problem.

5   Positively Crossing G–Supermodular Set Functions

A set function h is *positively crossing G–supermodular* if (1) holds for every crossing pair (X, Y) for which $h(X), h(Y) > 0$.

Let $M = (V; E, A)$ be a mixed graph, where E is the set of undirected edges and A is the set of directed edges. Then the task of finding a (k,l)-edge-connected orientation of M for a fix root s is equivalent to finding an orientation of the edges in E that covers the set function $\max\{h_{kl} - \varrho_A, 0\}$, where h_{kl} is defined in (19). This set function isn't crossing G–supermodular anymore, but it is positively crossing G–supermodular for any G. This motivates the study of the h-orientation problem for positively crossing G–supermodular set functions, and the corresponding augmentation problems.

The characterizations in this section involve some complicated set families. Every cross-free family $\mathcal{F}$ has a *tree-representation* (T, φ), where $T = (W, B)$ is a directed tree, and $\varphi : V \rightarrow B$ is a mapping such that $\{\varphi^{-1}(W_e) \mid e \in B\} = \mathcal{F}$, where W_e is the component of $T - e$ entered by e. A *tree-composition* of $\emptyset \neq X \subset V$ is a cross-free composition of X which has a tree-representation $(T = (W, B), \varphi)$ such that $\varphi^{-1}(w) \neq \emptyset$ for every $w \in W$. Equivalently, a tree-composition of X is a cross-free composition of X that contains no partitions and co-partitions of V. A partition or a co-partition of V will be regarded as a tree-composition of $\emptyset$.

In this section we solve the degree-specified augmentation problem, by mainly the same methods as in Sect. 3, but instead of relying on the properties of base polyhedra, we use the following extension of the classical result on the TDI-ness of the intersection of base polyhedra:

Lemma 4. *Let* $p : 2^V \rightarrow \mathbb{Z} \cup \{-\infty\}$ *be a fully supermodular set function, and let* $q : 2^V \rightarrow \mathbb{Z} \cup \{-\infty\}$ *be a set function that is supermodular on the crossing pairs* (X, Y) *for which* $p(X) < q(X)$ *and* $p(Y) < q(Y)$. *Then the system*

$$\{x \in \mathbb{R}^V \mid x(V) = p(V);\ x(Z) \geq p(Z),\ x(Z) \geq q(Z)\ \forall Z \subseteq V\} \qquad (23)$$

is TDI; it has a feasible solution if and only if

$$p(\overline{X}) + q(\mathcal{F}) \leq (\alpha + 1)p(V) \qquad (24)$$

for every $X \subset V$ *(including the empty set) and every tree-composition* $\mathcal{F}$ *of* X *with covering number* α.

Proof. To prove TDI-ness, we have to show that the dual system

$$\max \{y_1 p + y_2 q - \beta p(V) : (y_1 + y_2)A - \beta \mathbf{1} = c,\ y_1, y_2, \beta \geq 0\}$$

has an integral optimal solution for every integral c, where $y_1, y_2 : 2^V \rightarrow \mathbb{Q}_+$ are dual variables on the sets, y_1 corresponding to the inequalities featuring p, y_2 corresponding to those featuring q, $\beta \in \mathbb{Q}_+$ is the dual variable for the inequality $x(V) \leq p(V)$, and A is the incidence matrix of all subsets of V. The main observation is that we can assume that y_1 is positive on a chain and y_2 is positive on a cross-free family: this can be achieved by a slight modification of the usual uncrossing technique. Consider the following operations:

- If $y_1(X), y_1(Y) > 0$ and neither $X \subseteq Y$, nor $Y \subseteq X$, decrease y_1 on X and Y by $\min\{y_1(X), y_1(Y)\}$, and increase y_1 by the same amount on $X \cap Y$ and $X \cup Y$.
- If $y_2(X), y_2(Y) > 0$, $p(X) < q(X)$, $p(Y) < q(Y)$ and X, Y are crossing, then decrease y_2 on X and Y by $\min\{y_2(X), y_2(Y)\}$, and increase y_2 by the same amount on $X \cap Y$ and $X \cup Y$.
- If $y_2(X) > 0$ and $p(X) \geq q(X)$, then decrease y_2 on X to 0 and increase y_1 on X by the same amount.

These operations do not decrease $y_1 p + y_2 q - \beta p(V)$, and they maintain $(y_1 + y_2)A - \beta \mathbf{1} = c$. We show that by repeatedly applying these operations (in any order), in a finite number of steps we get an optimal dual solution (y_1', y_2', β) such that y_1' is positive on a chain and y_2' is positive on a cross-free family.

Since $y_1, y_2 \in \mathbb{Q}$, there is a positive integer ν such that νy_1 and νy_2 are integral. The sum

$$\nu \left(2 \sum_{y_1(X)>0} y_1(X)|X|^2 + \sum_{y_2(X)>0} y_2(X)|X|^2 \right)$$

increases by at least 1 with any of the above operations, and it is bounded from above by $2\nu|V|^2 (\beta + \max_{v \in V} c(v))$. Thus the procedure terminates after a finite number of steps.

We proved that there is an optimal dual solution (y_1', y_2', β) where y_1' is positive on a chain and y_2' is positive on a cross-free family; but this means that this is also an optimal solution of the dual of the system we get if we restrict p to the sets where y_1' is positive, and restrict q to the sets where y_2' is positive (changing their value to $-\infty$ on all other sets). This system is the intersection of two base polyhedra, so it has an integral optimal dual solution, which is in turn optimal for the dual of the system (23); therefore the system (23) is TDI.

The proof of the non-emptiness condition (24) is similar: the infeasibility of the system is equivalent to the feasibility of its dual by the Farkas Lemma; a feasible dual solution can be uncrossed as above, so dual feasibility implies the emptiness of the intersection of the two base polyhedra given by p and q restricted to the sets where y_1' and y_2' are positive. Thus the emptiness condition for the intersection of base polyhedra (which is of the form (24)) is sufficient for the infeasibility of the original system.   □

Theorem 6. *Let $G = (V, E)$ be a graph, $h : 2^V \to \mathbb{Z}_+$ a positively crossing G–supermodular set function on V, and $m : V \to \mathbb{Z}_+$ a degree specification with $m(V)$ even; let*

$$h_m(X) := h(X) + \max \left\{ 0, m(X) - \frac{m(V)}{2} \right\}.$$

There exists an undirected graph $G' = (V, E')$ such that $G + G'$ has an orientation covering h and $d_{G'}(v) = m(v)$ for all $v \in V$ if and only if

$$h_m(\mathcal{F}) + \max \left\{ 0, m(\overline{X}) - \frac{m(V)}{2} \right\} \le e_G(\mathcal{F}) + (\alpha + 1) \frac{m(V)}{2}$$

for every $X \subset V$ and for every tree-composition $\mathcal{F}$ of X with covering number α.

Proof. The necessity follows from the fact that if $\mathcal{F}'$ is a regular family with covering number $\alpha + 1$, then $\varrho_{\vec{G}}(\mathcal{F}') \le e_G(\mathcal{F}')$ for any orientation of G, and

$$\varrho_{\vec{G}'}(\mathcal{F}') \le (\alpha + 1) \frac{m(V)}{2} - \sum_{X \in \mathcal{F}'} \max \left\{ 0, m(X) - \frac{m(V)}{2} \right\}$$

for any orientation $\vec{G}'$ of a graph G' satisfying the degree specification.

The sufficiency can be proved in essentially the same way as in Theorem 2: define G_0 and h_0 similarly, and for $X \subseteq V$, let

$$p(X) := i_G(X) + \max\left\{0, m(X) - \frac{m(V)}{2}\right\},$$

$$q(X) := h(X) + i_G(X) + \max\left\{0, m(X) - \frac{m(V)}{2}\right\},$$

In this case Lemma 1 implies that an orientation of G_0 covering h_0 exists if and only if the polyhedron

$$\{x : V \to \mathbb{R} \mid x(V) = p(V); \ x(Z) \geq p(Z), \ x(Z) \geq q(Z) \ \forall Z \subseteq V\}$$

has an integral point.

Claim. The set function p is fully supermodular, and the set function q is supermodular on the crossing pairs (X, Y) for which $p(X) < q(X)$ and $p(Y) < q(Y)$.

Proof. The set function p is the sum of two fully supermodular functions, so it is fully supermodular. Since h is positively crossing G–supermodular, q is supermodular on the crossing pairs (X, Y) for which $h(X), h(Y) > 0$, and these are exactly the crossing pairs for which $p(X) < q(X)$ and $p(Y) < q(Y)$.   □

Lemma 4 implies that an orientation of G_0 covering h_0 exists if and only if

$$p(\overline{X}) + q(\mathcal{F}) \leq (\alpha + 1)p(V)$$

for every $X \subset V$ and every tree-composition $\mathcal{F}$ of X with covering number α. Using (5) this is equivalent to the condition of the theorem.

From here we can follow the line of the proof of Theorem 2. Let $\vec{G}_0$ be the orientation of G_0 covering h_0, and let $\vec{G}$ denote the edges of $\vec{G}_0$ not incident with z. Let $m_i(v)$ be the multiplicity of the edge zv in $\vec{G}_0$, and $m_o(v)$ the multiplicity of the edge vz in $\vec{G}_0$. Define the set function $h'(X) = h(X) - \varrho_{\vec{G}}(X)$; h' is positively crossing supermodular. As in the proof of Theorem 2, we can apply Lemma 2 (with the m_i, m_o and h' defined above) to obtain a directed graph D whose underlying undirected graph is a feasible augmentation.   □

Remark 3. We can use the ellipsoid method to prove that the above theorem gives rise to a polynomial algorithm. To prove that the optimization for (23) can be solved in polynomial time, we show that the separation can be solved for a vector x. We know that the separation algorithm works for supermodular functions. Thus we can determine if there is a set X with $x(X) < p(X)$. If not, then for the set function $q^*(X) = \max(x(X), q(X))$, $B(q^*)$ is a base polyhedron. Therefore we can solve the corresponding optimization problem, which implies the solvability of the separation problem; this is equivalent to the separation problem for q concerning x.

Remark 4. The condition involving tree-compositions may seem unfriendly, but it is unavoidable, even in the special case when the problem is to find an orientation of the undirected edges of a mixed graph such that the resulting digraph is k-edge-connected. This orientation problem was already considered in [3], where crossing G–supermodular set functions with possible negative values were studied. The following example shows that the positively G–supermodular case is more general, i. e. not every positively crossing G–supermodular set function h can be made crossing G–supermodular by decreasing the value of h on some of the sets where it is 0.

Let X_1, X_2, X_3 be three subsets of a ground set V in general situation. Let $h(X_i) = 1$, $h(X_i \cup X_j) = 2$ $(i \neq j)$, $h(X_1 \cup X_2 \cup X_3) = 4$, and $h(X) = 0$ on the remaining sets; this is a positively crossing supermodular function. The value of $X_1 \cap X_2$ cannot be decreased since

$$h(X_1 \cap X_2) \geq h(X_1) + h(X_2) - h(X_1 \cup X_2) = 0 \ .$$

Therefore it is impossible to correctly modify h so as to satisfy

$$h(X_1 \cap X_2) \leq h(X_1 \cap X_2 \cap X_3) + h(X_1 \cap X_2 \cup X_3) - h(X_3) \leq -1 \ .$$

References

1. Frank, A.: On the orientation of graphs. J. Combinatorial Theory B **28** No. 3 (1980) 251–261
2. Frank, A.: Connectivity augmentation problems in network design. In: Birge, J.R., Murty, K.G. (eds.): Mathematical Programming: State of the Art. Univ. of Michigan (1994) 34–63
3. Frank, A.: Orientations of graphs and submodular flows. Congressus Numerantium **113** (1996) 111–142
4. Frank, A., Király, Z.: Parity constrained k-edge-connected orientations. In: Cornuejols, G., Burkard, R., Woeginger, G.J. (eds.): Lecture Notes in Computer Science, Vol. 1610. Springer-Verlag, Berlin Heidelberg New York (1999) 191–201
5. Fujishige,S.: Structures of polyhedra determined by submodular functions on crossing families. Mathematical Programming **29** (1984) 125–141
6. Lovász, L.: Combinatorial Problems and Exercices. North-Holland (1979)
7. Mader,W.: Konstruktion aller n-fach kantenzusammenhängenden Digraphen. Europ. J. Combinatorics **3** (1982) 63–67
8. Nash-Williams, C.St.J.A.: On orientations, connectivity and odd vertex pairings in finite graphs. Canad. J. Math. **12** (1960) 555–567
9. Tutte,W.T.: On the problem of decomposing a graph into n connected factors. J. London Math. Soc. **36** (1961) 221–230
10. Watanabe, T., Nakamura, A.: Edge-connectivity augmentation problems. Computer and System Sciences **35** No. 1 (1987) 96–144

An Extension of a Theorem of Henneberg and Laman*

András Frank[1,2] and László Szegő[1]**

[1] Department of Operations Research, Eötvös University, Kecskeméti u. 10-12.,
Budapest, Hungary H-1053
[2] Ericsson Traffic Laboratory, Laborc u. 1., Budapest, Hungary H-1037
frank@cs.elte.hu, szego@cs.elte.hu

Abstract. We give a constructive characterization of graphs which are the union of k spanning trees after *adding* any new edge. This is a generalization of a theorem of Henneberg and Laman who gave the characterization for $k = 2$.

We also give a constructive characterization of graphs which have k edge-disjoint spanning trees after *deleting* any edge of them.

Keywords. graph, constructive characterization, rigidity, packing and covering by trees

AMS subject classification. 05C40

1 Introduction

The idea of constructive characterizations in graph theory is not new. The first example is the following theorem of Tutte [15] from 1966. A graph on more than k nodes is said to be k-**node-connected** if after deleting less than k nodes the graph remains connected.

Theorem 1 (Tutte). *An undirected graph $G = (V, E)$ is 3-node-connected if and only if G can be obtained from the complete graph on 4 nodes by the following two operations:*

(i) *add a new edge,*
(ii) *take a node z and replace it by two nodes z_1, z_2, put an edge between them, and put edges incident to z_1, z_2 such that the union of the neigbours of them are exactly the original neighbours of z and there are at least two neighbours of z_i for $i = 1, 2$.*

In 1976 Lovász [8] proved the following theorem. A graph is said to be k-**edge-connected** if after deleting less than k edges the graph remains connected.

---

* Research supported by the Hungarian National Foundation for Scientific Research Grant, OTKA T17580.
** This author is supported by the Siemens-ZIB Fellowship Program and FKFP grant no. 0143/2001.

K. Aardal, B. Gerards (Eds.): IPCO 2001, LNCS 2081, pp. 145–159, 2001.
© Springer-Verlag Berlin Heidelberg 2001

Theorem 2 (Lovász). *An undirected graph* $G = (V, E)$ *is $2k$-edge-connected if and only if G can be obtained from a single node by the following two operations:*

(i)   *add a new edge,*

(ii) *add a new node z, subdivide k existing edges by new nodes, then identify the k subdividing nodes with z.*

The (ii) operation in this theorem is called **pinching k edges (with z).**

Similar constructive characterizations for directed edge-connectivity in directed graphs exist due to Mader [9].

Theorem 3 (Mader). *A directed graph* $G = (V, E)$ *is k-edge-connected if and only if G can be obtained from a single node by the following two operations:*

(i)   *add a new edge,*

(ii) *pinch k existing directed edges.*

Another example of this kind of theorems concerns spanning trees. We call an undirected graph k-**tree-connected** if it contains k edge-disjoint spanning trees. It was observed in [3] that a combination of Mader's characterization and Tutte's disjoint tree theorem gives rise to the following.

Theorem 4. *An undirected graph* $G = (V, E)$ *is k-tree-connected if and only if G can be built from a single node by the following two operations:*

(i)   *add a new edge,*

(ii) *pinch i ($0 \leq i \leq k - 1$) existing edges with a new node z, and add $k - i$ new edges connecting z with existing nodes.*

For a direct proof see Tay [13].

An undirected graph $G = (V, E)$ is said to be k-**stiff** if it is the union of k edge-disjoint spanning trees after *adding* any new edge, that is, $G + e$ is the union of k edge-disjoint spanning trees for every possible new edge $e = uv$ ($u, v \in V$) (multiple edges are permitted).

2-stiff or, as often called, minimal generically rigid graphs are important in statics. A framework in the plane is statically rigid if and only if its graph has a minimal generically rigid subgraph. This was proved by Laman [7]. A framework consists of rigid rods and rotatable joints. Its underlying graph is the natural one: the node set is the set of the joints and there is an edge between two nodes if there is a rod between the corresponding two joints. A consequence of the theorem of Laman is that the notion of rigidity is a property of the graph and not only its embedding into the plane with real rods and joints.

According to a theorem of Nash-Williams [12], a graph $G = (V, E)$ is k-stiff if and only if $|E| = k(|V| - 1) - 1$ and $\gamma_G(X) \leq k(|X| - 1) - 1$ for every subset $X \subseteq V$ with $|X| \geq 2$ (where $\gamma_G(X)$ denotes the number of the edges of G whose two end-nodes are in X). By combining theorems of Henneberg [5] and of Laman [7], one obtains the following constructive characterization of 2-stiff graphs.

Theorem 5 (Henneberg and Laman). *A graph G is 2-stiff if and only if G can be constructed from one (non-loop) edge by the following two operations:*

(i)   *add a new node z and connect z to two distinct existing nodes,*
(ii) *subdivide an existing edge uv by a node z and connect z to an existing node distinct from u and v.*

In this paper we give the generalization of this theorem for arbitrary k. The difficulties which come to the picture for k greater than 2 will be presented in the next section. We note however that 3-stiff graphs have no direct meaning in 3-dimensional rigidity.

In Sect. 3 we will give a corresponding constructive characterization of graphs which have k edge-disjoint spanning trees after *deleting* any edge of them.

For the sake of completeness we give the original theorem of Henneberg [5] and Laman [7].

Theorem 6 (Henneberg). *A framework in the plane is minimally rigid if and only if it can be constructed from one rod by the following two operations:*

(i)   *add a new joint z and connect z to two distinct existing joints by rods,*
(ii) *subdivide an existing rod uv by a node z and connect z to an existing joint distinct from u and v.*

Theorem 7 (Laman). *A framework is minimally rigid if and only if its underlying graph $G = (V, E)$ has the following property: $|E| = k|V| - (k + 1)$, $\gamma_G(X) \le k|X| - (k + 1)$ for all $X \subseteq V, |X| \ge 2$.*

2   Construction of k-Stiff Graphs

Let k be an integer not less than 2. Let K_2^{k-1} denote the graph on two nodes with $k - 1$ parallel edges.

Theorem 8. *$G = (V, E)$ is a graph. The following are equivalent:*

(1) *G k-stiff.*
(2) *$|E| = k|V| - (k + 1)$ and $\gamma_G(X) \le k|X| - (k + 1)$ for all subsets $X \subseteq V$, $|X| \ge 2$.*
(3) *G can be built from K_2^{k-1} by applying the following operation:*
     Choose a subset F of i existing edges ($0 \le i \le k - 1$), pinch the elements of F with a new node z, and add $k - i$ new edges connecting z with other nodes so that there are no k parallel edges in the resulting graph.

The equivalence of (1) and (2) is straightforward by a theorem of Nash-Williams [12]. The fact that (3) implies (2) is an easy exercise.

(2) implies (3). This is the main point of this section.

After some definitions and lemmas, we give a necessary and sufficient condition when the inverse of operation (3) is applicable at node s, which is important for an inductive proof.

A graph which satisfies the conditions in (2) is called a **Laman-graph**.

A graph D on node-set U is called an **admissible graph** if it satisfies the following property:

(4) $\gamma_D(X) \le k|X| - (k+1)$ for all subsets $X \subseteq U, |X| \ge 2$.

In the graph $G = (V, E)$ **splitting off a pair of edges** at node s means the operation of replacing su and sv by a new edge connecting u and v.

At node s with degree $k + i$ $(0 \le i \le k - 1)$ **admissible splitting off** j $(1 \le j \le k - 1)$ pairs of edges means j number of splitting off a pair of edges such that the resulting induced subgraph on $V - s$ is an admissible graph (we often leave out the word admissible). If $j = i$, then it is called a **complete splitting off**.

Our goal is to find a node s with degree $k + i$ $(0 \le i \le k - 1)$ in the Laman-graph G, such that i pairs of edges can be split off, that is, the inverse operation of (3) can be applied at the node s in such a way that the resulting graph is also a Laman-graph. This will give our inductive proof.

G'_s will denote the graph that we obtained by splitting off some pairs of edges incident to s. We will use the term **split edge** in G'_s for an edge uv which comes from splitting off edges su and sv.

Definition 1. *Let b_G denote the following function on the subsets of V with cardinality at least 2:*

$$b_G(X) := k|X| - (k+1) - \gamma_G(X).$$

By this definition we have the following: for a graph G, property (2) holds if and only if $b_G(V) = 0$ and $b_G(X) \ge 0$ for all subsets $X \subseteq V, |X| \ge 2$. The graph $G = (U, F)$ is an admissible graph if and only if $b_G(X) \ge 0$ for all subsets $X \subseteq U, |X| \ge 2$.

If $b_G(V) = 0$, then X is said to be a G-**tight** set. From now on we leave G out if it is unambiguous.

All the lemmas below are about admissible graphs.

Lemma 1. *Let X and $Y \subseteq V$ and $|X \cap Y| \ge 2$. Then*

$$b(X) + b(Y) = b(X \cap Y) + b(X \cup Y) + d(X, Y).$$

Proof. $b(X) + b(Y) = k|X| - (k+1) - \gamma_G(X) + k|Y| - (k+1) - \gamma_G(Y) = k(|X| + |Y|) - 2(k+1) - (\gamma_G(X \cap Y) + \gamma_G(X \cup Y) + d_G(X, Y)) = k|X \cap Y| - (k+1) - \gamma_G(X \cap Y) + k|X \cup Y| - (k+1) - \gamma_G(X \cup Y) + d_G(X, Y) = b(X \cap Y) + b(X \cup Y) + d(X, Y).$ □

Lemma 2. *If $X_1, X_2, X_3 \subseteq V$ and $|X_j \cap X_l| = 1$ for all possible pairs and $|X_1 \cap X_2 \cap X_3| = 0$, then*

$$b\left(\bigcup_{j=1}^{3} X_j\right) \le \sum_{j=1}^{3} b(X_j) - k + 2.$$

Proof. $b(\bigcup_{j=1}^3 X_j) = k|\bigcup_{j=1}^3 X_j| - (k+1) - \gamma_G(\bigcup_{j=1}^3 X_j) \le k(\sum_{j=1}^3 |X_j| - 3) - (k+1) - \sum_{j=1}^3 \gamma_G(X_j) = \sum_{j=1}^3 (k|X_j| - (k+1) - \gamma_G(X_j)) - k + 2 = \sum_{j=1}^3 b(X_j) - k + 2.$ □

Lemma 3. $X, Y \subseteq V$, $|X \cap Y| = 1$. *Then*

$$b(X) + b(Y) = b(X \cup Y) - 1 + d(X, Y).$$

Proof. $b(X) + b(Y) = k|X| - (k+1) - \gamma_G(X) + k|Y| - (k+1) - \gamma_G(Y) = k(|X| + |Y| - 1) - (k+1) - 1 - (\gamma_G(X) + \gamma_G(Y)) = k|X \cup Y| - (k+1) - 1 - (\gamma_G(X \cup Y) - d_G(X, Y)) = b(X \cup Y) - 1 + d(X, Y).$ □

From now on G is a graph which satisfies (2) in our theorem, that is, G is a Laman-graph, and not K_2^{k-1}. It is easy to see that there exists a node s with degree $d(s)$ such that $k \le d(s) \le 2k - 1$. It is also clear that the multiplicity of edge uv is at most $k - 1$ (by (2): $\gamma_G(\{u, v\}) \le k|\{u, v\}| - (k+1) = k - 1$).

Observation 1. *The edges su and sv cannot be split off (that is, adding the edge uv to the induced subgraph of G on $V - s$ does not result in an admissible graph) if and only if there exists a tight set in G which does not contain s but u and v.*

Observation 2. *By Lemma 2 a splitting off at node s cannot be kept on if and only if the remaining neighbours of s are in a tight set which does not contain s or there is only one remaining neighbour of s.*

Theorem 9. *Let G be a Laman-graph. If $s \in V$ has degree k or $k + 1$, then a complete splitting off is applicable at it.*

Proof. If s has degree k, then a complete splitting off means deleting it with all its adjacent edges. This results obviously in a Laman-graph.

If s has degree $k + 1$, then we should find a pair of edges su and sv with $u \ne v$ such that $G - s + uv$ is an admissible graph.

There is no tight set X not containing s which contains all the neighbours of s because, if there was one, then $b_G(X + s) < 0$ which contradicts to (2). Because of the fact that there are no edges with multiplicity greater than $k - 1$, the neighbour-set of s in G has at least two elements, so by Lemma 2 and Observation 1 there is an admissible splitting off.   □

If $k = 2$, then there is no other case, so we proved the theorem of Henneberg and Laman: every node with degree 2 or 3 admits a complete splitting off. If $k \ge 3$, then life is much more complicated, as was observed by Z. Király [6]. He found a graph for $k = 3$ in which there is no splitting off 2 pairs of edges (that we would need) at a node with degree 5.

Here we give a necessary and sufficient condition for a node with degree $k + i$ ($2 \le i \le k - 1$) (let us call a node like this a **small node**) which admits a complete splitting off. Let $\Gamma_G(v)$ denote the number of the nodes in graph G that are connected to node v by an edge.

Theorem 10. *At node s with $d_G(s) = k + i$ $(2 \leq i \leq k - 1)$ there exists a complete splitting off if and only if there do not exist the following subsets $X_1, X_2, \ldots, X_m \subset V - s$ such that the following holds:*

a) $X_j \cap X_l = \{t\}$ *with a fix node $t \in V - s$ for all possible pairs X_j, X_l,*
b) $b_G(X) < d_G(s, X_j - t)$ *for all possible j,*
c) $d_G(s,t) > (k - i) + d_G(s, V - s - \cup_{j=1}^m X_j) + \sum_{j=1}^m b_G(X_j)$.

Proof. Let us consider a small node s. The necessity of the condition is obvious, because the sets X_j give that the maximum number of edges between s and t which can be split off with other edges is at most $d_G(s, V - s - \cup_{j=1}^m X_j) + \sum_{j=1}^m b_G(X_j)$, but by c), for a complete splitting off, we would need more (i).

Sufficiency. Let us consider a maximal splitting off with respect to the number of split edges, moreover in the resulting graph G'_s the number of neighbours of s is maximal, that is, $|\Gamma_{G'_s}(s)|$ is maximal, moreover, if $|\Gamma_{G'_s}(s)| \geq 2$, then the tight set containing $|\Gamma_{G'_s}(s)|$ is maximal. By Observation 1 this tight set gives the fact that there is no more splitting off at s.

If we managed to split off i pairs of edges at s, then it is the inverse operation of (3), so there exists a complete splitting off. If not, then we will find the sets X_j.

Lemma 4. G'_s *is obtained by a maximal but not complete splitting off at s. If s has only one neighbour t in G'_s, then there exists a split edge which is disjoint from t.*

Proof. Let us suppose that we split maximum number l pairs of edges and one endnode of every split edge is t. Since this splitting off is not complete, $l < i$. Then in the original graph G:

$$d_G(s,t) = d_G(s) - l = k + i - l > k,$$

which contradicts the condition in (2) for the set $\{s,t\}$. □

Lemma 5. G'_s *is obtained by a maximal but not complete splitting off at s. If s has at least 2 neighbours in G'_s, then let P_{max} denote the maximal tight set which covers all the neighbours of s in G'_s.*

Then there exists a split edge which is disjoint from the nodes of P_{max}.

Proof. Let us consider a maximal splitting off at node s, and let j denote the number of split edges in P_{max} and let l denote the number of edges with exactly one endnode in P_{max}. Let us suppose that there are no other split edges. Then:

$$\gamma_G(P_{max}+s) = \gamma_{G'_s}(P_{max})+j+l+(k+i-2(j+l)) = \gamma_{G'_s}(P_{max})+k+(i-(j+l))$$

$$> k|P_{max}| - (k+1) + k = k|P_{max} + s| - (k+1).$$

□

Lemma 6. *Let G be an admissible graph and X is a maximal tight set in it which contains the distinct nodes c_1, c_2. Let d be a node in $V - X$. Then there is no tight set which contains c_i and d for $i = 1$ or 2.*

Proof. We may suppose that there is a tight set P containing c_1 and d. According to Lemma 1 $P \cap X = \{c_1\}$ because X is maximal. By Lemmas 1 and 2 we can see that there is no tight set containing c_2 and d. □

Lemma 7. *Let G'_s denote the graph obtained by splitting off some edges at s. Let $as, bs \in E(G'_s)$ $(a \neq b)$ and uv be a split edge in G'_s such that the maximal tight set P does not contain s, u, v but a, b.*

If sa and sb are both multiple edges, then instead of spliting off su, sv we can split off $(sa, su$ and $sa, sv)$ or $(sb, su$ and $sb, sv)$.

If there is a third distinct node c in P with edge sc in G'_s such that su, sc is not splittable, then instead of spliting off su, sv we can split off $(sa, su$ and $sb, sv)$ or $(sb, su$ and $sa, sv)$.

Proof. According to Lemma 6, we can see that there are no tight sets which would be obstacles to the 'splitting off's in every case, it is remained to see, that we can apply the corresponding two 'splitting off's at the same time. If not, then there is a set with too many induced edges, which contains the two new split edges. But it means, that, before this, there is a tight set containing a, b, u, v but not s, which contradicts the maximality of P according to Lemma 1. □

Case 1. Let us suppose that $|\Gamma_{G'_s}(s)| \geq 3$, let a_1, a_2, a_3 denote three of these nodes.

According to Observation 2, there exists a maximal tight set P containing $\Gamma_{G'_s}(s)$. By Lemma 5, there is a split edge uv disjoint from P. By Lemma 7, it follows that the splitting off we consider is not maximal, a contradiction.

Case 2. Let us suppose that $|\Gamma_{G'_s}(s)| = |\{t, z\}| = 2$. If $d_{G'_s}(s, t) \geq 2$ and $d_{G'_s}(s, z) \geq 2$, then, as above:

According to Observation 2 there exists a maximal tight set P containing $\Gamma_{G'_s}(s)$. By Lemma 5 there is a split edge uv disjoint from P. By Lemma 7, it follows that the splitting off we consider is not maximal, a contradiction.

We may suppose, that the multiplicity of edge sz is in G'_s exactly one, that is, it is not a multiple edge. We have: $d_{G'_s}(s, t) \geq k + i - 2(i - 1) - 1 = k - i + 1 \geq 2$.

Let $u \in V$ be an arbitrary node which is incident to some split edge which is disjoint from t. Let P_u be the maximal tight set which does not contain s but u and t and contains the minimal number of split edges that are disjoint from t. Let P_z be the maximal tight set which does not contain s but z and t and contains the minimal number of split edges that are disjoint from t. We will see that these sets give the setsystem in Theorem 10.

Case 3. Let us suppose that $|\Gamma_{G'_s}(s)| = |\{t\}| = 1$. We have: $d_{G'_s}(s, t) \geq k + i - 2(i - 1) = k - i + 2 \geq 3$.

Here, let us define the sets P_u as in the above case.

In the above two cases the setsystem of sets P_u is called the **flower** of node s. A set P_u is called a **petal** of the flower or node s.

Proposition 1. *There is no split edge in an arbitrary petal which is disjoint from t.*

Proof. Let us suppose on the contrary that there is a split edge ab.

Let us consider P_z. First let us suppose that a, b, z are three distinct nodes. Since splitting off st, sz intead of sa, sb would result in a maximal splitting off with three neighbours of s, there is a tight set X which is an obstacle to it, that is, it contains t, z and exactly one of a and b. $X \cap P_z$ contains a smaller number of split edges, which gives a contradiction. Now let us suppose that $a = z$. The justification is just the same as above.

Let us consider P_u. Now we have a split edge uv such that $v \notin P_u$ (if not, then split off st, su instead of su, sv results in a maximal splitting off with one more remaining neighbour of s). $P_v \cap P_u = \{t\}$ because of Lemma 1. Splitting off su, st and sv, sa instead of sa, sb and su, sv would result in a maximal splitting off with one more neighbour of s, so there exists an obstacle to it, that is, the set X containing a, u, v, t, not s which is tight in G'_s. But then $X \cap P_z$ contains a smaller number of split edges, which gives a contradiction.   □

Proposition 2. *Let us suppose we defined P_u and P_v for nodes u, v. Then $P_u = P_v$, or $P_u \cap P_v = \{t\}$.*

Proof. By Lemma 1 $P_u \subseteq P_v$ can not be. If $P_u \neq P_v$ and $|P_u \cap P_v| \geq 2$, then by Lemma 1 $d_{G'_s}(P_u, P_v) = 0$ and $P_u \cup P_v$ is tight. Since it does not contain any split edge disjoint from t, it contradicts the choice of P_u by maximality.   □

We state that the sets P_u satisfy the condition of the theorem. Now **a)** and **b)** follows. **c)** is implied by the fact that the maximal splitting off that we consider is not complete and the number of the neighbours of s after the maximal splitting off is $\Gamma_{G'_s}(s) \leq 2$. This is the end of the proof of Theorem 10.   □

From now on, if s is a small node which does not admit a complete splitting off, then we have a flower with it, and it can be the type of Case 2 (that is, which comes from a maximal splitting off with two remaining neighbours of s) then we will refer to it as a first type flower, or it can be the type of Case 3 (that is, s has one remaining neighbour), then we refer to it as a second type flower.

Let us fix the set-system $\mathcal{P}_s$ that we defined to every small node without a complete splitting off. (That is, we consider a special flower whose number of its petals is minimal, moreover the petals are maximal sets.) Let $T(s)$ denote the node t for every s, and it is called the **blocking node** of s, that is, it is the center node of the flower of s.

It follows from Theorem 10 that every small node which does not admit a complete splitting off has a unique blocking node. But this is not important for our proof, let us fix one flower, and it has a unique centre.

We have the following.

Lemma 8 (number of petals). *For any small node s which does not admit a complete splitting off: $|\mathcal{P}_s| \geq 3$.*

Proof. If the flower of s is of first type, then there exists a maximal tight set P_{max} in graph G'_s which was obtained after a maximal splitting off which contains P_z as a subset. By Proposition 1, there exists a split edge ab disjoint from P_{max}. P_z together with P_a and P_b are three different petals.

If the flower of s is of second type, then it is enough to see, that there are at least two split edges not incident to t (the centre of the flower). By Lemma 4, there exists one split edge like this. Let us suppose that there is no other split edge disjoint from t. Then: let m be the number of split edges incident to t in set $P_u \cup P_v$, moreover let l be the number of split edges incident to t with the other endnode not in $P_u \cup P_v$. Since $P_u \cup P_v$ is a tight set in G'_s, $b_G(P_u \cup P_v) = m + 1$. As we have a maximal but not complete splitting off, $m + l + 1 < i$. So, $b_G(P_u \cup P_v + s) = b_G(P_u \cup P_v) + k - d_G(s, P_u \cup P_v) = m + 1 + k - (k + i - l) = m + l + 1 - i < 0$, which is a contradiction. □

Lemma 9 (flower-lemma). *If petal P of s contains the small node s' such that $T(s) = T(s')$ and P' is a petal of s' and $P' - P \neq \emptyset$, then $s \in P'$.*

Proof. Let us suppose that $s \notin P'$.
First case. $|P \cap P'| \geq 2$.

Let n be the number of split edges in $P \cap P'$ in graph $G'_{s'}$ (since this set is a subset of a petal, there cannot be split edges not incident to t), moreover let m be the number of split edges in $P' - P$ in the same graph. By a), $b_G(P') = n + m$ and $b_G(P \cap P') \geq n$, $d_G(P, P') \geq m$.

Now we have $b_G(P) \leq b_G(P \cup P') = b_G(P) + b_G(P') - b_G(P \cap P') - d_G(P, P') \leq b_G(P) + n + m - n - m = b_G(P)$. Which means that $P \cup P'$ contradicts to the maximality of petal P of s.
Second case. $|P \cap P'| = 1$.

Let m be the number of split edges in P' in graph $G'_{s'}$ (these are all incident to t since P' is a petal). This gives $b_G(P') = m$, $d_G(P, P') \geq m + 1$ (+1 follows from the fact that every petal P_0 of s' contains a node which is incident to a split edge disjoint from t in graph $G'_{s'}$ which is obtained by a maximal splitting off. That is, there is at least one more edge between s' and $P_0 - t$ in G.)

So, $b_G(P) \leq b_G(P \cup P') = b_G(P) + b_G(P') + 1 - d_G(P, P') \leq b_G(P) + m + 1 - (m + 1) = b_G(P)$. Which means that $P \cup P'$ contradicts to the maximality of petal P of s. □

Let us consider a petal P of the flower of a small node s. If there is a small node $s' \in P$ whose blocking node is also $T(s) = t$, then let us define the following (the flower of s' is denoted by $X_1, X_2, \ldots X_m$):

$$\tau(s') = \min_{l=1,2,\ldots,m} |\{s'' \in (\cup_{j=1}^m X_j - X_l) \cap P : T(s'') = t\}|.$$

According to the Flower lemma, there is a small node s_0 in P such that $\tau(s_0) = 0$. This means that s_0 has at least two petals that are entirely in P and do not contain any small node with blocking node $T(s) = t$.

It is clear that a tight set X either has two nodes with $k - 1$ parallel edges, or $d(v, X - v) \geq k$ for an arbitrary $v \in X$.

Proposition 3. *Let X_1 and X_2 be two petals of s_0 that do not contain any small node with blocking node t. Then $d_G(t, X_1 - t) + d_G(t, X_2 - t) \geq k$*

Proof. $d_G(t, X_1 - t) + d_G(t, X_2 - t) \geq d_{G'_s}(t, X_1 - t) + d_{G'_s}(t, X_2 - t) - (i - 1) \geq d_{G'_s}(t, X_1 - t) + d_{G'_s}(t, X_2 - t) - (k - 2) \geq 2(k - 1) - (k - 2) = k.$ □

Let us give a lower bound on the edges that are incident to some blocking node $T(s) = t$ for some s and whose other endnode is not a small node (that is, has degree at least $2k$) or a small node whose blocking node is also t. Let $\Delta(t)$ denote this number for blocking node t.

Let us consider an arbitrary blocking node t that is a blocking node of some small nodes. Let s be a small node like that. Let us consider three petals of s P_1, P_2, P_3 (they exist by lemma 'number of petals'). We may exchange some indices to get one of the following four cases.

Case A. There are no small nodes in the above petals blocking by t.

Then $\Delta(t) \geq \sum_{j=1}^{3} d_G(t, P_j - t) \geq \sum_{j=1}^{3} d_{G'_s}(t, P_j - t) - (i - 1) \geq 3(k - 1) - (k - 2) = 2k - 1.$

Case B. P_1 contains at least one small node blocking by t, P_2, P_3 do not.

By Proposition 3: $\Delta(t) \geq k + d_G(t, P_2 - t) + d_G(t, P_3 - t) \geq k + d_{G'_s}(t, P_2 - t) + d_{G'_s}(t, P_3 - t) - (i - 1) \geq k + (k - 1) + (k - 1) - (k - 2) = 2k.$

Case C. P_1, P_2 contains at least one small node blocking by t, P_3 does not.

By Proposition 3: $\Delta(t) \geq 2k + d_G(t, P_3 - t) \geq 2k + d_{G'_s}(t, P_3 - t) - (k - 2) \geq 2k + (k - 1) - (k - 2) = 2k + 1.$

Case D. P_1, P_2, P_3 contains at least one small node blocking by t.

Then: $\Delta(t) \geq 3k.$

We have in every case: $\Delta(t) \geq 2k - 1.$

We saw that, for a small node s, if $d(s) = k + i$ and $T(s) = t$, then $d_G(s, t) \geq k - i + 1$. Let n_{k+i} denote the number of nodes with degree $(k + i)$, $2 \leq i \leq k - 1$.

If there is a node with degree less than $k + 2$, than a complete splitting off is applicable at it by Theorem 9. To prove Theorem 8, let us suppose that every node has degree at least $k + 2$. Let $T \subseteq V$ be the set of the blocking nodes for every small node. Let us suppose that every small node has a blocking node, that is, it does not admit a complete splitting off.

Now we have:

$$2|E| = 2(k|V| - (k + 1)) = 2k|V| - 2k - 2 = 2|E| \geq$$

$$\geq \sum_{t \in T} d_G(t) + \sum_{i=2}^{k-1} (k + i)n_{k+i} + 2k(|V| - |T| - \sum_{i=2}^{k-1} n_{k+i}) \geq$$

$$\geq (2k - 1)|T| + \sum_{i=2}^{k-1} (k - i + 1)n_{k+i} + \sum_{i=2}^{k-1} (k + i)n_{k+i} + 2k(|V| - |T| - \sum_{i=2}^{k-1} n_{k+i}) =$$

$$= 2k|V| - |T| + \sum_{i=2}^{k-1} n_{k+i} \geq 2k|V|.$$

$\sum_{i=2}^{k-1} n_{k+i} \geq |T|$ holds because we fixed one blocking node to every small node. So we arrive at a contradiction which means there exists a small node at which a complete splitting off is applicable.   □

The following theorem can be proved by a slight modification of the above computation.

Theorem 11. *If G is k-stiff and is not K_2^{k-1}, then there are at least three nodes such that a complete splitting off is applicable at them.*

An open question is how we can quickly give the k edge-disjoint spanning trees after adding an arbitrary new edge if we know how the graph is built up by the operations.

The following theorem characterizes the connected graphs that are the union of k forests after adding an arbitrary edge.

Theorem 12. *Graph G is the union of k spanning trees after adding an arbitrary edge if and only if it is a connected subgraph of a k-stiff graph.*

Proof. It is straightforward that any connected subgraph of a k-stiff graph has this property.

By the theorem of Nash-Williams [12], $G = (V, E)$ is the union of k (not necessarily edge-disjoint) spanning trees after adding an arbitrary edge if and only if it is connected and $\gamma_G(X) \leq k|X| - (k+1)$ for all $X \subseteq V$. We claim that if $|E| < k|V| - (k+1)$, then we can add an edge e such that $G + e$ is also the union of k forests after adding an arbitrary edge. This will prove the theorem.

Let us consider a maximal tight set X and node $u \in X$, other node $v \notin X$. If we cannot add edge uv, then there exists a tight set Y containing u and v. According to Lemma 2, there is a node a in $X - Y$ and a node b in $Y - X$ such that we may add edge ab to G.   □

We remark about constructive characterizations in general that the fact that one obtains the required type of graphs by the operations is always much easier to prove than the other direction: in a graph which has the required property we can always find a node where we can perform an inverse operation such that after it we get a smaller graph which also has the required property.

In the proof of the theorem of Lovász or that of Mader one finds the following: an inverse operation can always be performed at a node with suitable degree such that it gives a smaller graph of the same kind.

This was also the case for $k = 2$ with k-stiff graphs, but not if k is greater than 2. Here we can easily find a node with the degree we are looking for (this is not the case in the theorems of Lovász and Mader) but in general there is no way to perform the inverse operation at that node.

(We remark that there are other constructive characterizations where the inverse operation is not only considering one single node and performing some operation with the edges but something different as in the case of Theorem 1.)

We finish this section by putting the question if there is a similar constructive characterization of graphs that are the union of k edge-disjoint spanning trees after adding arbitrary l number of edges. We mention that the basic lemmas we used in our proof are valid if and only if $2l \leq k$, furthermore there is no graph on three nodes that are the union of k edge-disjoint spanning trees after adding l number of arbitrary edges if $2l > k$.

3   Construction of $(k, 1)$-Edge-Connected Digraphs and $(k, 1)$-Partition-Connected Graphs

In a directed graph by **splitting off** a pair of edges $e = uz, f = zv$ we mean the operation of replacing e and f by a new directed edge from u to v. Suppose that the in-degree and the out-degree of z is the same, that is, $\varrho(z) = \delta(z)$. By a **complete splitting** at z we mean the following operation: pair the edges entering and leaving z and split off all these pairs.

For non-negative integers $l \leq k$, we call a digraph D (k, l)-**edge-connected** (in short, (k, l)-ec) if D has a node s so that there are k (resp., l) edge-disjoint paths from s to every other node (there are l edge-disjoint paths from every node to s). If there is an exceptional node z for which the existence of these edge-disjoint paths is not required, we say that D is (k, l)-**edge-connected apart from** z. When the role of s is emphasized, we say that D is (k, l)-ec with respect to root-node s. (k, k)-edge-connectivity is abbreviated by k-edge-connectivity and $(k, 0)$-edge-connectivity is sometimes called rooted k-edge-connectivity. Note that by Menger's theorem a digraph is (k, l)-ec if and only if

$$\varrho(X) \geq k \text{ for every subset } \emptyset \subset X \subseteq V - s \tag{1}$$

and

$$\delta(X) \geq l \text{ for every subset } \emptyset \subset X \subseteq V - s \tag{2}$$

where $\varrho(X) := \varrho_D(X)$ and $\delta(X) := \delta_D(X)$ denote the number of edges entering and leaving the subset X, respectively.

We say an undirected graph $G = (V, E)$ is (k, l)-**partition-connected** if there are at least $k(t - 1) + l$ edges connecting distinct classes of every partition of V into t $(t \geq 2)$ non-empty subsets.

The following result exhibits a link between the two concepts. It is a special case of a general orientation theorem appeared in [1].

Theorem 13. *Let $0 \leq l \leq k$ be integers. An undirected graph $G = (V, E)$ has a (k, l)-edge-connected orientation if and only if G is (k, l)-partition-connected.*

Mader's directed splitting off theorem [10] is as follows.

Theorem 14. *Let $D = (U + z, E)$ be a digraph which is k-edge-connected apart from z. If $\varrho(z) = \delta(z)$, then there is a complete splitting at z resulting in a k-ec digraph on node-set U.*

This result has been extended in [2] as follows.

Theorem 15. *Let $D = (U + z, E)$ be a digraph which is (k, l)-edge-connected apart from z. If $\varrho(z) = \delta(z)$, then there is a complete splitting at z resulting in a (k, l)-ec digraph on node-set U.*

We need the following corollary of Theorem 14.

Theorem 16. *Let $D = (U + z, E)$ be a digraph which is*

$$(k, 0) \text{ -ec apart from } z \ (k \geq 1) \text{ with respect to a root node } s \in V. \qquad (3)$$

If $\varrho(z) > \delta(z)$, then there are $\varrho(z) - \delta(z)$ edges entering z so that (3) continues to hold after discarding these edges. If $\varrho(z) = \delta(z)$, then there is a complete splitting at z preserving (3).

Proof. For every node $v \in U + z$ for which $\varrho(v) > \delta(v)$, add $\varrho(v) - \delta(v)$ parallel edges from v to s. In the resulting digraph D' clearly $\varrho'(v) \leq \delta'(v)$ holds for every node $v \in U - s$. Hence $\delta'(X) \geq \varrho'(X) = \varrho(X) \geq k$ holds for every subset $X \subseteq V - s, X \neq \{z\}$, that is, D' is k-ec apart from z.

By Theorem 14 there is a complete splitting at z resulting in a k-ec digraph. It follows that in case $\varrho(z) = \delta(z)$ this complete splitting, when applied to D, preserves (3). If $\varrho(z) > \delta(z)$, then there are $\varrho(z) - \delta(z)$ edges entering z such that their pairs at the complete splitting are necessarily newly added edges from z to s. Therefore these edges can be deleted from D without destroying (3). $\quad\square$

W. Mader used Theorem 14 to derive Theorem 3 on the constructive characterization of k-ec digraphs. Analogously, Theorem 16 may be used to derive the following.

Theorem 17. *A directed graph $D = (V, E)$ is $(k, 0)$-edge-connected if and only if D can be obtained from a single node by the following two operations: (i) Add a new edge, (ii) pinch j $(0 \leq j \leq k - 1)$ existing edges with a new node z, and add $k - j$ new edges entering z.*

Given these constructive characterizations of (k, k)-ec and $(k, 0)$-ec digraphs, one may formulate the following general conjecture.

Conjecture 1. A directed graph D is (k, l)-edge-connected $(0 \leq l \leq k - 1)$ if and only if it can be built up from a node by the following two operations: (j) add a new edge, (jj) pinch i $(l \leq i \leq k - 1)$ existing edges with a new node z, and add $k - i$ new edges entering z and leaving existing nodes.

Conjecture 2. An undirected graph G is (k, l)-partition-connected if and only if it can be built up from a node by the following two operations: (j) add a new edge, (jj) pinch i $(l \leq i \leq k - 1)$ existing edges with a new node z, and add $k - i$ new edges connecting z with existing nodes.

By Theorem 13 the second conjecture follows from the first one. Theorem 17 asserts the truth of this conjecture for $l = 0$. The conjecture was proved for $l = k - 1$ in [4]. Here we verify the conjecture for $l = 1$. The proof relies on the following lemma.

Lemma 10. *Let $D = (V, E)$ be a $(k, 1)$-edge-connected digraph which is minimal in the sense that the deletion of any edge destroys $(k, 1)$-edge-connectivity ($k \geq 2, |V| \geq 2$). Then D has a node z with $k = \varrho(z) > \delta(z)$ for which there is a set F of $\varrho(z) - \delta(z)$ edges entering z so that $D - F$ is $(k, 1)$-edge-connected apart from z.*

Proof. By (2), there is an edge e entering s. Since (1) cannot break down by deleting e, it follows from the minimality of D that e leaves a subset $X \subset V - s$ for which $\delta(X) = 1$. Since $\varrho(X) \geq k \geq 2$, there must be a node z in X for which $\varrho(z) > \delta(z)$. Let us choose such a node z so that the distance of s from z is as large as possible.

Proposition 4. *Let F be a subset of at most $k - 1$ edges entering z. Then $D' := D - F$ satisfies (2).*

Proof. Assume indirectly that there is a subset $X \subseteq V - s$ for which $\delta_{D'}(X) = 0$. As $\delta(X) \geq 1$, the elements of set of edges of D leaving X are all in F. Therefore $\delta(X) \leq |F| < k$ and, by $\varrho(X) \geq k$, X must contain a node z' for which $\varrho(z') > \delta(z')$. Since the head of each edge leaving X is z, we obtain that each path from z' to s must go through z contradicting the maximal-distance choice of z.   □

Proposition 5. *$\varrho(z) = k$.*

Proof. By Proposition 4 property (2) cannot break down when an edge entering z is left out. Hence the minimality of D implies that every edge entering z enters a subset $X \subseteq V - s$ for which $\varrho(X) = k$. If X and Y are two subsets of V containing z for which $k = \varrho(X) = \varrho(Y)$, then $\varrho(X) + \varrho(Y) \geq \varrho(X \cap Y) + \varrho(X \cup Y) \geq k + k$ from which $\varrho(X \cap Y) = k$ follows. This implies that there is a unique smallest subset Z containing z for which $\varrho(Z) = k$ such that every edge entering z enters Z as well. But then the in-degree of z cannot exceed k and hence $\varrho(z) = k$ as D is $(k, 1)$-ec.   □

By Theorem 16 there is a subset F of edges of D entering z for which $|F| = \varrho(z) - \delta(z) < k$ and the digraph $D - F$ is $(k, 0)$-ec. Now Proposition 4 implies that $D - F$ is actually $(k, 1)$-ec, completing the proof of the lemma.   □

Theorem 18. *A digraph $D_0 = (V, E)$ is $(k, 1)$-edge-connected if and only if D_0 can be built up from a node by the following two operations: (j) add a new edge, (jj) pinch i ($1 \leq i \leq k - 1$) existing edges with a new node z, and add $k - i$ new edges entering z and leaving existing nodes.*

Proof. It is straightforward to see that the two operations preserve $(k, 1)$-edge-connectivity. To prove the reverse direction we use induction on the number of edges. If there is an edge e whose deletion preserves $(k, 1)$-edge-connectivity, then $D_0 - e$ has a required construction by the inductive hypothesis from which the construction of D_0 can be obtained by giving back e (operation (j)).

Therefore we may assume that D_0 is minimally $(k, 1)$-edge-connected with respect to edge deletion. We are done if $|V| = 1$ so assume that $|V| \geq 2$.

By Lemma 10 there is a node z with $k = \varrho(z) > \delta(z)$ for which there is a subset F of $\varrho(z) - \delta(z)$ edges entering z so that the digraph $D_0 - F$ is $(k,1)$-ec apart from z By Theorem 15 there is a complete splitting at z so that the resulting digraph $D_1 = (V - z, E_1)$ is $(k,1)$-ec. By the inductive hypothesis D_1 can be constructed from a node by the two given operations. But then D_0 is also constructible this way as D_0 arises from D_1 by operation (ii).   □

By combining this result with Theorem 13 we obtain the the following special case of Conjecture 2.

Theorem 19. *An undirected graph G is $(k,1)$-partition-connected if and only if it can be built up from a node by the following two operations: (j) add a new edge, (jj) pinch i $(1 \leq i \leq k - 1)$ existing edges with a new node z, and add $k - i$ new edges connecting z with existing nodes.*

Acknowledgement. The authors thank Zoltán Király for fruitful discussions on the topic.

References

1. A. Frank: On the orientation of graphs. J. Combinatorial Theory, Ser. B, Vol. **28**, No. 3 (1980), 251–261
2. A. Frank: Connectivity augmentation problems in network design, in: Mathematical Programming: State of the Art 1994, eds. J.R. Birge and K.G. Murty, The University of Michigan, (1994) 34–63
3. A. Frank: Connectivity and network flows, in: Handbook of Combinatorics (eds. R. Graham, M. Grötschel and L. Lovász), Elsevier Science B.V. (1995) 111–177.
4. A. Frank and Z. Király: Graph orientations with edge-connection and parity constraints. Combinatorica (to appear)
5. L. Henneberg: Die graphische Statik der starren Systeme. Leipzig (1911)
6. Z. Király, personal communication (2000)
7. G. Laman: On graphs and rigidity of plane skeletal structures. J. Engineering Math. **4** (1970) 331–340
8. L. Lovász: Combinatorial Problems and Exercises. North-Holland (1979)
9. W. Mader: Ecken vom Innen- und Aussengrad k in minimal n-fach kantenzusammenhängenden Digraphen. Arch. Math. 25 (1974), 107–112
10. W. Mader: Konstruktion aller n-fach kantenzusammenhängenden Digraphen. Europ. J. Combinatorics **3** (1982) 63–67
11. C. St. J. A. Nash-Williams: Edge-disjoint spanning trees of finite graphs. J. London Math. Soc. **36** (1961) 445–450
12. C. St. J. A. Nash-Williams: Decomposition of finite graphs into forests. J. London Math. Soc. **39** (1964) 12
13. T.-S. Tay: Henneberg's method for bar and body frameworks. Structural Topology **17** (1991) 53–58
14. W. T. Tutte: On the problem of decomposing a graph into n connected factors. J. London Math. Soc. **36** (1961) 221–230
15. W. T. Tutte: Connectivity in Graphs. Toronto University Press, Toronto (1966)

Bisubmodular Function Minimization

Satoru Fujishige[1] and Satoru Iwata[2]

[1] Graduate School of Engineering Science
Osaka University, Toyonaka, Osaka 560-8531, Japan
fujishig@sys.es.osaka-u.ac.jp
[2] Graduate School of Information Science and Technology
University of Tokyo, Tokyo 113-8656, Japan
iwata@sr3.t.u-tokyo.ac.jp

Abstract. This paper presents the first combinatorial, polynomial-time algorithm for minimizing bisubmodular functions, extending the scaling algorithm for submodular function minimization due to Iwata, Fleischer, and Fujishige. A bisubmodular function arises as a rank function of a delta-matroid. The scaling algorithm naturally leads to the first combinatorial polynomial-time algorithm for testing membership in delta-matroid polyhedra. Unlike the case of matroid polyhedra, it remains open to develop a combinatorial strongly polynomial algorithm for this problem.

1 Introduction

Let V be a finite nonempty set of cardinality n and 3^V denote the set of ordered pairs of disjoint subsets of V. Two binary operations $\sqcup$ and $\sqcap$ on 3^V are defined by

$$(X_1, Y_1) \sqcup (X_2, Y_2) = ((X_1 \cup X_2)\backslash(Y_1 \cup Y_2), (Y_1 \cup Y_2)\backslash(X_1 \cup X_2))$$
$$(X_1, Y_1) \sqcap (X_2, Y_2) = (X_1 \cap X_2, Y_1 \cap Y_2).$$

A function $f : 3^V \to \mathbf{R}$ is called *bisubmodular* if it satisfies

$$f(X_1, Y_1) + f(X_2, Y_2) \geq f((X_1, Y_1) \sqcup (X_2, Y_2)) + f((X_1, Y_1) \sqcap (X_2, Y_2))$$

for any (X_1, Y_1) and (X_2, Y_2) in 3^V. This paper presents the first combinatorial polynomial-time algorithm for minimizing bisubmodular functions.

A bisubmodular function generalizes a submodular function as follows. Let 2^V denote the family of all the subsets of V. A function $g : 2^V \to \mathbf{R}$ is called *submodular* if it satisfies

$$g(Z_1) + g(Z_2) \geq g(Z_1 \cup Z_2) + g(Z_1 \cap Z_2)$$

for any $Z_1, Z_2 \subseteq V$. For a submodular function g, we define a bisubmodular function $f : 3^V \to \mathbf{R}$ by

$$f(X, Y) = g(X) + g(V - Y) - g(V).$$

K. Aardal, B. Gerards (Eds.): IPCO 2001, LNCS 2081, pp. 160–169, 2001.
© Springer-Verlag Berlin Heidelberg 2001

If (X, Y) is a minimizer of f, then both X and $V - Y$ are minimizers of g. Thus, bisubmodular function minimization is a generalization of submodular function minimization.

The first polynomial-time algorithm for submodular function minimization is due to Grötschel–Lovász–Schrijver [16]. They also give the first strongly polynomial algorithms in [17]. Their algorithms rely on the ellipsoid method, which is not efficient in practice. Recently, two combinatorial strongly polynomial algorithms are devised independently by Schrijver [22] and Iwata–Fleischer–Fujishige [18]. Both of these new algorithms are based on a combinatorial pseudopolynomial time algorithm of Cunningham [8]. The algorithm of Schrijver [22] directly achieves the strong polynomiality, whereas Iwata–Fleischer–Fujishige [18] develop a scaling algorithm with weakly polynomial time complexity and then convert it to a strongly polynomial one.

In the present paper, we extend the scaling algorithm of Iwata–Fleischer–Fujishige [18] to solve the minimization problem for integer-valued bisubmodular functions. The resulting algorithm runs in $O(n^5 \log M)$ time, where M designates the maximum value of f. This bound is weakly polynomial, and it remains open to develop a combinatorial strongly polynomial algorithm.

2 Delta-Matroids

A bisubmodular function arises as a rank function of a delta-matroid introduced independently by Bouchet [3] and Chandrasekaran–Kabadi [6]. A delta-matroid is a set system $(V, \mathcal{F})$ with $\mathcal{F}$ being a nonempty family of subsets of V that satisfies the following exchange property:

$$\forall F_1, F_2 \in \mathcal{F}, \forall v \in F_1 \triangle F_2, \exists u \in F_1 \triangle F_2 : F_1 \triangle \{u, v\} \in \mathcal{F},$$

where $\triangle$ denotes the symmetric difference. A slightly restricted set system with an additional condition $\emptyset \in \mathcal{F}$ had been introduced by Dress–Havel [11]. A member of $\mathcal{F}$ is called a feasible set of the delta-matroid. Note that the base and the independent-set families of a matroid saitisfy this exchange property. Thus, a delta-matroid is a generalization of a matroid.

Chandrasekaran–Kabadi [6] showed that the rank function $\varrho : 3^V \to \mathbf{Z}$ defined by

$$\varrho(X, Y) = \max\{|X \cap F| - |Y \cap F| \mid F \in \mathcal{F}\}$$

is bisubmodular. The convex hull of the characteristic vectors of the feasible sets is described by

$$P(\varrho) = \{x \mid x \in \mathbf{R}^V, x(X) - x(Y) \le \varrho(X, Y)\},$$

which is called the delta-matroid polyhedron. This fact follows from the greedy algorithm [3,6] for optimizing a linear function over the feasible sets.

Given a vector $x \in \mathbf{R}^V$, one can test if x belongs to $P(\varrho)$ by minimizing a bisubmodular function $f(X, Y) = \varrho(X, Y) - x(X) + x(Y)$. Even for such a special case of bisubmodular function minimization, no combinatorial algorithm

was known to run in polynomial time. This is in contrast with the matroid polyhedron, for which Cunningham [7] devised a combinatorial strongly polynomial
algorithm for testing membership.

A simple example of a delta-matroid is a matching delta-matroid [4], whose
feasible sets are the perfectly matchable vertex subsets of an undirected graph.
The delta-matroid polyhedron is the matchable set polytope [2]. For this special
case, Cunningham–Green-Krótki [10] developed an augmenting path algorithm
for solving the separation problem in polynomial time with the aid of the scaling
technique.

3 Bisubmodular Polyhedra

As a generalization of the delta-matroid polyhedron, a *bisubmodular polyhedron*

$$P(f) = \{x \mid x \in \mathbf{R}^V, \forall (X,Y) \in 3^V : x(X) - x(Y) \le f(X,Y)\}$$

is associated with a general bisubmodular function $f : 3^V \to \mathbf{R}$, where we assume
$f(\emptyset, \emptyset) = 0$. For a vector $x \in \mathbf{R}^V$, we denote $\|x\| = \sum_{v \in V} |x(v)|$. The following
min-max relation characterizes the minimum value of f.

Theorem 1 ([15]). *For any bisubmodular function f,*

$$\min\{f(X,Y) \mid (X,Y) \in 3^V\} = \max\{-\|x\| \mid x \in P(f)\}.$$

$\square$

The linear optimization problem over the bisubmodular polyhedron can be
solved by the following greedy algorithm, which was first introduced by Dunstan–
Welsh [12].

Let $\sigma : V \to \{+, -\}$ be a sign function. For any subset $U \subseteq V$, we denote by
$U|\sigma$ the pair $(X,Y) \in 3^V$ with $X = \{u \mid u \in U, \sigma(u) = +\}$ and $Y = \{u \mid u \in U, \sigma(u) = -\}$. We also write $f(U|\sigma) = f(X,Y)$ for any function $f : 3^V \to \mathbf{R}$,
and $x(U|\sigma) = x(X) - x(Y)$ for any vector $x \in \mathbf{R}^V$.

Let $L = (v_1, \cdots, v_n)$ be a linear ordering of V. For each $j = 1, \cdots, n$, let
$L(v_j) = \{v_1, \cdots, v_j\}$. The *greedy algorithm* with respect to L and a sign function
σ assigns $y(v_j) := \sigma(v_j)\{f(L(v_j)|\sigma) - f(L(v_{j-1})|\sigma)\}$ for each $j = 1, \cdots, n$. Then
the resulting vector $y \in \mathbf{R}^V$ is an extreme point of the bisubmodular polyhedron
$P(f)$.

Given a linear ordering $L = (v_1, \cdots, v_n)$ and a sign function σ, for a weight
function $w : V \to \mathbf{R}$ that satisfies $|w(v_1)| \ge \cdots \ge |w(v_n)|$ and $w(v) = \sigma(v)|w(v)|$
for each $v \in V$, the vector y generated by the greedy algorithm with respect to
L and σ maximizes the linear function $\sum_{v \in V} w(v)y(v)$ over the bisubmodular
polyhedron $P(f)$. See [14, §3.5 (b)] for a survey on bisubmodular polyhedron
including the validity of the greedy algorithm (also see [1]).

Based on the validity of this greedy algorithm, Qi [21] established a connection between bisubmodular functions and their convex extensions. This is a
generalization of a result of Lovász [19] on submodular functions, and it leads to

a polynomial-time algorithm for bisubmodular function minimization using the ellipsoid method.

The concept of bisubmodular polyhedron is extended to that of jump system by Bouchet–Cunningham [5]. A jump system is a set of lattice points satisfying a certain axiom. Examples include the set of degree sequences of a graph [9]. The lattice points contained in an integral bisubmodular polyhedron form a jump system, called a convex jump system, and conversely the convex hull of a jump system is an integral bisubmodular polyhedron. Recently, Lovász [20] investigated the membership problem in jump systems and proved a min-max theorem for a fairly wide class of jump systems. This result contains many interesting combinatorial theorems including Theorem 1. The present paper provides an algorithmic approach to this membership problem in convex jump systems.

4   Scaling Algorithm

This section presents a scaling algorithm for minimizaing an integer-valued bisubmodular function $f : 3^V \to \mathbf{Z}$, provided that an oracle for evaluating the function value is available.

The scaling algorithm works with a positive parameter δ. The algorithm keeps a vector $x \in \mathrm{P}(f)$ as a convex combination of extreme points of $\mathrm{P}(f)$. Namely, $x = \sum_{i \in I} \lambda_i y_i$ with $\lambda_i > 0$ for each $i \in I$ and $\sum_{i \in I} \lambda_i = 1$. Each extreme point y_i is generated by the greedy algorithm with respect to L_i and σ_i. It also keeps a pair of functions $\varphi : V \times V \to \mathbf{R}$ and $\psi : V \times V \to \mathbf{R}$. The function φ is skew-symmetric, i.e., $\varphi(u,v) + \varphi(v,u) = 0$ for any $u, v \in V$, while ψ is symmetric, i.e., $\psi(u,v) = \psi(v,u)$ for any $u, v \in V$. These functions are called δ-*feasible* if they satisfy $-\delta \le \varphi(u,v) \le \delta$ and $-\delta \le \psi(u,v) \le \delta$ for any $u, v \in V$. The boundaries $\partial \varphi$ and $\partial \psi$ are defined by $\partial \varphi(u) = \sum_{v \in V} \varphi(u,v)$ and $\partial \psi(u) = \sum_{v \in V} \psi(u,v)$.

The algorithm starts with an extreme point $x \in \mathrm{P}(f)$ generated by the greedy algorithm with respect to a linear ordering L and a sign function σ. The initial value of δ is given by $\delta := \|x\|/n^2$.

Each scaling phase starts by cutting the value of δ in half. Then it modifies φ and ψ to make them δ-feasible. This can be done by setting each $\varphi(u,v)$ and $\psi(u,v)$ to the closest values in the interval $[-\delta, \delta]$. The rest of the scaling phase aims at decreasing $\|z\|$ for $z = x + \partial \varphi + \partial \psi$.

Given δ-feasible φ and ψ, the algorithm constructs an auxiliary directed graph $G(\varphi, \psi)$ as follows. Let V^+ and V^- be the copies of V. For each $v \in V$, we denote its copies by $v^+ \in V^+$ and $v^- \in V^-$. The vertex set of $G(\varphi, \psi)$ is $V^+ \cup V^-$. The arc set $A(\varphi, \psi) = A(\varphi) \cup A(\psi)$ of $G(\varphi, \psi)$ is defined by

$$A(\varphi) = \{(u^+, v^+) \mid u \ne v, \, \varphi(u,v) \le 0\} \cup \{(u^-, v^-) \mid u \ne v, \, \varphi(u,v) \ge 0\},$$
$$A(\psi) = \{(u^+, v^-) \mid \psi(u,v) \le 0\} \cup \{(u^-, v^+) \mid \psi(u,v) \ge 0\}.$$

For any subset $U \subseteq V$, we also denote its copies by $U^+ \subseteq V^+$ and $U^- \subseteq V^-$.

Let $S = \{v \mid v \in V, z(v) \le -\delta\}$ and $T = \{v \mid v \in V, z(v) \ge \delta\}$. A simple directed path in $G(\varphi, \psi)$ from $S^+ \cup T^-$ to $S^- \cup T^+$ is called a δ-*augmenting path*. If there exists a δ-augmenting path P, the algorithm applies the following δ-augmentation to φ and ψ.

Augment$(\delta, P, \varphi, \psi)$:

- For each (u^+, v^+) in P, $\varphi(u,v) := \varphi(u,v) + \delta/2$ and $\varphi(v,u) := \varphi(v,u) - \delta/2$.
- For each (u^-, v^-) in P, $\varphi(u,v) := \varphi(u,v) - \delta/2$ and $\varphi(v,u) := \varphi(v,u) + \delta/2$.
- For each (u^+, v^-) in P, $\psi(u,v) := \psi(u,v) + \delta/2$ and $\psi(v,u) := \psi(v,u) + \delta/2$.
- For each (u^-, v^+) in P, $\psi(u,v) := \psi(u,v) - \delta/2$ and $\psi(v,u) := \psi(v,u) - \delta/2$.

As a result of a δ-augmentation, $\|z\|$ decreases by δ.

After each δ-augmentation, the algorithm computes an expression of x as a convex combination of affinely independent extreme points of $P(f)$ chosen from among $\{y_i \mid i \in I\}$. This can be done by a standard linear programming technique using Gaussian elimination.

If there is no δ-augmenting path, let $X^+ \subseteq V^+$ and $Y^- \subseteq V^-$ be the sets of vertices reachable by directed paths from $S^+ \cup T^-$. Then we have $S \subseteq X$, $T \subseteq Y$, and $X \cap Y = \emptyset$. For each $i \in I$, consider a pair of disjoint subsets $W_i = \{u \mid u^{\sigma_i(u)} \in X^+ \cup Y^-\}$ and $R_i = \{u \mid u^{\sigma_i(u)} \in X^- \cup Y^+\}$. We now introduce two procedures Double-Exchange and Tail-Exchange.

Procedure Double-Exchange(i, u, v) is applicable if u immediately succeeds v in L_i and either $u \in W_i$ and $v \notin W_i$ or $u \notin R_i$ and $v \in R_i$ hold. Such a triple (i, u, v) is called *active*. The first step of the procedure is to compute

$$\beta := f(L_i(u)\backslash\{v\}|\sigma_i) - f(L_i(u)|\sigma_i) + \sigma_i(v)y_i(v).$$

Then it interchanges u and v in L_i and updates y_i as $y_i := y_i + \beta(\sigma_i(u)\chi_u - \sigma_i(v)\chi_v)$. The resulting y_i is an extreme point generated by the new linear ordering L_i and sign function σ_i.

If $\lambda_i\beta \leq \delta$, Double-Exchange$(i, u, v)$ is called *saturating*. Otherwise, it is called *nonsaturating*. In the nonsaturating case, the procedure adds to I a new index k with y_k, σ_k and L_k being the previous y_i, σ_i and L_i, and assigns $\lambda_k := \lambda_i - \delta/\beta$ and $\lambda_i := \delta/\beta$. In both cases, x moves to $x := x + \alpha(\sigma_i(u)\chi_u - \sigma_i(v)\chi_v)$ with $\alpha = \min\{\delta, \lambda_i\beta\}$. In order to keep z invariant, the procedure finally modifies φ or ψ appropriately. If $\sigma_i(u) = \sigma_i(v)$, it updates $\varphi(u,v) := \varphi(u,v) - \sigma_i(u)\alpha$ and $\varphi(v,u) := \varphi(v,u) + \sigma_i(u)\alpha$. On the other hand, if $\sigma_i(u) \neq \sigma_i(v)$, then $\psi(u,v) := \psi(u,v) - \sigma_i(u)\alpha$ and $\psi(v,u) := \psi(v,u) - \sigma_i(u)\alpha$. A formal description of Double-Exchange is given in Figure 1.

Procedure Tail-Exchange(i, v) is applicable if v is the last element in L_i and $v \in R_i$. Such a pair (i, v) is also called *active*. The first step of the procedure is to reverse the sign $\sigma_i(v)$. It then computes

$$\beta := f(V|\sigma_i) - f(V\backslash\{v\}|\sigma_i) - \sigma_i(v)y_i(v).$$

and updates $y_i := y_i + \sigma_i(v)\beta\chi_v$. The resulting y_i is an extreme point generated by L_i and the new σ_i.

If $\lambda_i\beta \leq \delta$, Tail-Exchange$(i, v)$ is called *saturating*. Otherwise, it is called *nonsaturating*. In the nonsaturating case, the procedure adds to I a new index

Double-Exchange(i, u, v);
$\beta := f(L_i(u)\backslash\{v\}|\sigma_i) - f(L_i(u)|\sigma_i) + \sigma_i(v)y_i(v)$;
$\alpha := \min\{\delta, \lambda_i\beta\}$;
If $\alpha < \lambda_i\beta$ **then**
$\qquad k \leftarrow$ a new index;
$\qquad I := I \cup \{k\}$;
$\qquad \lambda_k := \lambda_i - \alpha/\beta$;
$\qquad \lambda_i := \alpha/\beta$;
$\qquad y_k := y_i$;
$\qquad L_k := L_i$;
Update L_i by interchanging u and v;
$y_i := y_i + \beta(\sigma_i(u)\chi_u - \sigma_i(v)\chi_v)$;
$x := x + \alpha(\sigma_i(u)\chi_u - \sigma_i(v)\chi_v)$;
If $\sigma_i(u) = \sigma_i(v)$ **then**
$\qquad \varphi(u, v) := \varphi(u, v) - \sigma_i(u)\alpha$;
$\qquad \varphi(v, u) := \varphi(v, u) + \sigma_i(u)\alpha$;
Else
$\qquad \psi(u, v) := \psi(u, v) - \sigma_i(u)\alpha$;
$\qquad \psi(v, u) := \psi(v, u) - \sigma_i(u)\alpha$.

Fig. 1. Algorithmic description of Procedure Double-Exchange(i, u, v).

k with y_k, σ_k and L_k being the previous y_i, σ_i and L_i, and assigns $\lambda_k := \lambda_i - \delta/\beta$ and $\lambda_i := \delta/\beta$. In both cases, x moves to $x := x + \sigma_i(v)\alpha\chi_v$ with $\alpha = \min\{\delta, \lambda_i\beta\}$. In order to keep z invariant, the procedure finally modifies ψ as $\psi(v, v) := \psi(v, v) - \sigma_i(v)\alpha$. A formal description of Tail-Exchange is given in Figure 2.

If there is no δ-augmenting path in $G(\varphi, \psi)$ and neither Double-Exchange nor Tail-Exchange is applicable, the algorithm terminates the δ-scaling phase by cutting the value of δ in half.

A formal description of our scaling algorithm BFM is now given in Figure 3.

5   Validity and Complexity

This section is devoted to the analysis of our scaling algorithm. We first discuss the validity.

Lemma 1. *At the end of the δ-scaling phase, the current $(X, Y) \in 3^V$ and $z = x + \partial\varphi + \partial\psi$ satisfy $\|z\| \leq 2n\delta - f(X, Y)$.*

Proof. At the end of the δ-scaling phase, we have $y_i(X) - y_i(Y) = f(X, Y)$ for each $i \in I$. Hence, x satisfies $x(X) - x(Y) = f(X, Y)$. By the definition of (X, Y), we immediately have $\partial\varphi(X) > 0$, $\partial\varphi(Y) < 0$, $\partial\psi(X) > 0$, and $\partial\psi(Y) < 0$, where note that $\partial\varphi(X) = \sum\{\varphi(u, v) \mid u \in X, v \in V\backslash X\}$ with $\varphi(u, v) > 0$

$$
\begin{array}{l}
\text{Tail-Exchange}(i, v); \\[2pt]
\sigma_i(v) := -\sigma_i(v); \\[2pt]
\beta := f(V|\sigma_i) - f(V\backslash\{v\}|\sigma_i) - \sigma_i(v)y_i(v); \\[2pt]
\alpha := \min\{\delta, \lambda_i\beta\}; \\[2pt]
\textbf{If } \alpha < \lambda_i\beta \textbf{ then} \\[2pt]
\qquad k \leftarrow \text{ a new index}; \\[2pt]
\qquad I := I \cup \{k\}; \\[2pt]
\qquad \lambda_k := \lambda_i - \alpha/\beta; \\[2pt]
\qquad \lambda_i := \alpha/\beta; \\[2pt]
\qquad y_k := y_i; \\[2pt]
\qquad L_k := L_i; \\[2pt]
y_i := y_i + \sigma_i(v)\beta\chi_v; \\[2pt]
x := x + \sigma_i(v)\alpha\chi_v; \\[2pt]
\psi(v, v) := \psi(v, v) - \sigma_i(v)\alpha.
\end{array}
$$

Fig. 2. Algorithmic description of Procedure Tail-Exchange(i, v).

$(u \in X, v \in V \setminus X)$ and similarly the other inequalities. Since $S \subseteq X$, $T \subseteq Y$, and $X \cap Y \neq \emptyset$, we have $z(v) \leq \delta$ for $v \in X$ and $z(v) \geq -\delta$ for $v \in Y$. Therefore, we have $\|z\| \leq -z(X) + z(Y) + 2n\delta \leq -x(X) + x(Y) + 2n\delta = -f(X, Y) + 2n\delta$.

Theorem 2. *The algorithm obtains a minimizer of f at the end of the δ-scaling phase with $\delta < 1/(3n^2)$.*

Proof. Since $|\partial\varphi(v)| \leq (n-1)\delta$ and $|\partial\psi(v)| \leq n\delta$ for each $v \in V$, it follows from Lemma 1 that $\|x\| \leq (2n^2+n)\delta - f(X, Y) < 1 - f(X, Y)$. For any $(X', Y') \in 3^V$, we have $f(X', Y') \geq -\|x\| > f(X, Y) - 1$. Hence (X, Y) is a minimizer of the integer-valued function f.

We now give a running time bound of our algorithm.

Lemma 2. *Each scaling phase performs $O(n^2)$ augmentations.*

Proof. At the beginning of the δ-scaling phase, the algorithm modifies φ and ψ to make them δ-feasible. This changes $\|z\|$ by at most $2n^2\delta$. Therefore, by Lemma 1, the pair (X, Y) must satisfy $\|z\| \leq 2n^2\delta + 4n\delta - f(X, Y)$ after updating φ and ψ at the beginning of the δ-scaling phase. On the other hand, we have $\|z\| \geq -z(X) + z(Y) \geq -x(X) + x(Y) - 2n^2\delta = -f(X, Y) - 2n^2\delta$. Thus $\|z\|$ decreases by at most $4n\delta + 4n^2\delta$ until the end of the δ-scaling phase. Since each δ-augmentation decreases $\|z\|$ by δ, the number of δ-augmentations in the δ-scaling phase is at most $4n^2 + 4n$, which is $O(n^2)$.

Lemma 3. *The algorithm performs Procedure Double-Exchange $O(n^3)$ times and Tail-Exchange $O(n^2)$ times between δ-augmentations. Moreover, the running time between δ-augmentations is $O(n^3)$.*

BFM(f):
Initialization:
    $L \leftarrow$ a linear ordering on V;
    $\sigma \leftarrow$ a sign function on V;
    $x \leftarrow$ an extreme vector in $\mathrm{P}(f)$ generated by L and σ;
    $I := \{i\}$, $y_i := x$, $\lambda_i := 1$, $L_i := L$;
    $\varphi := \mathbf{0}$, $\psi := \mathbf{0}$;
    $\delta \leftarrow \|x\|/n^2$;
While $\delta \geq 1/(3n^2)$ **do**
    $\delta := \delta/2$;
    For $(u, v) \in V \times V$ **do**
        Change $\varphi(u,v)$ and $\psi(u,v)$ to the closest values in the interval $[-\delta, \delta]$;
    $S := \{v \mid x(v) + \partial\varphi(v) \leq -\delta\}$;
    $T := \{v \mid x(v) + \partial\varphi(v) \geq \delta\}$;
    $X^+ \leftarrow$ the set of vertices in V^+ reachable from $S^+ \cup T^-$ in $G(\varphi, \psi)$;
    $Y^- \leftarrow$ the set of vertices in V^- reachable from $S^+ \cup T^-$ in $G(\varphi, \psi)$;
    $Q \leftarrow$ the set of active triples and active pairs;
    While $\exists\delta$-augmenting path or $Q \neq \emptyset$ **do**
        If $\exists P$: δ-augmenting path **then**
            Augment(δ, P, φ, ψ);
            Update S, T, X^+, Y^-, Q;
            Express x as $x = \sum_{i \in I} \lambda_i y_i$ by possibly smaller affinely independent
                subset I and positive coefficients $\lambda_i > 0$ for $i \in I$;
        Else
            While $\not\exists\delta$-augmenting path and $Q \neq \emptyset$ **do**
                Find an active $(i, u, v) \in Q$ or active $(i, v) \in Q$;
                Apply Double-Exchange(i, u, v) or Tail-Exchange(i, v);
                Update X^+, Y^-, Q;
Return (X, Y);
End.

Fig. 3. A scaling algorithm for bisubmodular function minimization.

Proof. Procedure Double-Exchange moves a vertex of W_i towards the head of L_i and/or a vertex in R_i towards the tail of L_i. Procedure Tail-Exchange changes a vertex of R_i to W_i. No vertex goes out of W_i. A vertex of R_i can be switched to W_i by Tail-Exchange. However, it does not go out of $R_i \cup W_i$. Thus, for each $i \in I$, after at most $O(n^2)$ applications of Double-Exchange and $O(n)$ applications of Tail-Exchange to $i \in I$, the subset R_i is empty and $W_i = L_i(w)$ holds for some $w \in V$. At this point, neither Double-Exchange nor Tail-Exchange is applicable to $i \in I$.

After each δ-augmentation, the algorithm updates the convex combination $x = \sum_{i \in I} \lambda_i y_i$ so that $|I| \leq n + 1$. A new index is added to I as a result of nonsaturating Double-Exchange(i, u, v) and Tail-Exchange(i, v). In both cases, v joins W_i. This can happen at most $n - 1$ times before the algorithm finds a δ-

augmenting path or finishes the δ-scaling phase. Hence, $|I|$ is always $O(n)$, and the algorithm performs Double-Exchange $O(n^3)$ times and Tail-Exchange $O(n^2)$ times between δ-augmentations.

It should be noted that each nonsaturating Double-Exchange or Tail-Exchange requires $O(n)$ time and each saturating one requires $O(1)$ time. Moreover, there are $O(n^3)$ saturating and $O(n)$ nonsaturating Double-Exchange or Tail-Exchange between δ-augmentations, as shown above. Also note that updating the convex combination $x = \sum_{i \in I} \lambda_i y_i$ is performed in $O(n^3)$ time by Gaussian elimination. Hence, the running time between δ-augmentations is $O(n^3)$.

Let M be the maximum value of f. Since $f(\emptyset, \emptyset) = 0$, the maximum value M is nonnegative.

Theorem 3. *The scaling algorithm finds a minimizer of f in* $O(n^5 \log M)$ *time.*

Proof. For the initial $x \in P(f)$, let $B = \{v \mid x(v) > 0\}$ and $C = \{v \mid x(v) < 0\}$. Then we have $\|x\| = x(B) - x(C) \leq f(B, C) \leq M$. Hence the algorithm performs $O(\log M)$ scaling phases. It follows from Lemmas 2 and 3 that each scaling phase performs $O(n^5)$ function evaluations and arithmetic operations. Therefore the total running time is $O(n^5 \log M)$.

6 Conclusion

We have described a polynomial-time algorithm for minimizing integer-valued bisubmodular functions. If we are given a positive lower bound ϵ for the difference between the minimum and the second minimum value of f, a variant of the present algorithm works for any real-valued bisubmodular function f. The only required modification is to change the stopping rule $\delta < 1/(3n^2)$ to $\delta < \epsilon/(3n^2)$. The running time is $O(n^5 \log(M/\epsilon))$. Thus we obtain a polynomial-time algorithm for testing membership in delta-matroid polyhedra. One can make this algorithm strongly polynomial with the aid of a generic preprocessing technique of Frank–Tardos [13] using simultaneous Diophantine approximation. However, a more natural strongly polynomial algorithm is desirable.

References

1. K. Ando and S. Fujishige: On structures of bisubmodular polyhedra, *Math. Programming*, 74 (1996), 293–317.
2. E. Balas and W. R. Pulleyblank: The perfectly matchable subgraph polytope of an arbitrary graph, *Combinatorica*, 9 (1989), 321–337.
3. A. Bouchet: Greedy algorithm and symmetric matroids, *Math. Programming*, 38 (1987), 147–159.
4. A. Bouchet: Matchings and $\triangle$-matroids, *Discrete Appl. Math.*, 24 (1989), 55–62.
5. A. Bouchet and W. H. Cunningham: Delta-matroids, jump systems and bisubmodular polyhedra, *SIAM J. Discrete Math.*, 8 (1995), 17–32.
6. R. Chandrasekaran and S. N. Kabadi: Pseudomatroids, *Discrete Math.*, 71 (1988), 205–217.

7. W. H. Cunningham: Testing membership in matroid polyhedra, *J. Combin. Theory*, B36 (1984), 161–188.
8. W. H. Cunningham: On submodular function minimization, *Combinatorica*, 5 (1985), 185–192.
9. W. H. Cunningham and J. Green-Krótki: *b*-matching degree sequence polyhedra, *Combinatorica*, 11 (1991), 219–230.
10. W. H. Cunningham and J. Green-Krótki: A separation algorithm for the matchable set polytope, *Math. Programming*, 65 (1994), 139–150.
11. A. Dress and T. F. Havel: Some combinatorial properties of discriminants in metric vector spaces, *Adv. Math.*, 62 (1986), 285–312.
12. F. D. J. Dunstan and D. J. A. Welsh: A greedy algorithm solving a certain class of linear programmes, *Math. Programming*, 5 (1973), 338–353.
13. A. Frank and É. Tardos: An application of simultaneous Diophantine approximation in combinatorial optimization, *Combinatorica*, 7 (1987), 49–65.
14. S. Fujishige: *Submodular Functions and Optimization*, North-Holland, 1991.
15. S. Fujishige: A min-max theorem for bisubmodular polyhedra, *SIAM J. Discrete Math.*, 10 (1997), 294–308.
16. M. Grötschel, L. Lovász, and A. Schrijver: The ellipsoid method and its consequences in combinatorial optimization, *Combinatorica*, 1 (1981), 169–197.
17. M. Grötschel, L. Lovász, and A. Schrijver: *Geometric Algorithms and Combinatorial Optimization*, Springer-Verlag, 1988.
18. S. Iwata, L. Fleischer, and S. Fujishige: A combinatorial strongly polynomial algorithm for minimizing submodular functions, *J. ACM*, to appear.
19. L. Lovász: Submodular functions and convexity. *Mathematical Programming — The State of the Art*, A. Bachem, M. Grötschel and B. Korte, eds., Springer-Verlag, 1983, 235–257.
20. L. Lovász: The membership problem in jump systems, *J. Combin. Theory*, Ser. B, 70 (1997), 45–66.
21. L. Qi: Directed submodularity, ditroids and directed submodular flows, *Math. Programming*, 42 (1988), 579–599.
22. A. Schrijver: A combinatorial algorithm minimizing submodular functions in strongly polynomial time, *J. Combin. Theory*, Ser. B, 80 (2000), 346–355.

On the Integrality Gap of a Natural Formulation of the Single-Sink Buy-at-Bulk Network Design Problem*

Naveen Garg[1], Rohit Khandekar[1], Goran Konjevod[2], R. Ravi[3], F.S. Salman[4], and Amitabh Sinha[3]

[1] Department of Computer Science and Engineering,
Indian Institute of Technology, New Delhi, India.
{naveen, rohitk}@cse.iitd.ernet.in

[2] Department of Computer Science and Engineering,
Arizona State University, Tempe, AZ 85287, USA.
goran@asu.edu

[3] GSIA, Carnegie Mellon University,
Pittsburgh PA 15213-3890, USA.
{ravi, asinha}@andrew.cmu.edu

[4] Krannert School of Management,
Purdue University, West Lafayette, IN 47907, USA.
salmanf@mgmt.purdue.edu

Abstract. We study two versions of the single sink *buy-at-bulk* network design problem. We are given a network and a single sink, and several sources which demand a certain amount of flow to be routed to the sink. We are also given a finite set of cable types which have different cost characteristics and obey the principle of economies of scale. We wish to construct a minimum cost network to support the demands, using our given cable types. We study a natural integer program formulation of the problem, and show that its integrality gap is $O(k)$, where k is the number of cable types. As a consequence, we also provide an $O(k)$-approximation algorithm.

1 Introduction

1.1 Motivation

We study two network design problems which often arise in practice. Consider a network consisting of a single server and several clients. Each client wishes to route a certain amount of flow to the server. The cost per unit flow along an edge is proportional to the edge length. However, we can reduce the cost per unit flow of routing by paying a certain fixed cost (again proportional to the length

---

* This research was supported by a faculty development grant awarded to R. Ravi by the Carnegie Bosch Institute, Carnegie Mellon University, Pittsburgh PA 15213-3890.

K. Aardal, B. Gerards (Eds.): IPCO 2001, LNCS 2081, pp. 170–184, 2001.

of the edge). We call the problem of finding a minimum cost network supporting the required flow the *deep-discount* problem.

Alternatively, at each edge we might be able to pay for and install a certain capacity selected from a set of allowable discrete capacity units, and then route flow (up to the installed capacity) for free. The problem of finding a minimum cost network in this scenario is called the *buy-at-bulk* network design problem [14].

Both the above problems reflect economies of scale, in the cost of routing unit flow as the installation cost increases (the deep-discount problem), and in the cost per unit capacity as the capacity increases (the buy-at-bulk problem).

1.2   Our Results

The two problems are in fact equivalent up to a small loss in the value of the solution. In this paper, we focus on the deep-discount problem. We study the structure of the optimum solution, and show that an optimal solution exists which is a tree. We provide a natural IP formulation of the problem, and show that it has an integrality gap of the order of the number of cables. We also provide a polynomial time approximation algorithm by rounding the LP relaxation.

1.3   Previous Work

Mansour and Peleg [11] gave an $O(\log n)$-approximation for the single sink buy-at-bulk problem with a single cable type (only one discrete unit of capacity allowable) for a graph on n nodes. They achieved this result by using a low-weight, low-stretch spanner construction [2].

Designing networks using facilities that provide economies of scale has attracted interest in recent years. Salman *et al* [14] gave an $O(\log D)$ approximation algorithm for the single sink buy-at-bulk problem in Euclidean metric spaces, where D is the total demand. Awerbuch and Azar [4] gave a randomized $O(\log^2 n)$ approximation algorithm for the buy-at-bulk problem with many cable types and many sources and sinks, where n is the number of nodes in the input graph. This improves to $O(\log n \log \log n)$ using the improved tree metric construction of Bartal [5]. For the single sink case with many cable types, an $O(\log n)$ approximation was obtained by Meyerson, Munagala and Plotkin [12] based on their work on the Cost-Distance two metric network design problem. Salman *et al* also gave a constant approximation in [14] for the single cable type case using a LAST construction [10] in place of the spanner construction used in [11]. The approximation ratio was further improved by Hassin, Ravi and Salman [9].

Andrews and Zhang [3] studied a special case of the single-sink buy-at-bulk problem which they call the *access network design* problem and gave an $O(k^2)$ approximation, where k is the number of cable types. As in the deep-discount problem, they use a cost structure where each cable type has a buying and a routing cost, but they assume that if a cable type is used, the routing cost is at least a constant times the buying cost.

An improved approximation to the problem we study was obtained simultaneously but independently by Guha, Meyerson and Munagala [7], who designed a constant-factor approximation algorithm. Their algorithm is combinatorial and is based on their prior work on the access network design problem [8], as opposed to our focus on the LP relaxation and its integrality gap.

1.4   Outline of the Paper

In the next section, we define the deep-discount problem formally and show its relation to the k-cable buy-at-bulk problem. In Sections 3 through 6, we introduce and study our integer program formulation and show that it has low integrality gap. We conclude with time complexity issues and open questions in Section 7.

2   Problem Definition and Inter-reductions

2.1   The Deep-Discount Problem

Let $G = (V, E)$ be a graph with edge-lengths $l : E \rightarrow I\!R^+$. Let $d(u, v)$ denote the length of the shortest path between vertices u and v. We are given sources $\{v_1, \ldots, v_m\} = S$ which want to transport $\{\mathsf{dem}_1, \ldots, \mathsf{dem}_m\}$ units of flow respectively to a common sink $t \in V$. We also have a set of k discount types $\{\kappa_0, \kappa_1, \ldots, \kappa_{k-1}\}$ available for us to purchase and install. Each cable κ_i has an associated fixed cost p_i and a variable cost r_i. If we install cable κ_i at edge e and route f_e flow through it, the contribution to our cost is $l_e(p_i + f_e r_i)$. We may therefore view the installation of cable κ_i at an edge as paying a fixed cost $p_i l_e$ in order to obtain a discounted rate r_i of routing along this edge. The problem of finding an assignment of discount types to the edges and routing all the source demands to the sink at minimum total cost is the deep discount problem with k discount types (DD for short).

Let us order the rates as $r_0 > r_1 > \ldots > r_{k-1}$. The rate $r_0 = 1$ and the price $p_0 = 0$ correspond to not using any discount. (It is easy to see that we may scale our cost functions so that this is true in general.) Observe that if $p_i \geq p_{i+1}$ for some i then κ_i will never be used. Therefore, without loss of generality, we can assume $p_0 < p_1 < \ldots < p_{k-1}$.

2.2   The Buy-at-Bulk Problem with k-Cable Types

In an edge-weighted undirected graph, suppose we are given a set of sources $\{v_1, \ldots, v_m\}$ which want to transport $\{\mathsf{dem}_1, \ldots, \mathsf{dem}_m\}$ units of flow respectively to a common sink t. We have available to us k different cable types, each having capacity u_i and cost c_i. We wish to buy cables such that we have enough capacity to support the simultaneous flow requirements. We are allowed to buy multiple copies of a cable type. There is no flow cost; our only cost incurred is the purchase price of cables. The problem of finding a minimum cost feasible network is the buy-at-bulk problem with k cable types (BB for short). It is **NP**-Hard even when $k = 1$ [14].

2.3   Approximate Equivalence of BB and DD

Suppose we are given a BB instance $BB = (G, c, u)$ on a graph G with k cable types having costs and capacities $(1, 1), (c_1, u_1), \ldots, (c_{k-1}, u_{k-1})$. We transform it into an instance of DD by setting edge costs (fixed and per-unit) $(0, 1), (c_1, \frac{c_1}{u_1})$, $\ldots, (c_{k-1}, \frac{c_{k-1}}{u_{k-1}})$, and call this $DD(BB)$.

Conversely, given a DD instance $DD = (G, p, r)$ on a graph G with k discount types having prices and variable costs $(0, 1), (p_1, r_1), \ldots, (p_{k-1}, r_{k-1})$, we transform it into a BB instance $BB(DD)$ with cable types having costs and capacities $(1, 1), (p_1, \frac{p_1}{r_1}), \ldots, (p_{k-1}, \frac{p_{k-1}}{r_{k-1}})$.

It is easy to see that $BB(DD(BB)) = BB$ and $DD(BB(DD)) = DD$; i.e., the two transformations are inverses of each other. For a problem instance X, we abuse notation to let X also denote the cost of a feasible solution to it. Let X^* denote the cost of an optimal (integer) solution to X.

Lemma 1. $BB \leq DD^*(BB)$

Proof. Consider an edge e and let the flow on e in $DD^*(BB)$ be x_e. If the solution uses discount type 0 on e, then the BB solution does not pay any more than the routing cost already paid. If it uses a discount type $i > 0$, we install $\lceil \frac{x_e}{u_i} \rceil$ copies of cable type i at this edge. Clearly this gives us a feasible solution to BB. For this edge $DD^*(BB)$ has routing cost $\frac{l_e x_e c_i}{u_i}$ and building cost $l_e c_i$, hence a total cost of $l_e c_i (1 + \frac{x_e}{u_i})$. The BB solution has cable cost $l_e c_i \lceil \frac{x_e}{u_i} \rceil$ on edge e, which is no more than the total cost incurred by edge e in $DD^*(BB)$.

Lemma 2. $DD \leq 2BB^*(DD)$

Proof. We will initially allow for multiple discount types on each edge and pay for it all. A pruned solution will satisfy all the desired properties and cost only less. Hence let x_e^i be the number of copies of cable type i used at edge e. We only consider edges with non-zero flow, that is, where $x_e^i > 0$. Note that the flow on e is at most $x_e^i p_i / r_i$. We purchase a discount type i cable on edge e and route the flow through it on this discounted cable. We pay no more than $p_i l_e + \frac{x_e^i p_i}{r_i} r_i l_e \leq (x_e^i + 1) p_i l_e$, which is no more than two times the cost $x_e^i p_i l_e$ already paid by $BB(DD)$, since $x_e^i \geq 1$.

Together, the above two Lemmas imply that $BB^*(DD) \leq BB(DD) \leq DD^* \leq DD \leq 2BB^*(DD)$, so that a ρ approximation algorithm for BB gives a 2ρ approximation algorithm for DD. Similarly, a ρ approximation to DD is a 2ρ approximation to BB.

Given the above relations, we focus on the deep-discount formulation in this paper. One reason for choosing to work with this version is presented in Section 3 where we show there are always acyclic optimal solutions for the deep-discount problem, while this is not always the case for the buy-at-bulk problem (see, e.g., [14]). However, we continue to use the term "cable type" to refer to the discount type used in an edge from time to time, even though a solution to the deep-discount problem involves choices of discount types on edges rather than installation of cables.

3   Structure of an Optimum Solution to the Deep-Discount Problem

Let us look at how an optimal solution allocates the discount types to edges and routes the flow to the sink. Clearly an edge will use only one type of discount. Suppose in an optimum, an edge e uses discount-i. Define a new length function $l'_e := r_i l_e$. Clearly once the fixed cost for the discount is paid, the routing cost is minimized if we route along a shortest path according to the length function l'. Therefore, there is an optimum which routes along shortest paths according to such a length function l'. As a result, we can also assume that the flow never splits. That is, if two commodities share an edge then they share all the subsequent edges on their paths to t. This is because we can choose a shortest path tree in the support graph; flow along this tree to the root will never split.

The cost of routing f units of flow on an edge e using discount-i is $l_e(p_i + r_i f)$. So the discount type corresponding to minimum cost depends only on f and is given by $\mathtt{type}(f) := \mathtt{minarg}_i\{p_i + r_i f \mid 0 \le i < k\}$.

Lemma 3. *The function* $\mathtt{type}(f)$ *defined above is non-decreasing in* f.

Proof. Consider any two discount types, say i and j with $i < j$. We know that $r_i > r_j$ and $p_i < p_j$. As the cost functions $p + rf$ are linear, there is a critical value f of flow such that, $p_i + r_i f = p_j + r_j f$. If the flow is smaller than f, discount-i is better than discount-j. And if the flow is more than f, discount-j is better than discount-i. This proves the lemma.

Suppose the optimal flow is along a path P from v_j to t. As the flow never splits, the flow along P is non-decreasing as we go from v_j to t. Hence from the above lemma, the discount type never decreases as we go from v_j to t. We call this property *path monotonicity*.

Summarizing, we have the following.

Theorem 1. *There exists an optimum solution to the deep-discount problem which satisfies the following properties.*

- *The support graph of the solution is a tree.*
- *The discount types are non-decreasing along the path from a source to the root.*

Similar results were proved independently (but differently) in [3] and [7]. Figure 1 illustrates the structure of an optimum assuming 3 discount types.

4   Linear Program Formulation and Rounding

4.1   Overview of the Algorithm

First we formulate the deep-discount problem as an integer program. We then take the linear relaxation of the IP, and solve it to optimality. Clearly an optimal solution to the LP is a lower bound on the optimal solution to the IP.

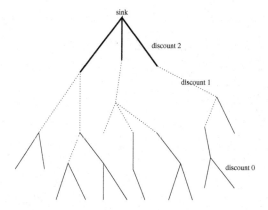

Fig. 1. Structure of an optimum solution to the deep-discount problem with three discount types.

We now use the LP solution to construct our solution. We have already seen that there is an integer optimal solution which is a layered tree. We construct such a tree in a *top-down* manner, starting from the sink. We iteratively augment the tree by adding cables of the next available lower discount type. At each stage we use an argument based on the values of the decision variables in an optimal LP solution to charge the cost of our solution to the LP cost. We thus bound the cost of building our tree. We next bound the routing costs by an argument which essentially relies on the fact that the tree is layered and that distances obey the triangle inequality.

4.2   Integer Program Formulation

We now present a natural IP formulation of the deep-discount problem. As is usual for flow problems, we replace each undirected edge by a pair of anti-parallel directed arcs, each having the same length as the original (undirected) edge. We introduce a variable z_e^i for each $e \in E$ and for each $0 \le i < k$, such that, $z_e^i = 1$ if we are using discount-i on edge e and 0 otherwise. The variable $f_{e;i}^j$ is the flow of commodity j on edge e using discount-i. For a vertex set S (or a singleton vertex v), we define $\delta^+(S)$ to be the set of arcs leaving S. That is, $\delta^+(S) = \{(u,v) \in E : u \in S, v \notin S\}$. Analogously, $\delta^-(S) = \{(u,v) \in E : u \notin S, v \in S\}$. The formulation is given in Figure 2.

The first term in the objective function is the cost of purchasing the various discount types at each edge; we call this the *building* cost. The second term is the total cost (over all vertices v_j) of sending $\mathbf{dem}_j$ amount of flow from vertex v_j to the sink; we call this the *routing* cost of the solution. These two components of the cost of an optimal solution are referred to as OPT_{build} and OPT_{route} respectively.

$$\min \sum_{e \in E} \sum_{i=0}^{k-1} p_i z_e^i l_e \quad + \sum_{v_j \in S} \sum_{e \in E} \sum_{i=0}^{k-1} \text{dem}_j f_{e;i}^j r_i l_e$$

subject to:

$$(i) \quad \sum_{e \in \delta^+(v_j)} \sum_{i=0}^{k-1} f_{e;i}^j \geq 1 \qquad\qquad \forall v_j \in S$$

$$(ii) \quad \sum_{e \in \delta^-(v)} \sum_{i=0}^{k-1} f_{e;i}^j = \sum_{e \in \delta^+(v)} \sum_{i=0}^{k-1} f_{e;i}^j \qquad\qquad \forall v \in V \setminus \{v_j, t\}, 1 \leq j \leq m$$

$$(iii) \quad \sum_{e \in \delta^-(v)} \sum_{i=q}^{k-1} f_{e;i}^j \leq \sum_{e \in \delta^+(v)} \sum_{i=q}^{k-1} f_{e;i}^j \qquad\qquad 0 \leq q < k, \forall v \in V \setminus \{v_j, t\},$$

$$\qquad\qquad\qquad\qquad\qquad\qquad\qquad\qquad 1 \leq j \leq m$$

$$(iv) \quad f_{e;i}^j \qquad\qquad \leq z_e^i \qquad\qquad \forall e \in E, 0 \leq i < k$$

$$(v) \quad \sum_{i=0}^{k-1} z_e^i \qquad\qquad \geq 1 \qquad\qquad \forall e \in E$$

$$(vi) \quad z, f \qquad\qquad \text{non-negative integers}$$

Fig. 2. Integer program formulation of the deep-discount problem.

The first set of constraints ensures that every source has an outflow of one unit which is routed to the sink. The second is the standard flow conservation constraints, treating each commodity separately. The third set of constraints enforces the path monotonicity discussed in the preceding Section, and is therefore valid for the formulation. The fourth simply builds enough capacity, and the fifth ensures that we install at least one cable type on each arc. Note that this is valid and does not add to our cost since we have the default cable available for installation at zero fixed cost.

Relaxing the integrality constraints (vi) to allow the variables to take real non-negative values, we obtain the LP relaxation. This LP has a polynomial number of variables and constraints, and can be therefore solved in polynomial time. The LP relaxation gives us a lower bound which we use in our approximation algorithm.

5   The Rounding Algorithm

5.1   Pruning the Set of Available Cables

We begin by pruning our set of available cables, and we show that this does not increase the cost by more than a constant factor. This pruning is useful in the analysis.

The following lemma shows that the cost of solution does not increase by a large factor if we restrict ourselves to rates that are sufficiently different, that is, they decrease by a constant factor.

Let OPT be the optimum value with rates $r_0, r_1, \ldots, r_{k-1}$ and corresponding prices $p_0, p_1, \ldots, p_{k-1}$. Let $\epsilon \in (0,1)$ be a real number. Assume that $\epsilon^{l-1} \geq r_{k-1} > \epsilon^l$. Now, let us create a new instance as follows. Let the new rates be $1, \epsilon, \ldots, \epsilon^{l-1}$. For each i, let the price corresponding to ϵ^i be p_j, where r_j is the largest rate not bigger than ϵ^i. Let OPT' be the optimum value of this new problem.

Lemma 4. $OPT'_{route} \leq \frac{1}{\epsilon} OPT_{route}$

Proof. Consider an edge e which uses discount-j in OPT. In the solution of the new problem, change its discount type to ϵ^i such that $\epsilon^i \geq r_j > \epsilon^{i+1}$. Thus, for this edge, the price does not increase and the routing cost increases by a factor at most $1/\epsilon$.

Since $OPT'_{build} \leq OPT_{build}$, we have as a consequence that $OPT' \leq \frac{1}{\epsilon} OPT$. Hereafter we assume that the rates $r_0, r_1, \ldots, r_{k-1}$ decrease by a factor at least ϵ for some $0 < \epsilon < 1$, thereby incurring an increase in cost by a factor of at most $1/\epsilon$.

5.2   Building the Solution: Overview

Recall that G is our input graph, and k is the number of cable types. We also have a set of parameters $\{\alpha, \beta, \gamma, \delta\}$, all of which are fixed constants and whose effect will be studied in the analysis in Section 6.

We build our tree in a top-down manner. We begin by defining T_k to be the singleton vertex $\{t\}$, the sink. We then successively augment this tree by adding cables of discount type i to obtain T_i, for i going down $k-1, k-2, \ldots, 1, 0$. Our final tree T_0 is the solution we output. Routing is then trivial – simply route along the unique path from each source to the sink in the tree.

Our basic strategy for constructing the tree T_i from T_{i-1} is to first identify a subset of demand sources that are not yet included in T_{i-1} by using information from their contributions to the routing cost portion of the LP relaxation. In particular, we order these candidate nodes in non-decreasing order of the radius of a ball that is defined based on the routing cost contribution of the center of the ball. We then choose a maximal set of non-overlapping balls going forward in this order. This intuitively ensures that any ball that was not chosen can be charged for their routing via the smaller radius ball that overlapped with it that is included in the current level of the tree. After choosing such a subset of as yet unconnected nodes, we build an approximately minimum building cost Steiner tree with these nodes as terminals and the (contracted) tree T_{i-1} as the root. The balls used to identify this subset now also serve a second purpose of relating the building cost of the Steiner tree to the fractional optimum. Finally, in a third step, we convert the approximate Steiner tree rooted at the contracted T_{i-1} to a LAST (light approximate shortest-path tree [10]) which intuitively ensures that all nodes in the tree are within a constant factor of their distance from the root T_{i-1} in this LAST without increasing the total length (and hence the building cost) of the tree by more than a constant factor. This step is essential to guarantee that the routing cost via this level of the tree does not involve long paths and thus can be charged to within a constant factor of the appropriate LP routing cost bound.

Algorithm Deep-discount$(G, K, \alpha, \beta, \gamma, \delta)$
   G: input graph
   K: set of cables
   $\alpha, \beta, \gamma, \delta$: parameters (fixed constants)
1. Prune the set of available cables as described in 5.1.
2. Solve the LP relaxation of the IP described in 4.2.
3. $T_k = \{t\}$.
4. For $i = k - 1, k - 2, \ldots, 1$:
      Define $S_i := \emptyset$
      $\forall v_j \notin T_{i+1}$:
         $R_j^i = \dfrac{C_j^0 + \ldots + C_j^{i-1}}{r_{i-1}}$.
      If $T_{i+1} \cap B(v_j, \gamma R_j^i) \neq \emptyset$,
         proxy$_i(v_j)$:= any (arbitrary) vertex in $T_{i+1} \cap B(v_j, \gamma R_j^i)$
         $S_i := S_i \cup \{v_j\}$.
      Order the remaining vertices $L_i = V \setminus (T_{i+1} \cup S_i)$ in nondecreasing
         order of their corresponding ball radii.
      While $L_i \neq \emptyset$:
         Let $B(v_j, \gamma R_j^i)$ be the smallest radius ball in L.
         $\forall u \in L \cap B(v_j, \gamma R_j^i)$:
            proxy$_i(u) := v_j$
            $L := L \setminus \{u\}$
         $L := L \setminus \{v_j\}$
         $S_i := S_i \cup \{v_j\}$
      Comment: S_i is the set of sources chosen to be connected at this level.
      Contract T_{i+1} to a singleton node t_{i+1}.
      Build a Steiner tree ST_i with $S_i \cup \{t_{i+1}\}$ as terminals (Elaborated in
         the text below – the parameter δ is used here).
         Use discount type i for these edges.
      Convert ST_i into an (α, β)-LAST rooted at t_{i+1}, denoted $LAST_i$.
      Define $T_i := T_{i+1} \cup LAST_i$.
5. For every source vertex $v_j \notin T_1$:
      Compute a shortest path P from v_j to any node in T_1.
      Augment T_1 by including the edges in P.
      Use cable type 0 on the edges in P.
      $T_0 := T_1$.
6. Route along shortest paths in T_0. This is the solution we output.

Fig. 3. The algorithm

5.3   Building the Solution: Details

The details of the algorithm are presented in Figure 3. Let C_j^i denote the fraction of the routing cost for routing unit flow from v_j to t corresponding to discount-i, that is, $C_j^i = \sum_e f_{e;i}^j r_i l_e$. Hence $\sum_{0 \le i < k} C_j^i$ is the total routing cost for vertex v_j. For a vertex v and a positive number R, let $B(v, R) = \{u \in V : d(u, v) \le R\}$ denote the ball of radius R centered at vertex v.

Selecting vertices for inclusion in the current level. The bulk of the work is done in Step 4. We first choose a certain set of vertices (S_i at level i), and then build a Steiner tree connecting the chosen vertices to the root component (T_{i+1}). We note that this step is somewhat similar to the "tour ball" construction in [13].

Building the Steiner tree. We build balls $B(v_j, \delta R_j^i)$ around each selected vertex v_j. We note that we will choose $\delta < \gamma$, where γ is the dilation parameter for the radius of the balls used in the vertex selection step. We then build an approximately minimum Steiner tree which connects these selected balls to T_{i+1}. More formally, we contract each ball and introduce a new node for it. We also contract T_{i+1} and introduce a node for it. Then we run an approximation algorithm to find a Steiner tree connecting all the selected nodes that has cost at most twice the value of a fractional Steiner tree, i.e., within twice the cost of an LP relaxation for the Steiner tree problem on these nodes (See, e.g., [1]). Then we un-contract the balls and extend the edges of the resulting forest incident on the boundary of $B(v_j, \delta R_j^i)$ with direct edges to the center v_j. Thus we have a tree connecting all the selected vertices v_j to T_{i+1}.

Converting the Steiner tree to a LAST. The Steiner tree constructed so far may have a very large diameter, since we have not taken the routing into consideration so far. Hence it may lead to very high routing costs in the solution. To get around this, we use a construction due to Khuller, Raghavachari and Young [10] which achieves short paths from a root node while being light.

Definition 1 (Light approximate shortest-path tree). *Let $G = (V, E)$ be a graph with a length function $l : E \to \mathbb{R}^+$ and let $t \in V$ be a root vertex. Let $\alpha, \beta > 1$ be real numbers. An (α, β)-LAST rooted at t is a tree T in G such that the total length of T is at most α times the length of an MST of G, and for any vertex $v \in V$, the length of the (v, t) path along T is at most β times the length of a shortest (v, t) path in G.*

The Steiner tree constructed can now be transformed into an (α, β)-LAST rooted at t_{i+1} (the contracted version of T_{i+1}) for some constants $\alpha, \beta > 1$ using the algorithm of [10]. The edges in the LAST will use discount-i. We then un-contract the root component. This breaks up the LAST into a forest where each subtree is rooted at some vertex in the un-contracted tree T_{i+1}. Define T_i to be the union of T_{i+1} and this forest.

In the last stage, we connect each source v_j not in T_1 to T_1 by a shortest path, using discount-0, thereby extending T_1 to include all remaining source vertices in T_0.

6   Analysis

We use the LP optimum OPT as a lower bound on the integer optimum. Let $OPT_{build} = \sum_e \sum_i p_i z_e^i l_e$ denote the total price paid for purchasing cables of all discount types in an LP optimum and let $OPT_{route} = \sum_{v_j} \sum_e \sum_i \text{dem}_j f_{e;i}^j r_i l_e$ be the total routing cost paid in that optimum. Thus $OPT = OPT_{build} + OPT_{route}$ is a lower bound on the total cost. In this section, we prove that, for our algorithm, the total building cost is $O(k \cdot OPT_{build})$ and the total routing cost is $O(k \cdot OPT_{route})$. Thus we establish that the integrality gap of the formulation is no more than $O(k)$.

6.1   Building Cost

We analyze the total price paid for installing discount-i cables when we augment the tree T_{i+1} to T_i.

Note that in building an (α, β)-LAST from the tree, we incur a factor of at most α in the building cost. We argue that the cost of building the tree at the current stage is $O(OPT_{build})$. Then, summing over all k stages, we get that the total building cost is $O(k \cdot OPT_{build})$.

For any source vertex v, the following Lemma proves that there is sufficient fractional z-value crossing a ball around v to allow us to pay for an edge crossing the ball. Since the LP optimum pays for this z, we can charge the cost of our edge to this fractional z and hence obtain our approximation guarantee.

Lemma 5. *Let $S \subset V$ be a set of vertices such that $t \notin S$ and $B(v_j, \delta R_j^i) \subset S$. Then,*

$$\sum_{q=i}^{k-1} \sum_{e \in \delta^+(S)} z_e^q \geq 1 - \frac{1}{\delta}.$$

Proof. Assume for the sake of contradiction that the sum is less than $1 - 1/\delta$. So the total flow starting from the source v_j which crosses S using discount types smaller than i is more than $1/\delta$. As it pays at least r_{i-1} per unit distance per unit flow, the total routing cost is more than

$$\frac{\delta R_j^i r_{i-1}}{\delta} = \frac{C_j^0 + \ldots + C_j^{i-1}}{r_{i-1}} r_{i-1} = C_j^0 + \ldots + C_j^{i-1}$$

This is a contradiction, as the total cost spent in discount types smaller than i is exactly $C_j^0 + \ldots + C_j^{i-1}$.

We built a LAST which used discount-i. So the building cost of the LAST is p_i times the length of the LAST. The following Lemma gives a bound on this cost.

Lemma 6. *The cost of the LAST built at any stage is $O(OPT_{build})$.*

Proof. If we scale up the z-values in the optimum by a factor $\delta/(\delta - 1)$, Lemma 5 indicates that we have sufficient z-value of types i or higher to build a Steiner tree connecting the balls $B(v_j, \delta R_j^i)$ to T_{i+1}. If we use the primal dual method [1], we incur an additional factor of 2 in the cost of the Steiner tree as against the LP solution z-values. Thus, its cost will be at most

$$2\frac{\delta}{\delta - 1}p_i \sum_{q=i}^{k-1}\sum_e z_e^q \leq 2\frac{\delta}{\delta - 1}OPT_{build}.$$

After un-contracting the balls, we extended the forest to centers v_l by direct edges between the boundaries of forest edges in $B(v_l, \delta R_l^i)$ and v_l. We can account for this extension by using the following observation. For a center v_l, the cost of extension is at most $\frac{\delta}{\gamma - \delta}$ times the cost of the forest inside $B(v_l, \gamma R_l^i)$. Furthermore, during the selection of the vertices, we ensured that for any two selected vertices v_l and v_j, the balls $B(v_l, \gamma R_l^i)$ and $B(v_j, \gamma R_j^i)$ are disjoint. Thus the total cost of the extended tree is at most $1 + \frac{\delta}{\gamma - \delta}$ times the cost of the previous forest. Hence cost of the Steiner tree built is at most $2\frac{\gamma}{\gamma - \delta}\frac{\delta}{\delta - 1}OPT_{build}$. Subsequently, the cost of the LAST built from this tree is at most $2\alpha\frac{\gamma}{\gamma - \delta}\frac{\delta}{\delta - 1}OPT_{build}$ [10]. For fixed constants α, δ, γ with $\gamma > \delta$, this is $O(OPT_{build})$ and completes the proof.

The total building cost is the sum of building costs at each stage, and we have k such stages. Thus, we have the following.

Lemma 7. *The total building cost is $O(k \cdot OPT_{build})$.*

6.2   Routing Cost

After constructing the tree, for each source vertex v_j, we route the corresponding commodity along the unique (v_j, t) path on the tree. Let $OPT_j = \sum_i C_j^i$ denote the routing cost per unit flow for v_j in the optimum. We prove that the routing cost for a source v_j is $O(k)$ times OPT_j. Thus the total routing cost is $O(k \cdot OPT_{route})$.

Refer to Figure 4 for the following analysis.

Lemma 8. *For any source vertex v_j, the cost of routing unit amount of its corresponding commodity is $O(k \cdot OPT_j)$.*

Proof. Let the (v_j, t) path along T_0 be $v_j = u_0, u_1, \ldots, u_k = t$ such that the sub-path (u_i, u_{i+1}) uses discount-i for $0 \leq i < k$. Note that if discount-i is not

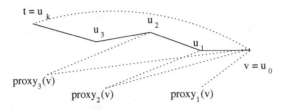

Fig. 4. Analysis for routing cost.

used, then $u_i = u_{i+1}$. Let $d_T(u_i, u_{i+1})$ be the distance between u_i and u_{i+1} in the tree T_0. Then, for v_j, the routing cost per unit flow is $\sum_i r_i d_T(u_i, u_{i+1})$.

For $1 \leq i < k$, let $\mathtt{proxy}_i(v_j)$ denote the proxy of v_j in stage $k-i$. Moreover, for all j, define $\mathtt{proxy}_k(v_j) = t$. We have $d(v_j, \mathtt{proxy}_{i+1}(v_j)) \leq 2\gamma \frac{C_j^0 + \ldots + C_j^i}{r_i} \leq 2\gamma \frac{OPT_j}{r_i}$. We also know that $r_i d_T(u_i, u_{i+1}) \leq \beta \cdot r_i d(u_i, \mathtt{proxy}_{i+1}(v_j))$ because when we constructed the LAST in stage $k-i$, $d(u_i, u_{i+1})$ was at most β times the shortest path connecting u_i to T_{i+1}. Also this shortest path is shorter than $d(u_i, \mathtt{proxy}_{i+1}(v_j))$, as $\mathtt{proxy}_{i+1}(v_j)$ was in T_{i+1}.

By induction on i we prove that, $r_i d_T(u_i, u_{i+1}) \leq M \cdot OPT_j$ for some constant M.

For the base case when $i = 0$, v_j was connected by a shortest path to T_1. Hence $r_0 d_T(u_0, u_1) \leq r_0 d(v_j, \mathtt{proxy}_1(v_j)) \leq r_0 \cdot 2\gamma \frac{C_j^0}{r_0} \leq 2\gamma OPT_j \leq M \cdot OPT_j$ for sufficiently large M.

Now assume $r_l d_T(u_l, u_{l+1}) \leq M \cdot OPT_j$ for all $l < i$. Using triangle inequality and the induction hypothesis, we get

$$r_i \cdot d_T(u_i, u_{i+1}) \leq \beta \cdot r_i \cdot d(u_i, \mathtt{proxy}_{i+1}(v_j))$$

$$\leq \beta \cdot r_i \sum_{q=0}^{i-1} d(u_q, u_{q+1}) + \beta \cdot r_i \cdot d(u_0, \mathtt{proxy}_{i+1}(v_j))$$

$$= \beta \sum_{q=0}^{i-1} \frac{r_i}{r_q} \cdot r_q \cdot d(u_q, u_{q+1}) + \beta \cdot r_i \cdot d(v_j, \mathtt{proxy}_{i+1}(v_j))$$

$$\leq \beta \sum_{q=0}^{i-1} \epsilon^{i-q} \cdot M \cdot OPT_j + \beta \cdot 2\gamma OPT_j$$

$$\leq (\frac{\beta\epsilon}{1-\epsilon} M + 2\beta\gamma) OPT_j$$

$$\leq M \cdot OPT_j$$

for $M \geq \frac{2\beta\gamma(1-\epsilon)}{1-\epsilon(1+\beta)}$. This completes the induction. Summing over all edges in the path from v_j to t, we get the statement of the lemma.

Summing the routing cost bound over all source vertices v_j, we obtain that the total routing cost is no more than $O(k \cdot OPT_{route})$.

7   Conclusion

The exact approximation factor of our algorithm depends on the parameters. If we set $(\alpha, \beta, \gamma, \delta, \epsilon)$ to be $(7, \frac{4}{3}, 3, 2, \frac{1}{5})$ respectively, we obtain an approximation factor of $60k$ for both components of the cost function. The running time of our algorithm is dominated by the time to solve an LP with $O(mnk)$ constraints and variables.

7.1   Recent Work

The work of Guha *et al* [7] is combinatorial, and they build their tree in a bottom up manner. Their approach is to gather demand from nodes by means of Steiner trees until it is more profitable to use the next higher type of cable available. They then connect such trees using shortest path trees that gather sufficient demand to use the next cable type. They iteratively do this until all nodes are connected to the sink. Their algorithm being purely combinatorial has a much better running time. However, their approximation ratio is a large constant, roughly 2000. We can contrast this with our approximation factor, which is $60k$ with k being the number of cables after pruning.

After learning about their work, we have been able to tighten the ratio of the building cost component of our solution to the analogous component in the LP relaxation (OPT_{build}) to a constant. We show how to do this in the extended version of our paper [6]. Essentially, we prune the set of available cables so as to get a sufficient (geometric) increase in the fixed cost of higher index cables. Subsequently, if our LP has purchased a certain amount of a certain cable, we allow ourselves to purchase the same amount of all cables of lower index. Given the geometric costs, this only results in a constant factor dilation of the LP lower bound. We show that this solution is near-optimal, and we compare ourselves against it. Our algorithm can then charge the building cost of cable type i to what the augmented LP paid for cable type i only, instead of the entire building cost of the LP. This enables us to prove that the integrality gap of the building cost component is low. However, we do not see yet how to improve the routing cost component of our solution.

7.2   Open Questions

The main open question from our work is the exact integrality gap of this problem, whether it is a constant, $O(k)$, or something in between. The question of getting even better approximation ratios for this problem remains open. The problem can be generalized to allow different source-sink pairs; for this problem the current state of the art is a polylogarithmic approximation [4].

References

1. Agrawal, A., Klein, P., Ravi, R.: When trees collide: An approximation algorithm for the generalized Steiner problem on networks. SIAM Journal of Computing,24(3):440-456, 1995.
2. Althöfer, I., Das, G., Dobkin, D., Joseph, D., Soares, J.: On sparse spanners of weighted graphs. Discrete and Computational Geometry, 9:81-100, 1993.
3. Andrews, M., Zhang, L.: The access network design problem. Proc. of the 39th Ann. IEEE Symp. on Foundations of Computer Science, 42-49, October 1998.
4. Awerbuch, B., Azar, Y.: Buy at bulk network design. Proc. 38th Ann. IEEE Symposium on Foundations of Computer Science, 542-547, 1997.
5. Bartal, Y.: On approximating arbitrary metrics by tree metrics. Proc. 30th Ann. ACM Symposium on Theory of Computing, 1998.
6. Garg, N., Khandekar, R., Konjevod, G., Ravi, R., Salman, F.S., Sinha, A.: A mathematical formulation of a transportation problem with economies of scale. Carnegie Bosch Institute Working Paper 01-1, 2001.
7. Guha, S., Meyerson, A., Munagala, K.: Improved combinatorial algorithms for single sink edge installation problems. To appear in Proc. 33rd Ann. ACM Symposium on Theory of Computing, 2001.
8. Guha, S., Meyerson, A., Munagala, K.: Heirarchical placement and network design problems. Proc. 41st Ann. IEEE Symposium on Foundations of Computer Sciece, 2000.
9. Hassin, R., Ravi, R., Salman, F.S.: Approximation algorithms for a capacitated network design problem. Proc. of the APPROX 2000, 167-176, 2000.
10. Khuller, S., Raghavachari, B., Young, N.E.: Balancing minimum spanning and shortest path trees. Algorithmica, **14**, 305-322, 1993.
11. Mansour, Y., Peleg, D.: An approximation algorithm for minimum-cost network design. The Weizman Institute of Science, Rehovot, 76100 Israel, Tech. Report CS94-22, 1994; Also presented at the DIMACS workshop on Robust Communication Networks, 1998.
12. Meyerson, A., Munagala, K., Plotkin, S.: Cost-distance: Two metric network design. Proc. 41st Ann. IEEE Symposium on Foundations of Computer Science, 2000.
13. Ravi, R., Salman, F.S.: Approximation algorithms for the traveling purchaser problem and its variants in network design. Proc. of the European Symposium on Algorithms, 29-40, 1999.
14. Salman, F.S., Cheriyan, J., Ravi R., Subramanian, S.: Approximating the single-sink link-installation problem in network design. SIAM Journal of Optimization 11(3):595-610, 2000.

Circuit Mengerian Directed Graphs

Bertrand Guenin

Department of Combinatorics and Optimization, University of Waterloo

Abstract. A feedback set of a digraph $D = (V, E)$ is a set of vertices and edges such that its removal makes the digraph acyclic. Let $w : V \cup E \to Z_+$ be a non-negative cost function. We say that D is Circuit Mengerian if, for every non-negative cost function w, the minimum weight feedback set is equal to the cardinality of the largest collection of circuits $\mathcal{F}$ with the property that, for every element $t \in V \cup E$, no more than $w(t)$ circuits of $\mathcal{F}$ use t. This property is closed under digraph minors, thus Circuit Mengerian digraphs can be characterized by a list of minor minimal non Circuit Mengerian digraphs. In this paper we give such an excluded minor characterization.

1 Introduction

We use the convention that paths and circuits have no repeated vertices, and that in digraphs they are directed. We say that t is an *element* of a digraph if t is either a vertex or an edge. Given a digraph D and a set of elements S we write $D \setminus S$ for the digraph obtained by deleting elements of S. A set S of elements of D is a *feedback set* if $D \setminus S$ is acyclic, i.e. it does not have (directed) circuits. Let us assign to every element t of D a non-negative value $w(t)$. The *weight* $w(S)$ of a set of elements S is the value $\sum_{t \in S} w(t)$. We write $\tau_w(D)$ for the minimum weight of a feedback set. A family $\mathcal{F}$ of circuits form a *w-matching* if for every element t of D there are at most $w(t)$ circuits of $\mathcal{F}$ using t. The cardinality of the largest w-matching of circuits is denoted $\nu_w(D)$. Observe that $\tau_w(D) \geq \nu_w(D)$. We say that a digraph D is *Circuit Mengerian* (CM) if for every cost function $w : V(D) \cup E(D) \to Z_+$, we have $\tau_w(D) = \nu_w(D)$. In this paper we give an excluded minor characterization of CM digraphs.

We say that an edge e of a digraph D with head v and tail u is *special* if either e is the only edge of D with head v, or it is the only edge of D with tail u, or both. We say that a digraph D is a *minor* of a digraph D' if D can be obtained from a subdigraph of D' by repeatedly contracting special edges. Suppose that D is CM. Note that deleting an element t of D is equivalent to choosing $w(t) = 0$. Observe that contracting a special edge uv (where say uv is the only edge leaving u) is equivalent to choosing $w(u) = +\infty$ and $w(uv) = +\infty$. These observations imply that,

Remark 1. The class of CM digraphs is closed under digraph minors.

By an *odd double circuit* we mean the digraph obtained from an undirected circuit of odd length at least three, by replacing each edge by a pair of directed edges,

K. Aardal, B. Gerards (Eds.): IPCO 2001, LNCS 2081, pp. 185–195, 2001.

one in each direction. A *chain* is the digraph obtained from an undirected path, with at least one edge, by replacing every edge by a pair of directed edges, one in each direction. The *endpoints* of the chain are the vertices which correspond to the endpoints of the path. A chain is odd (resp. even) if it has an odd (resp. even) number of vertices. Given an integer n we write $[n]$ for the set $\{1, \ldots, n\}$. Consider the digraph D with vertices $u, v_1, \ldots, v_k$, where $k \geq 3$, and edges: $v_i v_{i+1}$ for all $i \in [k-1]$ when i is odd; $v_{i+1} v_i$ for all $i \in [k-1]$ when i is even; uv_i for all $i \in [k]$ when i is odd; $v_i u$ for all $i \in [k]$ when i is even. A digraph isomorphic to D or to the reverse of D is a Δ-*digraph*. Vertex u is called the *center* of D and v_1, v_k are the endpoints of D. The parity of a Δ-digraph is the parity of k. A Δ_1-*chain* is the digraph obtained by identifying the endpoints of an even Δ-digraph to the endpoints of an odd chain. Consider an even Δ-digraph D with endpoints v_1, v_2 and center u and an odd Δ-digraph D' with endpoints v_1', v_2' and center u'. A Δ_2-*chain* is the digraph obtained by identifying u and u', identifying v_1, v_2' to the endpoints of an odd chain, and identifying v_1', v_2 to the endpoints of another odd chain.

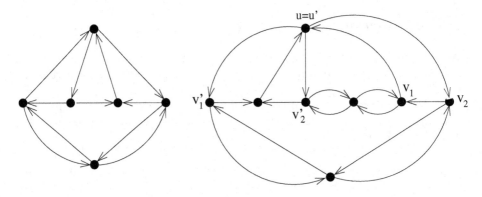

Fig. 1. A Δ_1-chain (left). A Δ_2-chain (right)

Next we present the main result of this paper.

Theorem 1. *D is CM if and only if it has no odd double circuit, no Δ_1-chain, and no Δ_2-chain minor.*

1.1   Related Results

Suppose we assign to each vertex v a non-negative value $w(v)$. Clearly, the weight of the minimum weight vertex feedback set is at least as large as the cardinality of the largest collection of circuits with the property that for every vertex v no more than $w(v)$ circuits of the collection use v. A natural problem is to try to characterize which digraphs have the property that equality holds in the aforementioned relation for every non-negative assignments of weights. Digraphs with this property are closed under induced subdigraphs but not under digraph minors and there is a long list of obstructions to this property. However the problem has been solved for the special case when D is a tournament [9] or a bipartite tournament [10]. When $w(t) = 1$ for every element t of D, we write $\tau(D)$ for $\tau_w(D)$ and $\nu(D)$ for $\nu_w(D)$. Digraph D for which $\tau(D') = \nu(D')$ for every subdigraph D' of D are said to *pack*. The digraph F_7 consists of vertices $\{0, \dots, 6\}$ and edges $i(i + 1), i(i + 5)$ for all $i \in \{0, \dots 6\}$ (where additions are taken modulo 7).

Theorem 2 (Guenin and Thomas [5]). *D packs if and only if D has no odd double circuit and no F_7 minor.*

It follows from Remark 1 that CM digraphs pack. However, not every digraph that packs is CM as the Δ_1-chain of Figure 1 illustrates.

1.2   Clutters

A *clutter* $\mathcal{H}$ is a finite family of sets, over some finite ground set $E(\mathcal{H})$, with the property that no set of $\mathcal{H}$ contains, or is equal to, another set of $\mathcal{H}$. A $0, 1$ matrix, whose columns are indexed by $E(\mathcal{H})$ and whose rows are the characteristic vectors of the sets in $\mathcal{H}$, is denoted by $M[\mathcal{H}]$. A clutter $\mathcal{H}$ is *ideal* if $\{x \geq \mathbf{0} : M[\mathcal{H}]x \geq \mathbf{1}\}$ is an integral polyhedron, that is its extreme points have $0, 1$ coordinates. The *blocker* $b(\mathcal{H})$ of $\mathcal{H}$ is the clutter defined as follows: $E\big(b(\mathcal{H})\big) := E(\mathcal{H})$ and $b(\mathcal{H})$ is the set of inclusion-wise minimal members of $\{U : U \cap C \neq \emptyset, \forall C \in \mathcal{H}\}$. A clutter $\mathcal{H}$ is *Mengerian* if the system $M[\mathcal{H}]x \geq \mathbf{1}, x \geq \mathbf{0}$ is totally dual integral. Let w be a vector with $|E(\mathcal{H})|$ non-negative components. We define $\tau_w(\mathcal{H}) = \min\{wx : M[\mathcal{H}]x \geq \mathbf{1}, x \geq \mathbf{0}\}$, $\nu_w(\mathcal{H}) = \max\{\mathbf{1}y : yM[\mathcal{H}] \leq w, y \in Z_+^{|\mathcal{H}|}\}$, and $\nu_w^*(\mathcal{H}) = \max\{\mathbf{1}y : yM[\mathcal{H}] \leq w, y \geq \mathbf{0}\}$. If all entries of w are 1 we write $\tau(\mathcal{H})$ for $\tau_w(\mathcal{H})$, $\nu(\mathcal{H})$ for $\nu_w(\mathcal{H})$, and $\nu^*(\mathcal{H})$ for $\nu_w^*(\mathcal{H})$. Note that a clutter $\mathcal{H}$ is Mengerian if and only for all $w \in Z_+^{|E(\mathcal{H})|}$, $\tau_w(\mathcal{H}) = \nu_w(\mathcal{H})$. Similarly, $\mathcal{H}$ is ideal if and only for all $w \in Z_+^{|E(\mathcal{H})|}$, $\tau_w(\mathcal{H}) = \nu_w^*(\mathcal{H})$. Thus Mengerian clutters are ideal but the converse is not true in general. Let $D = (V, E)$ be a digraph. The *clutter of circuits* of D, denoted $\mathcal{C}(D)$, is the clutter with ground set $V \cup E$ and where $T \in \mathcal{C}(D)$ if and only if T is the set of vertices and edges of some circuit of D. Observe that a digraph D is CM if and only if the clutter $\mathcal{C}(D)$ is Mengerian. We say that a digraph D is *Circuit Ideal* (CI) if the clutter $\mathcal{C}(D)$ is ideal. Similarly, as for CM digraphs we have,

Remark 2. The class of CI digraphs is closed under digraph minors.

Let $\mathcal{H}$ be a clutter and $i \in E(\mathcal{H})$. The *contraction* $\mathcal{H}/i$ and *deletion* $\mathcal{H} \setminus i$ are clutters with ground set $E(\mathcal{H}) - \{i\}$ where: $\mathcal{H}/i$ is the set of inclusion-wise minimal members of $\{S - \{i\} : S \in \mathcal{H}\}$ and; $\mathcal{H} \setminus i := \{S : i \notin S \in \mathcal{H}\}$. Contractions and deletions can be performed sequentially, and the result does not depend on the order. A clutter obtained from $\mathcal{H}$ by a sequence of deletions I_d and a sequence of contractions I_c ($I_c \cap I_d = \emptyset$) is called a *minor* of $\mathcal{H}$ and is denoted $\mathcal{H} \setminus I_d/I_c$. It is well known that Mengerian and ideal clutters are closed under minor. To distinguish between minor operations for digraphs and clutters we will sometimes talk of *digraph-minors* or of *clutter-minors*. There is a one-to-one correspondence between deletions on the digraph and deletions on the associated clutter.

Remark 3. Let D be a digraph and t an element of D. Then $\mathcal{C}(D \setminus t) = \mathcal{C}(D) \setminus t$.

An *odd hole* is a clutter, denoted $\mathcal{C}_k^2$, such that k is odd where and $M[\mathcal{C}_k^2]$ is a $k \times k$ non-singular matrix, with exactly 2 non-zero entries in every row and every column. Since $\tau(\mathcal{C}_k^2) = \lceil k/2 \rceil > \lfloor k/2 \rfloor = \nu(\mathcal{C}_k^2)$, odd holes are not Mengerian. Note that if D is an odd double circuit then $\mathcal{C}(D)/E(D)$ is an odd hole. Suppose D is a Δ_1, or Δ_2-chain. Let u denote the center of the Δ-digraphs. Then $\mathcal{C}(D)/(E(D) \cup \{u\})$ is an odd hole. Hence,

Remark 4. If D is an odd double circuit, a Δ_1-chain, or a Δ_2-chain then $\mathcal{C}(D)$ has an odd hole minor.

Combining the previous observation with theorem 1 implies,

Theorem 3. *Let D be a digraph. Then $\mathcal{C}(D)$ is Mengerian if and only if it has no odd hole minor.*

1.3   Outline of the Paper

Let D be a digraph. Theorem 1 follows from the following two results.

Lemma 1. *D is CI if and only if it has no odd double circuit, no Δ_1-chain, and no Δ_2-chain minor.*

Lemma 2. *D is CM if and only if it is CI.*

A sketch of the proof of Lemma 1 is given in Section 2. In Section 3 we show properties of CI digraphs which are not CM. Finally, Section 4 outlines the proof of Lemma 2.

2   Circuit Ideal Digraphs

2.1   Lehman's Theorem

A clutter is said to be *minimally non-ideal* (mni) if it is not ideal but all its proper minors are ideal. The clutter $\mathcal{J}_t$, for $t \geq 1$, is defined as follows: $E(\mathcal{J}_t) := \{0, \dots, t\}$ and $\mathcal{J}_t := \{\{1, \dots, t\}\} \cup \{\{0, i\} : i = 1, \dots, t\}$. By abuse of notation we shall identify a clutter $\mathcal{H}$ with its matrix representation $M[\mathcal{H}]$.

Theorem 4 (Lehman [6]). *Let A be a minimally non-ideal matrix with n columns and let B its blocker. Then B is minimally non-ideal. If A is distinct from $\mathcal{J}_t$ then A (resp. B) has a square, non-singular row submatrix $\bar{A}$ (resp. $\bar{B}$) with r (resp. $\bar{r}$) non-zero entries in every row and column. In addition $r\bar{r} > n$ and rows of A (resp. B) not in $\bar{A}$ (resp. $\bar{B}$) have at least $r + 1$ (resp. $\bar{r} + 1$) non-zero entries. Moreover, $\bar{A}\bar{B}^T = J + (r\bar{r} - n)I$, where J denotes an $n \times n$ matrix filled with ones.*

Given a clutter $\mathcal{H}$, we write $\bar{\mathcal{H}}$ for the clutter which contains only the sets of minimum cardinality of $\mathcal{H}$. Let $\mathcal{H}$ be mni and $\mathcal{K} = b(\mathcal{H})$. Define $A := M[\mathcal{H}], B := M[\mathcal{K}]$ and $\bar{A} := M[\bar{\mathcal{H}}], \bar{B} := M[\bar{\mathcal{K}}]$. Consider the element $L \in \bar{\mathcal{H}}$ (resp. $U \in \bar{\mathcal{K}}$) that corresponds to the ith row of $\bar{A}$ (resp. $\bar{B}$). We call U the *mate* of L. If $\mathcal{H} \neq \mathcal{J}_t$ then $\bar{A}\bar{B}^T = J + (r\bar{r} - n)I$. Hence U intersects every element of $\bar{\mathcal{H}}$ exactly once; except for L which is intersected $q = r\bar{r} - n + 1 \geq 2$ times. If $\mathcal{H} = \mathcal{J}_t$ then $b(\mathcal{H}) = \mathcal{J}_t$, and U intersects every element of $\bar{\mathcal{H}}$ exactly once; except for L which is intersected $q = 2$ times or $q = t \geq 2$ times.

2.2   Chains

Given a digraph D we write $t \in D$ to mean t is an element of D (a vertex or an edge). Given two subdigraphs D_1, D_2 we denote the intersection of the elements of D_1 and D_2 by $D_1 \cap D_2$. A collection of circuits $\mathcal{F} = \{C_1, C_2, \ldots, C_n\}$ is a *closed chain* if $n \geq 3$ is odd and the following properties are satisfied: (1) for all $i, j \in [n]$ with $i \neq j$ either $C_i \cap C_j = \emptyset$ or C_i and C_j intersect in a unique path (which may consists of a single vertex); (2) for all $i \in [n]$, there is an element in $C_i \cap C_{i+1}$ (where $n + 1 = 1$) which is not an element of any circuit in $\mathcal{F} - \{C_i, C_{i+1}\}$. A closed chain is minimal if the digraph induced by the elements of the chain does not contain a closed chain with fewer circuits. Observe that the "only-if" direction of Lemma 1 follows immediately from Remark 4. Let D be any digraph. The other direction follows from the next two propositions.

Proposition 1. *If $M(D)$ is not ideal then D has a closed chain.*

Proposition 2. *A minimal closed chain is either an odd double circuit, a Δ_1-chain, or a Δ_2-chain.*

The proof of Proposition 2 is a lengthy case analysis and hence it is not included in this paper.

Proof. (Sketch for proposition 1) Suppose $\mathcal{C}(D)$ is not ideal and let $\mathcal{A}$ be a mni minor of $\mathcal{C}(D)$. Because of Remark 3 we may assume $\mathcal{A} = \mathcal{C}(D)/I$. Let r be the cardinality of the elements of $\bar{\mathcal{A}}$.

Claim. Some submatrix of $M[\bar{\mathcal{A}}]$ is an odd hole.

Proof. (of Claim) If $\bar{\mathcal{A}} = \mathcal{J}_t$ then the first three rows and columns of $M[\mathcal{J}_t]$ contain an odd hole. Otherwise we know from Theorem 4 that $M[\bar{\mathcal{A}}]$ is a square non-singular matrix with exactly r ones on every row. It follows that $\frac{1}{r}\mathbf{1}$ is an extreme point of the polyhedron $\{x \geq \mathbf{0} : M[\bar{\mathcal{A}}]x \geq \mathbf{1}\}$. Hence $M[\bar{\mathcal{A}}]$ is not ideal. But matrices which are not ideal contain an odd hole as a submatrix (see Berge [1], Fulkerson et al. [4]). $\diamond$

We may assume that the odd hole of Claim 1 is contained in rows and columns 1 to k of $M[\bar{\mathcal{A}}]$. Let $L_1, \ldots, L_k$ be the sets of $\bar{\mathcal{A}}$ corresponding to rows 1 to k of $M[\bar{\mathcal{A}}]$. We denote by $\mathcal{F}$ a collection of circuits $C_1, \ldots, C_k$ where $L_i = \big(E(C_i) \cup V(C_i)\big) - I$ for all $i \in [k]$. There are exactly two circuits of $\mathcal{F}$ which contain any element of D corresponding to a column $i \in [k]$ of $M[\bar{\mathcal{A}}]$. This implies that property (2) of chains is satisfied. Consider any collection $\mathcal{F}$ and a pair of circuits $C, C' \in \mathcal{F}$ where C, C' both contain a same pair of distinct vertices v and w. Let P be the vw-path of C, let Q be the wv-path of C, let P' be the vw-path of C', and let Q' be the wv-path of C'. We need to show that $\mathcal{F}$ can be chosen so that in addition property (1) also holds, i.e. that $P = P'$ or that $Q = Q'$. Here we only indicate how to show the weaker property that $P - I = P' - I$ or $Q - I = Q' - I$. Given a subdigraph D' we write $l(D')$ for $|(V(D') \cup E(D')) - I|$. Let $\hat{C}$ denote the circuit included in $P + Q'$ and let $\bar{C}$ denote the circuit included in $Q + P'$. By definition of $\bar{N}$, $l(C) = l(C') = r$ and $l(\hat{C}) \geq r, l(\bar{C}) \geq r$. It follows that $l(P) = l(P')$ and $l(Q) = l(Q')$. Hence, $\hat{C} \in \bar{\mathcal{A}}$ and $\bar{C} \in \bar{\mathcal{A}}$. Moreover, every element of $(P \cup Q') - I$ (resp. $(P' \cup Q) - I$) is an element of $\hat{C}$ (resp. $\bar{C}$). Suppose for a contradiction $P - I \neq P' - I$ and $Q - I \neq Q' - I$. Then $C, C', \hat{C}, \bar{C}$ are all distinct elements of $\bar{\mathcal{A}}$. Let U be the mate of C. Then $|U \cap (P \cup Q)| \geq 2$. Since $\hat{C} \in \mathcal{B}$ we have $|U \cap \hat{C}| = 1$ and in particular $|U \cap P| \leq 1$. Similarly, $\bar{C}$ implies $|U \cap Q| \leq 1$. It follows that there are element $p, q \in U$ where $p \in P$ and $q \in Q$. Since $|U \cap C'| = 1$ there is an element $f \in C' \cap U$. By symmetry we may assume $f \in Q'$ but then $U \cap \hat{C} \supseteq \{p, f\}$ a contradiction since $|U \cap \hat{C}| = 1$. $\square$

3   Properties of CI But Non CM Digraphs

A clutter (or matrix) is *minimally non-Mengerian* if it is not Mengerian but all its proper clutter-minors are Mengerian. A digraph is *minimally non CM*, is it is not CM but all its digraph-minors are CM. We shall need the following observations.

Proposition 3 (Cornuéjols et al. [3]). *Let M be an ideal, minimally non-Mengerian $0, 1$ matrix.*

(i) for every column i of M, $\tau(M \setminus i) = \tau(M) - 1$,
(ii) for every column i of M, $\tau(M/i) = \tau(M)$,
(iii) every row of M has at least two non-zero entries.

The following weak form of the complementary slackness theorem [12] will also be required.

Remark 5. Let x^* be an optimal solution to the linear program $\min\{cx : Mx \geq \mathbf{1}, x \geq \mathbf{0}\}$ and let y^* be an optimal solution to $\max\{y\mathbf{1} : yM \leq c, y \geq \mathbf{0}\}$. If $M_{i.}x > 1$ then $y_i^* = 0$ and if $M_{.j}y < c_j$ then $x_j^* = 0$.

Proposition 4. *Let D be a minimally non CM digraph which is CI. Then there is a subset I of elements of D such that $M(D)/I$ is minimally non-Mengerian. Moreover, $V(D) \subseteq I$.*

Proof. (Sketch) Let $M := M(D)$ and let N be a minimally non-Mengerian matrix. Because of Remark 3 we may assume that $N = M/I$. We only present the proof for the case where $\tau(N) > \nu(N)$ (this is equivalent to assuming correctness of the replication conjecture [2]). Let i denote a column of N, we will show that i corresponds to an edge of D. Note,

$$\min\{\mathbf{1}x : Mx \geq \mathbf{1}, x \in \{0,1\}^n\} = \tag{1}$$
$$\min\{\mathbf{1}x : Mx \geq \mathbf{1}, x > \mathbf{0}\} = \tag{2}$$
$$\max\{y\mathbf{1} : yM \leq \mathbf{1}, y \geq \mathbf{0}\}. \tag{3}$$

The first equality holds because M is ideal. The second equality follows from linear programming duality. It follows from Proposition 3(i) that there is an optimal solution to (1) with $x_i^* = 1$. Let y^* be an optimal solution to (3). Remark 5 implies that $(y^*M)_i = 1$. Hence in particular there is a row m of M with $m_i = 1$ and $y_m^* > 0$. Proposition 3(iii) implies that for some column $j \neq i$ we have $m_j = 1$. Suppose that $\tau(N \setminus \{i,j\}) = \tau(N) - 2$. Then there is an optimal solution to (1) with $x_i^* = x_j^* = 1$. But then Remark 5 would imply that $y_m^* = 0$, a contradiction. Hence $\tau(N \setminus \{i,j\}) = \tau(N) - 1$. Because N is minimally non-Mengerian, $\nu(M \setminus \{i,j\}) = \tau(N) - 1$. Thus there exists a collection $\mathcal{R}$ of $\tau(N) - 1$ circuits of D which do not use elements i or j and which pairwise intersect in at most I. Proposition 3(ii) states that $\tau(N/j) = \tau(N)$. Because N is minimally non-Mengerian, $\nu(N/j) = \tau(N)$. Thus there exists a collection $\mathcal{B}$ of $\tau(N)$ circuits of D which pairwise intersect in at most $I \cup \{j\}$.

Claim: Every column of N is an element of a circuit of $\mathcal{R} \cup \mathcal{B}$.

Proof. (of Claim) Suppose for a contradiction there is a column k of N which is not in any circuit of $\mathcal{R} \cup \mathcal{B}$. Let T be a minimal set of columns which intersect all rows of $N \setminus k$. Suppose T contains j. Then the collection $\mathcal{R}$ imply that $|T - \{j\}| \geq |\mathcal{R}| = \tau(N) - 1$. Suppose T does not contain j. Then the collection $\mathcal{B}$ imply that $|T| \geq |\mathcal{B}| = \tau(N)$. Hence, $\tau(N \setminus k) = \tau(N)$. Because N is minimally non-Mengerian $\tau(N \setminus k) = \nu(N \setminus k)$, i.e. there exists $\tau(N \setminus k) = \tau(N)$ rows of N which are pairwise disjoint. This implies $\nu(N) = \tau(N)$ a contradiction. $\diamond$

Suppose for a contradiction that i corresponds to a vertex of D. Then i is in a unique circuit of $\mathcal{B}$ and in no circuit of $\mathcal{R}$. It follows from the Claim that i must have in- and out-degree one. Then it is routine to check that we may replace the two series edges incident to i by a single edge with weight one. The resulting digraph D' is a digraph minor of D and $\tau(D') > \nu(D')$. Hence D' is not CM, a contradiction. $\square$

Suppose a minimally non CM digraph D is strongly connected, but not strongly 2-connected, Thus there is a vertex v such that $D \setminus v$ is not strongly connected. Then there is a partition of $V(D) - \{v\}$ into non-empty sets X_1, X_2 such that all edges with endpoints in both X_1 and X_2 have tail in X_1 and head in X_2. Let F be the set of all these edges. For $i = 1, 2$ let D_i be the digraph obtained from D by deleting all edges with both endpoints in $X_{3-i} \cup \{v\}$ and identifying all vertices of $X_{3-i} \cup \{v\}$. Thus edges of F belong to both D_1 and D_2; in D_1 they have head v and in D_2 they have tail v. We say that D is a *1-sum* of D_1 and D_2. Let $w : E(D) \rightarrow Z_+$ and let $w' : V(D) \cup E(D) \rightarrow Z_+$ be defined as follows: $w'(t) = +\infty$ if $t \in V$ and $w'(t) = w(t)$ otherwise. We write $\bar{\tau}_w(D)$ for $\tau_{w'}(D)$, $\bar{\nu}_w(D)$ for $\nu_{w'}(D)$, $\bar{\nu}^*_w(D)$ for $\nu^*_{w'}(D)$. When $w(e) = 1$ for all $e \in E$ then $\bar{\tau}(D) = \bar{\tau}_w(D), \bar{\nu}(D) = \bar{\nu}_w(D), \bar{\nu}^*(D) = \bar{\nu}^*_w(D)$.

Proposition 5. *Let D be a minimally non CM digraph which is CI. Then D is strongly 2-connected.*

Proof. It is not hard to show that D must be strongly connected. Because D is CI and minimally non CM, it follows from Proposition 4 that for some $w : E(D) \rightarrow Z_+$, $\bar{\tau}_w(D) > \bar{\nu}_w(D)$. Because we allow parallel edges we may assume $w(e) = 1$ for all $e \in E(D)$. Suppose for a contradiction D is not strongly 2-connected. We will show $\bar{\tau}(D) = \bar{\nu}(D)$ contradicting the fact that D is not CM. Let $M = M(D)$. Since D is CI, M is ideal. It follows from linear programming duality that there is a edge feedback set T of D and a vector y^* such that $|T| = y^* \mathbf{1}$ where $y^* M \leq \mathbf{1}$ and $y^* \geq \mathbf{0}$. It follows from Proposition 3(i) and Remark 5 that $y^* M = \mathbf{1}$. Since D is not strongly 2-connected it is the 1-sum of some digraphs D_1 and D_2. Let F be the set of edges which belong to both D_1 and D_2. Throughout this proof i shall denote an arbitrary element of $[2]$. Let $\mathcal{L}_i$ be the family consisting of circuits C of D which are circuits of $D_i \setminus F$ and for which $y_C > 0$. Let $\mathcal{F}$ be the family consisting of circuits C of D which use some edge of F and for which $y_C > 0$. Let $l_i = \sum_{C \in \mathcal{L}_i} y^*_C$. Since for all $f \in F$, $(y^* M)_f = 1$, $\sum_{C \in \mathcal{F}} y_C = |F|$. Observe that y^* and the families $\mathcal{L}_i$ and $\mathcal{F}$ imply that $\bar{\nu}^*(D_i) = l_i + |F|$. Because D is strongly connected, each D_i is a digraph-minor of D. Since D is minimally non CM it follows that each D_i is CM. Hence $\bar{\nu}(D_i) \geq \lceil l_i \rceil + |F|$. Let $\hat{\mathcal{L}}_i$ be the corresponding family of circuits. Construct a family of circuits $\mathcal{L}$ as follows: (i) include each circuit of $\hat{\mathcal{L}}_1$ and $\hat{\mathcal{L}}_2$ which does not use edges of F; (ii) for every edge $f \in F$ for which there exist circuits $C_1 \in \hat{\mathcal{L}}_1, C_2 \in \hat{\mathcal{L}}_2$ and $f \in C_1 \cap C_2$, include the circuit $C_1 \cup C_2 - \{f\}$ in $\mathcal{L}$. Let q_i be the number edges of F which are not in circuits of $\hat{\mathcal{L}}_i$. Since $\hat{\mathcal{L}}_i \geq \lceil l_i \rceil + |F|$, there are $\lceil l_i \rceil + q_i$ circuits of $\hat{\mathcal{L}}_i$ not using any edge of F. Hence, there are $\lceil l_1 \rceil + q_1 + \lceil l_2 \rceil + q_2$ circuits included in (i) and at least $|F| - q_1 - q_2$ circuits included in (ii). Thus $|\mathcal{L}| \geq \lceil l_1 \rceil + \lceil l_2 \rceil + |F|$ and $\bar{\nu}(D) \geq \bar{\tau}(D)$, a contradiction. $\qquad\square$

4   Circuit Mengerian Digraphs

4.1   Pfaffian Orientations

Let G be a bipartite graph with bipartition (A, B), and let M be a perfect matching in G. We denote by $D(G, M)$ the digraph obtained from G by directing every edge of G from A to B, and contracting every edge of M. It is clear that every digraph is isomorphic to $D(G, M)$ for some bipartite graph G and some perfect matching M. A graph G is k-extendable, where k is an integer, if every matching in G of size at most k can be extended to a perfect matching. A

2-extendable bipartite graph is called a *brace*. The following straightforward relation between k-extendability and strong k-connectivity will be required [14].

Proposition 6. *Let G be a bipartite graph, let M be a perfect matching in G, and let $k \geq 1$ be an integer. Then G is k-extendable if and only if $D(G, M)$ is strongly k-connected.*

A circuit C of G is *central* if $G \backslash V(C)$ has a perfect matching. Let D be an orientation of G. A circuit C of G is *oddly oriented* if an odd number of its edges are directed in the direction of each orientation of C. A bipartite graph G is *Pfaffian* is there exists an orientation of G such that every central circuit is oddly oriented. Let G be a bipartite graph and M a perfect matching in G such that $D(G, M)$ is isomorphic to F_7. This defines G uniquely up to isomorphism, and the graph so defined is called the *Heawood graph*. Let G_0 be a bipartite graph, let C be a central circuit of G_0 of length 4, and let G_1, G_2 be subgraphs of G_0 such that $G_1 \cup G_2 = G_0, G_1 \cap G_2 = C$, and $V(G_1) - V(G_2) \neq \emptyset \neq V(G_2) - V(G_1)$. Let G be obtained from G_0 by deleting all the edges of C. In this case we say that G is the *4-sum* of G_1 or G_2 along C. This is a slight departure from the definition in [14], but the class of simple graphs obtainable according to our definition is the same, because we allow parallel edges.

Proposition 7 (Robertson et al. [14]). *Let G be a 4-sum of G_1 and G_2. Then G is a Pfaffian brace if and only if both G_1 and G_2 are Pfaffian braces.*

Theorem 5. *Let G be a brace, and let M be a perfect matching in G. Then the following conditions are equivalent.*

(i)   *G is Pfaffian,*
(ii) *$D(G, M)$ has no minor isomorphic to an odd double circuit.*
(iii) *either G is isomorphic to the Heawood graph, or G can be obtained from planar braces by repeatedly applying the 4-sum operation,*

Proof. The equivalence of (i) and (iii) is the main result of [11] and [14]. Condition (ii) is equivalent to the other results of Little [7] and Seymour and Thomassen [13]. See also [11].   □

Using this structural decomposition we will deduce.

Proposition 8. *Let G be a Pfaffian brace which is not isomorphic to the Heawood graph, let M be a perfect matching, and let $D = D(G, M)$. For every $w : E(D) \rightarrow Z_+$, $\bar{\tau}_w(D) = \bar{\nu}_w(D)$.*

This last proposition will be sufficient to show that every CI digraph is CM.

Proof. (of Lemma 2) Let D be a CI digraph which is minimally non CM. It follows from Proposition 4 that $\bar{\tau}_w(D) > \bar{\nu}_w(D)$ for some $w : E(D) \rightarrow Z_+$. Proposition 5 implies that D is strongly 2-connected. Recall $D = D(G, M)$ for some bipartite graph G with perfect matching M. Note that G is not isomorphic

to the Heawood graph; for otherwise D is isomorphic to F_7 which is not CI (it contains a Δ_1-chain as a minor). It follows from Proposition 6 that G is 2-extendable. Since D is CI it has no odd double circuit minors. It follows from Theorem 5 that G is Pfaffian. Hence Proposition 8 implies that $\bar{\tau}_w(D) = \bar{\nu}_w(D)$, a contradiction.
$\square$

Proof. (Sketch for Proposition 8) Suppose for a contradiction the result does not hold and let us assume that G is a counterexample with fewest number of vertices. Lucchesi and Younger [8](Theorem B) proved that for planar digraph, $\bar{\tau}_w(D) = \bar{\nu}_w(D)$ for any $w : E(D) \to Z_+$. Hence, we may assume D is not planar and thus that G is not planar either. It follows from Theorem 5 that G is the 4-sum along a central circuit C of length four of some bipartite graphs G_1 and G_2. For $i = 1, 2$, let $M_i = M \cap E(G_i)$ and $D_i = D(G_i, M_i)$. We know from Proposition 7 that G_1 and G_2 are Pfaffian braces. Moreover, G_1 is contained in G. Robertson et al. [14][Theorem 6.7] showed that if a Pfaffian brace G contains the Heawood graph then G is isomorphic to the Heawood graph. It follows that G_1, and similarly G_2, are not isomorphic to the Heawood graph. Moreover, G_1 and G_2 have fewer vertices than G so it follows that for D_i and all its minors D_i', $\bar{\tau}_w(D_i') = \bar{\nu}_w(D_i')$ for all weights $w : E(D_i') \to Z_+$.

There are several cases depending on how many edges of M have both endpoints in $V(C)$. We shall outline how to dispose of the case where two edges of M have both endpoints in C, the other cases can be dealt with using similar, but more involved, arguments. In this case $|V(D_1) \cap V(D_2)| = 2$ and let $V(D_1) \cap V(D_2) = \{u_1, u_2\}$. Moreover, $E(D_1) \cap E(D_2) = \{u_1 u_2, u_2 u_1\}$. Define $D_i' = D_i \backslash \{u_1 u_2, u_2 u_1\}$. Let $w' : E(D) \cup \{u_1 u_2, u_2 u_1\} \to Z_+$ where $w'(u_1 u_2) = w'(u_2 u_1) = +\infty$ and $w'(e) = w(e)$ for all $e \in E(D)$. Let $i, j, k \in [2]$ where $j \neq k$. Define $r_i^{jk} = \bar{\tau}_{w'}(D_i' + u_j u_k) - \bar{\tau}_{w'}(D_i')$.

Claim: For $i = 1, 2$, $r_i^{12} = 0$ or $r_i^{21} = 0$.

Proof. (of Claim) Let j, k denote distinct elements of [2]. If $r_i^{jk} > 0$ there is no set of edges $T \subseteq E(D_i')$ with $w(T) = \bar{\tau}_w(D_i')$ which intersects all circuits of D_i' and all paths from u_k to u_j. Hence, if $r_i^{12} > 0$ and $r_i^{21} > 0$ then there is no set of edges $T \subseteq E(D_i')$ with $w(T) = \bar{\tau}_w(D_i')$ which intersects all circuits of D_i', a contradiction.
$\diamond$

Because of the Claim we may assume (after possibly relabeling D_1, D_2 and u_1, u_2) that $r_1^{21} = 0$ and $r_1^{12} \geq r_2^{21}$. For $i \in [2]$, let $\hat{w}_i : E(D) \cup \{u_1 u_2, u_2 u_1\} \to Z_+$ where $\hat{w}_i(u_1 u_2) = r_i^{12}$, $\hat{w}_i(u_2 u_1) = r_i^{21}$, and $\hat{w}_i(e) = w(e)$ for $e \in D(D)$. Note for $i, j, k \in [2]$ where $j \neq k$, $\bar{\tau}_{\hat{w}_i}(D_i' + u_j u_k) = \bar{\tau}_{w'}(D_i' + u_j u_k)$. Since $D_1' + u_1 u_2$ is a minor of D_1 it follows that $\bar{\tau}_{\hat{w}_i}(D_1' + u_1 u_2) = \bar{\nu}_{\hat{w}_i}(D_1' + u_1 u_2)$. Hence there is a collection of $\bar{\tau}_{w'}(D_1')$ circuits $\mathcal{C}_1$ and r_1^{12} $u_2 u_1$-paths $\mathcal{P}_1$ which form a $\hat{w}_1$-matching. Similarly, there is a collection of $\bar{\tau}_{w'}(D_2')$ circuits $\mathcal{C}_2$ and r_2^{21} $u_1 u_2$-paths $\mathcal{P}_2$ which form a $\hat{w}_2$-matching. Combining $\mathcal{C}_1 \cup \mathcal{C}_2$ with the circuits obtained by pairing as many paths of $\mathcal{P}_1$ and $\mathcal{P}_2$ as possible, we get a collection of $\bar{\tau}_{w'}(D_1') + \bar{\tau}_{w'}(D_2') + r_2^{21}$ circuits which form a w-matching. Since $r_1^{12} = 0$ there exists a set $T_1 \subseteq E(D_1')$ with $w'(T_1) = \bar{\tau}_{w'}(D_1')$ which intersects all circuits of D_1 and all $u_1 u_2$-paths.

There exists a set $T_2 \subseteq E(D'_2)$ with $w'(T_2) = \bar{\tau}_{w'}(D'_2) + r_2^{21}$ which intersects all circuits of D_2 and all $u_1 u_2$-paths. Then $T_1 \cup T_2$ intersect all circuits of D and $\bar{\tau}_w(D) = \bar{\nu}_w(D)$, a contradiction. $\qquad\qquad\square$

References

1. C. Berge. Balanced matrices. *Math. Programming*, 2:19–31, 1972.
2. M Conforti and G. Cornuéjols. Clutters that pack and the max-flow min-cut property: A conjecture. In W.R. Pulleyblank and F.B. Shepherd, editors, *The Fourth Bellairs Workshop on Combinatorial Optimization*, 1993.
3. G. Cornuéjols, B. Guenin, and F. Margot. The packing property. *Math. Program., Ser. A*, 89:113–126, 2000.
4. R. Oppenheim D. R. Fulkerson, A. Hoffman. On balanced matrices. *Math. Programming Study*, 1:120–132, 1974.
5. B. Guenin and R. Thomas. Packing directed circuits exactly. 1999. Preprint.
6. A. Lehman. On the width-length inequality and degenerate projective planes. In W. Cook and P.D. Seymour, editors, *Polyhedral Combinatorics*, volume 1 of *DIMACS Series in Discrete Math. and Theoretical Computer Science*, pages 101–105, 1990.
7. C. H. C. Little. A characterization of convertible $(0, 1)$-matrices. *J. of Combin. Theory Ser. B*, 18:187–208, 1975.
8. C.L. Lucchesi and D.H. Younger. A minimax relation for directed graphs. *J. London Math. Soc.*, 17(2):369–374, 1978.
9. W. Zang M. Cai, X. Deng. A tdi system and its application to approximation algorithm. In *Proc. 39th IEEE Symposium on Foundations of Computer Science*, 1998.
10. W. Zang M. Cai, X. Deng. A min-max theorem on feedback vertex sets. In *Integer Programming and Combinatorial Optimization*, Lecture Notes in Computer Science. 7th IPCO Conference, Graz, Austria, 1999.
11. W. McCuaig. Pólya's permanent problem, manuscript (81 pages). June 1997.
12. A. Schrijver. *Theory of Linear and Integer Programming*. Wiley-Interscience, 1986. ISBN 0-471-90854-1.
13. P. D. Seymour and C. Thomassen. Characterization of even directed graphs. *J. of Combin. Theory Ser. B*, 42:36–45, 1987.
14. R. Thomas, N. Robertson, and P.D. Seymour. Permanents, pfaffian orientations, and even directed circuits. *Ann. Math.*, 150:929–975, 1999.

Integral Polyhedra Related to Even Cycle and Even Cut Matroids*

Bertrand Guenin

Department of Combinatorics and Optimization, University of Waterloo

Abstract. A family of sets $\mathcal{H}$ is ideal if the polyhedron $\{x \geq \mathbf{0} : \sum_{i \in S} x_i \geq 1, \forall S \in \mathcal{H}\}$ is integral. Consider a graph G with vertices s, t. An *odd st-walk* is either: an odd st-path; or the union of an even st-path and an odd circuit which share at most one vertex. Let T be a subset of vertices of even cardinality. An *st-T-cut* is a cut of the form $\delta(U)$ where $|U \cap T|$ is odd and U contains exactly one of s or t. We give excluded minor characterizations for when the families of odd st-walks and st-T-cuts (represented as sets of edges) are ideal. As a corollary we characterize which extensions and coextensions of graphic and cographic matroids are 1-flowing.

1   Introduction

A *clutter* $\mathcal{H}$ is a finite family of sets, over some finite ground set $E(\mathcal{H})$, with the property that no set of $\mathcal{H}$ contains, or is equal to, another set of $\mathcal{H}$. A $0, 1$ matrix, whose columns are indexed by $E(\mathcal{H})$ and whose rows are the characteristic vectors of the sets in $\mathcal{H}$, is denoted by $M[\mathcal{H}]$. A clutter $\mathcal{H}$ is *ideal* if $\{x \geq \mathbf{0} : M[\mathcal{H}]x \geq \mathbf{1}\}$ is an integral polyhedron, that is its extreme points have $0, 1$ coordinates. The *blocker* $b(\mathcal{H})$ of $\mathcal{H}$ is the clutter defined as follows: $E(b(\mathcal{H})) := E(\mathcal{H})$ and $b(\mathcal{H})$ is the set of inclusion-wise minimal members of $\{U : U \cap C \neq \emptyset, \forall C \in \mathcal{H}\}$. It is well known that $b(b(\mathcal{H})) = \mathcal{H}$. Hence we say that $\mathcal{H}, b(\mathcal{H})$ form a *blocking pair* of clutters. Lehman [11] showed that if a clutter is ideal, then so is its blocker. A clutter is said to be *binary* if, for any $S_1, S_2, S_3 \in \mathcal{H}$, their symmetric difference $S_1 \triangle S_2 \triangle S_3$ contains, or is equal to, a set of $\mathcal{H}$. Equivalently, $\mathcal{H}$ is binary if, for every $S \in \mathcal{H}$ and $S' \in b(\mathcal{H})$, $|S \cap S'|$ is odd. Let $\mathcal{H}$ be a clutter and $i \in E(\mathcal{H})$. The *contraction* $\mathcal{H}/i$ and *deletion* $\mathcal{H} \setminus i$ are clutters with ground set $E(\mathcal{H}) - \{i\}$ where: $\mathcal{H}/i$ is the set of inclusion-wise minimal members of $\{S - \{i\} : S \in \mathcal{H}\}$ and; $\mathcal{H} \setminus i := \{S : i \notin S \in \mathcal{H}\}$. Observe that $b(\mathcal{H} \setminus i) = b(\mathcal{H})/i$ and $b(\mathcal{H}/i) = b(\mathcal{H}) \setminus i$. Contractions and deletions can be performed sequentially, and the result does not depend on the order. A clutter obtained from $\mathcal{H}$ by a sequence of deletions I_d and a sequence of contractions I_c ($I_c \cap I_d = \emptyset$) is called a *minor* of $\mathcal{H}$ and is denoted $\mathcal{H} \setminus I_d/I_c$. It is a *proper* minor if $I_c \cup I_d \neq \emptyset$. A clutter is said to be *minimally non-ideal* (mni) if it is not ideal but all its proper minors are ideal. The clutter $\mathcal{O}_{K_5}$ is defined as follows: $E(\mathcal{O}_{K_5})$ is the set of 10 edges of K_5 and $\mathcal{O}_{K_5}$ is the set of odd circuits of K_5 (the

---

* This work supported by the Fields Institute and the University of Waterloo.

triangles and the circuits of length 5). The 10 constraints corresponding to the triangles define a fractional extreme point $(\frac{1}{3}, \frac{1}{3}, \ldots, \frac{1}{3})$ of the associated polyhedron. Consequently, $\mathcal{O}_{K_5}$ is not ideal and neither is its blocker. The ground set of $\mathcal{L}_{F_7}$ are the elements $\{1, 2, 3, 4, 5, 6, 7\}$ of the Fano matroid and the sets in $\mathcal{L}_{F_7}$ are the circuits of length three (the lines) of the Fano matroid, i.e.,

$$\mathcal{L}_{F_7} := \{\{1, 3, 5\}, \{1, 4, 6\}, \{2, 3, 6\}, \{2, 4, 5\}, \{1, 2, 7\}, \{3, 4, 7\}, \{5, 6, 7\}\}.$$

The fractional point $(\frac{1}{3}, \frac{1}{3}, \ldots, \frac{1}{3})$ is an extreme point of the associated polyhedron, hence $\mathcal{L}_{F_7}$ is not ideal. The blocker of $\mathcal{L}_{F_7}$ is $\mathcal{L}_{F_7}$ itself. The following excluded minor characterization is predicted,

Seymour's Conjecture [Seymour [14] p. 200, [15] (9.2), (11.2)]
A binary clutter is ideal if and only if it has no $\mathcal{L}_{F_7}$, no $\mathcal{O}_{K_5}$, and no $b(\mathcal{O}_{K_5})$ minor.

1.1 Clutters of Odd st-Walks

Consider a graph G and a subset of its edges Σ. The pair (G, Σ) is called a *signed graph*. We say that a subset of edges L is *odd* (resp. *even*) if $|L \cap \Sigma|$ is odd (resp. even). Let s, t be vertices of G. We call a subset of edges of (G, Σ) an *odd st-walk* if it is an odd st-path; or it is the union of an even st-path P and an odd circuit C, where P and C share at most one vertex. We will see in Section 2 that the family of odd st-walks form a binary clutter. Figure 1 gives a representations of the clutter $\mathcal{L}_{F_7}$ as a clutter of odd st-walks. The dashed lines correspond to edges in Σ.

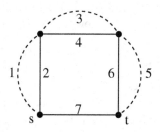

Fig. 1. The clutter $\mathcal{L}_{F_7}$ represented as an st-walk.

Theorem 1. *A clutter of odd st-walks is ideal if and only if it has no $\mathcal{L}_{F_7}$ and no $\mathcal{O}_{K_5}$ minor.*

A signed graph is said to be *weakly bipartite* if its odd circuits form an ideal clutter (where the ground set is the edge set). If $s = t$ there exist no odd st-paths in (G, Σ). Hence, in that case, the clutter of odd st-walks is the clutter of odd circuits. It is easy to verify that $\mathcal{L}_{F_7}$ is not a clutter of odd circuits. Moreover, the family of clutters of odd circuits is closed under taking minors. Consequently the following theorem is a corollary of Theorem 1.

Theorem 2 (Guenin [7]). *A signed graph is weakly bipartite if and only if the clutter of its odd circuits has no $\mathcal{O}_{K_5}$ minor.*

1.2   Clutters of st-T-Cuts

Consider a graph G and a subset T of its vertices with $|T|$ even. A *T-join* is an inclusion-wise minimal set of edges J such that T is the set of vertices of odd degree of the edge-induced subgraph $G[J]$. A *T-cut* is an inclusion-wise minimal set of edges $\delta(U) := \{(u, v) : u \in U, v \notin U\}$, where U is a set of vertices of G that satisfies $|U \cap T|$ odd. T-joins and T-cuts generalize many interesting special cases. If $T = \{s, t\}$, then the T-joins (resp. T-cuts) are the st-paths (resp. st-cuts) of G. If $T = V$, then the T-joins of size $|V|/2$ are the perfect matchings of G. The case where T is identical to the set of odd-degree vertices of G is known as the Chinese postman problem [9,4]. The families of T-joins and T-cuts form a blocking pair of clutters. Let s, t be vertices of G. An *st-T-cut* [6] is a T-cut $\delta(U)$, where U contains exactly one of s or t. We will see in Section 2 that the family of st-T-cuts form a binary clutter. In Figure 2 we represent $\mathcal{L}_{F_7}$ as an st-T-cut. In Figure 3 we represent $b(\mathcal{O}_{K_5})$ as an st-T-cut. The solid vertices correspond to the vertices in T.

Theorem 3. *A clutter of st-T-cuts is ideal if and only if it has no $\mathcal{L}_{F_7}$ and no $b(\mathcal{O}_{K_5})$ minor.*

Suppose t is an isolated vertex. Then the st-T-cuts of G are the T-cuts of G. Since $\mathcal{L}_{F_7}$ and $b(\mathcal{O}_{K_5})$ are not clutters of T-cuts, and since this class of clutters is closed under taking minors, the following theorem is a corollary of Theorem 3,

Theorem 4 (Edmonds [4]). *Clutters of T-cuts and T-joins are ideal.*

If T consists of two vertices s', t' then the st-T-cuts are known as *2-commodity cuts* [8]. The blocker of the clutter of 2-commodity cuts is the set of all st-paths and $s't'$-paths. It is referred to as a clutter of *2-commodity flow*. Since $\mathcal{L}_{F_7}$ and $b(\mathcal{O}_{K_5})$ are not clutters of 2-commodity flows, and since this class of clutters is closed under taking minors, the following theorem is a corollary of Theorem 3,

Theorem 5 (Hu [8]). *Clutters of 2-commodity flows and 2-commodity cuts are ideal.*

Recall that $\mathcal{L}_{F_7}, \mathcal{O}_{K_5}$ are clutters of odd st-walks and $\mathcal{L}_{F_7}, b(\mathcal{O}_{K_5})$ are clutters of st-T-cuts. Together with theorems 1 and 3 this implies the following corollary which suggests a new way of attacking Seymour's conjecture,

Corollary 1. *The following statements are equivalent:*

(i)   Seymour's conjecture holds;

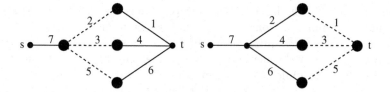

Fig. 2. The clutter $\mathcal{L}_{F_7}$ represented as an st-T-cut.

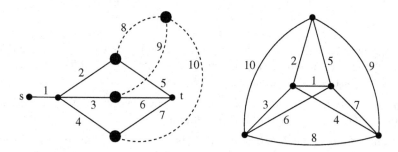

Fig. 3. The clutter $b(\mathcal{O}_{K_5})$ represented as an st-T-cut.

(*ii*) Let $\mathcal{H}$ be a mni binary clutter, then $\mathcal{H}$ or $b(\mathcal{H})$ is a clutter of odd st-walks;
(*iii*) Let $\mathcal{H}$ be a mni binary clutter, then $\mathcal{H}$ or $b(\mathcal{H})$ is a clutter of st-T-cuts;
(*iv*) A mni binary clutter is either a clutter of odd st-walks or a clutter of st-T-cuts.

1.3 1-Flowing Matroids

Let M be a matroid. The set of its elements is denoted $E(M)$ and the set of its circuits by $\Omega(M)$. A matroid M is said to be e-*flowing* [15] if for every weights $w \in Z_+^{|E(M)|-1}$ the following equality holds:

$$\min\Big\{ \sum_{i\in E(M)-\{e\}} w_i x_i : \sum_{i\in C-\{e\}} x_i \geq 1,$$

$$\forall C \text{ with } e \in C \in \Omega(M), x \in \{0,1\}^{|E(M)|-1}\Big\}$$

$$= \max\Big\{ \sum_{C:e\in C\in\Omega(C)} y_C : \sum_{C:\{e,j\}\subseteq C\in\Omega(C)} y_C \leq w_j,$$

$$\text{for all } j \in E(M) - \{e\}, y \geq 0\Big\}.$$

The matroid M is said to be 1-flowing if it is e-flowing for all its elements e. Consider the special case where M is a graphic matroid, and let s, t be the endpoints of edge e of the underlying graph. Then $C - \{e\}$ is an st-path. The solution to the minimization problem becomes a minimum weight st-cut, while the solution to the maximization problem becomes a maximum fractional st-flow. We know from the max-flow min-cut theorem [5] that equality holds in this case. Hence graphic matroids are 1-flowing. Next we define the matroids $AG(3,2)$ and T_{11} by giving their partial representations,

$$AG(3,2) \begin{bmatrix} 0\,1\,1\,1 \\ 1\,0\,1\,1 \\ 1\,1\,0\,1 \\ 1\,1\,1\,0 \end{bmatrix} \qquad T_{11} \begin{bmatrix} 1\,1\,0\,0\,1\,1 \\ 1\,1\,1\,0\,0\,1 \\ 0\,1\,1\,1\,0\,1 \\ 0\,0\,1\,1\,1\,1 \\ 1\,0\,0\,1\,1\,1 \end{bmatrix}. \tag{1}$$

Observe that $AG(3,2)$ is a one element extension of the Fano matroid and T_{11} is a one element extension of R_{10}. We write T_{11}^* for the dual of T_{11}. The following is a special case of Seymour's conjecture.

Conjecture 1 (Seymour[15] (11.2)). A binary matroid is 1-flowing if and only if it has no $AG(3,2)$, no T_{11}, and no T_{11}^* minor.

The rank 2 uniform matroid on 4 elements is written $U_{2,4}$. The next result implies Conjecture 1 for extensions and coextensions of graphic and cographic matroids.

Theorem 6. (i)   Suppose M is a coextension of a graphic matroid. Then M is 1-flowing if and only if it has no $AG(3,2)$, no T_{11}^*, and no $U_{2,4}$ minor.

(ii)   Suppose M is a coextension of a cographic matroid. Then M is 1-flowing if and only if it has no $AG(3,2)$, no T_{11}, and no $U_{2,4}$ minor.

(iii)   Suppose M is an extension of a cographic matroid. Then M is 1-flowing if and only if it has no $AG(3,2)$, no T_{11}, and no $U_{2,4}$ minor.

(iv)   Suppose M is an extension of a graphic matroid. Then M is 1-flowing if and only if it has no $AG(3,2)$, no T_{11}^*, and no $U_{2,4}$ minor.

1.4   Outline of the Paper

The paper is organized as follows. Section 2 constructs binary clutters from binary matroids. Special cases of this construction include odd st-walks and st-T-cuts. Section 3 reviews properties of minimally non-ideal binary clutters. The proof of Theorem 1 is found in Section 4. The proof of Theorem 6(i),(iii) is in Section 5. We have omitted the proof of theorem 4.

2   From Binary Matroids to Binary Clutters

Throughout this section, M will denote a binary matroid and $\Sigma \subseteq E(M)$ a subset of its element. The pair (M, Σ) is called a *signed matroid*. Recall that a set $L \subseteq E(M)$ is odd (resp. even) if $|L \cap \Sigma|$ is odd (resp. even). The dual of M is written M^*. Let B be the $|\Omega(M^*)| + 1 \times |E(M)|$ matrix, where the first $|\Omega(M^*)|$ rows are the incidence vectors of the cocircuits of M, and the last row is the incidence vector of Σ. The binary matroid with representation B is denoted $Even(M, \Sigma)$.

Proposition 1. *Let $N := Even(M, \Sigma)$.*

(i)   *A circuit of N is either: an even circuit of M; or the disjoint union of two odd circuits of M.*

(ii) *A cocircuit of N is either: a cocircuits of M; or of the form $C \triangle \Sigma$, where C is a cocycle of M.*

Proof. Let L be a circuit of $Even(M, \Sigma)$. By definition L intersects every cocircuit of M with even parity. Hence L is a cycle of M and it can be expressed as the disjoint union of a collection S of circuits of M. As L is even there is an even number of odd circuits in S. Since L is a circuit either S consists of a single even circuit or S consists of the union of two odd circuits. Finally, (ii) follows from the fact that the cocycle space of $Even(M, \Sigma)$ is spanned by the rows of B. □

Let e be an element of M. The family of sets $\{S - \{e\} : e \in S \in \Omega(M)\}$ is denoted $Port(M, e)$. Note that $Port(M, e)$ has ground set $E(M) - \{e\}$. The following results have appeared explicitly or implicitly in the literature, see for instance [2].

Proposition 2. *$Port(M, e)$ is a binary clutter.*

Proposition 3. *$Port(M, e)$ and $Port(M^*, e)$ form a blocking pair.*

The next proposition is an immediate corollary of propositions 1 and 3.

Proposition 4. *Let $e \in E(M) \cap \Sigma$ and let $\mathcal{H} := Port\big(Even(M, \Sigma), e\big)$.*

(i)   *An element of $\mathcal{H}$ is either: (1) an odd set $C - \{e\}$, where C is a circuit of M using e; or (2) the union of an even set $C - \{e\}$ and an odd set C', where C, C' are disjoint circuits of M and $e \in C$.*

(ii) *An element of $b(\mathcal{H})$ is either of the form: (1) $C - \{e\}$ where C is a cocircuit of M using e; or (2) of the form $(C \triangle \Sigma) - \{e\}$, where C is a cocycle of M avoiding e.*

2.1   Even Cycle Matroids

Here we consider the case where M is the graphic matroid of a graph G. In that case $Even(M, \Sigma)$ is called an *even cycle matroid*. Let $e = (s, t) \in \Sigma$ be an edge of G. Let $\Sigma' := \Sigma - \{e\}$ and let G' be the graph obtained by deleting edge e of G. Consider $\mathcal{H}$ from Proposition 4. In the subsequent discussion odd and even are meant with respect to Σ'. From (i) we know that an element of $\mathcal{H}$ is either: (1) an odd st-path, or (2) the edge-disjoint union of an even st-path P and an odd circuit C' of G'. Moreover, if P and C' share more than one vertex then $P \cup C$ contains, as a proper subset, an odd st-path; a contradiction since $\mathcal{H}$ is a clutter. Hence the elements of $\mathcal{H}$ are the odd st-walks of (G', Σ'). Note that if C is a cut of G avoiding edge (s, t) then it is a cut of G' where s and t are on the same shore. Statement (ii) thus implies,

Corollary 2. *An element in the blocker of the clutter of odd st-walks of (G', Σ') is either: an st-cut; or a set of the form $C \triangle \Sigma'$, where C is a cut with s and t on the same shore.*

2.2   Even Cut Matroids

Here we consider the case where M is the cographic matroid of a graph G. In that case $Even(M, \Sigma)$ is called an *even cut matroid*. Let $e = (s, t) \in \Sigma$ be an edge of G. Let $\Sigma' := \Sigma - \{e\}$ and let G' be the graph obtained by deleting edge e of G. Define T to be the set of vertices of odd degree in the edge-induced subgraph $G'[\Sigma']$. Consider $\mathcal{H}$ from Proposition 4. If C (resp. $C \cup C'$) is a cut of G using e then $C - \{e\}$ (resp. $(C \cup C') - \{e\}$) is a cut of G' where s and t are on distinct shores. Since the odd cuts of (G', Σ') are the T-cuts of G', (i) implies that $\mathcal{H}$ is the clutter of st-T-cuts of (G', Σ'). From (ii) we know that an element of $b(\mathcal{H})$ is either: (1) an st-path; or (2) is of the form $C \triangle \Sigma'$ (where C is a cycle of G') i.e., a T-join. Consequently,

Corollary 3. *An element in the blocker of the clutter of st-T-cuts of G' is either an st-path or a T-join.*

3   Properties of Minimally Non-ideal Clutters

In this section we review properties of mni clutters. The clutter $\mathcal{J}_t$, for $t \geq 1$, is defined as follows: $E(\mathcal{J}_t) := \{0, \dots, t\}$ and $\mathcal{J}_t := \{\{1, \dots, t\}\} \cup \{\{0, i\} : i = 1, \dots, t\}$. By abuse of notation we shall identify a clutter $\mathcal{H}$ with its matrix representation $M[\mathcal{H}]$.

Theorem 7 (Lehman [12]). *Let A be a minimally non-ideal matrix with n columns and let B its blocker. Then B is minimally non-ideal. If A is distinct from $\mathcal{J}_t$ then A (resp. B) has a square, non-singular row submatrix $\bar{A}$ (resp. $\bar{B}$) with r (resp. $\bar{r}$) non-zero entries in every row and column. In addition $r\bar{r} > n$ and rows of A (resp. B) not in $\bar{A}$ (resp. $\bar{B}$) have at least $r + 1$ (resp. $\bar{r} + 1$) non-zero entries. Moreover, $\bar{A}\bar{B}^T = J + (r\bar{r} - n)I$, where J denotes an $n \times n$ matrix filled with ones.*

Note that $\mathcal{J}_t$ is not binary. Bridges [1] proved that matrices $\bar{A}, \bar{B}$ such that, $\bar{A}\bar{B}^T = J + (r\bar{r} - n)I$, commute. So $\bar{A}\bar{B}^T = \bar{A}^T\bar{B}$ and in particular for each column j of $\bar{B}$ we have $\bar{A}^T col(\bar{B}, j) = 1 + (r\bar{r} - n)e_j$, where e_j is the jth unit vector. Combinatorially this can be expressed as follows,

Remark 1. Let j be the index of a column of $\bar{B}$. Let $L_1, \dots, L_{\bar{r}}$ (resp. $U_1, \dots, U_r$) be the characteristic sets of the rows of $\bar{A}$ (resp. $\bar{B}$) whose indices are given by the characteristic set of column j of $\bar{B}$ (resp. $\bar{A}$). Then $L_1, \dots, L_{\bar{r}}$ (resp. $U_1, \dots, U_r$) pairwise intersect in at most j and exactly $q := r\bar{r} - n + 1$ of these sets contain j.

Which implies,

Remark 2. Let $\mathcal{H}$ be a mni clutter distinct from $\mathcal{J}_t$ and consider $L_1, L_2 \in \bar{\mathcal{H}}$ with $e \in L_1 \cap L_2$. The mates U_1, U_2 of L_1, L_2 satisfy $U_1 \cap U_2 \subseteq \{e\}$.

Given a clutter $\mathcal{H}$, we write $\bar{\mathcal{H}}$ for the clutter which contains only the sets of minimum cardinality of $\mathcal{H}$. Let $\mathcal{H}, \mathcal{K}$ be a blocking pair of mni clutters. Define $A := M[\mathcal{H}], B := M[\mathcal{K}]$ and $\bar{A} := M[\bar{\mathcal{H}}], \bar{B} := M[\bar{\mathcal{K}}]$. Consider the element $C \in \bar{\mathcal{H}}$ (resp. $U \in \bar{\mathcal{K}}$) that corresponds to the ith row of $\bar{A}$ (resp. $\bar{B}$). Because, $\bar{A}\bar{B}^T = J + (r\bar{r} - n)I$, C intersects every element of $\bar{\mathcal{K}}$ exactly once; except for U which is intersected $q = r\bar{r} - n + 1 \geq 2$ times. We call U the *mate* of C. Thus every element of $\bar{\mathcal{H}}$ is paired with an element of $\bar{\mathcal{K}}$. In binary clutters mates intersect with odd parity. Hence,

Remark 3. Let $\mathcal{H}$ be a mni binary clutter, then $q \geq 3$. In particular $r \geq 3, \bar{r} \geq 3$.

We shall also need the following results.

Proposition 5 (Guenin [7]). *Let $\mathcal{H}$ be a mni binary clutter and $\mathcal{K}$ its blocker. For any $e \in E(\mathcal{H})$ there exist $L_1, L_2, L_3 \in \bar{\mathcal{H}}$ and $U_1, U_2, U_3 \in \bar{\mathcal{K}}$ such that*

(i) $L_1 \cap L_2 = L_1 \cap L_3 = L_2 \cap L_3 = \{e\}$.
(ii) $U_1 \cap U_2 = U_1 \cap U_3 = U_2 \cap U_3 = \{e\}$.
(iii) *For all $i, j \in \{1, 2, 3\}$ we have: $L_i \cap U_j = \{e\}$ if $i \neq j$ and $|L_i \cap U_j| = q \geq 3$ if $i = j$.*

Moreover, let $L \in \bar{\mathcal{H}}$ and let U be its mate; if $e \in L \cap U$ then we may assume $L = L_1$.

Proposition 6 (Cornuéjols et al. [3], Theorem 6.1, Claim 2). *Let $\mathcal{H}$ be a mni clutter. There exist no set T such that $|T \cap L| = 1, \forall L \in \bar{\mathcal{H}}$.*

Proposition 7 (Guenin [7]). *Let $\mathcal{H}$ be a mni binary clutter and $L_1, L_2 \in \bar{\mathcal{H}}$. If $L \subseteq L_1 \cup L_2$ and $L \in \mathcal{H}$ then either $L = L_1$ or $L = L_2$.*

4   Clutters of Odd st-Walks

Proof. (of Theorem 1) Clearly, if a clutter of odd st-walks has an $\mathcal{O}_{K_5}$ or $\mathcal{L}_{F_7}$ minor then it cannot be ideal. It remains to show the converse. Throughout this proof, $\mathcal{H}$ will denote a mni clutter of odd st-walks. The underlying signed graph is (G, Σ). The blocker of $\mathcal{H}$ is written $\mathcal{K}$. The cardinality of the elements of $\bar{\mathcal{H}}$ (resp. $\bar{\mathcal{K}}$) is r (resp. $\bar{r}$). Suppose $s = t$. Then $\mathcal{H}$ is a clutter of odd circuits (see Section 1.1). It then follows from Theorem 2 that $\mathcal{H} = \mathcal{O}_{K_5}$. Consequently, we can assume that s and t are distinct vertices. An odd st-walk is minimum if it is in $\bar{\mathcal{H}}$. We will say that an odd circuit C is minimum if there exists an edge disjoint even st-path P such that $C \cup P \in \bar{\mathcal{H}}$. Similarly, we define minimum even st-paths. The set of edges incident to a vertex v is denoted $\delta(v)$.

Claim. There is a minimum odd circuit C_s containing vertex s but avoiding vertex t. Similarly, there is a minimum odd circuit C_t containing vertex t but avoiding vertex s.

Proof. (of Claim) We know from Corollary 2 that $\delta(s) \in \mathcal{K}$. Hence in particular $\delta(s)$ intersects every element of $\bar{\mathcal{H}}$ at least once. It follows from Proposition 6 that $\delta(s)$ intersects some element of $\bar{\mathcal{H}}$ more than once. Thus there exists a minimum odd circuit C_s using vertex s. Suppose for a contradiction that C_s uses vertex t as well. Then C_s can be partitioned into st-paths P and P'. As C_s is odd, one of these paths (say P) is odd. Hence $P \in \mathcal{H}$, a contradiction as $\mathcal{H}$ is a clutter. The same argument applies to vertex t. ◇

Claim. There exists a unique minimum even st-path.

Proof. (of Claim) Let P_s (resp. P_t) be a minimum even st-path such that $P_s \cup C_s$ (resp. $P_t \cup C_t$) is in $\bar{\mathcal{H}}$. The set $P_s \cup C_t$ contains an odd st-walk. It follows from Proposition 7 that $P_s = P_t$. Consider $L \in \bar{\mathcal{H}}$ of the form $P \cup C$, where P is a minimum even st-path and C a minimum odd circuit. We may assume C is distinct from C_s. The set $P \cup C_s$ contains an odd st-walk. It follows from Proposition 7 that $P = P_s$. ◇

The unique minimum even st-path of Claim 2 is denoted P_{even}.

Claim. Let W be a minimum odd st-walk which shares a vertex $v \neq s$ with C_s. Then either W is an odd st-path and it uses an edge of C_s incident to s; or W is of the form $P_{even} \cup C$ where C is a minimum odd circuit. Moreover, if C shares another vertex v' with C_s as well, then C and C' are not edge disjoint.

Proof. (of Claim) Consider first the case where W is an odd st-path P. We can partition C_s into paths Q_1 and Q_2 with endpoints s, v. Since C_s is odd we may assume that Q_1 is odd and Q_2 is even. We can partition P into paths P_1 and P_2, where P_1 has endpoints s, v and P_2 endpoints t, v. We may assume P_1 is distinct from Q_1 and Q_2 since otherwise we are done for this case. If P_1 is odd then let $L := Q_1 \cup P_2$. If P_1 is even then let $L := Q_2 \cup P_2$. In either cases L contains an odd st-path distinct from W and $P_{even} \cup C_s$; a contradiction to Proposition 7. Consider now the case where the minimum odd st-walk W is of the form $P_{even} \cup C$, where C is a minimum odd circuit using distinct vertices v, v' of C_s. We can partition C_s into paths Q_1 and Q_2 with endpoints v, v'. We can partition C into paths P_1 and P_2 with endpoints v, v'. We may assume that C and C_s are edge disjoint since otherwise we are done. Since C_s (resp. C) is odd exactly one of Q_1, Q_2 (resp. P_1, P_2) is odd. If P_1 and Q_2 have distinct parity, then let $L := P_1 \cup Q_2 \cup P_{even}$. If P_1 and Q_2 have the same parity, then let $L := P_1 \cup Q_1 \cup P_{even}$. In either cases L contains an odd st-walk distinct from W and $P_{even} \cup C_s$; a contradiction to Proposition 7. ◇

The mate of $P_{even} \cup C_s$ is denoted U_s.

Claim. The intersection of $P_{even} \cup C_s$ and U_s consists of three edges; namely the two edges of C_s incident to s and one edge of P_{even}.

Proof. (of Claim) Suppose for a contradiction there exists an edge $e \in (C_s \cap U_s) - \delta(s)$. Let $L_1 := P_{even} \cup C_s$. Since $e \in L_1 \cap U_s$, it follows from Proposition 5 that there exists $L_2 \in \bar{\mathcal{H}}$ such that $L_1 \cap L_2 = \{e\}$. Because of Claim 2, L_2 must be an st-path; a contradiction to Claim 3. Let T be the set of edges in $U_s \cap (P_{even} \cup C_s)$. We proved that $T \subseteq \delta(s) \cup P_{even}$. As $|U_s \cap (P_{even} \cup C_t)| = 1$, we have $|T \cap P_{even}| = 1$. Remark 3 states $q := |T| \geq 3$. Thus $|T| = 3$ and the result follows.   ◇

By Claim 4 there is an edge e_s in $U_s \cap P_{even}$. Remark 1 states that there exists a collection of minimum odd st-walks, $\mathcal{L} := \{L_1, \ldots, L_{\bar{r}}\}$ which pairwise intersect at most in e_o. Moreover, because $e_s \subset U_s$, we may assume that $L_1 = P_{even} \cup C_s$. Because of Claim 4, C_s is not a loop, hence it has a vertex v which is distinct from s.

Claim. Let $v \neq s$ be a vertex of C_s. There exists a set $L \in \mathcal{L} - \{L_1\}$ of the form $P_{even} \cup C$, where C is a minimum odd circuit which uses v.

Proof. (of Claim) The minimum odd st-paths in $\mathcal{L}$ are edge disjoint from C_s. By Claim 3, none of these st-paths use edges incident to v. Suppose the claim does not hold. Then none of the elements in $\mathcal{L} - \{L_1\}$ contain any edge incident to v. We know from Remark 1 that every edge of G is contained in some set of $\mathcal{L}$. Thus vertex v of G has degree two. Consider the two edges e, e' that are incident to v. Every odd st-walk uses either both, or none of e, e'. Equivalently, the columns of $M[\mathcal{H}]$ indexed by e, e' are identical. But this contradicts Theorem 7 since $M[\mathcal{H}]$ should be non-singular.   ◇

If $|P_{even}| \geq 2$ then sets of $\mathcal{L} - \{L_1\}$ are odd st-paths; a contradiction to the previous claim. Thus

Claim. The path P_{even} consists of a single edge.

The edge of Claim 6 is denoted e_o. Recall (Remark 1) that sets of $\bar{\mathcal{H}}$ are in $\mathcal{L}$ if and only if their mates contain e_o. Suppose there exist 4 sets in $\mathcal{L}$ which are not odd st-paths. Then each of these sets, and their mates, contain e_o. Thus $q := col(M[\mathcal{H}], e_o)col(M[\bar{\mathcal{K}}], e_o) \geq 4$. Since $\bar{A}^T \bar{B} = \bar{A}\bar{B}^T = J + (q-1)I$, it follows that mates intersect in at least four element; a contradiction to Claim 4. Thus there exist at most three sets in $\mathcal{L}$ of the form $P_{even} \cup C$. We know that $L_1 = P_{even} \cup C_s$. By symmetry we can assume that $L_2 = P_{even} \cup C_t$. We know from Claim 3 that there exists a set $L_3 \in \mathcal{L} - \{L_1, L_2\}$ of the form $P_{even} \cup C$. Hence L_1, L_2, L_3 are the only odd st-walks of $\mathcal{L}$ which are not odd st-paths. Let v_s be any vertex of C_s distinct from s and v_t be any vertex of C_t distinct from t. It follows from Claim 5, that the odd circuit C uses vertex v_s. By symmetry C uses vertex v_t as well. Since v_s and v_t were chosen arbitrarily, and since C is edge disjoint from C_s, it follows from Claim 3 that v_s is the unique vertex of C_s distinct from s. Hence, $r = 3$ and let $I_d = E(\mathcal{H}) - \{P_{even} \cup C_s \cup C_t \cup C\}$. Then $\mathcal{H} \setminus I_d$ is $\mathcal{L}_{F_7}$ (see Figure 1). Moreover, since $\mathcal{H}$ is mni, $I_d = \emptyset$.   □

5   1-Flowing Matroids

Throughout this section M will denote a binary matroid. The next proposition shows that there exists a one to one correspondence between ideal binary clutters and e-flowing matroid. The result is easy and is part of the folklore so we have omitted the proof.

Proposition 8. *Let $e \in E(M)$. Then $Port(M, e)$ is ideal if and only if M is e-flowing. In particular, M is 1-flowing if and only if $Port(M, e)$ is ideal for all $e \in E(M)$.*

We omit the proof of the next result as well,

Remark 4. Let $e \in E(M)$ and $I_c, I_d \subseteq E(M) - \{e\}$ with $I_c \cap I_d = \emptyset$. Then

$$Port(M, e) \setminus I_d / I_c = Port(M \setminus I_d / I_c).$$

Let C be a cycle of M. Then $e \in C$ if and only if $|C \cap \{e\}|$ is odd. Let D be any cocircuit of M using e and let $\Sigma := D - \{e\}$. Then $|C \cap \{e\}|$ is odd if and only if $|C \cap (D \triangle \{e\})| = |C \cap \Sigma|$ is odd. Thus the odd (resp. even) cycles of M are the cycles using (avoiding) e. This proves (i) of the next remark. Suppose C is a cycle of $M \setminus e$, then equivalently C is a cycle of M with $e \notin C$. Hence $|C \cap \Sigma|$ is even. As $e \notin C$, C is a cycle of M/e i.e., C is a cycle of $Even(M/e, \Sigma)$. This shows (ii) of the next remark.

Remark 5. Let $e \in E(M)$, let D be a cocircuit of M using e, and let $\Sigma := D - \{e\}$.

(i) $Port(M, e)$ is the set of odd circuits of $(M/e, \Sigma)$.
(ii) $Even(M/e, \Sigma) = M \setminus e$.

Consider a connected binary matroid M and $e \in E(M)$. Then for all circuits $C \in \Omega(M)$ either $e \in C$ or $C = C_1 \triangle C_2$ where $C_1, C_2 \in \Omega(M)$ and $e \in C_1 \cap C_2$ [10]. So in particular,

Remark 6. Let $e \in E(M)$. The connected component of M using e is uniquely determined by $Port(M, e)$.

Remark 7.

(i)   For all elements e of $AG(3, 2)$ we have $Port(AG(3, 2), e) = \mathcal{L}_{F_7}$.
(ii)   There is a unique element e of T_{11} such that $Port(T_{11}, e) = b(\mathcal{O}_{K_5})$.
(iii)   There is a unique element e of T_{11}^* such that $Port(T_{11}, e) = \mathcal{O}_{K_5}$.

For (ii),(iii) the element corresponds to the last column of the representation of T_{11} given in (1). A clutter $\mathcal{H}$ is said to have the *Max Flow-Min Cut* property (MFMC) if the system $M[\mathcal{H}] \geq 1, x \geq 0$ is Totally Dual Integral (see [13]). The clutter $\mathcal{Q}_6$ is the clutter of odd circuits of the complete graph K_4.

Theorem 8 (Seymour [14]). *A binary clutter has the MFMC property if and only it has no $\mathcal{Q}_6$ minor.*

We shall also need the following result,

Theorem 9 (Cornuéjols et al. [3], Theorem 1.8). *Let $\mathcal{H}$ be a mni clutter, let $L \in \bar{\mathcal{H}}$, and let U be the mate of L. Suppose $q = |L \cap U| \geq 3$. Then $\mathcal{H} \setminus e$ does not have the MFMC property for any $e \in E(\mathcal{H})$.*

Consider a mni binary clutter $\mathcal{H}$. Then (Remark 3) $q \geq 3$. Let e be any element of $E(\mathcal{H})$. It follows from Theorem 9 that the minor $\mathcal{H} \setminus e$ does not have the MFMC property. Theorem 8 implies that $\mathcal{H} \setminus e$ has a $\mathcal{Q}_6$ minor. Since the blocker of a mni clutter is mni, we get the following corollary,

Corollary 4. *Let $\mathcal{H}$ be a mni binary clutter and $e \in E(\mathcal{H})$. Then $\mathcal{H}/e$ has a $b(\mathcal{Q}_6)$ minor.*

Consider a graph G with a pair of distinct vertices s and t. The clutter $\mathcal{H}$ where $E(\mathcal{H})$ is the edge set of G and $\mathcal{H}$ is the set of all st-paths (resp. st-bonds) is called a *clutter of st-path* (resp. *st-bonds*). The following is well known and easy to check,

Remark 8. The clutters of st-paths and st-bonds have no $b(\mathcal{Q}_6)$ minors.

The fractional Max-Flow Min-Cut theorem [5] can be stated by saying that the clutter of st-paths is ideal. Since clutters of st-paths and st-bonds form a blocking pairs this implies the next remark. Observe that this also follows from Remark 8 and Theorem 8.

Remark 9. Clutters of st-paths and st-bonds are ideal.

Proposition 8 and Remark 7 imply that none of $AG(3,2), T_{11}$, and T_{11}^* are 1-flowing. It is easy to check that the matroid $U_{2,4}$ is not 1-flowing either. Seymour [15] showed that the class of 1-flowing matroids is closed under minors. Thus it suffices to show the "if" direction in Theorem 6. Since non-binary matroids have a $U_{2,4}$ minor (see [16]) it suffices to prove Theorem 6 for binary matroids. It follows from Proposition 8, Proposition 3, and the fact that the blocker of an ideal clutter is ideal, that the 1-flowing property is closed under duality (see also [15]). Hence, to prove Theorem 6 is suffices to show (i) and (ii). The next result proves (i). The proof of (ii) is similar to that of (i) but uses Theorem 3 instead of Theorem 1.

Proposition 9. *Let M be a binary coextension of a graphic matroid. Then M is 1-flowing if it has no $AG(3,2)$ and no T_{11}^* minor.*

Proof. Suppose M is not 1-flowing. We will show that M has an $AG(3,2)$ or T_{11}^* minor. It is trivial to check that a matroid M is 1-flowing if and only if all its components are 1-flowing so we will assume that M is connected. Since M is a coextension of a graphic matroid, there is an element e of M such that M/e is the graphic matroid of a graph G. Proposition 8 implies that for some element f of M, $Port(M, f)$ is non-ideal. Consider first the case where $e = f$. Because M is connected f is not a loop and there exists a cocircuit D of M which contains f. It follows from Remark 5(i) that $Port(M, f)$ is the clutter of odd circuits of $(G, D - \{f\})$. Theorem 2 implies $Port(M, f) \setminus I_d/I_c = \mathcal{O}_{K_5}$ for some $I_c, I_d \subseteq E(M) - \{f\}$. By Remark 4, $\mathcal{O}_{K_5} = Port(M \setminus I_d/I_c, f)$. It follows from remarks 6 and 7 that the connected component of $M \setminus I_d/I_c$ using f is T_{11}^*. Hence T_{11}^* is a minor of M. Thus we may assume that e and f are distinct elements. Let s, t be the endpoints of edge f of G. Let $\mathcal{K} := Port(M, f) \setminus I_d/I_c$ be a mni minor. As $I_d \cap I_c = \emptyset$ it suffices to consider the following cases,

Case 1 $e \notin I_c \cup I_d$.

Because of Remark 4, $\mathcal{K}/e = Port(M/e, f) \setminus I_d/I_c$. Since M/e is the graphic matroid of graph G, it follows that $Port(M/e, f)$ is the clutter of st-paths of G. Hence $\mathcal{K}/e$ is a clutter of st-paths. It follows from Remark 8 that $\mathcal{K}/e$ has no $b(\mathcal{Q}_6)$ minor, a contradiction to Corollary 4.

Case 2 $e \in I_c$.

Because of Remark 4, $\mathcal{K} = Port(M/e, f) \setminus I_d/(I_c - \{e\})$. Since $Port(M/e, f)$ is the clutter of st-paths of G, it follows that $\mathcal{K}$ is a clutter of st-paths as well. Remark 9 implies $\mathcal{K}$ is ideal, a contradiction.

Case 3 $e \in I_d$.

Since M is connected, so is M^* and thus there is a cocircuit D of M using both e and f. Remark 5(ii) states that $M \setminus e = Even(M/e, D - \{e\})$. As $f \in D - \{e\}$ it follows (see Section 2.1) that $Port(M \setminus e, f) = Port(Even(M/e, D - \{e\}))$ is the clutter of odd st-walks of $(G, D - \{e\})$. Because of Remark 4 $\mathcal{K} = Port(M \setminus e, f) \setminus (I_d - \{e\})/I_c$. So in particular, $\mathcal{K}$ is a clutter of odd st-walks. It follows from Theorem 1 that $\mathcal{K}$ is equal to $\mathcal{L}_{F_7}$ or to $\mathcal{O}_{K_5}$. By Remark 4, $\mathcal{O}_{K_5}$ or $\mathcal{L}_{F_7}$ is equal to $Port(M \setminus I_d/I_c, f)$. It follows from remarks 6 and 7 that the connected component of $M \setminus I_d/I_c$ using f is $AG(3, 2)$ or T_{11}^*. Hence $AG(3, 2)$ or T_{11}^* is a minor of M. □

Acknowledgments: I would like to thank Gérard Cornuéjols for helpful discussions in the early stages of this research.

References

1. W. G. Bridges and H. J. Ryser. Combinatorial designs and related systems. *J. Algebra*, 13:432–446, 1969.
2. G. Cornuéjols and B. Guenin. Ideal binary clutters, connectivity and a conjecture of Seymour. Submitted to SIAM J. on Discrete Mathematics, 1999.
3. G. Cornuéjols, B. Guenin, and F. Margot. The packing property. *Math. Program., Ser. A*, 89:113–126, 2000.
4. J. Edmonds and E. L. Johnson. Matching, euler tours and the chinese postman. *Math. Prog.*, 5:88–124, 1973.
5. L.R. Ford and D.R. Fulkerson. *Flows in Networks*. Princeton University Press, 1962.
6. M.X. Goemans and V.S. Ramakrishnan. Minimizing submodular functions over families of sets. *Combinatorica*, 15:499–513, 1995.
7. B. Guenin. A characterization of weakly bipartite graphs. In R. E. Bixby, E. A. Boyd, and R. Z. Ríos-Mercado, editors, *Integer Programming and Combinatorial Optimization*, volume 1412 of *Lecture Notes in Computer Science*, pages 9–22. 6th International IPCO Conference, Houston, Texas, Springer, June 1998. Submitted to J. of Comb. Theory Ser. B.
8. T. C. Hu. Multicommodity network flows. *Operations Research*, 11:344–360, 1963.

9. Mei-Ko Kwan. Graphic programming using odd or even points (in chinese). *Acta Mathematica Sinica*, 10:263–266, 1960.

10. A. Lehman. A solution of the Shannon switching game. *J. SIAM*, 12(4):687–725, 1964.

11. A. Lehman. On the width-length inequality. *Mathematical Programming*, 17:403–417, 1979.

12. A. Lehman. On the width-length inequality and degenerate projective planes. In W. Cook and P.D. Seymour, editors, *Polyhedral Combinatorics*, volume 1 of *DIMACS Series in Discrete Math and Theoretical Computer Science*, pages 101–105, 1990.

13. A. Schrijver. *Theory of Linear and Integer Programming*. Wiley-Interscience, 1986. ISBN 0-471-90854-1.

14. P.D. Seymour. The matroids with the max-flow min-cut property. *J. Comb. Theory Ser. B*, 23:189–222, 1977.

15. P.D. Seymour. Matroids and multicommodity flows. *European J. of Combinatorics*, pages 257–290, 1981.

16. W. T. Tutte. A homotopy theorem for matroids, i,ii. *Amer. Math. Soc.*, 88:144–174, 1958.

A Unified Framework for Obtaining Improved Approximation Algorithms for Maximum Graph Bisection Problems

Eran Halperin and Uri Zwick

Department of Computer Science, Tel-Aviv University, Tel-Aviv 69978, Israel.
E-mail: {`heran,zwick`}@post.tau.ac.il.

Abstract. We obtain improved semidefinite programming based approximation algorithms for *all* the natural maximum bisection problems of graphs. Among the problems considered are: MAX $\frac{n}{2}$-BISECTION – partition the vertices of a graph into two sets of equal size such that the total weight of edges connecting vertices from different sides is maximized; MAX $\frac{n}{2}$-VERTEX-COVER – find a set containing half of the vertices such that the total weight of edges touching this set is maximized; MAX $\frac{n}{2}$-DENSE-SUBGRAPH – find a set containing half of the vertices such that the total weight of edges connecting two vertices from this set is maximized; and MAX $\frac{n}{2}$-UnCUT – partition the vertices into two sets of equal size such that the total weight of edges that do *not* cross the cut is maximized. We also consider the directed versions of these problems, MAX $\frac{n}{2}$-DIRECTED-BISECTION and MAX $\frac{n}{2}$-DIRECTED-UnCUT. These results can be used to obtain improved approximation algorithms for the unbalanced versions of the partition problems mentioned above, where we want to partition the graph into two sets of size k and $n - k$, where k is not necessarily $\frac{n}{2}$. Our results improve, extend and unify results of Frieze and Jerrum, Feige and Langberg, Ye, and others.

1 Introduction

Goemans and Williamson [GW95] use semidefinite programming to obtain a 0.878-approximation algorithm for the MAX CUT problem: Given an undirected graph $G = (V, E)$ with nonnegative edge weights, partition the vertex set V into two sets S and $T = V - S$ such that the total weight of edges in the cut (S, T) is maximized. The MAX $\frac{n}{2}$-BISECTION problem is a variant of the MAX CUT problem in which the sets S and T are required to be of size $\frac{n}{2}$, where $n = |V|$ is assumed to be even. In other words, we are required to partition the vertices of the graph into two blocks of equal cardinality, such that the total weight of edges connecting vertices from different blocks is maximized. Frieze and Jerrum [FJ97] use extensions of the ideas set forth by Goemans and Williamson [GW95] to obtain a 0.651-approximation algorithm for the MAX $\frac{n}{2}$-BISECTION problem. Ye [Ye99] recently improved this result and obtained a 0.699-approximation algorithm for the problem. As one of our results here, we

K. Aardal, B. Gerards (Eds.): IPCO 2001, LNCS 2081, pp. 210–225, 2001.
© Springer-Verlag Berlin Heidelberg 2001

obtain another small improvement to the performance ratio that can be obtained for MAX $\frac{n}{2}$-BISECTION.

An improved approximation algorithm for the MAX $\frac{n}{2}$-BISECTION problem for *regular* graphs was recently obtained by Feige *et al.* [FKL00b].

Several graph bisection problems that are similar to MAX $\frac{n}{2}$-BISECTION problem were also considered in the literature. In Sect. 2, we consider all possible bisection problems for both undirected and directed graphs. For undirected graphs, there are three other natural bisection problems, which are the MAX $\frac{n}{2}$-VERTEX-COVER, the MAX $\frac{n}{2}$-DENSE-SUBGRAPH, and the MAX $\frac{n}{2}$-UnCUT problems. (These problems were defined in the abstract and will be defined again in Sect. 2). Approximation algorithms for these problems were obtained by Feige and Langberg [FL99], Ye and Zhang [YZ99] and Han *et al.* [HYZ00], [HYZZ00]. We obtain improved performance ratios for all three problems. We also consider maximum bisection problems of directed graphs, namely, the MAX $\frac{n}{2}$-DIRECTED BISECTION, which was considered by Ageev, Hassin and Sviridenko [AHS00], and the MAX $\frac{n}{2}$-DIRECTED-UnCUT, which to the best of our knowledge was not considered before. We obtain improved approximation algorithms also for these problems. A table comparing our results to the previously known results is given in Sect. 2.

Our results and techniques may be used to obtain improved approximation algorithms for the unbalanced version of all these problems. See Feige and Langberg [FL99] and Han, Ye and Zhang [HYZ00] for details.

To obtain our results we use a collection of old and new techniques. We show, for the first time, that adding the 'triangle inequalities' to the semidefinite relaxations of maximum bisection problems provably helps. To demonstrate this, we use a *global* argument, as opposed to local arguments used in previous results. We also introduce the somewhat counterintuitive idea of 'flipping' the bisection obtained by rounding the solution of the semidefinite programming relaxation.

2   Graph Bisection Problems

Let $G = (V, E)$ be a directed graph. Let $n = |V|$. Whenever needed, we assume that n is even. Let $S \subset V$ be such that $|S| = \frac{n}{2}$ and let $T = V - S$. The partition of the vertices of G into S and T partitions the edges of G into the four classes:

$$A = \{ (i, j) \in E \mid i, j \in S \},$$
$$B = \{ (i, j) \in E \mid i \in S, j \in T \},$$
$$C = \{ (i, j) \in E \mid i \in T, j \in T \},$$
$$D = \{ (i, j) \in E \mid i, j \in T \}.$$

(See the right-hand size of Fig. 1). If we ignore the direction of the edges (or if the graph is initially undirected), then the edges fall into the classes A, $B \cup C$ and D, as shown on the left-hand size of Fig. 1.

We let $w_{ij} \geq 0$ denote the weight of the edge (i, j). (If $(i, j) \notin E$, then $w_{ij} = 0$.) For a subset of edges $F \subseteq E$, we let $w(F) = \sum_{(i,j) \in F} w_{ij}$. The MAX

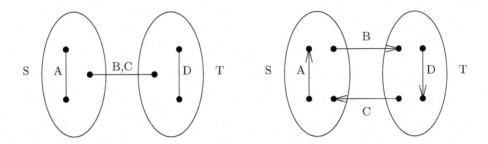

Fig. 1. Types of edges obtained by partitioning the vertices of a graph into two sets in the undirected case (on left-hand side) and the directed case (on right-hand side).

(a, b, c, d)-BISECTION problem calls for maximizing $a\,w(A) + b\,w(B) + c\,w(C) + d\,w(D)$. When a, b, c, d range over 0 and 1 we obtain all the natural bisection problems. When $b = c$, the direction of the edges makes no difference and the resulting problem is a bisection problem of undirected graphs. If, for example, $a = d = 0$ and $b = c = 1$, then the resulting problem is the well studied MAX $\frac{n}{2}$-BISECTION. (It is possible to consider more general bisection problems in which a, b, c and d assume arbitrary nonnegative values, but the problems obtained are less natural. Our techniques could be used to devise approximation algorithms also for such problems.)

All the interesting bisection problems that can be obtained in this way are listed in Table 1. MAX (a, b, c, d)-BISECTION problems, with $a, b, c, d \in \{0, 1\}$ that are not listed in Table 1 are either equivalent to one of the problems listed there, or can be trivially solved in polynomial time. (For example, the optimal solution of the MAX $\frac{n}{2}$-DIRECTED-VERTEX-COVER, corresponding to $a = b = 1$ and $c = d = 0$, is obtained by letting S be the set the $\frac{n}{2}$ vertices with the largest outdegree.) For each of the bisection problems listed in Table 1 we obtain an improved performance ratio. For some problems the improvement is quite substantial. For others, such as MAX $\frac{n}{2}$-BISECTION, the improvement is quite small. Even this improvement is important, however, as it is obtained by adding the 'triangle inequalities' to the semidefinite programming relaxation of the problem, and using a global argument to show their contribution to the relaxation. The question of whether these inequalities can be used to obtain an improved performance ratio for the related MAX CUT problem is still open (see [GW95],[FG95],[Zwi99],[FKL00a],[FS01]).

3   Techniques Used by Our Approximation Algorithms

Our algorithms are based first and foremost on the ideas of Frieze and Jerrum [FJ97]. To these we add some ideas used by Feige and Langberg [FL99] and by Ye [Ye99], and some ideas that were not used before. Each one of these ingredients is briefly discussed below.

Table 1. Maximum bisection problems of undirected and directed graphs and the performance ratios obtained for them here and in previous works.

ABCD	problem	our result	previous result
0 1 1 0	MAX $\frac{n}{2}$-BISECTION	0.7016	0.6993 [Ye99]
1 0 0 0	MAX $\frac{n}{2}$-DENSE-SUBGRAPH	0.6221	0.5866 [YZ99]
1 0 0 1	MAX $\frac{n}{2}$-UNCUT	0.6436	0.6024 [YZ99]
1 1 1 0	MAX $\frac{n}{2}$-VERTEX-COVER	0.8452	0.8113 [HYZZ00]

Undirected bisection problems

ABCD	problem	our result	previous result
0 1 0 0	MAX $\frac{n}{2}$-DIRECTED-BISECTION	0.6440	0.5000 [AHS00]
1 0 1 1	MAX $\frac{n}{2}$-DIRECTED-UNCUT	0.8118	0.7500

Directed bisection problems

3.1   Using the 'Triangle Inequalities'

All our algorithms start by solving a semidefinite programming relaxation of the problem at hand. To obtain our improved results, we strengthen the semidefinite programming relaxations by adding the so called 'triangle inequalities', i.e., the inequalities $v_i \cdot v_j + v_i \cdot v_k + v_j \cdot v_k \geq -1$, for $0 \leq i, j, k \leq n$. These inequalities were used already in some algorithms for maximum bisection problems (e.g., [FL99]), but were previously analyzed using local arguments. Here we show for the first time that a global argument that considers all triples (i, j, k) together can be used to improve the analysis of the algorithm.

3.2   Using Outward Rotations

By solving the semidefinite programming relaxation of the problem considered, we obtain an (almost) optimal solution $v_0, v_1, \ldots, v_n$. We then have to round these vectors to obtain a partition of the graph. The rounding procedure picks a random hyperplane, which partitions the vectors into two parts (as in [GW95]). Before rounding the vectors $v_0, v_1, \ldots, v_n$, we subject them to a rotation process. Rotations were used before by Feige and Langberg [FL99] and by Ye [Ye99]. Feige and Langberg [FL99] use Feige-Goemans [FG95] rotations. Ye [Ye99] uses a different type of rotations that is also used by Nesterov [Nes98] and by Zwick [Zwi99]. Following [Zwi99], we call these *outward rotations*. Outward rotations are also used by us here as they seem to be better suited for maximum bisection problems than Feige-Goemans rotations. Although such rotations were used before, they were not used yet in conjunction with the other ingredients.

3.3   Adding Linear Randomized Rounding

For asymmetric problems we obtain better performance ratios by combining the random hyperplane rounding method with a rounding procedure similar to the one used when rounding solutions of linear programming relaxations. The relaxations of asymmetric problems contain a vector v_0 whose role is to break the symmetry. In the linear rounding technique, vertex i is placed in S with probability $(1 - v_0 \cdot v_i)/2$, and in T with probability $(1 + v_0 \cdot v_i)/2$. The choices made for the different vertices are *independent*.

3.4   'Flipping' the Bisection

The rounding procedure returns a cut (S, T). Usually, this cut is not a bisection, i.e., $|S| \neq \frac{n}{2}$. A bisection (S', T') is then obtained by 'mending' the cut (S, T). Some of the maximum bisection problems are symmetric, i.e., the value of a cut (S', T') is equal to the value of a cut (T', S'). Some, like MAX $\frac{n}{2}$-VERTEX-COVER and $\frac{n}{2}$-DENSE-SUBGRAPH are not. For such asymmetric problems, our algorithms compare the values of (S', T') and of (T', S') and return the bisection with the larger value. Interestingly, this improves the performance ratio achieved. This is somewhat counterintuitive, at least at first, as all previous steps of the algorithm attempt to make (S', T'), and not (T', S'), a 'heavy' bisection.

4   A Generic Approximation Algorithm

We now describe a generic approximation algorithm that can be turned into an approximation algorithm for any problem from the family MAX (a, b, c, d)-BISECTION. Our algorithms for the different bisection problems are obtained by setting various parameters used by these generic algorithm.

$$
\text{Max} \sum_{i,j \geq 1} w_{ij} \frac{\left\{ \begin{array}{c} (a + b + c + d) - (a + b - c - d)(v_0 \cdot v_i) \\ -(a - b + c - d)(v_0 \cdot v_j) + (a - b - c + d)(v_i \cdot v_j) \end{array} \right\}}{4}
$$

$$
\text{s.t.} \sum_{i,j \geq 1} v_i \cdot v_j = 0
$$

$$
\begin{aligned}
v_i \cdot v_j + v_i \cdot v_k + v_j \cdot v_k &\geq -1 &,\quad 0 \leq i, j, k \leq n \\
v_i \cdot v_j - v_i \cdot v_k - v_j \cdot v_k &\geq -1 &,\quad 0 \leq i, j, k \leq n \\
- v_i \cdot v_j - v_i \cdot v_k + v_j \cdot v_k &\geq -1 &,\quad 0 \leq i, j, k \leq n \\
- v_i \cdot v_j + v_i \cdot v_k - v_j \cdot v_k &\geq -1 &,\quad 0 \leq i, j, k \leq n \\
v_i \cdot v_i &= 1 &,\quad 0 \leq i \leq n
\end{aligned}
$$

Fig. 2. A generic semidefinite programming relaxation of a weighted instance of the MAX (a, b, c, d)-BISECTION problem.

ALGORITHM LARGE-(a, b, c, d)-BISECTION:

Input: A weighted (directed) graph $G = (V, E)$.
Output: A large (a, b, c, d)-bisection.

1. **(Initialization)** Choose parameters $0 \leq \tau_1, \tau_2$, a rotation coefficient $0 \leq \rho \leq 1$, a probability $0 \leq \nu \leq 1$, and a tolerance $\epsilon > 0$. Let $\mu = \alpha_{abcd}(\rho, \nu) + \tau_1 \beta(\rho, \nu) + \tau_2 \gamma(\rho, \nu)$.

2. **(Solving the relaxation)** Solve the semidefinite programming relaxation of the instance given in Fig. 2 and obtain an (almost) optimal solution $v_0, v_1, \ldots, v_n$. Let z^* be the value of this solution.

3. **(Rotating the vectors)** Construct unit vectors $v'_0, v'_1, \ldots, v'_n$ such that $v'_i \cdot v'_j = \rho (v_i \cdot v_j)$, for every $0 \leq i < j \leq n$.

4. **(Find initial partition)** Repeat
   a) Choose a random vector r and let $S = \{ i \mid \text{sgn}(v'_i \cdot r) \neq \text{sgn}(v'_0 \cdot r) \}$ and $T = V - S$.
   b) With probability ν override the choice of S made above, and for each $i \in V$, put i in S, independently, with probability $(1 - v_0 \cdot v_i)/2$, and in T, otherwise.
   c) Let $X = value_{abcd}(S, T)$, $Y_1 = |T|$, $Y_2 = |S||T|$ and $Z = \frac{1}{z^*}X + \frac{2\tau_1}{n}Y_1 + \frac{4\tau_2}{n^2}Y_2$.
   Until $Z \geq (1 - \epsilon)\mu$.

5. **(Constructing a bisection)** If $|S| \geq \frac{n}{2}$, let S' be a random subset of S of size $\frac{n}{2}$ and let $T' = V - S'$. If $|S| \leq \frac{n}{2}$, let T' be a random subset of T of size $\frac{n}{2}$, and let $S' = V - T'$.

6. **(Flipping the bisection)** If $value_{abcd}(S', T') \geq value_{abcd}(T', S')$ then output (S', T'). Otherwise, output (T', S').

Fig. 3. A generic approximation algorithm for MAX (a, b, c)-BISECTION.

The generic algorithm is given in Fig. 3. It starts by setting some parameters. (The meaning of these parameters is explained later.) It then solves the semidefinite programming relaxation of the problem described in Fig. 2. The relaxation used is the standard relaxation of these problems, augmented by the 'triangle inequalities'. Recall that $w_{ij} \geq 0$ is the weight of the edge (i, j). Each vertex $i \in V$ has a unit vector associated with it. There is also a unit vector v_0 that plays an important role for asymmetric problems. To see that this is indeed a relaxation, note that if (S, T) is a bisection of the graph then by letting $v_i = -v_0$, if $i \in S$, and $v_i = v_0$, if $i \in T$, where v_0 is an arbitrary unit vector, we obtain a feasible solution of the relaxation whose value is $value_{abcd}(S, T) = a\, w(A) + b\, w(B) + c\, w(C) + d\, w(D)$. This follows from the fact that each term of the objective function can be written in the following way:

$$a \frac{1 - v_0 \cdot v_i - v_0 \cdot v_j + v_i \cdot v_j}{4} + b \frac{1 - v_0 \cdot v_i + v_0 \cdot v_j - v_i \cdot v_j}{4}$$
$$+ c \frac{1 + v_0 \cdot v_i - v_0 \cdot v_j - v_i \cdot v_j}{4} + d \frac{1 + v_0 \cdot v_i + v_0 \cdot v_j + v_i \cdot v_j}{4},$$

and the term multiplied by a, for example, assumes the value 1 only if $v_i = v_j = -v_0$, i.e., if $(i,j) \in A$. Note that if we set the vectors in this way then $\sum_{i=1}^{n} v_i = 0$, as $|S| = |T|$, and thus $\sum_{i,j\geq 1} v_i \cdot v_j = (\sum_{i=1}^{n} v_i) \cdot (\sum_{j=1}^{n} v_j) = 0$. Also note that the constraint $\sum_{i,j\geq 1} v_i \cdot v_j = 0$ implies that $\sum_{i=1}^{n} v_i = 0$.

The relaxation given in Fig. 2 is equivalent to a semidefinite program. An (almost) optimal solution $v_0, v_1, v_2, \ldots, v_n$ of it can therefore be found in polynomial time using either the Ellipsoid algorithm [GLS81], or using a more efficient interior-point method algorithm (see [WSV00]). We let z^* denote the value of this (almost) optimal solution.

The algorithm then subjects the vectors $v_0, v_1, v_2, \ldots, v_n$ to outward-rotation, and obtains new vectors $v_0', v_1', v_2', \ldots, v_n'$ such that $v_i' \cdot v_j' = \rho(v_i \cdot v_j)$, where $0 \leq \rho \leq 1$ is a rotation parameter. (See [Zwi99] for how this can be done.) A different value of ρ is used for each bisection problem, but this value is used for all the instances of that problem. We may assume, without loss of generality, that $v_0', v_1', v_2', \ldots, v_n' \in \mathrm{IR}^{n+1}$.

Following Frieze and Jerrum [FJ97], who follow Goemans and Williamson [GW95], the algorithm then chooses a random vector r according to the $(n+1)$-dimensional normal distribution. This vector r defines a hyperplane that partitions the vertices of the graph into the sets $S = \{i \in V \mid \mathrm{sgn}(v_i' \cdot r) \neq \mathrm{sgn}(v_0' \cdot r)\}$ and $T = V - S$. Then, with probability ν, this choice of S is ignored and a new set S is constructed as follows: each vertex $i \in V$ is placed in S, independently, with probability $(1 - v_0 \cdot v_i)/2$, and in T, otherwise. (This is what we referred to in Sect. 3.3 as linear rounding.)

The partition (S,T) determines the quantities $X = value_{abcd}(S,T)$, $Y_1 = |T|$, $Y_2 = |S||T|$ and $Z = \frac{1}{z^*}X + \frac{2\tau_1}{n}Y_1 + \frac{4\tau_2}{n^2}Y_2$. Here, X is the value of this partition (which is not necessarily a bisection yet), while Y_1 and Y_2 measure, in some way, how close this partition is to a bisection. Finally, Z is a weighted combination of X, Y_1 and Y_2, where τ_1 and τ_2 are problem specific parameters, and z^* is the value of the optimal solution of the semidefinite program.

The functions $\alpha_{abcd}(\rho,\nu)$, $\beta(\rho,\nu)$ and $\gamma(\rho,\nu)$, to be defined later, are such that $E[X] \geq \alpha_{abcd}(\rho,\nu)z^*$, $E[Y_1] \geq \beta(\rho,\nu)\frac{n}{2}$, and $E[Y_2] \geq \gamma(\rho,\nu)\frac{n^2}{4}$. (The quantities X, Y_1 and Y_2 are random variables as they are determined the partition (S,T), which is determined in turn by the random vector r.) Thus, $\mu = \alpha_{abcd}(\rho,\nu) + \tau_1\beta(\rho,\nu) + \tau_2\gamma(\rho,\nu)$, computed at the initialization phase of the algorithm, is a lower bound on $E[Z]$.

If $Z \geq (1 - \epsilon)\mu$, then the partition (S,T) obtained using the random vector r, or using the linear rounding, is accepted. Otherwise, a new partition is generated. Here, $\epsilon > 0$ is a tolerance parameter set at the initialization phase of the algorithm. As $E[Z] \geq \mu$, $Z \leq 1 + \tau_1 + \tau_2$, and $\epsilon > 0$, it is easy to see, using Markov's inequality, that the expected number of random vectors that have to be chosen until $Z \geq (1 - \epsilon)\mu$ is only $O(\frac{1}{\epsilon})$.

The partition (S,T) obtained in this way is usually not a bisection. It is mended is the following way. If $|S| \geq \frac{n}{2}$, we let S' be a random subset of S of size $\frac{n}{2}$ and then let $T' = V - S'$. Otherwise, if $|S| \leq \frac{n}{2}$, we let T' be a random subset of T of size $\frac{n}{2}$, and then let $S' = V - T'$. The partition (S', T') obtained in this way is now a bisection.

Finally, the algorithm compares $value_{abcd}(S', T')$ and $value_{abcd}(T', S')$, the values of the bisections (S', T') and (T', S') and outputs the better bisection. This step is helpful only for asymmetric problems.

5 The Functions $\alpha_{abcd}(\rho, \nu)$, $\beta(\rho, \nu)$ and $\gamma(\rho, \nu)$

The functions $\alpha_{abcd}(\rho, \nu)$, $\beta(\rho, \nu)$ and $\gamma(\rho, \nu)$ play a central role in our algorithms and in their analysis. The function $\alpha_{abcd}(\rho, \nu)$ is problem specific. The functions $\beta(\rho, \nu)$ and $\gamma(\rho, \nu)$ are the same for all the bisection problems. As mentioned in the previous section, these functions should satisfy $E[X] \geq \alpha_{abcd}(\rho, \nu)z^*$, $E[Y_1] \geq \beta(\rho, \nu)\frac{n}{2}$, and $E[Y_2] \geq \gamma(\rho, \nu)\frac{n^2}{4}$. The larger we could make these functions, while satisfying the required conditions, the larger the performance ratios obtained by our algorithms would be.

The bounds on $\alpha_{abcd}(\rho, \nu)$ and $\beta(\rho, \nu)$ are obtained using standard local ratio techniques (see [GW95], [FG95] and [Zwi99]). Thus, for example,

$$\alpha_{0110}(\rho, 0) = \min_{-1 \leq x < 1} \frac{2}{\pi} \frac{\arccos(\rho x)}{1 - x} .$$

We specify the function $\alpha_{0110}(\rho, \nu)$ only for $\nu = 0$, as $\nu = 0$ is the optimal choice for symmetric problems, i.e., problems with $abcd = dcba$. A full description of how the functions $\alpha_{abcd}(\rho, \nu)$ are computed is given in the full version.

Let β'' and γ'' be the expectations of $\frac{2Y_1}{n}$ and $\frac{4Y_2}{n^2}$, respectively, given that step 4(b) is executed, i.e., when vertex i is independently put in S, with probability $\frac{1 - v_0 \cdot v_i}{2}$. Note that $\beta'' = \beta(\rho, 1)$, $\gamma'' = \gamma(\rho, 1)$, for any $0 \leq \rho \leq 1$. Also let $\beta'(\rho) = \beta(\rho, 0)$ and $\gamma'(\rho) = \gamma(\rho, 0)$. It is easy to see that we can let:

$$\beta(\rho, \nu) = (1 - \nu)\beta'(\rho) + \nu\beta'' ,$$
$$\gamma(\rho, \nu) = (1 - \nu)\gamma'(\rho) + \nu\gamma'' .$$

We next show that we may assume that $\beta'' = \gamma'' = 1 - O(1/n)$. This follows as

$$\beta'' = \frac{2}{n} \sum_{i=1}^{n} \Pr[i \in T] = \frac{2}{n} \sum_{i=1}^{n} \frac{1 + v_0 \cdot v_i}{2} = 1 .$$

(Recall that $\sum_{i,j} v_i \cdot v_j = 0$ implies $\sum_{i=1}^{n} v_i = 0$, and therefore $\sum_{i=1}^{n} v_0 \cdot v_i = 0$.)
Next, we have

$$\gamma'' = \frac{4}{n^2} \sum_{i,j} \Pr[i \in S , j \in T] = \frac{4}{n^2} \sum_{i=1}^{n} \frac{1 - v_0 \cdot v_i}{2} \sum_{j \neq i} \frac{1 + v_0 \cdot v_j}{2} \geq 1 - \frac{2}{n} ,$$

as for any i, we have $\sum_{j \neq i} \frac{1 + v_0 \cdot v_j}{2} = \sum_{j=1}^{n} \frac{1 + v_0 \cdot v_j}{2} - \frac{1 + v_0 \cdot v_i}{2} \geq \frac{n}{2} - 1$. In the full version of the paper we show that

$$\beta'(\rho) = \min_{-1 \leq x \leq 0} \frac{2}{\pi} \frac{\arccos(\rho x) - x \arccos(\rho)}{1 - x} ,$$

is a valid choice for $\beta'(\rho)$.

It is not difficult to verify that $\gamma'(\rho) = \beta'(\rho)$ is a valid choice for $\gamma'(\rho)$. One of the contributions of this paper is a demonstration that

$$\gamma'(\rho) = \min_{-1 \leq x \leq -\frac{1}{3}} \frac{1}{\pi} \left(\arccos(\rho x) + \frac{3(x+1)}{4} \arccos(-\frac{\rho}{3}) + \frac{1-3x}{4} \arccos(\rho) \right)$$

is also a valid choice. This is a better choice as it is not difficult to check that $\beta'(\rho) < \gamma'(\rho)$, for every $0 < \rho \leq 1$. The proof that this choice is valid in given in Appendix A. We show there that obtaining good choices for $\gamma'(\rho)$ is closely related to the behavior of the MAX CUT algorithms of Goemans and Williamson [GW95] and of Zwick [Zwi99], strengthened by the 'triangle inequalities', on *complete* graphs. To show that the above choice of $\gamma'(\rho)$ is valid, we rely on Turán's theorem that states that any undirected graph on n vertices with more than $\frac{n^2}{4}$ edges contains a triangle (Turán [Tur41]). Thus, if $v_0, v_1, \ldots, v_n$ is a feasible solution of the relaxation given in Fig. 2, then for at most $\frac{n^2}{4}$ pairs (i, j), where $1 \leq i < j \leq n$, we can have $v_i \cdot v_j < -\frac{1}{3}$. We *conjecture* that the function

$$\bar{\gamma}'(\rho) = \frac{1}{2\pi} \left(3 \arccos(-\frac{\rho}{3}) + \arccos(\rho) \right) ,$$

which satisfies $\gamma'(\rho) < \bar{\gamma}'(\rho)$, for any $0 < \rho \leq 1$, is also a valid choice for $\gamma'(\rho)$. If true, then further improvements to all the bounds given in Table 1 would be obtained. Furthermore, $\bar{\gamma}'(\rho)$ would then be the optimal choice for $\gamma'(\rho)$.

6   The Performance Ratio of the Generic Algorithm

We continue with the analysis of the generic approximation algorithm for bisection problems given in Fig. 3. We have already reached a stage in which the algorithm finds a partition (S, T) of the graph for which $Z \geq (1 - \epsilon)\mu$. Recall that $X = value_{abcd}(S, T)$, $Y_1 = |T|$, $Y_2 = |S||T|$, that $Z = \frac{1}{z^*}X + \frac{2\tau_1}{n}Y_1 + \frac{4\tau_2}{n^2}Y_2$, and that $\mu = \alpha_{abcd}(\rho, \nu) + \tau_1\beta(\rho, \nu) + \tau_2\gamma(\rho, \nu)$. Let $\lambda = \frac{X}{z^*}$ and $\delta = \frac{|S|}{n}$, so that $X = \lambda z^*$, $|S| = \delta n$, $|T| = (1 - \delta)n$, and thus $Y_1 = (1 - \delta)n$ and $Y_2 = \delta(1 - \delta)n^2$. To simplify the analysis, we assume that $\epsilon = 0$. As ϵ may be arbitrarily small, we may get arbitrarily close to the ratio obtained by this analysis. Thus,

$$\lambda + 2(1 - \delta)\tau_1 + 4\delta(1 - \delta)\tau_2 \geq \alpha_{abcd}(\rho, \nu) + \tau_1\beta(\rho, \nu) + \tau_2\gamma(\rho, \nu) .$$

The partition (S, T) is usually not a bisection. To covert it into a bisection, we choose either a random subset of T of size $\frac{n}{2}$, if $\delta \leq \frac{1}{2}$, and call it T', or a random subset of S of size $\frac{n}{2}$, and call it S', if $\delta \geq \frac{1}{2}$. We let (S', T') be the bisection obtained in this way.

Suppose, at first, that $\delta \geq \frac{1}{2}$. Each element of S has a probability of $p = \frac{1}{2\delta}$ of being chosen to S'. However, as we are choosing a subset of size exactly $\frac{n}{2}$, these events are *not* independent. But, as n may be assumed to be large, these events are almost independent. To simplify the analysis, we assume that these events are independent. This can be justified as follows. We first construct an almost-bisection (S', T'), where S' is obtained by choosing each element of S independently with probability p. (We still assume that $\delta \leq \frac{1}{2}$.) With over-whelming probability $|S'| = \frac{n}{2} + O(\sqrt{n})$. We then convert (S', T') into an exact bisection (S'', T'') by choosing subsets of size exactly $\frac{n}{2}$. It is easy to see that the loss incurred by moving from (S', T') to (S'', T'') is negligible. Similarly, if $\delta \leq \frac{1}{2}$, we assume that each element of T is chosen to be in T' independently with probability $q = \frac{1}{2(1-\delta)}$.

We find it more convenient now to replace the names A, B, C, D and a, b, c, d by indexable names. We let $(F_1, F_2, F_3, F_4) = (A, B, C, D)$ and $(a_1, a_2, a_3, a_4) = (a, b, c, d)$. With our new notation, we have $value_{abcd}(S, T) = \sum_{i=1}^{4} a_i w(F_i)$. It is easy to check that if F_1, F_2, F_3, F_4 is the partition of the edges induced by (S, T), and F_1', F_2', F_3', F_4' is the partition induced by (S', T'), then

$$
\begin{pmatrix} E[w(F_1')] \\ E[w(F_2')] \\ E[w(F_3')] \\ E[w(F_4')] \end{pmatrix} = \begin{pmatrix} p^2 & 0 & 0 & 0 \\ p(1-p) & p & 0 & 0 \\ p(1-p) & 0 & p & 0 \\ (1-p)^2 & 1-p & 1-p & 1 \end{pmatrix} \begin{pmatrix} w(F_1) \\ w(F_2) \\ w(F_3) \\ w(F_4) \end{pmatrix} \quad , \qquad \text{if } \delta \geq \tfrac{1}{2} \, ,
$$

$$
\begin{pmatrix} E[w(F_1')] \\ E[w(F_2')] \\ E[w(F_3')] \\ E[w(F_4')] \end{pmatrix} = \begin{pmatrix} 1 & 1-q & 1-q & (1-q)^2 \\ 0 & q & 0 & q(1-q) \\ 0 & 0 & q & q(1-q) \\ 0 & 0 & 0 & q^2 \end{pmatrix} \begin{pmatrix} w(F_1) \\ w(F_2) \\ w(F_3) \\ w(F_4) \end{pmatrix} \quad , \qquad \text{if } \delta \leq \tfrac{1}{2} \, .
$$

Denote the matrices appearing in these expressions by $M_1(p)$ and $M_2(q)$. Let

$$
M(\delta) = \begin{cases} M_1(\frac{1}{2\delta}) & \text{if } \delta \geq \tfrac{1}{2} \, , \\ M_2(\frac{1}{2(1-\delta)}) & \text{if } \delta \leq \tfrac{1}{2} \, . \end{cases}
$$

Recall, again, that $X = \sum_{i=1}^{4} a_i w(F_i) = \lambda z^*$. Let λ_i, for $1 \leq i \leq 4$, be such that $w(F_i) = \lambda_i z^*$. Then, we have $\sum_{i=1}^{4} a_i \lambda_i = \lambda$. Also, as $\sum_{i=1}^{4} w(F_i) = w(E) \geq z^*$, we also have $\sum_{i=1}^{4} \lambda_i \geq 1$. Define

$$
f_{abcd}(\delta, \lambda_1, \lambda_2, \lambda_3, \lambda_4) = (a_1 \; a_2 \; a_3 \; a_4) \, M(\delta) \, (\lambda_1 \; \lambda_2 \; \lambda_3 \; \lambda_4)^T \, .
$$

We are finally able to state the performance ratio of our algorithm:

Theorem 1. *The performance ratio of the approximation algorithm* LARGE-(a, b, c, d)-PARTITION, *given in Fig. 3, when run with the parameters* τ_1, τ_2 *and* ρ *and with the functions* $\alpha_{abcd}(\rho, \nu)$, $\beta(\rho, \nu)$ *and* $\gamma(\rho, \nu)$, *is at least*

$$g_{abcd}(\tau_1, \tau_2, \rho, \nu) =$$

$$
\begin{bmatrix} \min & \max\{ f_{abcd}(\delta, \lambda_1, \lambda_2, \lambda_3, \lambda_4) \, , \; f_{dcba}(\delta, \lambda_1, \lambda_2, \lambda_3, \lambda_4) \} \\ \text{s.t.} & \lambda + 2(1-\delta)\tau_1 + 4\delta(1-\delta)\tau_2 \geq \alpha_{abcd}(\rho, \nu) + \tau_1 \beta(\rho, \nu) + \tau_2 \gamma(\rho, \nu) \\ & a_1\lambda_1 + a_2\lambda_2 + a_3\lambda_3 + a_4\lambda_4 = \lambda \\ & \lambda_1 + \lambda_2 + \lambda_3 + \lambda_4 \geq 1 \\ & 0 \leq \delta \leq 1 \, , \; 0 \leq \lambda, \lambda_1, \lambda_2, \lambda_3, \lambda_4 \end{bmatrix} .
$$

Proof. It is not difficult to verify that the expected value of the bisection (S', T') constructed by the algorithm is $f_{abcd}(\delta, \lambda_1, \lambda_2, \lambda_3, \lambda_4)\, z^*$, and that the expected value of the flipped bisection (T', S') is $f_{dcba}(\delta, \lambda_1, \lambda_2, \lambda_3, \lambda_4)\, z^*$. (Note that flipping the bisection is equivalent to changing $abcd$ to $dcba$.) As z^* is the optimal value of the relaxation, it is an upper bound on the value of the optimal bisection. Thus, $\max\{\, f_{abcd}(\delta, \lambda_1, \lambda_2, \lambda_3, \lambda_4)\,,\; f_{dcba}(\delta, \lambda_1, \lambda_2, \lambda_3, \lambda_4)\,\}$ is the performance ratio achieved when $|S| = \delta n$ and $w(F_i) = \lambda_i z^*$, for $1 \le i \le 4$. Finally, it was argued above that $\delta, \lambda, \lambda_1, \lambda_2, \lambda_3, \lambda_4$ satisfy all the stated inequalities. The claim of the theorem follows. $\qquad\Box$

Although the functions $f_{abcd}(\delta, \lambda_1, \lambda_2, \lambda_3, \lambda_4)$ and $g_{abcd}(\tau_1, \tau_2, \rho, \nu)$ are fairly complicated functions, it is possible to evaluate them numerically for any value of their arguments. To see this, let us write the function $g_{abcd}(\tau_1, \tau_2, \rho, \nu)$ in the following form:

$$
g_{abcd}(\tau_1.\tau_2, \rho, \nu) = \min_{0 \le \delta \le 1}
\begin{bmatrix}
\min & z \\
\text{s.t.} & \lambda \ge h(\tau_1, \tau_2, \rho, \nu, \delta) \\
& a_1\lambda_1 + a_2\lambda_2 + a_3\lambda_3 + a_4\lambda_4 = \lambda \\
& \lambda_1 + \lambda_2 + \lambda_3 + \lambda_4 \ge 1 \\
& z \ge f_{abcd}(\delta, \lambda_1, \lambda_2, \lambda_3, \lambda_4) \\
& z \ge f_{dcba}(\delta, \lambda_1, \lambda_2, \lambda_3, \lambda_4) \\
& 0 \le \lambda, \lambda_1, \lambda_2, \lambda_3, \lambda_4
\end{bmatrix},
$$

where

$$
h(\tau_1, \tau_2, \rho, \nu, \delta) = \alpha_{abcd}(\rho, \nu) + \tau_1\beta(\rho, \nu) + \tau_2\gamma(\rho, \nu) - (2(1 - \delta)\tau_1 + 4\delta(1 - \delta)\tau_2).
$$

Thus, for every fixed $\tau_1, \tau_2, \rho, \delta$, we get a linear program whose optimal value is attained at one of its vertices. Thus, for any τ_1, τ_2, and ρ, the performance of the algorithm is easy to compute. 'Flipping' the bisection will improve the performance of the algorithm if two of the inequalities defining the optimal value of the linear program are $z = f_{1110}(\delta, \lambda_1, \lambda_2, \lambda_3)$, and $z = f_{0111}(\delta, \lambda_1, \lambda_2, \lambda_3)$.

To obtain our best approximation algorithms for the various problems we choose the parameters τ_1, τ_2, ρ and ν that maximize the function $g_{abcd}(\tau_1, \tau_2, \rho, \nu)$. The computations here were done numerically. The performance guarantees obtained in this way may be made rigorous with the help of a rigorous global optimization system. See Zwick [Zwi00] for further details.

7   Parameter Setting

The setting of the parameters τ_1, τ_2, ρ and ν using which we obtain the results stated in Table 1 are given in Table 2. It is interesting to note that $\rho < 1$, i.e., outward rotations are used, for all the bisection problems. Also linear rounding is used, i.e., $\nu > 0$, by the algorithms for all the *asymmetric* problems. Finally, $\tau_1 > 0$ only for the asymmetric undirected problems.

Table 2. The settings of the parameters τ_1, τ_2, ρ and ν using which the results stated in Table 1 are obtained.

ABCD	problem	ratio	τ_1	τ_2	ρ	ν
0 1 1 0	MAX $\frac{n}{2}$-BISECTION	0.7016	0	1.83	0.89	0
1 0 0 0	MAX $\frac{n}{2}$-DENSE-SUBGRAPH	0.6221	0.69	2.15	0.87	0.23
1 0 0 1	MAX $\frac{n}{2}$-UnCUT	0.6436	0	3.86	0.81	0
1 1 1 0	MAX $\frac{n}{2}$-VERTEX-COVER	0.8452	0.5	0.37	0.88	0.06
0 1 0 0	MAX $\frac{n}{2}$-DIRECTED-BISECTION	0.6440	0	1.61	0.93	0.16
1 0 1 1	MAX $\frac{n}{2}$-DIRECTED-UnCUT	0.8118	0	2.10	0.74	0.04

8   Concluding Remarks

We presented a unified approach for developing approximation algorithms for all maximum bisection problems of directed and undirected graphs. Our approach uses almost all previously used techniques together with some techniques that are used here for the first time. Using this unified approach we obtained improved approximation algorithms for *all* the maximum bisection problems.

It is possible to obtain some small improvements to the performance ratios reported here. These improvements are obtained by adding some additional features to the generic approximation algorithm presented in Fig. 3. Some improvements may be obtained, for some of the problems, by combining outward rotations with Feige-Goemans rotations [FG95]. The improvements, however, are quite small while the analysis becomes substantially more complicated.

Obtaining substantially improved ratios, especially for MAX $\frac{n}{2}$-BISECTION, which is perhaps the most natural bisection problem, remains a challenging open problem. It would also be interesting to prove, or disprove, our conjecture, made in Sect. 5, regarding the function $\bar{\gamma}(\rho)$.

References

AHS00.   A. Ageev, R. Hassin, and M. Sviridenko. An 0.5-approximation algorithm for MAX DICUT with given sizes of parts. Manuscript, 2000.

FG95.   U. Feige and M.X. Goemans. Approximating the value of two prover proof systems, with applications to MAX-2SAT and MAX-DICUT. In *Proceedings of the 3nd Israel Symposium on Theory and Computing Systems, Tel Aviv, Israel*, pages 182–189, 1995.

FJ97.   A.M. Frieze and M. Jerrum. Improved approximation algorithms for MAX k-CUT and MAX BISECTION. *Algorithmica*, 18:67–81, 1997.

FKL00a. U. Feige, M. Karpinski, and M. Langberg. Improved approximation of Max-Cut on graphs of bounded degree. Technical report, E-CCC Report number TR00-043, 2000.

FKL00b. U. Feige, M. Karpinski, and M. Langberg. A note on approximating MAX-BISECTION on regular graphs. Technical report, E-CCC Report number TR00-043, 2000.

FL99.   U. Feige and M. Langberg. Approximation algorithms for maximization problems arising in graph partitioning. Manuscript, 1999.

FS01.   U. Feige and G. Schechtman. On the integrality ratio of semidefinite relaxations of MAX CUT. In *Proceedings of the 33th Annual ACM Symposium on Theory of Computing, Crete, Greece*, 2001. To appear.

GLS81.   M. Grötschel, L. Lovász, and A. Schrijver. The ellipsoid method and its consequences in combinatorial optimization. *Combinatorica*, 1:169–197, 1981.

GW95.   M.X. Goemans and D.P. Williamson. Improved approximation algorithms for maximum cut and satisfiability problems using semidefinite programming. *Journal of the ACM*, 42:1115–1145, 1995.

HYZ00.   Q. Han, Y. Ye, and J. Zhang. An improved rounding method and semidefinite programming relaxation for graph partition. Manuscript, 2000.

HYZZ00.   Q. Han, Y. Ye, H. Zhang, and J. Zhang. On approximation of Max-Vertex-Cover. Manuscript, 2000.

Nes98.   Y. E. Nesterov. Semidefinite relaxation and nonconvex quadratic optimization. *Optimization Methods and Software*, 9:141–160, 1998.

Tur41.   P. Turán. On an extremal problem in graph theory. *Mat. Fiz. Lapok*, 48:436–452, 1941.

WSV00.   H. Wolkowicz, R. Saigal, and L. Vandenberghe, editors. *Handbook of semidefinite programming*. Kluwer, 2000.

Ye99.   Y. Ye. A 0.699-approximation algorithm for Max-Bisection. Manuscript, 1999.

YZ99.   Y. Ye and J. Zhang. Approximation of dense-$\frac{n}{2}$-subgraph and the complement of min-bisection. Manuscript, 1999.

Zwi99.   U. Zwick. Outward rotations: a tool for rounding solutions of semidefinite programming relaxations, with applications to max cut and other problems. In *Proceedings of the 31th Annual ACM Symposium on Theory of Computing, Atlanta, Georgia*, pages 679–687, 1999.

Zwi00.   U. Zwick. Analyzing the MAX 2-SAT and MAX DI-CUT approximation algorithms of Feige and Goemans. Manuscript, 2000.

A   The Function $\gamma'(\rho)$

We show that the choice of $\gamma'(\rho)$ given in Sect. 5 is valid, i.e., that $E[\|S\|\|T\|] \geq \gamma'(\rho)\frac{n^2}{4}$, for any sequence of unit vectors $v_0, v_1, \ldots, v_n$ that satisfies the constraints of the generic semidefinite programming relaxation given in Fig. 2. Note that these constraints do not depend on the specific problem considered. By the linearity of the expectation, we have $E[\|S\|\|T\|] = \frac{1}{\pi}\sum_{i<j}\arccos(\rho(v_i \cdot v_j))$. Let

$$\gamma^*(\rho) = \begin{bmatrix} \min & \frac{4}{\pi n^2}\sum_{i<j}\arccos(\rho x_{ij}) \\ \text{s.t.} & X \succeq 0 \\ & x_{ii} = 1 \qquad\qquad , 1 \leq i \leq n \\ & \sum_{i<j}x_{ij} = -\frac{n}{2} \\ & x_{ij} + x_{ik} + x_{jk} \geq -1 , 1 \leq i,j,k \leq n \\ & -1 \leq x_{ij} \leq 1 \qquad , 1 \leq i,j \leq n \end{bmatrix},$$

where $X \succeq 0$ indicates that the matrix $X = (x_{ij})$ is a positive semidefinite matrix. It is clear from the definition of $\gamma^*(\rho)$ that $E[\|S\|\|T\|] \geq \gamma^*(\rho)\frac{n^2}{4}$, as for any feasible solution $v_0, v_1, \ldots, v_n$ of the semidefinite program, the matrix $X = (x_{ij})$, where $x_{ij} = v_i \cdot v_j$, is a feasible solution of the optimization problem appearing in the definition of $\gamma^*(\rho)$. It is enough to show, therefore, that $\gamma^*(\rho) \geq \gamma'(\rho)$. If $X = (x_{ij})$ is a feasible point of the problem defining $\gamma^*(\rho)$, we let

$$\mathcal{A} = \{(i,j) \mid x_{ij} < -\frac{1}{3}, \ 1 \leq i < j \leq n\},$$
$$\mathcal{B} = \{(i,j) \mid -\frac{1}{3} \leq x_{ij} < 0, \ 1 \leq i < j \leq n\},$$
$$\mathcal{C} = \{(i,j) \mid 0 \leq x_{ij}, \ 1 \leq i < j \leq n\}.$$

Consider the subgraph $G' = (V, \mathcal{A})$. As for every $(i,j) \in \mathcal{A}$ we have $x_{ij} < -\frac{1}{3}$, and as for every $1 \leq i, j, k \leq n$ we have $x_{ij} + x_{ik} + x_{jk} \geq -1$, we get that the graph $G' = (V, \mathcal{A})$ is *triangle-free*. A theorem of Turán [Tur41] states that a triangle-free n-vertex graph can have at most $\frac{n^2}{4}$ edges. It follows, therefore, that $|\mathcal{A}| \leq \frac{n^2}{4}$. We now define the following function:

$$\gamma''(\rho) = \begin{bmatrix} \min & \frac{4}{\pi n^2} \sum_{i<j} \arccos(\rho x_{ij}) \\ \text{s.t.} & \sum_{i<j} x_{ij} = -\frac{n}{2} \\ & -1 \leq x_{ij} \leq 1, \ 1 \leq i, j \leq n \\ & |\{(i,j) \mid x_{ij} \leq -\frac{1}{3}\}| \leq \frac{n^2}{4} \end{bmatrix}.$$

As all the feasible points of the minimization problem defining $\gamma^*(\rho)$ are also feasible points of the minimization problem defining $\gamma''(\rho)$, we get that $\gamma''(\rho) \leq \gamma^*(\rho)$. We show next that $\gamma''(\rho) = \gamma'(\rho) + O(\frac{1}{n})$, for every $0 \leq \rho \leq 1$.

The proof relies on the convexity/concavity properties of the function $f(x) = \arccos(\rho x)$ (see Fig. 4). It is easy to see that $f(x)$ is concave in $[0,1]$ and convex in $[-1, 0]$. As an immediate corollary, we get:

(i) If $0 < x - \epsilon \leq x \leq y \leq y + \epsilon \leq 1$, then $f(x-\epsilon) + f(y+\epsilon) \leq f(x) + f(y)$.
(ii) If $-1 \leq x \leq x + \epsilon \leq y - \epsilon \leq y \leq 0$, then $f(x + \epsilon) + f(y - \epsilon) \leq f(x) + f(y)$.

Thus, if $X = (x_{ij})$ is an optimal solution of the minimization problem that appears in the definition of $\gamma''(\rho)$, and if $(i,j), (k,l) \in \mathcal{A}$, or $(i,j), (k,l) \in \mathcal{B}$, then $x_{ij} = x_{kl}$, otherwise, we can decrease the objective function by moving x_{ij} and x_{kl} closer together. Similarly, if $(i,j), (k,l) \in \mathcal{C}$, and $x_{ij} \leq x_{kl}$, then either $x_{ij} = 0$, or $x_{kl} = 1$, otherwise, we can decrease the objective function by moving x_{ij} and x_{kl} further apart. There can therefore be only a single point $(i,j) \in \mathcal{C}$ for which $x_{ij} \neq 0$ and $x_{ij} \neq 1$. As each point makes only an $O(\frac{1}{n^2})$ difference, we assume, for simplicity, that there is no such point. Next note that if $(i,j) \in \mathcal{B}$, $(k,l) \in \mathcal{C}$, and $x_{kl} = 0$, then we can move x_{ij} and x_{kl} closer together. It follows, therefore, that there exist $x < -\frac{1}{3}$ and $-\frac{1}{3} \leq y \leq 0$ (see Fig. 4) such that for

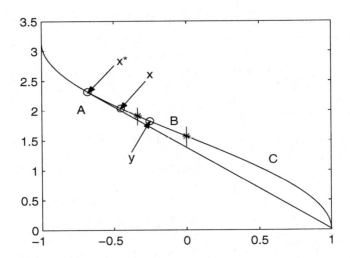

Fig. 4. The function $\arccos(x)$, the sets $\mathcal{A}, \mathcal{B}, \mathcal{C}$, and the points x^*, x and y.

every $(i, j) \in \mathcal{A}$ we have $x_{ij} = x$, for every $(i, j) \in \mathcal{B}$ we have $x_{ij} = y$, and for every $(i, j) \in \mathcal{C}$ we have $x_{ij} = 1$.

We next show that $\mathcal{A}$ is not empty. If $\mathcal{A}$ is empty and n is large enough, then $\mathcal{B}$ must contain at least four points. It is easy to check that for every $-\frac{1}{3} \le y \le 0$ we have $4f(y) > 3f(\frac{4y-1}{3}) + f(1)$, thus we can decrease the objective function by replacing the four points y, y, y, y by the points $\frac{4y-1}{3}, \frac{4y-1}{3}, \frac{4y-1}{3}, 1$. Using a very similar argument we in fact get that $|\mathcal{A}| = \frac{n^2}{4}$.

As $\mathcal{A}$ is not empty, we must have $y = -\frac{1}{3}$, as otherwise we can decrease the objective function by moving a point from $\mathcal{B}$ and a point from $\mathcal{A}$ closer together.

Let $a = |\mathcal{A}| = \frac{n^2}{4}$, $b = |\mathcal{B}|$ and $c = |\mathcal{C}|$. Then, we must have

$$a \cdot x + b \cdot (-\tfrac{1}{3}) + c = -\tfrac{n}{2} \quad , \quad a + b + c = \binom{n}{2} .$$

Solving these equations, we get $a = \frac{n^2}{4}$, $b = \frac{3n^2(1+x)}{16}$, $c = \frac{n^2(1-3x)}{16} - \frac{n}{2}$. Thus,

$$\gamma''(\rho) = \min_{-1 \le x \le -\frac{1}{3}} \frac{4}{\pi n^2} \left(\frac{n^2}{4} f(x) + \frac{3n^2}{16}(1+x)f(-\tfrac{1}{3}) + (\frac{n^2}{16}(1-3x) - \frac{n}{2})f(1) \right)$$

$$= \min_{-1 \le x \le -\frac{1}{3}} \frac{1}{\pi} \left(\arccos(\rho x) + \frac{3(x+1)}{4} \arccos(-\tfrac{\rho}{3}) + (\frac{1-3x}{4} - \frac{2}{n}) \arccos(\rho) \right) ,$$

as required.

As mentioned in Sect. 5, we conjecture that $\gamma^*(\rho) = \bar{\gamma}'(\rho) + O(\frac{1}{n})$, where

$$\bar{\gamma}'(\rho) = \tfrac{1}{2\pi} \left(3 \arccos(-\tfrac{\rho}{3}) + \arccos(\rho) \right) ,$$

and that the optimal solution for the optimization problem defining $\gamma^*(\rho)$ is attained, when $4|n$, at

$$x_{ij} = \begin{cases} -\frac{1}{3} & \text{if } i \not\equiv j \,(\mathrm{mod}\,4), \\ 1 & \text{otherwise.} \end{cases}$$

This correspond to the following spatial configuration: The unit vectors v_1, v_2, v_3 and v_4 point to the corners of regular tetrahedron, and thus $v_i \cdot v_j = -\frac{1}{3}$, for $i \neq j$, and each one of these vectors is duplicated $\frac{n}{4}$ times. We have numerical evidence that supports this conjecture, but not a proof.

Finally, we explain the comment made in Sect. 5, that obtaining a good choice for $\gamma'(\rho)$ is closely related to the study of the MAX CUT algorithms of Goemans and Williamson [GW95] and of Zwick [Zwi99], strengthened by the 'triangle inequalities', on *complete* graphs. In the semidefinite programming relaxation given in Fig. 2, we have the constraint $\sum_{i,j \geq 0} v_i \cdot v_j = 0$. As for every sequence of vectors $v_1, v_2, \ldots, v_n$ we have $\sum_{i,j \geq 0} v_i \cdot v_j = (\sum_{i=1}^{n} v_i) \cdot (\sum_{i=1}^{n} v_i) \geq 0$, we get that any solution of our semidefinite programming relaxation is an *optimal* solution of the semidefinite programming relaxation of the MAX CUT problem of the complete graph on n vertices, whose objective function in $\sum_{i<j} \frac{1 - v_i \cdot v_j}{2}$.

Synthesis of 2-Commodity Flow Networks

Refael Hassin and Asaf Levin

Department of Statistics and Operations Research, Tel-Aviv University, Tel-Aviv
69978, Israel. {hassin,levinas}@post.tau.ac.il

Abstract. We investigate network planning and design under volatile
conditions of link failures and traffic overload. Our model is a non-
simultaneous 2-commodity problem. We characterize the feasible solutions and using this characterization we reduce the size of the LP program. For the case that all non-zero requirements are equal we present a
closed fractional optimal solution, a closed integer (where the capacities
of the solution network are integer) optimal solution and we investigate
the integral case for which an integer 2-commodity flow must exist for
every pair of requirements.

1 Introduction

Consider an undirected graph $G = (V, E)$ with $V = \{1, ..., n\}$. Let r_e, $e \in E$,
be given nonnegative *requirements*. We assume that for every $e = (i, j) \in E$
there is currently an edge with capacity r_e connecting i and j. Thus, these
edges have sufficient capacity to enable *simultaneous flows* which satisfy all the
requirements. The k-PROTECTION NETWORK PROBLEM is to build an undirected
capacitated *protection network* of minimum total edge capacity, such that its
edge capacities $b_{i,j}$ $i, j \in V$ together with the original non-failed edges will be
capable of satisfying simultaneously the required flows, in case that at most k
edges of E fail (are disconnected).

An alternative interpretation is that we are given a network which is capable
of carrying the regular level of requirements. However, at different times there
are picks in the flow and the additional requirement is denoted by r_e. It is
assumed that the probability of more than k picks is negligible. The goal is to
build extra capacity of minimum total value that will be capable of carrying any
pick requirements of at most k pairs of terminals.

The k-PROTECTION NETWORK PROBLEM is a special case of the problem
of synthesis of a time-varying requirements multicommodity flow network in
which the requirement between every pair of vertices is time-dependent, and
the network designed to meet all requirements during all time periods. This
problem was solved via a specialized linear programming method by Gomory
and Hu ([2], see also [3] pages 197-213). We generalize some of our results to
the time varying requirements 2-commodity flow network in which we are given
a family Γ of *scenarios* each contains a pair of edges e_i and e_j. We have to
build an undirected network of minimum total edge capacity, such that its edge
capacities together with the original non-failed edges will be capable of satisfying

K. Aardal, B. Gerards (Eds.): IPCO 2001, LNCS 2081, pp. 226–235, 2001.
© Springer-Verlag Berlin Heidelberg 2001

simultaneously the required flows, in case that at the edges of any scenario fail. Each scenario is composed of a single edge failure or a failure of a pair of edges. It generalizes the 2-PROTECTION NETWORK PROBLEM as there may be pairs of edges that are not in Γ.

The 1-PROTECTION NETWORK PROBLEM has an efficient algorithm by Gomory and Hu ([1], see also [3] pages 142-148). Sridhar and Chandrasekaran solve efficiently (see [7]) the 1-PROTECTION NETWORK INTEGRAL PROBLEM in which capacities are restricted to integer values. It is well known that in such a case also integer flows can be maintained.

The subject of this paper is the 2-PROTECTION NETWORK PROBLEM. Tham [8] provided a heuristic for this problem. Ouveysi and Wirth [4] and [5] provided heuristics for the problem for the case that the protection network should be 1 or 2 hop network. Sastry [6] solved the problem when only one pair of edges need to be satisfied i.e., $|\Gamma| = 1$. We will prove that a protection network is feasible if and only if it satisfies the subset of requirements which correspond to a union of 2 disjoint spanning trees with maximum sum of requirements. This observation reduces the size of a linear programming formulation of the problem. For the important special case with *unit requirements* that is $r_e = 1 \; \forall e \in E$, we present a closed solution. For this case we present a closed solution also to the integer capacities case (when all the capacities in the solution must be integer). For the case where the flows in the protection network must be integer we present a $\frac{7}{6}$-approximation algorithm and closed solutions to some special cases.

For a function f over E and a subset $E' \subset E$ we denote $f(E') = \sum_{e \in E'} f_e$.

The 2-PROTECTION NETWORK PROBLEM is to compute capacities b which minimize $\sum_{i,j \in V} b_{ij}$ so that there exist nonnegative flow functions y, z which satisfy for every $i, j \in V$ and $(k, l), (p, q) \in E$,

$$y_{ijklpq} + y_{jiklpq} + z_{ijklpq} + z_{jiklpq} \leq b_{ij},$$

and for every $j \in V$ and $(k, l), (p, q) \in E$

$$\sum_i y_{ijklpq} - \sum_i y_{jiklpq} = \begin{cases} -r_{kl} & j = k \\ r_{kl} & j = l \\ 0 & \text{otherwise,} \end{cases}$$

$$\sum_i z_{ijklpq} - \sum_i z_{jiklpq} = \begin{cases} -r_{pq} & j = p \\ r_{pq} & j = q \\ 0 & \text{otherwise.} \end{cases}$$

In this formulation b_{ij} denotes the capacity of the edge between i and j in the protection network, y_{ijklpq} and z_{ijklpq} denote the $k - l$ and $p - q$ flows, respectively through (i, j) in the i to j direction, when the edges $(k, l), (p, q)$ fail. This is a linear program with $O(n^6)$ variables and $O(n^6)$ constrains. Therefore, an optimal solution can be found in polynomial time. However, even for medium sized instances the resulting problem may be too large for a practical solution.

2   Protection Equivalence

There are $O(n^4)$ pairs of edges in G, and the protection network must be able to satisfy the flow requirements of every such pair.

A MAXIMUM WEIGHT UNION OF TWO SPANNING TREES is a subgraph of G consisting of a union of two disjoint spanning trees with maximum total value of requirements over its edges.

We will prove that it is sufficient to find a solution to the problem for a maximum weight union of two spanning trees. Consequently, the size of the linear program reduces to $O(n^4)$ constraints and variables.

Two requirement vectors $\mathbf{R} = (r_e\ e \in E)$ and $\mathbf{R}' = (r'_e\ e \in E)$ are k-protection equivalent if a solution of the k-protection network of one is a solution to the problem with respect to the other and vice versa.

For a class R of protection equivalent vectors, $\mathbf{R} \in R$ is minimal if $\mathbf{R} = (r_e \in E)$ and there is no $\mathbf{R}' \in R$ $\mathbf{R}' = (r'_e\ e \in E)$ such that $r'_e \leq r_e\ \forall e \in E$ and for some e $r'_e < r_e$.

A Max Bi-flow Min Bi-cut theorem of Hu ([3] p. 180) states that there exist a 2 commodity flow of size f_1, f_2 between the two pairs $\{i_1, j_1\}$ and $\{i_2, j_2\}$, respectively, if and only if for every cut the size of the cut is at least: $f_1 + f_2$ if it separates both $\{i_1, j_1\}$ and $\{i_2, j_2\}$, f_1 if it only separates $\{i_1, j_1\}$ but not $\{i_2, j_2\}$, f_2 if it only separates $\{i_2, j_2\}$ but not $\{i_1, j_1\}$.

For the 1-PROTECTION NETWORK PROBLEM Gomory and Hu [1] proved that it is sufficient to duplicate a maximum spanning tree. This is not sufficient for the failure of two edges as can be seen by the following example:

Example 2.1 Let G be the triangle with a unit requirement for every edge. Any pair of edges is a maximum spanning tree, but a copy of the tree (which is a protection against the failure of 2 edges of the tree) won't protect against a failure of two edges including the non-tree edge.

Theorem 2.2 Let $T_1 \cup T_2$ be a maximum weight union of two spanning trees of G. Let $\mathbf{R}' = (r'_e\ e \in E)$ where $r'_e = r_e$ for $e \in T_1 \cup T_2$ and $r'_e = 0$ otherwise. Then $\mathbf{R}'$ is 2-protection equivalent to $\mathbf{R}$.

Proof: Consider a 2-protection network H with respect to $\mathbf{R}'$. We will prove that H is a 2-protection network with respect to $\mathbf{R}$ as well. The other direction is obvious as for every $e \in E$ $r_e \geq r'_e$ by definition.

We need to prove that H will survive the failure of any pair of edges $(i, j), (k, l)$. If both (i, j) and (k, l) belong to $T_1 \cup T_2$ then by definition H will survive their failure. Therefore, we can assume without loss of generality that $(k, l) \notin T_1 \cup T_2$ and furthermore that $(i, j) \notin T_1$. Let P_1 be the unique $k - l$ path in T_1. Clearly, for each $(u, v) \in P_1$ $r_{uv} \geq r_{kl}$ (otherwise, (k, l) could replace (u, v) in T_1 to obtain a higher capacity union of two trees – a contradiction). Similarly, let P_2 be the unique $i = j$ path in T_2 (if $(i, j) \in T_2$ then $P_2 = (i, j)$). As above, for each $(u, v) \in P_2$ $r_{uv} \geq r_{ij}$.

For every cut that contains both edges there are edges (k', l') from P_1 and (i', j') from P_2 in this cut. As H is a 2-protection network with respect to $\mathbf{R}'$ it survives the failure of $(i', j'), (k', l')$. Therefore, by Hu's Theorem it has a capacity which is at least $r_{i'j'} + r_{k'l'}$ in this cut. If only one edge of the pair, say (i, j), is separated by the cut, then define (i', j') as before. As H is a 2-protection network with respect to $\mathbf{R}'$, it survives the failure of (i', j'). Therefore, by Hu's theorem, the capacity of H in this cut is at least r_{ij}. This proves, again by Hu's theorem, that H will survive the failure of (i, j) and (k, l).   ∎

Theorem 2.2 states that it is possible to consider only the function $\mathbf{R}'$ induced by a maximum requirement union of 2 spanning trees. However, this function is not minimal.

Theorem 2.3 *A maximum requirement union of 2 spanning trees is not minimal protection equivalent, that is, it contains an edge which can be removed without changing the optimal 2-protection network.*

We generalize Theorem 2.2 to the case that the set of scenarios doesn't include all the pairs of edges.

Two scenario sets Γ and Γ' are *equivalent* if a solution of the protection network of one is a solution to the problem with respect to the other and vice versa.

For $e \in E$ let $E_e = \{e' \in E | \{e, e'\} \in \Gamma\}$ and let F_e be a maximum weight forest in the graph $G' = (V, E_e)$.

Theorem 2.4 *Let $\Gamma' = \{\{e, e'\} | e \in E, \ e' \in F_e\}$. Then Γ' and Γ are equivalent.*

Proof: Consider a protection network H with respect to Γ'. We will prove that H is a protection network with respect to Γ as well. The other direction is obvious as $\Gamma' \subseteq \Gamma$ by definition.

We need to prove that H will survive the failure of edges $(i, j), (k, l)$ of a scenario in Γ. If $\{(i, j), (k, l)\} \in \Gamma'$ then by definition H will survive their failure. Therefore, w.l.o.g. we may assume that (k, l) is not in $F_{(i,j)}$. Denote by P the path in $F_{(i,j)}$ between k and l.

For every cut C that contains both (i, j) and (k, l) there is an edge (k', l') from P in this cut. Since $F_{(i,j)}$ is a maximum weight forest, $r_{k'l'} \geq r_{kl}$. As H is a protection network with respect to Γ' it survives the failure of $\{(i, j), (k', l')\}$. Therefore, by Hu's Theorem C has a capacity of at least $r_{ij} + r_{k'l'} \geq r_{ij} + r_{kl}$. If only one edge of the pair, say (i, j), is separated by C, then H survives the failure of (i, j) and therefore, has enough capacity in this cut. This proves, again by Hu's theorem, that H will survive the failure of (i, j) and (k, l).   ∎

Theorem 2.4 proves that in checking whether a protection network survives the failure of a collection of scenarios, one should check only $O(n^3)$ pairs of edges. In particular, Theorem 2.4 provides a way to reduce the number of constraints and variable in the LP program.

3 Lower Bounds

Define
$$R_i^1 = max\{r_{ij}|j \in V\},$$
$$R_i^2 = max\{r_{ij} + r_{ik}|j, k \in V, \ j \neq k\},$$
$$R_i^1(\Gamma) = max\{r_{ij}|\{(i,j)\} \in \Gamma\},$$
$$R_i^2(\Gamma) = max\{r_{ij} + r_{ik}|\{(i,j),(i,k)\} \in \Gamma\}.$$
Gomory and Hu [1] gave a lower bound for the 1-protection network problem:

Lemma 3.1 *Every solution to the 1-protection network problem must have total capacity of at least $\frac{1}{2}\sum_{i=1}^{n} R_i^1$.*

The following is an extension of Lemma 3.1 to the 2-protection network problem.

Lemma 3.2 *Every solution to the 2-protection network problem must have total capacity of at least $\frac{1}{2}\sum_{i=1}^{n} R_i^2$.*

Proof: For every vertex i the protection network must have capacities for edges incident to i which are at least the sum of the two maximum requirement of i. This is clearly necessary for the case in which the two edges fail. As every edge is incident in two vertices it is count exactly twice. Therefore, summing over all vertices result the correctness of the lemma.   ■

For a scenario set Γ the following lower bound is the extension of Lemma 3.2.

Lemma 3.3 *Every solution to the protection network problem with scenario set Γ must have total capacity of at least $\frac{1}{2}\sum_{v \in V} Max\{R_v^1(\Gamma), R_v^2(\Gamma)\}$.*

Proof: For every vertex v it must have capacity of adjacent edges (in the solution network) of at least $Max\{R_v^1(\Gamma), R_v^2(\Gamma)\}$. This is so as it must survive the failure of every single edge adjacent to v and every pair adjacent to v that is in Γ.   ■

As opposed to the 1-protection network for which the lower bound of Lemma 3.1 is achievable in all cases, it is not the case in the lower bound of Lemma 3.2 for the 2-protection network problem as seen by Remark 4.2 and by the following lemma for the star of 4 vertices (in this case the lower bound of Lemma 3.2 is not achievable).

Lemma 3.4 *The cost of 2-protection network of a graph with 4 vertices is at least the cost of its maximum requirement spanning tree.*

Proof: There are two non-isomorphic spanning trees in K_4: a path of 4 vertices and a star. We will show that in each case there exists a set of cuts such that (i) every edge in the graph participates in at most two cuts, (ii) every edge in the maximum spanning tree participate in exactly two cuts, and (iii) each cut contains (at most) two edges of the spanning tree.

Let C_i $i = 1, ..., l$ be such a set of cuts and let e_1^i, e_2^i be the edges in the tree that belong to C_i. Then

$$b(C_i) \geq r_1^i + r_2^i.$$

Summing over $i = 1, ..., l$ we get

$$2b(G) \geq \sum_i b(C_i) \geq \sum_i (r_1^i + r_2^i) = 2r(T),$$

where the first inequality follows from (i) and the equality follows from (ii). Thus, $b(G) \geq r(T)$. If the tree is a path $\{(1,2), (2,3), (3,4)\}$, then the following are the 3 cuts: $(\{1,4\}, \{2,3\}), (\{2\}, \{1,3,4\}), (\{3\}, \{1,2,4\})$. If the tree is a star with center 1, then the following are the 3 cuts: $(\{1,4\}, \{2,3\}), (\{1,2\}, \{3,4\}), (\{1,3\}, \{2,4\})$.
∎

4   Unit Requirements – Fractional Capacities

In this section we solve the case where $r_e = 1$ $e \in E$. Let R_1 (R_2) be the set of vertices with only one adjacent edge (at least two adjacent demand edges) in G, $k = |R_1|$ and $l = |R_2|$. W.l.o.g. we assume that $R_1 \cup R_2 = V$ (any connections made to a vertex with no adjacent demand edge can be made instead to any node in $R_1 \cup R_2$).

If there is a pair of R_1 vertices with a demand edge connecting them, we can delete these vertices from G, obtain a solution to the resulted network and then add an edge with capacity 1 between them to obtain an optimal solution to the original problem.

If G is a cycle or a path then E itself is an optimal solution as it obviously satisfies the requirements and its cost is equal to the lower bound of Lemma 3.2.

Algorithm 4.1

1. *If $R_2 = V$ then return a Hamiltonian cycle with capacity 1.*
2. *Suppose that $k \leq 2l$. If $l \geq 3$, construct a cycle with capacity 1 for all its edges over R_2. (If $R = \{a, b\}$ $(l = 2)$ use instead of the cycle the edge (a, b) with capacity $b_{a,b} = 2$). Map R_1's vertices to the cycle's edges so that every R_1 vertex is mapped to an edge and for every cycle edge there are at most two R_1 vertices that are mapped to it. Denote the mapping by m. For every R_1 vertex u let $m(u) = (v, w)$, then update the capacity function as $b_{u,v} = b_{u,w} = \frac{1}{2}$ and $b_{v,w} \leftarrow b_{v,w} - \frac{1}{2}$.*
3. *Suppose that $k \geq 2l$. For every edge (u, v) connecting an R_2 vertex and an R_1 vertex define $b_{u,v} = \frac{2}{k}$, and for every edge (v, w) connecting two R_1 vertices define $b_{v,w} = \frac{1 - \frac{2l}{k}}{k - 2}$.*

Remark 4.2 *The cost of the solution in Step 3 is larger than the lower bound of Lemma 3.2. The difference is $\frac{k-2l}{2(k-2)}$. This quantity equals $\frac{1}{2}$ when $l = 1$ and for every other value of l it is smaller than $\frac{1}{2}$. When l is constant the value is increasing with k. We note that the solution given in [8] even when applied to this case is not optimal.*

5   Unit Requirements – Integral Capacities

In this section we solve the case where $r_e = 1$ $\forall e \in E$ under the additional requirement that the solution b is integer.

The problem for the case of 1-protection network was solved optimally by Sridhar and Chandrasekaran [7] for general $\{r_e\}$.

We first assume that G is connected. In this case there are two possibilities: Either G is a spanning tree or it has a cycle.

Lemma 5.1 *If G is a spanning tree then the solution defined by $b_e = r_e$ $\forall e \in E$ is optimal.*

Proof: It is a feasible solution, trivially. To show optimality, if the graph with the positive capacities is not connected then there exists a pair of vertices from distinct connected components of G which have positive requirement between them, and this requirement is not satisfied. Therefore, the solution must contain a spanning tree and therefore, in this case the defined solution is optimal.   ∎

Lemma 5.2 *If G is connected and contains a cycle, then a Hamiltonian cycle over the vertex set is an optimal solution.*

Proof: If G contains a cycle then as we argue in the proof of Lemma 5.1 the solution must contain a spanning tree with capacity 1. However, a spanning tree T is not a feasible solution. If the cycle in G has the following vertices: $v_1, v_2, \ldots, v_{k-1}, v_k = v_0$ and the cycle edges are (v_i, v_{i+1}) $\forall k \geq i \geq 0$, there is a pair of adjacent edges $(v_i, v_{i+1}), (v_{i+1}, v_{i+2})$ such that the paths in the spanning tree T that connect v_i and v_{i+1} and connect v_{i+1} and v_{i+2} overlap in at least one edge. Then T will not survive the failure of the pair of edges $(v_i, v_{i+1}), (v_{i+1}, v_{i+2})$. Therefore, the solution must be with cost at least n. The solution defined by an Hamiltonian cycle with capacity 1 and the rest of the edges with 0 capacity is feasible and costs n, and therefore it is optimal.   ∎

Next we consider the case where G is composed of k connected components. Let $V_1, V_2, \ldots, V_i$ be components such that G is a spanning tree over each component, and let $V_{i+1}, \ldots, V_k$ be components such that G contains a cycle over each one. Let G^b be the union of a Hamiltonian cycle over the vertices in $V_{i+1} \cup V_{i+2} \cup \ldots \cup V_k$ and the subgraph induced by $V_1 \cup V_2 \cup \ldots \cup V_i$. The cost of this solution is $n - i$.

Theorem 5.3 *G^b is optimal.*

Proof: Let $G' = (V, E')$ be an optimal solution (if in the solution there is an edge with capacity 2 or more we use multiple copies of it, therefore, we assume all the capacities are 1). For every connected component of G all its vertices are in the same connected component of G'. This is so as otherwise, there exist adjacent vertices in G, u, v, that are in distinct connected components of G', therefore, G' won't survive the failure of the edge (u, v). For every $j \geq i$ G contains a cycle over V_j, therefore, the connected component of G' that contains V_j must contain a cycle (as in the proof of Lemma 5.2). If a connected component of G' contains a cycle the connected component can be replaced with a Hamiltonian cycle over its vertices without enlarging its cost. Therefore, the cost of G' is $n - i'$ where i' is the number of connected components of G' that it induce a spanning tree over them. Therefore, $i' \leq i$, and as a result the solution G^b is optimal. ∎

6 Unit Requirements – Integral Flows

In this section we require that for every pair of edges that fail there will be integer flows that backup it in the protection network. The complexity of this variant is still an open question.

We note that for the 1-protection network problem this type of integer requirement is the same as the requirement that b is integer. However, for the 2-protection network problem this is not the case.

6.1 Clique Requirements

We solve the problem for the case of clique: $r_e = 1$ $\forall e \in E$, $G = (V, E)$ is complete graph.

Suppose that $n = 6k$ for some integer k. We will show that the following graph $G_o = (V, E_o)$ is optimal for this case: The non-zero capacity edges have unit capacity and they consist of the set E_o as follows (see Figure 1 for $n = 18$): E_o contains a cycle C on $k = \frac{2}{3}n$ vertices $v_1, ..., v_k$. It also contains for every even $i = 2, 4, ..., \frac{k}{2}$ a *connecting path* (denoted CP) starting at v_i and ending at $v_{i+\frac{k}{2}}$ with edges $(v_i, u_i), (u_i, w_i), (w_i, v_{i+\frac{k}{2}})$.

This graph has $\frac{7n}{6}$ edges.

Theorem 6.1 G_o *is feasible and optimal.*

Theorem 6.1 provides us a nice approximation algorithm for the general requirement graph (we assume all the requirements are 0 or 1), as follows:

Algorithm 6.2

1. *In the graph of positive requirements find the partition of the connected components.*

2. *For every connected component which is a tree take this tree into the solution and remove it from the graph.*

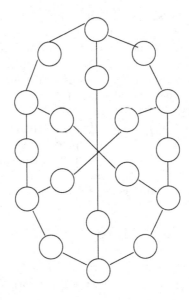

Fig. 1. An optimal solution for the integer flows problem

3. *Consider the vertices of the rest of the components. If their number of is not a multiple of 6 then add at most 5 new vertices called dummy vertices. Find a graph over this vertex set according to the solution G_o.*
4. *If the new solution has dummy vertices then contract them into an arbitrary regular vertex.*

This is a linear time algorithm.

Theorem 6.3 *Algorithm 6.2 is asymptotically $\frac{7}{6}$-approximation algorithm.*

Proof: The feasibility of the solution is straightforward as G_o provides a solution for an extended requirement network and each spanning tree satisfies the requirements of its component. For the $\frac{7}{6}$ ratio note the following: An optimal solution to the problem must also be feasible to the problem when we only require that the capacity will be integer. Therefore, it must include at least k edges for every connected component with k vertices which is not a tree. Therefore, any solution includes at least $n - n_T$ edges where n_T is the number of the connected components which are trees. As the solution has at most $\frac{7}{6}(n - n_T + 5)$ edges the asymptotic ratio of $\frac{7}{6}$ is proved.   ∎

References

[1] R.E. Gomory and T.C. Hu, "Multi-terminal network flows", *J. SIAM* **9**, 551-570, 1961.

[2] R.E. Gomory and T.C. Hu, "Synthesis of a communication network", *J. SIAM*, **12**, 348-369, 1964.

[3] T.C. Hu, *Integer Programming and Network Flows*, Addison Wesley Publishing, 1969.

[4] I. Ouveysi and A. Wirth, "On design of a survivable network architecture for dynamic routing: optimal solution strategy and an efficient heuristic", *European Journal of Operational Research*, **117**, 30-44, 1999.

[5] I. Ouveysi and A. Wirth, "Fast heuristics for protection networks for dynamic routing", *Journal of the operational research society*, **50**, 262-267, 1999.

[6] T. Sastry, "A characterization of the two-commodity network design problem", *Networks*, **36**, 9-16, 2000.

[7] S. Sridhar and R. Chandrasekaran, "Integer solution to synthesis of communication networks", *Mathematics of Operation Research*, **17**, 581-585, 1992.

[8] Y. K. Tham, "Network design for simultaneous traffic flow requirements", *IEICE Trans. Commun.*, **E80-B**, 930-938, 1997.

Bounds for Deterministic Periodic Routing Sequences

A. Hordijk and D.A. van der Laan

Mathematical Institute, Leiden University, P.O. Box 9512, 2300 RA Leiden,
The Netherlands

Abstract. We consider the problem of routing arriving jobs to parallel queues according to a deterministic periodic routing sequence. We introduce a combinatorial notion called the unbalance for such routing sequences. This unbalance is used to obtain an upper bound for the average waiting time of the routed jobs. The best upper bound for given (optimized) routing fractions is obtained when the unbalance is minimized. The problem of minimizing the unbalance is investigated and we show how to construct sequences with small unbalance.

1 Introduction and Notations

A stream of arriving customers are routed to parallel queues according to some deterministic periodic routing policy. The routing policy is characterised with an integer sequence $U = (U_1, U_2, \dots)$ where U_n is the queue to which the n-th arriving customer is routed. In this paper we derive bounds on the average waiting time of the arriving customers. U is called a routing sequence. Such a routing policy is also called a generalized round robin (GRR) routing policy (see [8] and [5]). The policy and corresponding integer sequence are periodic with period T if $U_n = U_{n+T}$ for every $n \in \mathbb{N}$. A GRR routing policy does not depend on dynamical information such as the number of customers in each queue or the remaining workload in each queue and is therefore a static routing policy and easy to implement. For homogeneous servers, the optimality of the round robin routing policy is proved in [12]. The problem of constructing the optimal GRR routing policy for heterogeneous servers is difficult. In this paper we define for periodic integer sequences U a combinatorial notion called the unbalance of U. Then we use the unbalance of U to derive bounds on the average waiting time of the arriving customers if they are routed according to routing sequence U. We use these bounds to construct routing sequences with average waiting time close to some lower bound.

This paper is a short version of [10], which is submitted for publication in a journal. In [10] complete proofs of the statements of this paper and additional results can be found.

If $\mathbf{A}$ is a finite set (a so called alphabet) and $I \subseteq \mathbb{Z}$ is an interval in $\mathbb{Z}$ then a mapping $\Lambda : I \to \mathbf{A}$ is called an I - word or just a word. Thus a word is a sequence of letters from the alphabet $\mathbf{A}$ and a word is (in)finite if I is (in)finite.

K. Aardal, B. Gerards (Eds.): IPCO 2001, LNCS 2081, pp. 236–250, 2001.
© Springer-Verlag Berlin Heidelberg 2001

A subword of Λ is the restriction of Λ to some interval $J \subseteq I$. For a finite (sub)word Λ we denote the length of Λ by $|\Lambda|$. Further for a letter $a \in A$ the number of a's in Λ is denoted by $|\Lambda|_a$. For an I -word Λ the support of letter a is the set $S_a = \{t \in I : \Lambda(t) = a\}$. For an I -word Λ we have that letter a has density d if for every non-decreasing sequence of finite subintervals I_i with union I we have that $\frac{|S_a \cap I_i|}{|I_i|}$ tends to d. An (infinite) I - word Λ is periodic if for some $q \in \mathbb{N}$ it holds that $\Lambda(t) = \Lambda(t+q)$ for every t for which $t, t+q \in I$. Then q is said to be a period of the word and any subword of length q is said to be a period cycle of the word. The period of Λ is defined as the minimum of all the periods of Λ. If q is the period of word Λ then a subword of length q is said to be a primitive period cycle of the word. Let $N \geq 2$ be the number of parallel queues and thus $U_n \in \{1, 2, \dots, N\}$ for every $n \in \mathbb{N}$. Then the routing sequence $U = (U_1, U_2, \dots)$ can be seen as an $\mathbb{N}$ - word on the alphabet $\{1, 2, \dots, N\}$. Further to every letter $i \in \{1, 2, \dots, N\}$ corresponds a sequence of zeros and ones $u^i = (u_1^i, u_2^i, \dots)$ via the support of the letter, i.e. $u_n^i = 1$ if and only if $U_n = i$. We call u^i the routing or splitting sequence for queue i (see for example [9], [14], [2] and [11]). In the sequel we identify (integer) sequences with words and sometimes it is useful to choose an appropriate domain I for the word the sequence is identified with. So, we consider sequences and words as the same thing and therefore we have the same notions and operations for words as for sequences. For example if $u = (u_1, u_2, \dots, u_n)$ and $v = (v_1, v_2, \dots, v_m)$ are finite sequences then $uv = (u_1, u_2, \dots, u_n, v_1, v_2, \dots, v_n)$ is the concatenation of the two sequences. For $k \in \mathbb{N}$ we denote by u^k the sequence $uu \dots u$ in which the number of u's is k and we say that u^k is a power of u. A finite sequence is primitive if it is not a power of some other sequence. We denote $uu \dots$, the infinite sequence with period cycle u, by u^∞. Further a sequence (word) of zeros and ones is said to have density d if the letter 1 has density d.

Definition 1. *Let u be an infinite sequence of zeros and ones and Λ be the corresponding word defined on an appropriate infinite interval I. We say that u is regular with density d if for every $N \in \mathbb{N}$ we have for every subword Λ' of Λ of length N that $|\Lambda'|_1$ equals $\lfloor Nd \rfloor$ or $\lceil Nd \rceil$. In that case we also say that the corresponding set S_1 is a regular set (of density d).*

Regular sequences are important when applying deterministic routing or admission policies. Namely, in [1] and [3] it is shown under suitable conditions that if (at least) a fraction d of the arriving customers has to be routed to some queue then it is optimal to do that according to a regular sequence of density d. For queues with exponential service times the result goes back to [9]. In [1] and [3] it is shown that stationarity of the service times is sufficient just as for the interarrival times of the arriving customers. In case of two parallel queues they show that there exist a regular routing sequence U which is optimal. For this U there exist some $d \in [0, 1]$ such that u^1 is regular with density d and u^2 is regular with density $1 - d$. For $N \geq 3$ queues one often wants to route to the queues according to some (optimized) fractions, where the sum of these fractions is 1. However, for such fractions it is not always possible to route in such a way that the routing sequence for every queue is regular, i.e there exist no regular word on the alphabet $\{1, 2, \dots, N\}$ with densities of the letters equal to the prescribed fractions (see [2] and [15]). For example it is easily seen that there exists no regular word on an alphabet of three letters with densities equal to $\frac{1}{2}, \frac{1}{3}$ and $\frac{1}{6}$. Therefore routing with such fractions always gives an irregular routing sequence for at least one of the queues. In this paper we compare the performance of irregular routing sequences with regular routing sequences.

We proceed as follows. We restrict to periodic routing policies and thus to routing sequences with rational density. We define the notion unbalance for periodic words by using the period cycle of the word. In fact it is possible to define a primal unbalance and a dual unbalance, but here we restrict to the former notion. The dual unbalance can be used to obtain similar results for polling systems. The unbalance is a measure for the irregularity of the word. The unbalance of a regular sequence will be 0. We first do this for sequences of zeros and ones corresponding to routing to a single server queue. Then we use the unbalance in a sample path argument to bound the difference in average waiting time for customers routed to that queue according to an arbitrary periodic routing sequence and regular routing with the same density. We extend the results to routing to parallel queues. Using so called billiard sequences we construct periodic routing policies such that the average waiting time of all the arriving customers is close (within some bound) to the "regular" lower bound. We give a Mathematical Programming Problem (MPP) such that for given rational densities a periodic word with minimal unbalance can be obtained from an optimal solution of this MPP. The MPP can be transformed to an Integer Linear Program (ILP), which is given in [11].

2   The Unbalance for Periodic Sequences of Zeros and Ones

In this section all the words are defined on the alphabet $\{0, 1\}$ and we define the unbalance for periodic sequences of zeros and ones. From the classification of balanced sequences in ([13]) and the fact that regular sequences are particular balanced sequences (see [15]) we have for a regular infinite I - word of density d that

$$S_1 = I \cap \{\lfloor k\frac{1}{d} + \varphi \rfloor\}_{k \in \mathbb{Z}} \text{ or } S_1 = I \cap \{\lceil k\frac{1}{d} + \varphi \rceil\}_{k \in \mathbb{Z}} \tag{1}$$

for some $\varphi \in \mathbb{R}$. From this it follows that if u is regular with rational density $d = \frac{p}{q}$ with $p, q \in \mathbb{N}$, $\gcd(p, q) = 1$ then u is periodic with period q.

Example 1. The sequence $u = (1, 1, 0, 1, 0, 1, 0)^\infty$ is regular with density $\frac{4}{7}$ and (indeed) u is periodic with period 7. Further if u is identified with an N-word we have for the support of the ones that

$$S_1 = \{1, 2, 4, 6, 8, 9, 11, 13, 15, 16, 18, 20, \dots\} = \mathbb{N} \cap \{\lfloor k \cdot \frac{7}{4} - \frac{3}{4}\rfloor\}_{k \in \mathbb{Z}},$$

which is indeed as in (1).

We first define the unbalance for finite sequences of zeros and ones and then we extend the notion to infinite periodic sequences by considering the period cycle of such a sequence. We first have some more notations and definitions. We say that a finite sequence of zeros and ones is regular if it is the period cycle of some infinite regular sequence. Two finite sequences u, u' are said to be conjugate if there exist sequences v and w (v or w may be empty) such that $u = vw$ and $u' = wv$. This is an equivalence relation since u is conjugate to u' if and only if u' is a cyclic permutation of u. If u and u' are conjugate then we write $u \sim u'$. The conjugacy class of all the cyclic permutations of a finite sequence u is denoted by $\tilde{u}$. Let $\mathcal{P}(T, k)$ be all sequences of zeros and ones of length T containing exactly k ones and $\mathcal{R}(T, k) \subseteq \mathcal{P}(T, k)$ be the subset of the regular sequences in $\mathcal{P}(T, k)$. Let $u \in \mathcal{P}(T, k)$ and Λ the corresponding I - word on the alphabet $\{0, 1\}$. We assume without loss of generality that $I = \{1, 2, \dots, T\}$ and we denote u as well as Λ by $(u_1, u_2, \dots, u_T)$ where $u_n = \Lambda(n)$ for $n = 1, 2, \dots, T$. Further we say for $l = 0, 1, \dots, T - 1$ that $(u_{1+l}, u_{2+l}, \dots, u_T, u_1, u_2, \dots, u_l)$ is the l -th cyclic permutation of u.

For sequence $u \in \mathcal{P}(T, k)$ we define the counting function $\kappa_u : \{0, 1, \dots, T\} \to \mathbb{Z}$ by $\kappa_u(n) = \sum_{t=1}^{n} u_t$. Thus $\kappa(n)$ counts the number of ones in the first n letters of the corresponding word Λ. We also define the discrepancy function $\chi_u : \{0, 1, \dots, T\} \to \mathbb{Q}$ by $\chi_u(n) = n \cdot \frac{k}{T} - \kappa_u(n)$ for $n = 0, 1, \dots, T$. We define a partial order $\preceq$ on $\mathcal{P}(T, k)$.

Definition 2. *For $u, v \in \mathcal{P}(T, k)$ we say that $u \preceq v$ if $\kappa_u(n) \leq \kappa_v(n)$ for $n = 1, 2, \dots, T$.*

Proposition 1. *Let $u \in \mathcal{R}(T, k)$. Then $u' \in \mathcal{R}(T, k)$ if and only if $u \sim u'$. Moreover the partial order $\preceq$ on $\mathcal{P}(T, k)$ induces a total order on $\mathcal{R}(T, k)$.*

Since $\mathcal{R}(T, k)$ is finite it follows from Proposition 1 that $\mathcal{R}(T, k)$ contains a greatest element for this order. This greatest element is the sequence in $\mathcal{R}(T, k)$ in which " the ones are as much to the left as possible ". This greatest element of $\mathcal{R}(T, k)$ is important for our definition of the (primal) unbalance of a sequence $u \in \mathcal{P}(T, k)$. We denote this greatest element by $\omega(T, k)$ or just ω if no confusion is possible. For example $\omega(7, 4)$ is the sequence $(1, 1, 0, 1, 0, 1, 0)$. We have the following lemma which can be used to determine $\omega(T, k)$ quickly.

Lemma 1. *For the support set S_1 of $\omega(T, k)$ we have that*

$$S_1 = \{\lfloor i \cdot \frac{T}{k} + 1 - \frac{T}{k}\rfloor\}_{i=1}^{k} \text{ and } \kappa_{\omega(T,k)}(n) = \lceil n \cdot \frac{k}{T}\rceil \text{ for } n = 0, 1, \dots, T.$$

We have seen in Proposition 1 that the regular sequences $\mathcal{R}(T, k)$ form a conjugacy class in $\mathcal{P}(T, k)$. We have the following theorem in which the partial order is used to give a characterising property of the conjugacy class $\mathcal{R}(T, k)$ of regular sequences in the set of all the conjugacy classes of $\mathcal{P}(T, k)$.

Theorem 1. *Every conjugacy class $\widetilde{u}$ of $\mathcal{P}(T,k)$ contains an upper bound of $\mathcal{R}(T,k)$, i.e for every $u \in \mathcal{P}(T,k)$ there exists a $v \in \mathcal{P}(T,k)$ such that $v \sim u$ and $v \succeq w$ for every $w \in \mathcal{R}(T,k)$.*

Outline of the proof. Let $u \in \mathcal{P}(T,k)$ and let $l = \text{argmax}_{n=0,1,\ldots,T-1}\chi_u(n)$. We claim that $u' :=$ the l-th cyclic permutation of u, is an upper bound of $\mathcal{R}(T,k)$. Since $u' \sim u$ this proves the theorem.

Remark 1. By Proposition 1 we have that $u' \in \mathcal{P}(T,k)$ is an upper bound of $\mathcal{R}(T,k)$ if and only if $u' \succeq \omega(T,k)$. By Lemma 1 this is the case if the discrepancy function $\chi_{u'}$ is never strictly positive. If $\gcd(k,T) = 1$ then it is easily seen for every $u \in \mathcal{P}(T,k)$ that χ_u is injective on the domain $\{0,1,\ldots,T-1\}$ and thus $\text{argmax}_{n=0,1,\ldots,T-1}\chi_u(n)$ is unique. A consequence is that for every $u \in \mathcal{P}(T,k)$ the upper bound of $\mathcal{R}(T,k)$ in the conjugacy class $\widetilde{u}$ is unique if $\gcd(k,T) = 1$. If $\gcd(k,T) > 1$ then the upper bound is in general not unique.

Definition 3. *For $u \in \mathcal{P}(T,k)$, let u' be an upper bound with respect to the partial order $\preceq$ of $\mathcal{R}(T,k)$ in the conjugacy class $\widetilde{u}$. Then the (primal) unbalance $I(u)$ of u is defined by*

$$I(u) = \frac{1}{T}\sum_{n=1}^{T}(\kappa_{u'}(n) - \kappa_{\omega}(n)), \ \text{where} \ \omega = \omega(T,k) \ . \tag{2}$$

It can be checked that the unbalance is well defined for finite sequences of zeros and ones, i.e for every pair u', u'' of upper bounds of $\mathcal{R}(T,k)$ with $u' \sim u''$ it holds that $\sum_{n=1}^{T}\kappa_{u'}(n) = \sum_{n=1}^{T}\kappa_{u''}(n)$. The following example illustrates these results and definitions.

Example 2. We calculate the unbalance of the sequence $u = (0,1,0,0,1,0,1,1,0,1,0) \in \mathcal{P}(11,5)$. First we have to find an upper bound with respect to the partial order $\preceq$ of $\mathcal{R}(11,5)$ in the conjugacy class $\widetilde{u}$. Following the outline of the proof of Theorem 1 we find that $\text{argmax}_{n=0,1,\ldots,10}\chi_u(n) = 4$. Thus the 4-th cyclic permutation of u, which is $u' = (1,0,1,1,0,1,0,0,1,0,0)$ should be upper bound of $\mathcal{R}(11,5)$. Indeed $u' \succeq \omega(11,5) = (1,0,1,0,1,0,1,0,1,0,0)$ and it follows that $I(u) = I(u') = \frac{2}{11}$.

After defining the unbalance and a partial order for finite sequences we extend these notions to infinite periodic sequences (words). First of all we extend in a natural way the partial order $\preceq$ to the infinite sequences $u = (u_1, u_2, \ldots)$ of zeros and ones corresponding to $\mathbb{N}$-words on the alphabet $\{0,1\}$. We define just as for finite sequences (words) the counting function $\kappa_u : \mathbb{Z}_{\geq 0} \to \mathbb{Z}$ by $\kappa_u(n) = \sum_{t=1}^{n} u_t$ and we say that $u \preceq v$ if $\kappa_u(n) \leq \kappa_v(n)$ for $n = 1,2,\ldots$. For a sequence $u = (u_1, u_2, \ldots)$ of zeros and ones of density d corresponding to an $\mathbb{N}$ - word we also define the discrepancy function $\chi_u : \mathbb{Z}_{\geq 0} \to \mathbb{R}$ by $\chi_u(n) = n \cdot d - \kappa_u(n)$ for $n \in \mathbb{Z}_{\geq 0}$. Further we have the following equivalence relation $\sim$ for infinite periodic sequences. In fact $\sim$ is an equivalence relation for all infinite periodic integer sequences and the corresponding words on a finite alphabet A and not only for the sequences of zeros and ones.

Let u and v be infinite periodic integer sequences. Then we say that u is equivalent to v, $u \sim v$ if there exists a finite sequence w such that w is a period cycle of both u and v.

In our application of routing sequences we have in general that equivalent infinite periodic sequences have the same performance. Let $\mathcal{P}$ be the set of infinite periodic sequences of zeros and ones (modulo the equivalence relation), $\mathcal{R} \subseteq \mathcal{P}$ the subset of regular periodic sequences and for $d \in \mathbb{Q}$, $0 \leq d \leq 1$ let $\mathcal{P}(d) \subseteq \mathcal{P}$ be the subset of sequences with density d. We denote the regular sequence in $\mathcal{P}(d)$ by $\omega(d)$ or just ω. Note that if $u' \in \mathcal{P}(T, k)$ is a period cycle of $u \in \mathcal{P}$ then $u \in \mathcal{P}(d)$ where $d = \frac{k}{T}$. We define the unbalance for infinite periodic sequences of zeros and ones as follows.

Definition 4. *Let $u \in \mathcal{P}$ and let $u' \in \mathcal{P}(T, k)$ be a period cycle of u. Then we define the (primal) unbalance of u as $I(u) := I(u')$.*

The unbalance is well defined on $\mathcal{P}$. Namely, if u' and u'' are both period cycles of u then $I(u') = I(u'')$. When the unbalance of an infinite periodic sequence has to be computed it is of course most practical to compute the unbalance of a primitive period cycle of the sequence.

3   Bounding the Difference in Expected Average Waiting Time of Routing Sequences for One Queue

In this section we give a bound for the expected average waiting time for customers routed to a single server queue according to a routing sequence of zeros and ones. This bound depends on the unbalance of the routing sequence. To state the results we first have some definitions and notations.

Let $\{T_i\}_{i=1,2,\dots}$ be a sequence of arrival times of customers, with the convention that $T_1 = 0$. Put $\delta_i := T_{i+1} - T_i$ for $i = 1, 2, \dots$. Then $\{\delta_i\}$ is the sequence of interarrival times. Further these arriving customers are routed to a server according to some routing sequence $u = (u_1, u_2, \dots)$ of zeros and ones. For such a routing sequence we have the counting function $\kappa_u(n) = \sum_{t=1}^{n} u_t$ that we used to define a partial order $\preceq$ for such routing sequences. Further we define the the following related function $\nu_u(i) : \mathbb{Z}_{\geq 0} \to \mathbb{Z}$ with $\nu_u(j) = \min\{n \in \mathbb{Z}_{\geq 0} : \kappa_u(n) = j\}$ and we put $\tau_u(j) = \sum_{i=\max(\nu_u(j-1),1)}^{\nu_u(j)-1} \delta_i$ for $j = 1, 2, \dots$. Then $\tau_u(j)$ is the time elapsed between the routing of the $j-1$-th and j-th customer to the server according to routing sequence u. If we put $\lambda_u(j) = \sum_{k=1}^{j} \tau_u(k)$ for $j = 1, 2, \dots$ then $\lambda_u(j)$ is the time at which the $j - th$ customer is routed to the server according to routing sequence u. We have a sequence of service times $\{\sigma_j\}_{j=1,2,\dots}$, where σ_j is the service time of the j-th customer that is routed to the server according to the routing sequence. Further we define $W_u(j)$ to be the workload for the server at the moment the j-th customer is routed to the server according to routing sequence u. In other words $W_u(j)$ is the waiting time for the j-th customer that is routed to the server. We assume that the server starts empty at $T_1 = 0$ and thus $W_u(1) = 0$. If the interarrival times $\{\delta_i\}$ and service times $\{\sigma_j\}$ are random variables then we say that $\overline{W}(u)$ is the almost sure long-run average waiting time of customers routed to the server according to routing sequence u if $\lim_{m \to \infty} \frac{1}{m} \cdot \sum_{j=1}^{m} W_u(j) = \overline{W}(u)$ with probability one. From ergodic theory we have the following theorem.

Theorem 2. *Suppose that the interarrival times $\{\delta_i\}$ of customers arriving at the system are independent and identically distributed (i.i.d.) random variables with mean δ and the service times $\{\sigma_j\}$ of the considered server are i.i.d. random variables with mean σ and independent of the interarrival times. Further let u' and u'' be routing sequences of zeros and ones that are both representatives of some $u \in \mathcal{P}(d)$ where $d \in \mathbb{Q}$ (thus they have a common period cycle) and $\frac{\sigma}{\delta} \cdot d < 1$. Then $\overline{W}(u')$ and $\overline{W}(u'')$ exist and are finite. Moreover $\overline{W}(u') = \overline{W}(u'')$.*

Let δ, σ and d be as in Theorem 2. Then we say that $\rho := \frac{\sigma}{\delta} \cdot d$ is the traffic intensity for the server. By Theorem 2 all the routing sequences which are representative of some $u \in \mathcal{P}(d)$ have the same long-run average waiting time if the stability condition $\rho < 1$ is fulfilled. Therefore in this case we denote this long-run average waiting time simply by $\overline{W}(u)$. The following theorem which is obtained by a sample path argument is the main result of this paper.

Theorem 3. *Let the interarrival times $\{\delta_i\}$ and the service times $\{\sigma_j\}$ be as in Theorem 2. Further let $u \in \mathcal{P}(d)$ for some $d \in \mathbb{Q}$ such that $\rho < 1$ and let $\omega = \omega(d) \in \mathcal{P}(d)$ be the regular sequence of density d. Then*

$$\overline{W}(u) \leq \overline{W}(\omega) + \frac{\delta}{d} \cdot I(u) \ . \tag{3}$$

Remark 2. From [1] and [3] we have that $\overline{W}(u) \geq \overline{W}(\omega)$.

We give an outline of the proof of Theorem 3. We first assume that the interarrival times $\{\delta_i\}_{i=1,2,\ldots}$ and service times $\{\sigma_j\}_{j=1,2,\ldots}$ are fixed sequences of non-negative real numbers. Then we have the following lemma.

Lemma 2. *Let $u = (u_1, u_2, \ldots)$ and $v = (v_1, v_2, \ldots)$ be routing sequences of zeros and ones and suppose that $u \preceq v$. Then $W_v(j) + \lambda_v(j) \leq W_u(j) + \lambda_u(j)$ for every $j \in \mathbb{N}$.*

Note that $W_u(j) + \lambda_u(j)$ is the moment that the server starts serving the j-th customer that is routed to the server according to u. So, Lemma 2 states that if $u \preceq v$ then the moment when the service of the j-th customer that is routed to the server starts, is for v not later than for u. This can be proved by induction to j.

For every $j \in \mathbb{N}$ we have that $\lambda_u(j) - \lambda_v(j) = \sum_{i=\nu_v(j)}^{\nu_u(j)-1} \delta_i$. This is a sum of interarrival times δ_i, where the number of terms in the sum is $\nu_u(j) - \nu_v(j)$. Therefore, if we put $N_{uv}(m) = \sum_{j=1}^{m} (\nu_u(j) - \nu_v(j))$ for $m = 1, 2, \ldots$ then we have the following Corollary by Lemma 2.

Corollary 1. *Let u and v be routing sequences with $u \preceq v$. For every $m \in \mathbb{N}$ we have that*

$$\sum_{j=1}^{m} W_v(j) \leq \sum_{j=1}^{m} W_u(j) + \sum_{j=1}^{m} \sum_{i=\nu_v(j)}^{\nu_u(j)-1} \delta_i . \tag{4}$$

In the double sum the number of terms is $N_{uv}(m)$.

Then we show that if $u = (u')^\infty$ and $v = (v')^\infty$ for some $u', v' \in \mathcal{P}(T, k)$ and $\omega(T, k) \preceq u' \preceq v'$ then

$$\lim_{m \to \infty} \frac{N_{uv}(m)}{m} = \frac{T}{k} \cdot (I(v) - I(u)) . \tag{5}$$

Let $u \in \mathcal{P}(d)$ be as in Theorem 3. Then by Theorem 1 there exist $T, k \in \mathbb{N}$ with $d = \frac{k}{T}$ and $u'' \in \mathcal{P}(T, k)$ such that $u' := (u'')^\infty$ is a representative of u and $\omega' \preceq u'$ where $\omega' = \omega(T, k)^\infty$ is a representative of $\omega(d)$. By Theorem 2 we have that both $\overline{W}(u) = \overline{W}(u')$ and $\overline{W}(\omega) = \overline{W}(\omega')$ exist. We make a coupling between the interarrival times for u' and ω' and we also do this for the service times. After the coupling we can apply Corollary 1 and (5) and it can be shown that

$$\overline{W}(u') - \overline{W}(\omega') \leq \frac{\delta \cdot T}{k} \cdot (I(u) - I(\omega)) = \frac{\delta}{d} \cdot I(u),$$

which proves Theorem 3.

Remark 3. From this proof it follows that it is possible to refine Theorem 3 as follows: Let $u, v \in \mathcal{P}(d)$ be as in Theorem 3 and let ω' be a representative of $\omega(d)$ as above. Then, if there exist representatives u' and v' for u and v respectively such that $\omega' \preceq u' \preceq v'$ then

$$\overline{W}(v) - \overline{W}(u) \leq \frac{\delta}{d}(I(v) - I(u)) . \tag{6}$$

Moreover, the condition $\rho < 1$ in Theorem 3 is just necessary to apply Theorem 2. However, if both the interarrival times and the service times are deterministic then Theorem 2 also holds if $\rho = 1$ (see [11]). Thus in that case Theorem 3 also holds if $\rho = 1$ and in fact we have for that case the following stronger result from which it follows that the bound of Theorem 3 is tight.

Theorem 4. *Suppose that the interarrival times and service times are deterministic and equal to δ respectively σ. Let $u \in \mathcal{P}(d)$ for some $d \in \mathbb{Q}$ such that $\delta = d\sigma$, hence $\rho = 1$, and let $\omega = \omega(d)$ be the regular sequence of density d. Then*

$$\overline{W}(u) = \overline{W}(\omega) + \frac{\delta}{d} \cdot I(u) . \tag{7}$$

Outline of the proof. Let ω' be a representative of $\omega(d)$ as above and let u' be a representative of u such that $\omega' \preceq u'$. Then it can be shown that $W_{u'}(j) + \lambda_{u'}(j) = W_{\omega'}(j) + \lambda_{\omega'}(j) = (j-1)\sigma$ for every $j \in \mathbb{N}$. Hence we have for every $m \in \mathbb{N}$ that (4) holds with equality for u' and ω' with $\delta_i = \delta$ for $i = 1, 2, \ldots$. From this it follows that (3) also holds with equality.

4   Routing to Parallel Queues

In this section we derive upper bounds for the expected average waiting time for routing arriving customers to $N \geq 2$ parallel queues according to some periodic routing policy corresponding to a word $U = (U_1, U_2, \dots)$. Extending the notion of unbalance for periodic sequences of zeros and ones we define a (total) unbalance $O(U)$ for periodic integer sequences and corresponding words on some finite alphabet. Then we extend the result of Theorem 3 for one queue to multiple queues by using the unbalance $O(U)$ of U.

Let $d_1, d_2, \dots , d_N \in \mathbb{Q}_{>0}$ with $\sum_{i=1}^{N} d_i = 1$. Then we denote by $\mathcal{Q}(d_1, d_2, \dots , d_N)$ all the infinite periodic $\mathbb{N}$ - words U on the alphabet $\{1, 2, \dots , N\}$ for which $u^i \in \mathcal{P}(d_i)$ for $i = 1, 2, \dots , N$. Further for $T, k_1, k_2, \dots , k_N \in \mathbb{N}$ with $\sum_{i=1}^{N} k_i = T$ we denote by $\mathcal{Q}(\{T\}, k_1, k_2, \dots , k_N)$ all the $\mathbb{N}$-words on the alphabet $\{1, 2, \dots , N\}$ for which every subword of length T contains exactly k_i letters i for $i = 1, 2, \dots , N$. Note that $\mathcal{Q}(\{T\}, k_1, k_2, \dots , k_N) \subseteq \mathcal{Q}(\frac{k_1}{T}, \frac{k_2}{T}, \dots , \frac{k_N}{T})$. For example if $U = (1, 2, 1, 2, 1, 3, 1, 2, 1, 1, 2, 3)^\infty$ then $U \in \mathcal{Q}(\{12\}, 6, 4, 2) \subseteq \mathcal{Q}(\frac{1}{2}, \frac{1}{3}, \frac{1}{6})$. Further $u^1 = (1, 0, 1, 0, 1, 0, 1, 0, 1, 1, 0, 0)^\infty \in \mathcal{P}(\frac{1}{2})$, $u^2 = (0, 1, 0, 1, 0, 0, 0, 1, 0, 0, 1, 0)^\infty \in \mathcal{P}(\frac{1}{3})$ and $u^3 = (0, 0, 0, 0, 0, 1)^\infty \in \mathcal{P}(\frac{1}{6})$.

Let ψ be a routing policy and U the corresponding routing sequence. Then for $t \in \mathbb{N}$ we define $W(t) = W_\psi(t)$ as the waiting time of the t -th arriving customer if policy ψ is applied, which is the remaining workload for server U_t at the moment that the t-th customer arrives. We assume that all the servers $i \in \{1, 2, \dots , N\}$ are empty at $T_1 = 0$ and thus $W(t) = 0$ if $t = \nu_{u^i}(1)$ for some $i \in \{1, 2, \dots , N\}$. If the interarrival times and service times of the various servers are random variables then we say that $\overline{W}(\psi) = \overline{W}(U)$ is the almost sure long-run average waiting time of the arriving customers routed according to policy ψ if $\lim_{\tau \to \infty} \frac{1}{\tau} \cdot \sum_{t=1}^{\tau} W_\psi(t) = \overline{W}(\psi)$ with probability one. Let $\{\delta_i\}$ be the sequence of interarrival times and let $\{\sigma_j^i\}$ be the sequence of service times of server i for $i = 1, 2, \dots , N$, i.e σ_j^i is the service time of the j - th customer that is routed to server i. We define $\overline{W}^i(\psi) = \overline{W}^i(u^i)$ as the almost sure long-run average waiting time of customers routed to server i if policy ψ is applied in the same way as we did in the previous section. The only differences are that routing sequence u is replaced by routing sequence u^i and the sequence of service times $\{\sigma_j\}$ is replaced with the sequence of service times $\{\sigma_j^i\}$. From Theorem 2 we obtain the following theorem.

Theorem 5. *Suppose that the interarrival times $\{\delta_i\}$ of customers arriving at the system are i.i.d. random variables with mean δ and for every $i \in \{1, 2, \dots , N\}$ the service times $\{\sigma_j^i\}$ are i.i.d random variables with mean σ_i, and independent of the interarrival times. Let ψ be a routing policy that corresponds to some word $U \in \mathcal{Q}(d_1, d_2, \dots , d_N)$ such that $\rho_i := \frac{\sigma_i}{\delta} \cdot d_i < 1$ for $i = 1, 2, \dots , N$. Then $\overline{W}(\psi)$ exists and is finite. Moreover $\overline{W}(\psi) = \sum_{i=1}^{n} d_i \cdot \overline{W}^i(\psi)$.*

Definition 5. *Let $U \in \mathcal{Q}(d_1, d_2, \ldots, d_N)$ for some $d_i \in \mathbb{Q}_{>0}$ with $\sum_{i=1}^{N} d_i = 1$. Then we define the (total) unbalance of U as*

$$O(U) := \sum_{i=1}^{N} I(u^i) . \tag{8}$$

If it is clear from the context what is meant we just say unbalance and not total unbalance. Let the interarrival times $\{\delta_i\}$ and the service times $\{\sigma_j^i\}$ for $i = 1, 2, \ldots, N$ be as in Theorem 5. Further let $d_i \in \mathbb{Q}_{>0}$ for $i = 1, 2, \ldots, N$ with $\sum_{i=1}^{N} d_i = 1$. Recall that $\omega(d_i)$ is the regular sequence of density d_i. If $\overline{W}^i(\omega(d_i))$ exists and is finite for $i = 1, 2, \ldots, N$ then we put

$$\widetilde{R} = \widetilde{R}(d_1, d_2, \ldots, d_N) := \sum_{i=1}^{N} d_i \cdot \overline{W}^i(\omega(d_i)) . \tag{9}$$

Remark 4. If the interarrival times and service times are random variables then for the existence of $\widetilde{R}$ it suffices that $\rho_i < 1$ for $i = 1, 2, \ldots, N$, while if the interarrival times and service times are deterministic then it suffices that $\rho_i \leq 1$ for $i = 1, 2, \ldots, N$. It is possible to extend the definition of $\widetilde{R}$ and some of the results to the case of irrational d_i. In general $\widetilde{R}$ depends on the distribution of the interarrival times and service times and in some cases it is possible to compute $\widetilde{R}$ explicitly. See [6] and [11] for an exact computation in case of deterministic interarrival and service times and see [14] for computations and bounds in general.

Combining Theorem 3 and Theorem 5 we obtain the following theorem.

Theorem 6. *Let the interarrival times $\{\delta_i\}$ and the service times $\{\sigma_j^i\}$ for $i = 1, 2, \ldots, N$ be as in Theorem 5. Let $d_i \in \mathbb{Q}_{>0}$ for $i = 1, 2, \ldots, N$ with $\sum_{i=1}^{N} d_i = 1$. Suppose that a routing policy ψ is applied with corresponding word $U \in \mathcal{Q}(d_1, d_2, \ldots, d_N)$ and $\rho_i < 1$ for $i = 1, 2, \ldots, N$. Then*

$$\overline{W}(\psi) - \widetilde{R} \leq \delta \cdot O(U).$$

From [1] and [3] we have that $\overline{W}(\psi) \geq \widetilde{R}$, which is the lower bound obtained by replacing the routing sequence to queue i by the regular routing sequence with the same density for any queue i. Hence we have the following bounds.

$$\widetilde{R} \leq \overline{W}(\psi) \leq \widetilde{R} + \delta \cdot O(U) . \tag{10}$$

Remark 5. Suppose that we have a queueing system where the arrivals are according to a Poisson process with parameter λ. Suppose that the service times of server i are exponentially distributed with parameter μ_i. If d_i, the fraction of jobs that is routed to server i, equals $\frac{1}{q_i}$ for some $q_i \in \mathbb{N}$, then $\overline{W}^i(\omega(d_i))$ can be calculated in the following way. For the routing sequence $\omega(d_i)$ we have that among every q_i arriving jobs exactly one job is routed to server i. So, the inter-arrival times at server i consist of q_i Poisson arrivals with parameter λ. Hence the interarrival times for the queue of server i are Erlang distributed, namely according to an $E_\lambda^{q_i}$ distribution. Thus $\overline{W}^i(\omega(d_i))$ is the same as the average waiting time for a $E_\lambda^{q_i}/M/1$ queue, where the parameter of the service times is μ_i. So (see [7]) if $x_i \in (0,1)$ is a solution of the equation

$$x = \left(\frac{\lambda}{\lambda + \mu_i - \mu_i \cdot x}\right)^{q_i} , \tag{11}$$

then

$$\overline{W}^i(\omega(d_i)) = \frac{x_i}{\mu_i \cdot (1 - x_i)} . \tag{12}$$

In the following example we have calculated $\overline{W}^i(\omega(d_i))$ for $i = 1, 2, 3, 4$ by applying (12). Further we have explicitly calculated the lower bound $\widetilde{R}$ and upper bound $\widetilde{R} + \delta \cdot O(U)$ of (10) for $\overline{W}(\psi)$, where U is the word corresponding to the applied routing policy ψ.

Example 3. We consider a queueing system with 4 parallel servers where the arrivals are according to a Poisson process with parameter $\lambda = 11$. Hence for the mean interarrival time δ we have that $\delta = \frac{1}{11}$. The arriving jobs are routed to the servers according to the policy ψ that corresponds to the word

$$U = (1, 2, 3, 1, 4, 2, 1, 3, 1, 2, 4, 3)^\infty \in \mathcal{Q}(\{12\}, 4, 3, 3, 2) \subseteq \mathcal{Q}(\frac{1}{3}, \frac{1}{4}, \frac{1}{4}, \frac{1}{6}) .$$

Then we have that $I(u^1) = \frac{1}{12}$, $I(u^2) = 0$, $I(u^3) = \frac{1}{12}$, $I(u^4) = 0$ and thus $O(U) = \frac{1}{6}$. For every $i \in \{1, 2, 3, 4\}$ the service times are exponentially distributed with parameter μ_i and $\mu_1 = 4$, $\mu_2 = \mu_3 = 3$ and $\mu_4 = 2$. Then we find by (12) that $\overline{W}^1(\omega(\frac{1}{3})) = 1.7792$ (rounded to 4 decimals) and thus by Corollary 3 we have that $\overline{W}^1(u^1) \leq 1.7792 + 3 \cdot \frac{1}{11} \cdot \frac{1}{12} = 1.8019$. Similarly we have that $\overline{W}^2(\omega(\frac{1}{4})) = \overline{W}^2(u^2) = 2.2105$, $\overline{W}^3(\omega(\frac{1}{4})) = 2.2105$, $\overline{W}^3(u^3) \leq 2.2408$ and $\overline{W}^4(\omega(\frac{1}{6})) = \overline{W}^4(u^4) = 3.0732$. Hence by (9)

$$\widetilde{R} = \frac{1}{3} \cdot \overline{W}^1(\omega(\frac{1}{3})) + \frac{1}{4} \cdot \overline{W}^2(\omega(\frac{1}{4})) + \frac{1}{4} \cdot \overline{W}^3(\omega(\frac{1}{4})) + \frac{1}{6} \cdot \overline{W}^4(\omega(\frac{1}{6})) = 2.2105$$

and $\widetilde{R} + \delta \cdot O(U) = 2.2105 + \frac{1}{11} \cdot \frac{1}{6} = 2.2257$. So, by (10) we have that $2.2105 \leq \overline{W}(\psi) \leq 2.2257$.

The following theorem provides conditions under which the right-hand side inequality of (10) actually holds with equality.

Theorem 7. *Let the interarrival times $\{\delta_i\}$ be deterministic equal to δ and the service times $\{\sigma_j^i\}$ be deterministic equal to σ_i for $i = 1, 2, \ldots, N$. Let $d_i \in \mathbb{Q}_{>0}$*

for $i = 1, 2, \ldots, N$ *with* $\sum_{i=1}^{N} d_i = 1$. *Let* $p_i, q_i \in \mathbb{N}$ *be such that* $d_i = \frac{p_i}{q_i}$ *with* $\gcd(p_i, q_i) = 1$ *for* $i = 1, 2, \ldots, N$. *Suppose that a routing policy* ψ *is applied with corresponding word* $U \in \mathcal{Q}(d_1, d_2, \ldots, d_N)$ *and* $\rho_i = 1$ *for* $i = 1, 2, \ldots, N$. *Then*

$$\overline{W}(\psi) = \widetilde{R} + \delta \cdot O(U) = \delta \cdot \left(\frac{1}{2} - \sum_{i=1}^{N} \frac{1}{2q_i} + O(U) \right).$$

By combining Theorem 4 and Theorem 5 we get the first equality of Theorem 7. See [11] for the second equality which follows from the computation of $\widetilde{R}$ for this case.

Example 4. We consider a queueing system with 3 parallel servers where the interarrival times are deterministic and equal to $\delta = 3$. The arriving jobs are routed to the servers according to the policy ψ that corresponds to the word

$$U = (1, 2, 1, 2, 1, 3, 1, 2, 1, 3)^{\infty} \in \mathcal{Q}(\{10\}, 5, 3, 2) \subseteq \mathcal{Q}(\frac{1}{2}, \frac{3}{10}, \frac{1}{5}) .$$

All the service times are deterministic and for server 1 they are equal to $\sigma_1 = 6$, for server 2 equal to $\sigma_2 = 10$ and for server 3 equal to $\sigma_3 = 15$. Hence $\rho_i = 1$ for $i = 1, 2, 3$. Further $I(u^1) = 0$, $I(u^2) = \frac{1}{10}$, $I(u^3) = \frac{1}{10}$ and thus $O(U) = \frac{1}{5}$. So, according to Theorem 7 we have that

$$\overline{W}(\psi) = 3 \cdot \left(\frac{1}{2} - (\frac{1}{4} + \frac{1}{20} + \frac{1}{10}) + \frac{1}{5} \right) = \frac{9}{10}$$

which can be checked by direct calculation.

5   Billiard Sequences and Routing Sequences

In this section we give some properties of sequences (words) and their implications for the corresponding routing policies. In particular we have some properties for billiard sequences. We have the following theorem.

Theorem 8. *Let* $d_1, d_2, \ldots, d_N \in \mathbb{Q}_{>0}$ *with* $\sum_{i=1}^{N} d_i = 1$ *for some* $N \geq 2$. *Then there exists a word* $U \in \mathcal{Q}(d_1, d_2, \ldots, d_N)$ *such that* $O(U) \leq \frac{N}{2} - 1$.

By combining Theorem 8 and Theorem 6 we obtain the following corollary.

Corollary 2. *Let the interarrival times* $\{\delta_i\}$, *the service times* $\{\sigma_j^i\}$ *and* d_i *for* $i = 1, 2, \ldots, N$ *be as in Theorem 5. Suppose that* $\rho_i < 1$ *for* $i = 1, 2, \ldots, N$. *Then there exists a routing policy* ψ *corresponding to a word* $U \in \mathcal{Q}(d_1, d_2, \ldots, d_N)$ *such that*

$$\overline{W}(\psi) \leq \widetilde{R}(d_1, d_2, \ldots, d_N) + \delta \cdot (\frac{N}{2} - 1).$$

We show how to construct a word $U \in \mathcal{Q}(d_1, d_2, \ldots, d_N)$ such that $O(U) \leq \frac{N}{2} - 1$. We use the following algorithm which we call the Special Greedy (SG) algorithm.

SG algorithm. Let $d_1, d_2, \ldots, d_N \in \mathbb{R}_{>0}$ with $\sum_{i=1}^{N} d_i = 1$ and $x_1, x_2, \ldots, x_N \in \mathbb{R}_{\geq 0}$ be given. Then $U = (U_1, U_2, \ldots)$ is determined inductively in the following way. Suppose that $U_1, U_2, \ldots, U_{n-1}$ have been determined and thus $\kappa_{u^i}(n-1)$ is known for $i = 1, 2, \ldots, N$. Choose U_n such that $U_n = i \in \{1, 2, \ldots, N\}$ for which $\frac{x_i + \kappa_{u^i}(n-1)}{d_i}$ is minimal.

If $0 \leq x_i < 1$ for $i = 1, 2, \ldots, N$ then a word U constructed by the SG algorithm is known in the literature as a billiard word or sequence (see for example [4]). Namely, if a billiard ball bounces in an N-dimensional cube and a sequence of the integers $\{1, 2, \ldots, N\}$ (the integers correspond to the sides of the cube and opposite sides correspond to the same integer) is constructed by writing down an integer if the ball bounces against the side of the cube corresponding to that integer then such a sequence is a billiard sequence. It is easily seen that the word constructed by the SG algorithm is a billiard sequence obtained by starting in position $(x_1, x_2, \ldots, x_N)$ and having initial velocity vector $(-d_1, -d_2, \ldots, -d_N)$ in the unit cube $[0, 1]^N$. In the SG algorithm the choice of U_n is not totally determined if $\frac{x_i + \kappa_{u^i}(n-1)}{d_i}$ is minimal for several $i \in \{1, 2, \ldots, N\}$. We assume that in such cases a consistent choice is made. For example always the smallest i is chosen for which $\frac{x_i + \kappa_{u^i}(n-1)}{d_i}$ is minimal. If this modified SG algorithm is used then the word obtained by the algorithm is unique for every $x_1, x_2, \ldots, x_N \in \mathbb{R}_{\geq 0}$. For billiard sequences this consistent choice means that if the billiard ball hits multiple sides at the same time then the order in which the letters corresponding to those sides appear in the billiard sequence is also prescribed. We will call such a billiard sequence a consistent billiard sequence. A billiard sequence that is not consistent can be aperiodic even if the densities $d_1, d_2, \ldots, d_N \in \mathbb{Q}_{>0}$. For consistent billiard sequences we have the following lemma.

Lemma 3. Let $d_1, d_2, \ldots, d_N \in \mathbb{Q}_{>0}$ with $\sum_{i=1}^{N} d_i = 1$. Let $p_i, q_i, k_i \in \mathbb{N}$ for $i = 1, 2, \ldots, N$ and $T \in \mathbb{N}$ be such that $d_i = \frac{p_i}{q_i} = \frac{k_i}{T}$ with $\gcd(p_i, q_i) = 1$ for $i = 1, 2, \ldots, N$ and $\gcd(k_1, k_2, \ldots, k_N) = 1$. Let U be a consistent billiard sequence. Then $U \in \mathcal{Q}(\{T\}, k_1, k_2, \ldots, k_N) \subseteq \mathcal{Q}(d_1, d_2, \ldots, d_N)$.

Construction to prove Theorem 8. Let d_i, p_i, q_i, k_i and T be as in Lemma 3, $N \geq 2$ and let $k \in \{1, 2, \ldots, N\}$ be such that $q_k = \max_{i \in \{1, 2, \ldots, N\}} q_i$. Then it can be proved that in the following way a word $U \in \mathcal{Q}(\{T\}, k_1, k_2, \ldots, k_N) \subseteq \mathcal{Q}(d_1, d_2, \ldots, d_N)$ is constructed with $O(U) = 0$ if $N = 2$ and $O(U) < \frac{N}{2} - 1$ if $N \geq 3$. Put $x_k = 0$ and $x_j = 1 - \frac{1}{q_j}$ for every $j \in \{1, 2, \ldots, N\}$ for which $j \neq k$. Construct a consistent billiard sequence starting in $(x_1, x_2, \ldots, x_N)$ with initial velocity vector $(-d_1, -d_2, \ldots, -d_N)$ in the unit cube $[0, 1]^N$.

Example 5. In this example we construct in this way a sequence $U \in \mathcal{Q}(d_1, d_2, d_3)$ where $d_1 = \frac{2}{5}$, $d_2 = \frac{7}{20}$ and $d_3 = \frac{1}{4}$. We see that $q_2 = 20 = \max_i q_i$.

Thus we put $x_2 = 0$, $x_1 = 1 - \frac{1}{5} = \frac{4}{5}$ and $x_3 = 1 - \frac{1}{4} = \frac{3}{4}$. So, we apply the SG algorithm with these x_i and d_i and we obtain the billiard sequence $U = (2, 1, 2, 3, 1, 2, 1, 3, 2, 1, 3, 2, 1, 2, 1, 3, 1, 2, 3, 1)^\infty$. For this U we have that $I(u^1) = \frac{3}{20}$, $I(u^2) = \frac{1}{20}$ and $I(u^3) = \frac{1}{10}$. Hence $O(U) = \frac{3}{10} < \frac{3}{2} - 1 = \frac{1}{2}$ as expected.

Remark 6. We conjecture that for every consistent periodic billiard sequence U it holds that $O(U) \leq \frac{N}{2} - 1$ if $N > 2$.

For given rational densities we can prove that there is a consistent billiard sequence U with these densities which minimizes the unbalance among all integer sequences with the same densities. We show how to obtain a suitable mathematical programming problem (MPP) for minimizing the unbalance for given densities. The billiard sequence U is obtained by applying the SG algorithm to an optimal solution of this MPP. Recall the definition of the discrepancy function χ for infinite sequences of zeros and ones from section 2. Let $U \in \mathcal{Q}(d_1, d_2, \ldots, d_N)$ for some $d_1, d_2, \ldots, d_N \in \mathbb{R}_{>0}$ with $\sum_{i=1}^{N} d_i = 1$. Then we define

$$c_i = c_i(U) = \sup_{n \in \mathbb{Z}_{\geq 0}} \chi_{u^i}(n) \tag{13}$$

for $i = 1, 2, \ldots, N$. It is easily seen that if $U \in \mathcal{Q}(\{T\}, k_1, k_2, \ldots, k_N)$ for some $T, k_1, k_2, \ldots, k_N \in \mathbb{N}$ then $c_i = \max_{n \in \{0, 1, \ldots, T-1\}} \chi_{u^i}(n)$ for $i = 1, 2, \ldots, N$. From the following theorem it follows that minimizing the total unbalance $O(U)$ for given denisities $d_1, d_2, \ldots, d_N$ is equivalent to minimizing $\sum_{i=1}^{N} c_i(U)$.

Theorem 9. *Let $d_i \in \mathbb{Q}_{>0}$ for $i = 1, 2, \ldots, N$ be as in Lemma 3 and let $U \in \mathcal{Q}(d_1, d_2, \ldots, d_N)$. Then*

$$O(U) = \sum_{i=1}^{N} c_i(U) - \frac{N}{2} + \sum_{i=1}^{N} \frac{1}{2q_i} .$$

For given densities d_i with $\sum_{i=1}^{N} d_i = 1$ there exist $U \in \mathcal{Q}(d_1, d_2, \ldots, d_N)$ with $c_i(U) \leq z_i$ for $i = 1, 2, \ldots, N$ if and only if

$$t \geq \sum_{i=1}^{N} \max(\lceil t \cdot d_i - z_i \rceil, 0) \tag{14}$$

for $t = 0, 1, 2, \ldots$ (see [11] and [10]). Moreover, if the d_i and T are as in Lemma 3 then it suffices that (14) holds for $t = 0, 1, \ldots, T - 1$. Hence for these densities $\sum_{i=1}^{N} c_i(U)$ (and thus the total unbalance) can be minimized by solving the following MPP, where $z_1, z_2, \ldots, z_N$ are non-negative real variables.

$$\begin{array}{ll} \text{minimize} & \sum_{i=1}^{N} z_i \\ \text{subject to:} & \\ \sum_{i=1}^{N} \max(0, \lceil t \cdot d_i - z_i \rceil) \leq t & \text{for } t = 0, 1, \ldots, T - 1 \end{array} \tag{15}$$

From an optimal solution $z_1^*, z_2^*, \ldots, z_N^*$ of (15) a sequence U with minimal unbalance can be obtained by applying the SG algorithm with $x_i = z_i^*$ for $i = 1, 2, \ldots, N$. For this U it holds that $c_i(U) = z_i^*$ for $i = 1, 2, \ldots, N$. In [11] it is shown that there exist an optimal solution of (15) with $0 \le z_i^* < 1$ for $i = 1, 2, \ldots, N$ and then the obtained U is a periodic billiard sequence with period T. So, we may assume that $0 \le z_i < 1$ for $i = 1, 2, \ldots, N$ and then (14) can be replaced with $t \ge \sum_{i=1}^N \lceil t \cdot d_i - z_i \rceil$. Moreover, if $d_i = \frac{p_i}{q_i}$ with $\gcd(p_i, q_i) = 1$ then for an optimal solution $z_1^*, z_2^*, \ldots, z_N^*$ of (15) it holds that $z_i^* = \frac{k}{q_i}$ for some non-negative integer k. This can be used to transform (15) into an equivalent ILP as is done in [11]. Note that from $O(U) \ge 0$ for every sequence U and Theorem 9 it follows that the minimal value of (15) and the corresponding ILP is greater or equal than $\frac{N}{2} - \sum_{i=1}^N \frac{1}{2q_i}$. The LP relaxation of the ILP yields the same lower bound $\frac{N}{2} - \sum_{i=1}^N \frac{1}{2q_i}$ (see [11]) and also solving the Lagrange dual of (15) gives this lower bound. Thus the minimal total unbalance for given densities $d_1, d_2, \ldots, d_N$ is equal to the duality gap for (15).

References

1. Altman, E., Gaujal, B., Hordijk, A.: Admission control in stochastic event graphs. IEEE Trans. Automat. Control **45** (2000) 854–867
2. Altman, E., Gaujal, B., Hordijk, A.: Balanced sequences and optimal routing, J. Assoc. Comput. Mach. **47** (2000) 752–775
3. Altman, E., Gaujal, B., Hordijk, A.: Multimodularity, convexity and optimization properties. Math. Oper. Res. **25** (2000) 324–347
4. Arnoux, P., Mauduit, C., Shiokawa, I., Tamura, J.: Complexity of sequences defined by billiards in the cube. Bull. Soc. Math. France **122** (1994) 1–12
5. Arian, Y., Levy, Y.: Algorithms for generalized round Robin routing. Oper. Res. Lett. **12** (1992) 313–319
6. Gaujal, B., Hyon, E.: Optimal routing policy in two deterministic queues. Technical Report INRIA RR-3997 (2000). To appear in Calculateurs Parallèles
7. Gross, D., Harris, C. M.: Fundamentals of queueing theory. Wiley Series in Probability and Mathematical Statistics (1974)
8. Hajek, B.: The Proof of a Folk Theorem on Queueing Delay with Applications to Routing in Networks. J. ACM **30** (1983) 834–851
9. Hajek, B.: Extremal splittings of point processes. Math. Oper. Res. **10** (1985) 543–556
10. Hordijk, A., van der Laan, D. A.: Periodic routing to parallel queues with bounds on the average waiting time. Report MI no. 2000-44 Leiden University (2000). Submitted to Journal of Scheduling
11. van der Laan, D. A.: Routing jobs to servers with deterministic service times. Report MI no. 2000-20 Leiden University (2000). Submitted to Math. Oper. Res.
12. Liu, Z., Righter, R.: Optimal load balancing on distributed homogeneous unreliable processors. Oper. Res. **46** (1998) 563–573
13. Morse, M., Hedlund, G. A.: Symbolic dynamics II. Amer. J. Math. **60** (1940) 1–42
14. Shirakawa, H., Mori, M., Kijima, M.: Evaluation of Regular Splitting Queues. Comm. Statist. Stochastic Models **5** (1989) 219–234
15. Tijdeman, R.: Fraenkel's conjecture for six sequences. Discrete Math. **222** (2000) 223–234

Cutting Planes for Mixed 0-1 Semidefinite Programs

G. Iyengar[1] and M.T. Çezik[2]

[1] IEOR Dept, Columbia University, NY, NY, USA. `garud@ieor.columbia.edu`
[2] Bell Laboratories, Lucent Technologies, Holmdel, NJ, USA. `cezik@lucent.com`

1 Introduction

Since the seminal work of Nemirovski and Nesterov [14], research on semidefinite programming (SDP) and its applications in optimization has been burgeoning. SDP has led to good relaxations for the quadratic assignment problem, graph partition, non-convex quadratic optimization problems, and the TSP. SDP-based relaxations have led to approximation algorithms for combinatorial optimization problems such as the MAXCUT and vertex coloring. SDP has also found numerous applications in robust control and, as a natural extension, in robust optimization for convex programs with uncertain parameters. For a recent survey of semidefinite techniques and applications see [20].

In several of the optimization applications of SDPs binary variables naturally emerge. SDP relaxations for combinatorial optimization problems are typically obtained by relaxing binary constraints. Robust formulations of uncertain discrete problems also result in SDPs with binary variables. Another source of SDPs with binary variables is the semidefinite lift-and-project relaxation for mixed 0-1 linear programs (LP) [3,12]. Mixed 0-1 SDPs have the following general form,

$$\begin{aligned}
\min \ & c^T x \\
\text{subject to } & F_0 + \sum_{i=1}^{n} x_i F_i \succeq 0, \\
& x_i \in \{0,1\}, \quad i = 1, \dots, p,
\end{aligned} \qquad (1)$$

where $x \in \mathbf{R}^n$, $F_i = F_i^T \in \mathcal{S}^m$, the set of all symmetric $m \times m$ matrices, and $A \succeq 0$ denotes that the matrix A is positive semidefinite. Usually there are linear constraints, but these can be easily reformulated as semidefinite constraints. We will assume that the inequalities $0 \le x_j \le 1$ for $1 \le j \le p$ are present in the semidefinite constraint. The affine semidefinite constraint in (1) is called a linear matrix inequality (LMI).

Currently, the mixed 0-1 SDP (1) is approximately solved by relaxing the binary constraints. This is very similar to the early attempts to approximately solve integer LPs by relaxing the integer constraints. As the techniques for solving LPs became more efficient, there was a concerted effort to develop systematic means of improving the relaxation. We expect a similar development in the case of mixed 0-1 SDPs. The results in this paper are a first step in this direction.

We propose to solve (1) to optimality by the usual 2-step procedure: in the first step, the binary constraints are relaxed and the resulting SDP solved to

K. Aardal, B. Gerards (Eds.): IPCO 2001, LNCS 2081, pp. 251–263, 2001.

optimality. If the solution $\bar{x}$ is feasible for (1) then it must be optimal. Otherwise, one generates a convex constraint that is valid for the feasible set of the mixed 0-1 SDP (1) but is violated by $\bar{x}$, i.e. a convex "cut". This constraint is added to the convex relaxation and the steps repeated. In most cases, it is unlikely that cuts would be sufficient to solve mixed SDPs to optimality, and therefore, partitioning into subproblems via branching will be required.

In this paper, we develop several methods for generating linear cutting planes for the mixed 0-1 SDP. In §2 we develop the analog of the Chvátal-Gomory (CG) cutting plane method for mixed 0-1 SDPs. We show that the usual definition of the rank extends to these CG cuts and that all valid linear inequalities are dominated by CG inequalities. We also show that the CG procedure can be used to characterize the relative strengths of semidefinite and polyhedral relaxations of several combinatorial optimization problems. Next, we introduce matrix super-additive functions and show that CG cuts are a special case of matrix super-additive cuts.

In §3 we develop disjunctive cutting planes for mixed 0-1 SDPs by extending the work of Balas, Ceria and Cornuèjols [3]. To be successful, any cutting plane method must be able to efficiently generate the violated inequality. Balas et al [3] showed that, in the case of mixed 0-1 LPs, disjunctive cut generation amounts to solving an LP. We show that in the case of mixed 0-1 SDPs the disjunctive cuts can be efficiently generated by solving an SDP. We also show that the cuts can be generated in the space of fractional variables and then lifted. This is particularly relevant when the cuts are used in a branch-and-cut framework.

This approach can be interpreted as lift-and-project cut generation [16,12]. The lift-and-project method was previously generalized to mixed 0-1 convex programs by Stubbs and Mehrotra [17]. However, in this general setting, cut generation is not efficient and, as a consequence, the method does not yield an efficient optimization procedure.

In §4 we comment on the computational performance of disjunctive cuts.

2   Chvátal-Gomory Cutting Planes for Mixed 0-1 SDP

In this section we extend the Chvátal-Gomory cut generation procedure, developed in the context of mixed integer programs, to mixed 0-1 SDPs. Analogous to the case of integer programs, we first consider the case of pure 0-1 SDPs and then extend the results to mixed 0-1 SDPs.

Let $\mathcal{K}^\circ$ be the feasible set of a pure 0-1 SDP, i.e.

$$\mathcal{K}^\circ = \left\{ x \in \{0,1\}^n \,\middle|\, F_0 + \sum_{j=1}^n x_j F_j \succeq 0 \right\}, \qquad F_i = F_i^T \in S^m, i = 0, \ldots, n.$$

The Chvátal-Gomory cutting planes for $\mathcal{K}^\circ$ are based on the following equivalent representations of the feasible set of an LMI.

$$\left\{ x \in \mathbf{R}^n \,\middle|\, F_0 + \sum_{i=1}^n x_i F_i \succeq 0 \right\} = \left\{ x \in \mathbf{R}^n \,\middle|\, Z \bullet F_0 + \sum_{i=1}^n x_i (Z \bullet F_i) \geq 0, \quad \forall Z \succeq 0 \right\},$$

where $A \bullet B = \mathbf{Tr}(AB)$ is the usual inner product on the space of symmetric matrices. The above equivalence follows from the fact that the set of semidefinite matrices is self-dual with respect to this inner product.

The Chvátal-Gomory (CG) cutting plane procedure for $\mathcal{K}^\circ$ is as follows.

Step 1. For any $Z \succeq 0$, we have $Z \bullet F_0 + \sum_{i=1}^n x_i (Z \bullet F_i) \geq 0$ for all $x \in \mathcal{K}^\circ$.
Step 2. Since $x_i \geq 0$, $F_0 \bullet Z + \sum_{i=1}^n x_i \lceil Z \bullet F_i \rceil \geq 0$ for all $x \in \mathcal{K}^\circ$.
Step 3. Since $x \in \{0,1\}$, $\lfloor F_0 \bullet Z \rfloor + \sum_{i=1}^n x_i \lceil Z \bullet F_i \rceil \geq 0$ for all $x \in \mathcal{K}^\circ$.

A valid inequality of the form

$$\lfloor Z \bullet F_0 \rfloor + \sum_{i=1}^n x_i \lceil Z \bullet F_i \rceil \geq 0, \tag{2}$$

is called a CG cut for $\mathcal{K}^\circ$. Analogous to the CG procedure for integer programs, a CG cut for $\mathcal{K}^\circ$ can be combined with other CG cuts and the original LMI to generate new CG cuts. One can analogously define the *rank* of an CG inequality and get a hierarchy of CG cuts. The following result is a simple extension of the corresponding result for integer programs.

Lemma 1. *Define* $\mathcal{K}^\circ = \{x \in \{0,1\}^n \mid F_0 + \sum_{i=1}^n x_i F_i \succeq 0\}$. *Then, every valid inequality for* $\mathbf{conv}\,(\mathcal{K}^\circ)$ *is equivalent to or dominated by a CG cut.*

Proof. The result is established by reducing this problem to the polyhedral case. Let $\mathcal{K}_1$ be the feasible set of all rank-1 CG cuts with respect to the LMI, i.e.

$$\mathcal{K}_1 = \left\{ x \in \mathbf{R}^n \middle| \lfloor F_0 \bullet Z \rfloor + \sum_{i=1}^n x_i \lceil Z \bullet F_i \rceil \geq 0, \forall Z \succeq 0 \right\}.$$

First we show that $\mathcal{K}_1 \cap \{0,1\}^n = \mathcal{K}^\circ$. Clearly $\mathcal{K}_1 \cap \{0,1\}^n \supseteq \mathcal{K}^\circ$. To prove the other direction, let $\bar{x} \in (\mathcal{K}_1 \cap \{0,1\}^n) \backslash \mathcal{K}^\circ$, i.e. $\bar{x} \notin \{x \in \mathbf{R}^n \mid F_0 + \sum_{i=1}^n x_i F_i \succeq 0\}$. Therefore, there exists $\bar{Z} \succeq 0$ such that $d = -(\bar{Z} \bullet F_0 + \sum_{i=1}^n \bar{x}_i (\bar{Z} \bullet F_i)) > 0$. Since $\bar{x} \in \mathcal{K}_1$ and $k\,\bar{Z} \succeq 0$ for all $k \geq 0$, it follows that for all $k \geq 0$,

$$\lfloor k\bar{Z} \bullet F_0 \rfloor + \sum_{i=1}^n \bar{x}_i \lceil k\bar{Z} \bullet F_i \rceil \geq 0.$$

On the other hand, we have

$$\lfloor k\bar{Z} \bullet F_0 \rfloor + \sum_{i=1}^n \bar{x}_i \lceil k\bar{Z} \bullet F_i \rceil = k \left(\bar{Z} \bullet F_0 + \sum_{i=1}^n \bar{x}_i (\bar{Z} \bullet F_i) \right)$$
$$- \left(((k\bar{Z}) \bullet F_0)_f - \sum_{i=1}^n \bar{x}_i (1 - ((k\bar{Z}) \bullet F_i)_f) \right),$$
$$\leq -kd + n,$$

where $(a)_f = a - \lfloor a \rfloor$ is the fractional part of a. A contradiction. Therefore, $\mathcal{K}_1 \cap \{0,1\}^n = \mathcal{K}^\circ$. Now, the lemma follows by recognizing that $\mathcal{K}_1$ has the same structure as the feasible set of all rank-1 CG cuts of an integer program. $\qquad\square$

The CG procedure has interesting consequences for comparing LP and SDP relaxations for combinatorial optimization problems. For instance, consider the traveling salesman problem (TSP). A simple extension of the results in [8] shows that the TSP on n nodes can be formulated as the following 0-1 SDP [6],

$$
\begin{array}{l}
\text{minimize } \sum_{i,j=1}^{n} C_{ij} A_{ij}, \\
\text{subject to } A\mathbf{1} = 2\mathbf{1}, \quad \mathbf{diag}(A) = \mathbf{0}, \\
\qquad h_n \mathbf{I} + \frac{2-h_n}{n} \mathbf{1}\mathbf{1}^T - A \succeq 0, \\
\qquad A \in \{0,1\}^{n \times n},
\end{array}
\tag{3}
$$

where A denotes the node-node adjacency matrix of a graph and $h_n = 2\cos\left(\frac{2\pi}{n}\right)$. The sub-tour elimination constraints for the TSP on n nodes are given by

$$
\sum_{i,j \in W} A_{ij} \le |W| - 1, \quad \forall\, W \subset \{1,\dots,n\}, \quad 3 \le |W| \le n-1. \tag{4}
$$

Lemma 2. *The sub-tour elimination constraints for the TSP on a graph $\mathcal{G} = (V, E)$, with $|V| > 4$, are rank-1 CG cuts with respect to the constraints,*

$$
h_n \mathbf{I} + \left(\frac{2-h_n}{n}\right) \mathbf{1}\mathbf{1}^T - A \succeq 0, \quad A\mathbf{1} = 2\mathbf{1}, \quad \mathbf{diag}(A) = \mathbf{0}, \quad 0 \le A \le \mathbf{1}\mathbf{1}^T. \tag{5}
$$

Proof. Define $Z = \frac{1}{2}\mathbf{1}_W \mathbf{1}_W^T \succeq 0$ where $\mathbf{1}_W$ is the indicator vector of the set W. On applying the CG procedure with Z, we have

$$
Z \bullet A = \sum_{i,j \in W} A_{ij} \le \left\lfloor Z \bullet \left(h_n I + \frac{2-h_n}{n} \mathbf{1}\mathbf{1}^T\right) \right\rfloor. \tag{6}
$$

The constant term

$$
\begin{aligned}
Z \bullet \left(h_n I + \frac{2-h_n}{n}\mathbf{1}\mathbf{1}'\right) &= h_n \frac{1}{2}|W| + \frac{2-h_n}{2n}|W|^2, \\
&= |W|\left[\left(1 - \frac{h_n}{2}\right)\frac{|W|}{n} + \frac{h_n}{2}\right], \\
&< |W|,
\end{aligned}
\tag{7}
$$

where (7) follows from the fact that $(1 - h_1/2) = 1 - \cos(2\pi/n) > 0$ and $|W| < n$. Therefore, the CG rank of the sub-tour inequality is at most 1.

Suppose the sub-tour elimination inequality I_W corresponding to a subset W is a rank-0 CG cut with respect to (5). Let A^i, $i = 1, 2$, be the adjacency matrices of two distinct tours that satisfy I_W with equality. Then so does $\bar{A} = \frac{1}{2}(A^1 + A^2)$.

Let $\lambda_{(i)}(A)$ denote the i-th smallest eigenvalue of $(2I - A)$. The linear constraints in (5) imply that $\lambda_{(1)}(2I - A) = \lambda_{\min}(2I - A) = 0$ with $\frac{1}{\sqrt{n}}\mathbf{1}$ as the corresponding eigenvector. As a consequence the LMI in (3) can be rewritten as

$$
L_2(2I - A) \stackrel{\Delta}{=} \lambda_{(1)}(2I - A) + \lambda_{(2)}(2I - A) \ge 2 - h_n.
$$

Since A^i are tours, we have $L_2(2I - A^i) = (2 - h_n)$ [11]. The function L_2 is a strictly concave [1], therefore $L_2(2I - \bar{A}) > (2 - h_n)$. Let $\alpha = \frac{(2 - h_n)}{L_2(2I - \bar{A})} \in (0, 1)$ and A^w denote the adjacency matrix of the graph consisting of tours on the sets W and $V \backslash W$. Clearly, A^w violates I_W and so does $A_\alpha = \alpha \bar{A} + (1 - \alpha) A^w$. However, A_α satisfies the linear constraints in (5) and, since $L_2(2I - A^w) = 0$, we have

$$L_2(2I - A_\alpha) \geq \alpha L_2(2I - \bar{A}) + (1 - \alpha) L_2(2I - A^w) = \alpha L_2(2I - \bar{A}) = 2 - h_n.$$

A contradiction. □

Lemma 2 shows that (3) is strictly weaker than the linear relaxation with all the sub-tour elimination inequalities. This result strengthens an earlier result of Goemans and Rendl [9]. The following results are established similarly.

Lemma 3. *The clique constraints, the odd hole constraints and odd anti-hole constraints are rank-1 CG cuts with respect to the LMI in $TH(G)$ [12].*

Lemma 4. *All triangular inequalities for max-cut are rank-1 CG cuts with respect to the LMI in the Goemans-Williamson relaxation [10].*

Lemmas 2–4 assert that a single LMI "closely" approximates exponentially many inequalities in the sense that rank-1 CG cuts of the LMI subsume all the inequalities. We expect that the feasible set of the LMI will be a good relaxation for the feasible set of the exponentially many linear inequalities. Such an LMI representation would be particularly useful if there was no polynomial time separation algorithm for the class of linear inequalities.

The CG cuts can be extended to mixed 0-1 SDPs as follows.

Lemma 5. *Let $\mathcal{K}^\circ = \left\{ x \in \{0, 1\}^n, y \in \mathbf{R}_+^p \middle| F_0 + \sum_{i=1}^n x_i F_i + \sum_{j=1}^p y_j G_j \succeq 0 \right\}$, where $F_i, G_j \in S^m$. For any $Z \succeq 0$, the linear inequality*

$$\lfloor Z \bullet F_0 \rfloor + \sum_{i=1}^n x_i \lceil Z \bullet F_i \rceil + \frac{1}{1 - f_0} \sum_{j \in J^+} y_j (Z \bullet G_j) \geq 0,$$

where $J^+ = \{j \mid (Z \bullet G_j) > 0\}$ and $f_0 = (Z \bullet F_0) - \lfloor Z \bullet F_0 \rfloor$, is valid for $\mathcal{K}^\circ$.

All known results for CG cuts, except those that relate to facets, extend to mixed 0-1 SDPs. Next, we generalize this procedure via *nondecreasing matrix-super-additive functions*.

A function $\mathcal{F} : S^m \to S^q$ is non-decreasing and super-additive if it has the following properties:

(a) $A \succeq B \Rightarrow \mathcal{F}(A) \succeq \mathcal{F}(B)$,
(b) for all $A, B \in S^n$, $\mathcal{F}(A + B) \succeq \mathcal{F}(A) + \mathcal{F}(B)$

Lemma 6. *Let $\mathcal{F}$ is nondecreasing and super-additive matrix function, then $\mathcal{F}(F_0) - \sum_{i=1}^{n} x_i \mathcal{F}(-F_i) \succeq 0$ is a valid linear matrix inequality for $\mathcal{K}^\circ = \{x \in \{0,1\}^n \mid F_0 + \sum_{i=1}^{n} x_i F_i \succeq 0\}$.*

The CG cuts correspond to $\mathcal{F}(A) = \lfloor Z \bullet A \rfloor$. Some other useful functions are $\mathcal{F}(A) = \lfloor \lambda_{\min}(A) \rfloor$ and $\mathcal{F}(A) = \lfloor \lambda_{\min}(A(\mathcal{B})) \rfloor$ where $A(\mathcal{B})$ is a diagonal block. Unfortunately the range of these function is $\mathbf{R}$, i.e. the resulting LMI is a linear inequality. Matrix super-additive functions that map to higher dimensional symmetric matrices are unknown.

The super-additive cuts can also be extended to mixed 0-1 SDPs. Corresponding to a super-additive function $\mathcal{F}$, define the positive homogeneous function $\bar{\mathcal{F}}$ as follows,

$$\bar{\mathcal{F}}(A) = \lim_{\lambda \searrow 0^+} \frac{\mathcal{F}(\lambda A)}{\lambda}.$$

The following lemma follows from corresponding result for mixed 0-1 LPs.

Lemma 7. *Let $\mathcal{K}^\circ = \left\{ x \in \{0,1\}^n, y \in \mathbf{R}_+^p \,\middle|\, F_0 + \sum_{i=1}^{n} x_i F_i + \sum_{j=1}^{p} y_j G_j \succeq 0 \right\}$, where $F_i, G_j \in S^m$. For any nondecreasing super-additive function $\mathcal{F}$, the linear inequality*

$$\mathcal{F}(F_0) - \sum_{i=1}^{n} x_i \mathcal{F}(-F_i) - \sum_{j=1}^{p} y_j \bar{\mathcal{F}}(-G_j) \succeq 0,$$

is valid for $\mathcal{K}^\circ$.

3   Disjunctive Cutting Planes for Mixed 0-1 SDP

Let $\mathcal{K}$ denote the convex set obtained by relaxing the 0-1 constraints in $\mathcal{K}^\circ$, i.e.

$$\mathcal{K} = \left\{ x \in \mathbf{R}^n \,\middle|\, F(x) = F_0 + \sum_{i=1}^{n} x_i F_i \succeq 0 \right\}.$$

A lift-and-project procedure for the mixed 0-1 SDP is as follows [3,16].

Step 0. Select an index $j \in \{1, \dots, p\}$.
Step 1. Multiply $F(x) \succeq 0$ by x_j and $(1 - x_j)$ to obtain the nonlinear system,

$$\begin{aligned} x_j(F_0 + \sum_{i=1}^{n} x_i F_i) &\succeq 0 \\ (1 - x_j)(F_0 + \sum_{i=1}^{n} x_i F_i) &\succeq 0 \end{aligned} \tag{8}$$

Step 2. Linearize (8) by substituting $y_i = x_j x_i$, for $i \neq j$, $y_j = 0$ and x_j for x_j^2. The linear system in (x, y) is as follows,

$$\begin{aligned} (F_0 + F_j)x_j + \sum_{i=1}^{n} y_i F_i &\succeq 0, \\ \sum_{i=1}^{n} x_i F_i - (F_0 + F_j)x_j - \sum_{i=1}^{n} y_i F_i + F_0 &\succeq 0. \end{aligned} \tag{9}$$

Denote the set $\{(x, y) \in \mathbf{R}^{2n} \mid (x, y) \text{ satisfy } (9)\}$ by $\mathcal{M}_j(\mathcal{K})$.

Step 3. Project the $M_j(\mathcal{K})$ onto the x space. Call the projection $P_j(\mathcal{K})$, i.e.

$$P_j(\mathcal{K}) = \{x \mid (x, y) \in M_j(\mathcal{K})\}. \tag{10}$$

The properties of this procedure are summarized in the lemma below.

Lemma 8. *For any* $j \in \{1, \ldots, p\}$, $P_j(\mathcal{K}) =$ **conv** $(\mathcal{K} \cap \{x \in \mathbf{R}^n \mid x_j \in \{0, 1\}\})$. *For all* $j_1, j_2 \in \{1, \ldots, n\}$, $j_1 \neq j_2$,

$$P_{j_1}(P_{j_2}(\mathcal{K})) = P_{j_2}(P_{j_1}(\mathcal{K})) = \mathbf{conv}\,(\mathcal{K} \cap \{x_{j_1}, x_{j_2} \in \{0, 1\}\}).$$

More generally, for $t \in \{1, \ldots, p\}$,

$$P_{j_1, \ldots, j_t}(\mathcal{K}) = \mathbf{conv}\,(\mathcal{K} \cap \{x \in \mathbf{R}^n \mid x_j \in \{0, 1\}, j = j_1, \ldots, j_t\}).$$

The following corollary immediately follows.

Corollary 1. $P_{1, \ldots, p}(\mathcal{K}) = \mathbf{conv}\,(\mathcal{K} \cap \{x \mid x_i \in \{0, 1\}, 1 \leq i \leq p\}) = $ **conv** $(\mathcal{K}^\circ)$.

These properties are established by extending the disjunctive programming results for linearly constrained sets [2,3] to sets described by LMIs.

Theorem 1. *Let* $\mathcal{K} = \{x \in \mathbf{R}^n \mid F(x) = F_0 + \sum_{i=1}^n x_i F_i \succeq 0\}$. *Then* $x \in P_j(\mathcal{K})$ *if and only if there exists* $(y, y_0), (z, z_0) \in \mathbf{R}^{n+1}$ *such that*

$$\begin{aligned}
y_0 F_0 + \sum_{i=1} y_i F_i &\succeq 0, \\
z_0 F_0 + \sum_{i=1} z_i F_i &\succeq 0, \\
y_j &= 0, \\
z_j &= z_0, \\
y_0 + z_0 &= 1.
\end{aligned}$$

Proof. This representation follows by suitably extending the disjunctive programming results in [2,3].

From Corollary 1 it follows that p iterations of the lift-and-project procedure gives the convex hull **conv** $(\mathcal{K}^\circ)$, i.e. the solution set of the mixed SDP. Therefore, in principle, this procedure solves the mixed 0-1 SDP. However, each projection step could introduce an exponential number of LMIs, and therefore, the method is not computationally viable.

The alternative is to use the disjunctive representation in Theorem 1 to generate cuts separating the current iterate $\bar{x}$ from **conv** $(\mathcal{K}^\circ)$. If $\bar{x}$ is an extreme point of $\mathcal{K}$ and $0 < \bar{x}_j < 1$ for some $1 \leq j \leq p$ then $\bar{x} \notin P_j(\mathcal{K})$; hence, there exists a linear inequality valid for $P_j(\mathcal{K}) \supseteq \mathbf{conv}\,(\mathcal{K}^\circ)$ that is violated by $\bar{x}$. We show that finding the deepest linear cut is equivalent to solving an SDP.

3.1   Disjunctive Cutting Planes for Mixed 0-1 SDP

For a cutting plane method to be successful, the violated inequalities have to efficiently generated. We show that the set $\mathcal{P}_j(\mathcal{K})^*$ of all linear inequalities valid for
the projected set $\mathcal{P}_j(\mathcal{K})$ is described by LMIs and, consequently, cut generation
reduces to an SDP.

Let $\mathbf{vec} : \mathbf{R}^{m \times m} \rightarrow \mathbf{R}^{m^2}$ denote the function that maps a matrix $A \in \mathbf{R}^{m \times m}$
to the vector $a \in \mathbf{R}^{m^2}$ obtained by stacking the columns of matrix into a single
long vector. Define $f_i = \mathbf{vec}(F_i)$ for $i = 0, \ldots, n$, and $F \in \mathbf{R}^{n \times m^2}$ as follows,

$$
F = \begin{bmatrix} f_1^T \\ \vdots \\ f_n^T \end{bmatrix}. \tag{11}
$$

Theorem 2. *If* $\mathrm{int}((\mathbf{conv}\,(\mathcal{K}^\circ))) \neq \emptyset$, *then a linear inequality* $\alpha^T x \geq \beta$ *is
valid for* $\mathcal{P}_j(\mathcal{K})$, *i.e.* $(\alpha, \beta) \in \mathcal{P}_j(\mathcal{K})^*$, *if and only if there exist* $u_0, v_0 \geq 0$, *and*
$U, V \succeq 0$, *such that*

$$
\begin{aligned}
\alpha - u_0 e_j \qquad -Fu \qquad\quad &= 0, \\
\alpha \qquad -v_0 e_j^T \qquad -Fv &= 0, \\
f_0^T u \qquad\quad &= \beta, \\
+v_0 \qquad -f_0^T v &= \beta,
\end{aligned} \tag{12}
$$

where $u = \mathbf{vec}(U)$, $v = \mathbf{vec}(V)$, $f_0 = \mathbf{vec}(F_0)$ *and* $F \in \mathbf{R}^{n \times m^2}$ *is given by (11).*

Proof. The result follows by semidefinite programming duality. The condition
$\mathrm{int}((\mathbf{conv}\,(\mathcal{K}^\circ))) \neq \emptyset$ ensures that strong duality holds.   □

In Theorem 2, the Slater-type condition $\mathrm{int}(\mathbf{conv}\,(\mathcal{K}^\circ)) \neq \emptyset$ can be replaced by
any condition that ensures strong duality.

If $\bar{x} \notin \mathcal{P}_j(\mathcal{K})$, then, by minimum norm duality, the separation of $\bar{x}$ from
$\mathcal{P}_j(\mathcal{K})$ is given by,

$$
\begin{aligned}
&\text{maximize } \beta - \alpha^T \bar{x}, \\
&\text{subject to } (\alpha, \beta) \in \mathcal{P}_j^*(\mathcal{K}), \\
&\qquad\qquad \|\alpha\| \leq 1,
\end{aligned} \tag{13}
$$

where $\|\alpha\| = \sqrt{\alpha^T \alpha}$. The optimal solution (α^*, β^*) of (13) is the "deepest" linear
cut. The norm constraint $\|\alpha\| \leq 1$ is equivalent to the following linear matrix
inequality,

$$
\begin{bmatrix} 1 & \alpha^T \\ \alpha & I \end{bmatrix} \succeq 0.
$$

Thus, Theorem 2 implies that (13) is equivalent to the following SDP,

$$
\begin{aligned}
\text{maximize } & \beta - \alpha^T \bar{x}, \\
\text{subject to } & \alpha - u_0 e_j & -Fu & = 0, \\
& \alpha \quad -v_0 e_j^T & -Fv &= 0, \\
& & f_0^T u &= \beta, \\
& +v_0 & -f_0^T v &= \beta, \\
& \begin{bmatrix} 1 & \alpha^T \\ \alpha & I \end{bmatrix} \succeq 0, \quad U \succeq 0, \quad V \succeq 0.
\end{aligned}
\tag{14}
$$

In our computations, we solved (14) as a combination of linear matrix inequalities and a second-order cone constraint [14].

In the case of mixed 0-1 LP, the iterate $\bar{x}$ is always an extreme point of the convex relaxation $\mathcal{K}$ and, therefore, if $0 < \bar{x}_j < 1$ then $\bar{x} \notin P_j(\mathcal{K})$. However, in the case of mixed 0-1 SDP, the convex relaxation is solved using an interior point method that stops before reaching the boundary, i.e. $\bar{x}$ is, typically, not extreme, and, therefore, one might be unable to produce any cuts. A remedy is to use Pataki's algorithm [15] to generate an extreme point. However, in our computational experiments, we worked with the approximate optimal solution $\bar{x}$ of the SDP relaxations and chose j to be any index such that $0 < \bar{x}_j < 1$ without encountering any numerical difficulties.

Unlike in mixed 0-1 LP, here the dual set $P_j(\mathcal{K})^*$ does not have a finite number of extreme points. Therefore, this cutting plane algorithm will not, in general, be finitely terminating and one would have to resort to branching.

3.2   Lifting from the Space of Fractional Variables

Consider a node of the branch-and-cut tree with E as the index set of fractional components of $\bar{x}$. Then, without loss of generality, one can assume that $\bar{x}_j = 0$ for all $j \notin E$. Modify the constraint matrices in the cut generation SDP (14) by excluding all variables except those in E, i.e. remove all rows from F corresponding to $f_k^T, k \in \{1, \ldots, n\} \setminus E$. Denote the new dual feasible set by $P_j^*(K)^E$. Then, the modified SDP is given by,

$$
\begin{aligned}
\text{maximize } & \beta - \alpha_E^T \bar{x}_E, \\
\text{subject to } & (\alpha_E, \beta) \in P_j^*(\mathcal{K})^E, \\
& \|\alpha_E\| \leq 1,
\end{aligned}
\tag{15}
$$

The following lemma establishes the main result of this section.

Lemma 9. *Let $(\bar{\alpha}^E, \bar{\beta}^E)$ be an optimal solution for the modified SDP (15). Then, $(\bar{\alpha}^E, \bar{\beta}^E)$ can be complemented to an optimal solution $(\bar{\alpha}, \bar{\beta})$ of the following SDP,*

$$
\begin{aligned}
\text{maximize } & \beta - \alpha^T \bar{x}, \\
\text{subject to } & (\alpha, \beta) \in P_j^*(\mathcal{K}), \\
& \|\alpha_E\| \leq 1,
\end{aligned}
\tag{16}
$$

Proof. The result follows by modifying the proof of Theorem 3.2 in [3]. □

Table 1. Computational results for the TSP

Instance	SDP gap (%)	(%) gap closed		(%) gap	
		25 cuts	50 cuts	B-and-B	B-and-C
burma14	4.45	20.3	25.0	0.00	0.00
ulysses16	7.30	5.7	6.3	2.92	1.42
gr17	13.2	12.0	13.8	9.88	9.06
ulysses22	9.78	7.70	7.85	7.27	0.00
gr24	3.30	19.05	21.43	23.42	7.20
fri26	5.12	5.26	7.02	17.71	2.45
bayg29	3.54	12.50	13.89	4.53	4.53
bays29	3.56	12.50	13.89	17.02	9.60
Random $n = 16$	11.27	31.25	31.25	4.93	0.00
$n = 16$	3.11	50.00	100.00	7.77	0.00
$n = 16$	1.33	100.00	100.00	0.00	0.00
Random $n = 20$	6.84	38.46	46.15	6.84	5.78
$n = 20$	6.70	42.86	42.86	18.66	6.70
$n = 20$	6.77	44.45	55.56	6.77	5.26
Random $n = 25$	3.39	50.00	66.67	16.95	0.00
$n = 25$	4.24	60.00	80.00	20.95	0.00
$n = 25$	3.39	80.00	90.00	4.75	2.03
Random $n = 30$	0.69	100.00	100.00	2.76	0.00
$n = 30$	0.86	100.00	100.00	2.59	0.00
$n = 30$	0.77	50.00	100.00	3.07	0.00
Random $n = 35$	1.04	100.00	100.00	7.81	0.00
$n = 35$	0.51	100.00	100.00	3.54	0.00
$n = 35$	1.39	75.00	100.00	21.53	0.00

Notice that the cut obtained by lifting, although a supporting hyper-plane, may not be optimal for (14). If α is normalized using the $\mathcal{L}_1$-norm the results of Balas et al [3,4] can be extended to establish that the lifted cut is optimal for the general problem.

4   Computational Results

In this section we comment on our preliminary computational experiments with solving mixed 0-1 SDPs using disjunctive cuts and a rudimentary branch-and-cut method. The branch-and-cut code does not employ warm starts – in fact it is not clear how to warm start SDP solvers – and it does not employ any cut lifting. The codes are written in MATLAB and use SeDuMi [18] to solve the semidefinite and second-order cone programs encountered in the branch-and-cut procedure. We tested the pure cutting plane and branch-and-cut algorithms on random instances of the TSP and MAXCUT problems, as well as instances of the TSP taken from the TSPLIB [19].

4.1   Computational Results for the Traveling Salesman Problem

We attempted to solve the 0-1 SDP formulation for the TSP given in (3) to optimality [8,6].

In the cutting plane algorithm, at every iteration of the cut generation procedure, we chose the "most" fractional variable, i.e. one closest to 0.5, as our disjunction variable and generated only one disjunctive cut in every iteration. The algorithm was run for a maximum of 50 cuts. The results of our experiments with TSP instances are summarized in Table 1. The first column lists the name of the TSP instance – the first eight are instances from the TSPLIB and the rest random. The second column lists the integrality gap of the SDP relaxation. Columns 3 and 4 list the fraction of the gap closed after the addition of 25 and 50 cuts respectively. There was, on average, a 48.47% reduction in the integrality gap with 25 cuts, whereas an additional 25 cuts led to an additional reduction of only 9.02%. The cuts were not facet defining for any choice of disjunctive variables. In fact, the cuts tend to become almost parallel to the objective function as the number of cuts increases, and the performance degrades rapidly. The cutting plane method performs considerable better on random instances in comparison to the benchmark instances from the TSPLIB.

Columns 5 and 6 display the relative gaps of the solutions obtained from a pure branch-and-bound (B-B) and the branch-and-cut (B-C) procedure respectively. The experiments were stopped after exactly 50 nodes were searched unless the optimal solution was encountered earlier. We employed a depth-first search branching rule and branched on the fractional variable closest to 1. At every node we also maintained a feasible tour by a simple rounding procedure. The gaps reported correspond to the best tour encountered in the course of the algorithm. In the B-C procedure we branched whenever the addition of the cut did not result in at least a 0.1% improvement in the objective. Even with a limit of 50 nodes, the B-C code produces good results – it solves almost all random instances and produces good results for the TSPLIB instances. Also, the B-C code are significantly better than the B-B code – especially in case of random TSP instances. Since the number of branching nodes was restricted, the performance of both the B-B and B-C is poor when the gap of the initial SDP relaxation is large. This partially explains the poor performance on the TSPLIB instances.

4.2   Computational Results for MAXCUT

We solved the following 0-1 SDP formulation of the MAXCUT to optimality using the disjunctive cutting plane method.

$$\text{maximize } \sum_{i,j=1}^{n} W_{ij}(1 - X_{ij}),$$
$$\text{subject to } \text{diag}(X) = 1,$$
$$X \succeq 0,$$
$$X_{ij} \in \{-1, +1\},$$

where W is the Laplacian matrix of weights [10]. The results of our computational experiments are summarized in Table 2. The MAXCUT instances were

Table 2. Computational results for MAXCUT.

Instance	SDP gap (%)	Number of cuts
Random $n = 5$	2.81	36
$n = 5$	2.13	21
Random $n = 10$	1.87	217
$n = 10$	1.29	153
$n = 10$	0.81	164
$n = 10$	1.63	49
Random $n = 15$	0.63	261
$n = 15$	1.28	375
Random $n = 20$	1.14	297
$n = 20$	2.14	334
Random $n = 40$	3.39	450
$n = 40$	4.40	532

generated randomly with the weights chosen independently and identically from a uniform distribution on $[1, 100]$. The first column in Table 2 lists the size of the instance and the second column displays the integrality gap of the SDP relaxation. Column 3 lists the number of disjunctive cuts needed to solve the instance to optimality. As can be seen from the table, all the instances were solved to optimality with relatively modest number of cuts. This is a consequence of the fact that a single disjunctive cut was often enough to drive the disjunction variable to one of its boundaries in the next iteration. This is almost never the case in TSP instances. The total number of cuts required to solve the problem goes up with the size of the problem and the integrality gap of the initial SDP relaxation. The rate of improvement in the objective function decreases as more cuts are generated and on termination only 40% of the cuts are active.

4.3   Comments on Computation

An obvious criticism of the computational results is the size of the test problems. The problem sizes are limited primarily because of two computational issues. The first is the complexity of the Cholesky factorization in the interior point solver. In the worst case, the complexity of this step is $O(n^6)$, where n is the number of nodes in the graph. However, for problems where the constraints are relatively sparse – as is often the case with combinatorial problems – one could use sparse matrix techniques and iterative methods to speed up this step [5,7]. Such techniques could allow one to solve large mixed 0-1 SDPs provided the constraint matrices are in the appropriate classes.

The second issue is the absence of warm start techniques for the SDP solver. This severely limits the number of nodes in the branch-and-cut tree which, in turn, limits the problem size, since the typically large integrality gap of the initial SDP relaxation for large problems requires one to search over a larger number of branch-and-cut nodes. There is some hope that the recent work on warm start methods for interior point solvers for LP [13] might lead to warm start methods for SDP solvers. Another possibility is to do away with the interior point methods altogether and develop a simplex-type method for conic programs [15]. In summary, these two computational issues have to be suitably resolved for a branch-and-cut method for mixed 0-1 SDPs to be successful.

References

1. F. Alizadeh. Interior point methods in semidefinite programming with applications to combinatorial optimization. *SIAM J. Optim.*, 5(1):13–51, 1995.
2. E. Balas. Disjunctive programming. *Annals of Discrete Mathematics*, 5:3–51, 1979.
3. E. Balas, S. Ceria, and G. Cornuèjols. A list-and-project cutting plane algorithm for mixed 0-1 programs. *Mathematical Programming*, 58:295–324, 1993.
4. E. Balas, S. Ceria, and G. Cornuèjols. Mixed 0-1 programming by lift-and-project in a branch-and-cut framework. *Management Science*, 42:1229–1246, 1996.
5. S. Benson, Y. Ye, and X. Zhang. Solving large-scale sparse semidefinite programs for combinatorial optimization. Tech. Rep., Dept of Mgmt Sc., Univ. of Iowa, 1997.
6. M. T. Çezik. *Semidefinite methods for the traveling salesman and other combinatorial problems*. PhD thesis, IEOR Dept., Columbia University, 2000.
7. C. Choi and Y. Ye. Solving sparse semidefinite programs using dual scaling algorithm with an iterative solver. Tech. Rep., Dept. of Mgmt. Sc., Univ. of Iowa, 2000.
8. D. Cvetković, M. Čangalović, and V. Kovačević-Vujčić. Semidefinite programming methods for the symmetric TSP. In *IPCO VII*, p. 126–136. Springer, Berlin, 1999.
9. M. Goemans and F. Rendl. Combinatorial optimization. In *Handbook of Semidefinite Programming: Theory, Algorithms and Applications*. Kluwer, 2000.
10. M. Goemans and D. P. Williamson. Improved approximation algorithms for maximum cut and satisfiability problems using semidefinite programming. *Journal of ACM*, 42:1115–1145, 1995.
11. L. Lovász. *Combinatorial Problems and Exercises*. North-Holland, 1979.
12. L. Lovász and A. Schrijver. Cones of matrices and set-functions and 0-1 optimization. *SIAM Journal of Control and Optimization*, 1(2):166–190, 1991.
13. J. E. Mitchell and B. Borchers. Solving linear ordering problems with a combined interior point/simplex cutting plane algorithm. In *High performance optimization*, pages 349–366. Kluwer Acad. Publ., Dordrecht, 2000.
14. Y. Nesterov and A. Nemirovskii. *Interior-point polynomial algorithms in convex programming*. SIAM, Philadelphia, 1993.
15. G. Pataki. Cone-lp's and semidefinite programs : geometry and simplex-type method. In *Lecture Notes in Computer Science*, v. 1084, p. 162–174, 1996.
16. H. Sherali and W. Adams. A heirarchy of relaxations and convex hull representations for mixed zero-one programming problems. Tech. Rep., Virginia Tech., 1989.
17. R. A. Stubbs and S. Mehrotra. A branch-and-cut method for 0-1 mixed convex programming. *Mathematical Programming*, 86:515–532, 1999.
18. J. Sturm. Using SeDuMi 1.02, a MATLAB toolbox for optimization over symmetric cones. *Optimization Methods and Software*, 11–12:625–653, 1999.
19. TSPLIB: A library of sample instances for the TSP and related problems. Available at http://www.crpc.rice.edu/softlib/tsplib/.
20. W. Wolkowicz, R. Saigal, and L. Vandenberghe, editors. *Handbook of Semidefinite Programming*. Kluwer Acad. Publ., 2000.

Independence Free Graphs and Vertex Connectivity Augmentation

Bill Jackson[1*] and Tibor Jordán[2**]

[1] Department of Mathematical and Computing Sciences, Goldsmiths College, London SE14 6NW, England. e-mail: b.jackson@gold.ac.uk
[2] Department of Operations Research, Eötvös University, Kecskeméti utca 10-12, 1053 Budapest, Hungary. e-mail: jordan@cs.elte.hu

Abstract. Given an undirected graph G and a positive integer k, the k-vertex-connectivity augmentation problem is to find a smallest set F of new edges for which $G + F$ is k-vertex-connected. Polynomial algorithms for this problem have been found only for $k \leq 4$ and a major open question in graph connectivity is whether this problem is solvable in polynomial time in general.

In this paper we develop the first algorithm which delivers an optimal solution in polynomial time for every fixed k. In the case when the size of an optimal solution is large compared to k, we also give a min-max formula for the size of a smallest augmenting set.

A key step in our proofs is a complete solution of the augmentation problem for a new family of graphs which we call k-independence free graphs. We also prove new splitting off theorems for vertex connectivity.

1   Introduction

An undirected graph $G = (V, E)$ is *k-vertex-connected* if $|V| \geq k + 1$ and the deletion of any $k - 1$ or fewer vertices leaves a connected graph. Given a graph $G = (V, E)$ and a positive integer k, the k-vertex-connectivity augmentation problem is to find a smallest set F of new edges for which $G' = (V, E \cup F)$ is k-connected. This problem (and a number of versions with different connectivity requirements and/or edge weights) is an important and well-studied optimization problem in network design. The complexity of the vertex-connectivity augmentation problem is one of the most challenging open questions of this area. It is open even if the graph G to be augmented is $(k - 1)$-vertex-connected. Polynomial algorithms have been developed only for $k = 2, 3, 4$ by Eswaran and Tarjan [2], Watanabe and Nakamura [13] and Hsu [6], respectively.

* This research was carried out while the first named author was visiting the Department of Operations Research, Eötvös University, Budapest, supported by an Erdős visiting professorship from the Alfréd Rényi Mathematical Institute of the Hungarian Academy of Sciences.
** Supported by the Hungarian Scientific Research Fund, grant no. T029772, T030059, F034930 and FKFP grant no. 0607/1999.

K. Aardal, B. Gerards (Eds.): IPCO 2001, LNCS 2081, pp. 264–279, 2001.
© Springer-Verlag Berlin Heidelberg 2001

In this paper we give an algorithm which delivers an optimal solution in polynomial time for any fixed $k \geq 2$. Its running time is bounded by $O(n^5) + O(f(k)n^3)$, where n is the size of the input graph and $f(k)$ is an exponential function of k. We also obtain a min-max formula which determines the size of an optimal solution when it is large compared to k. In this case the running time of the algorithm is simply $O(n^5)$. A key step in our proofs is a complete solution of the augmentation problem for a new family of graphs which we call k-independence free graphs. We follow some of the ideas of the approach of [8], which used, among others, the splitting off method. We further develop this method for k-vertex-connectivity. Due to space limitations we omit several proofs and most of the algorithmic details.

We remark that the other three basic augmentation problems (where one wants to make G k-edge-connected or wants to make a digraph k-edge- or k-vertex-connected) have been shown to be polynomially solvable. These results are due to Watanabe and Nakamura [12], Frank [3], and Frank and Jordán [4], respectively. For more results on connectivity augmentation see the recent survey by Nagamochi [11]. In the rest of the introduction we introduce some definitions and our new lower bounds for the size of an augmenting set which makes G k-vertex-connected. We also state our main results.

In what follows we deal with undirected graphs and k-connected refers to k-vertex-connected. For two disjoint sets of vertices X, Y in a graph $G = (V, E)$ we denote the number of edges from X to Y by $d_G(X, Y)$ (or simply $d(X, Y)$). We use $d(X) = d(X, V - X)$ to denote the $degree$ of X. For a single vertex v we write $d(v)$. Let $G = (V, E)$ be a graph with $|V| \geq k + 1$. For $X \subseteq V$ let $N(X)$ denote the set of $neighbours$ of X, that is, $N(X) = \{v \in V - X : uv \in E$ for some $u \in X\}$. Let $n(X)$ denote $|N(X)|$. We use X^* to denote $V - X - N(X)$. We call X a $fragment$ if $X, X^* \neq \emptyset$. Let $a_k(G)$ denote the size of a smallest augmenting set of G with respect to k. It is easy to see that every set of new edges F which makes G k-connected must contain at least $k - n(X)$ edges from X to X^* for every fragment X. By summing up these 'deficiencies' over pairwise disjoint fragments, we obtain a useful lower bound on $a_k(G)$, similar to the one used in the corresponding edge-connectivity augmentation problem. Let $t(G) = \max\{\sum_{i=1}^{r} k - n(X_i) : X_1, ..., X_r$ are pairwise disjoint fragments in $V\}$. Then

$$a_k(G) \geq \lceil t(G)/2 \rceil. \tag{1}$$

Another lower bound for $a_k(G)$ comes from 'shredders'. For $K \subset V$ let $b(K, G)$ denote the number of components in $G - K$. Let $b(G) = \max\{b(K, G) : K \subset V, |K| = k - 1\}$. We call a set $K \subset V$ with $|K| = k - 1$ and $b(K, G) = q$ a q-shredder. Since $G - K$ has to be connected in the augmented graph, we have the second lower bound:

$$a_k(G) \geq b(G) - 1. \tag{2}$$

These lower bounds extend the two natural lower bounds used e.g. in [2, 6,8]. Although these bounds suffice to characterize $a_k(G)$ for $k \leq 3$, there are examples showing that $a_k(G)$ can be strictly larger than the maximum of these lower bounds, consider for example the complete bipartite graph $K_{3,3}$ with target $k = 4$. We shall show in Subsection 4.1 that if G is $(k-1)$-connected and $a_k(G)$ is large compared to k, then $a_k(G) = \max\{b(G) - 1, \lceil t(G)/2 \rceil\}$. The same result is not valid if we remove the hypothesis that G is $(k-1)$-connected. To see this consider the graph G obtained from $K_{m,k-2}$ by adding a new vertex x and joining x to j vertices in the m set of the $K_{m,k-2}$, where $j < k < m$. Then $b(G) = m$, $t(G) = 2m + k - 2j$ and $a_k(G) = m - 1 + k - j$. We shall see in Subsection 4.2, however, that if we modify the definition of $b(G)$ slightly, then we may obtain an analogous min-max theorem for augmenting graphs of arbitrary connectivity. For a set $K \subset V$ with $|K| = k - 1$ we define $\delta(K) = \max\{0, \max\{k - d(x) : x \in K\}\}$ and $b^*(K) = b(K) + \delta(K)$. We let $b^*(G) = \max\{b^*(K) : K \subset V, |K| = k - 1\}$. It is easy to see that $a_k(G) \geq b^*(G) - 1$. We shall prove in Subsection 4.2 that if $a_k(G)$ is large compared to k, then $a_k(G) = \max\{b^*(G) - 1, \lceil t(G)/2 \rceil\}$.

2   Preliminaries

In the so-called 'splitting off method' one extends the input graph G by a new vertex s and a set of appropriately chosen edges incident to s and then obtains an optimal augmenting set by splitting off pairs of edges incident to s. This approach was initiated by Cai and Sun [1] for the k-edge-connectivity augmentation problem and further developed and generalized by Frank [3]. In [7] we adapted the method to vertex-connectivity and proved several basic properties of the extended graph as well as the splittable pairs.

Given the input graph $G = (V, E)$, an *extension* $G + s = (V + s, E + F)$ of G is obtained by adding a new vertex s and a set F of new edges from s to V. In $G + s$ we define $\bar{d}(X) = n_G(X) + d(s, X)$ for every $X \subseteq V$. We say that $G + s$ is (k, s)-*connected* if

$$\bar{d}(X) \geq k \text{ for every fragment } X \subset V, \tag{3}$$

and that it is a k-*critical extension* if F is an inclusionwise minimal set with respect to (3). The minimality of F implies that every edge su in a critical extension is k-*critical*, that is, deleting su from $G + s$ destroys (3). (Thus an edge su is k-critical if and only if there exists a fragment X in V with $u \in X$ and $\bar{d}(X) = k$.) A fragment X with $d(s, X) \geq 1$ and $\bar{d}(X) = k$ is called *tight*. A fragment X with $d(s, X) \geq 2$ and $\bar{d}(X) \leq k + 1$ is called *dangerous*. Observe that if G is l-connected then for every $v \in V$ we have $d(s, v) \leq k - l$ in any k-critical extension of G. We shall use the following submodular inequalities for n and $\bar{d}$ from [7].

Proposition 1. *[7, Proposition 7] Let $G + s$ be an extension of a graph $G = (V, E)$ and $X, Y \subseteq V$. Then*

$$\bar{d}(X) + \bar{d}(Y) \geq \bar{d}(X \cap Y) + \bar{d}(X \cup Y) + |(N_G(X) \cap N_G(Y)) - N_G(X \cap Y)| +$$
$$+ |(N_G(X) \cap Y) - N_G(X \cap Y)| + |(N_G(Y) \cap X) - N_G(X \cap Y)|, \tag{4}$$

$$\bar{d}(X) + \bar{d}(Y) \geq \bar{d}(X \cap Y^*) + \bar{d}(Y \cap X^*) + d(s, X - Y^*) + d(s, Y - X^*). \quad (5)$$

Proposition 2. *[7, Proposition 8] Let $G + s$ be an extension of a graph $G = (V, E)$ and $X, Y, Z \subseteq V$. Then*

$$\bar{d}(X) + \bar{d}(Y) + \bar{d}(Z) \geq \bar{d}(X \cap Y \cap Z) + \bar{d}(X \cap Y^* \cap Z^*) + \bar{d}(X^* \cap Y^* \cap Z) +$$
$$+ \bar{d}(X^* \cap Y \cap Z^*) - |N_G(X) \cap N_G(Y) \cap N_G(Z)| + 2d(s, X \cap Y \cap Z). \quad (6)$$

Lemma 1. *[7, Lemma 10] Let $G + s$ be a critical extension of a graph G. Then $\lceil d(s)/2 \rceil \leq a_k(G) \leq d(s) - 1$.*

It follows from (3) that we can construct an augmenting set for G by adding edges between the vertices in $N(s)$ for any (k, s)-connected extension $G + s$ of G. This observation motivates the following definitions. *Splitting off* two edges su, sv in $G + s$ means deleting su, sv and adding a new edge uv. Such a split is *admissible* if the graph obtained by the splitting also satisfies (3). Notice that if $G + s$ has no edges incident to s then (3) is equivalent to the k-connectivity of G. Hence it would be desirable to know, when $d(s)$ is even, that there is a sequence of admissible splittings which isolates s. In this case, using the fact that $d(s) \leq 2a_k(G)$ by Lemma 1, the resulting graph on V would be an *optimal* augmentation of G with respect to k. This approach works for the k-edge-connectivity augmentation problem [3] but does not always work in the vertex connectivity case. The reason is that such 'complete splittings' do not necessarily exist. On the other hand, we shall prove results which are 'close enough' to yield an optimal algorithm for k-connectivity augmentation, using the splitting off method, which is polynomial for k fixed.

3   Independence Free Graphs

Let $G = (V, E)$ be a graph and k be an integer. Let X_1, X_2 be disjoint subsets of V. We say (X_1, X_2) is a k-*deficient pair* if $d(X_1, X_2) = 0$ and $|V - (X_1 \cup X_2)| \leq k - 1$. We say two deficient pairs (X_1, X_2) and (Y_1, Y_2) are *independent* if for some $i \in \{1, 2\}$ we have either $X_i \subseteq V - (Y_1 \cup Y_2)$ or $Y_i \subseteq V - (X_1 \cup X_2)$, since in this case no edge can simultaneously connect X_1 to X_2 and Y_1 to Y_2. We say G is k-*independence free* if G does not have two independent k-deficient pairs. (Note that if G is $(k-1)$-connected and (X_1, X_2) is a k-deficient pair then $X_2 = X_1^*$ and $V - (X_1 \cup X_2) = N(X_1) = N(X_2)$.) For example (a) $(k-1)$-connected chordal graphs and graphs with minimum degree $2k - 2$ are k-independence free; (b) all graphs are 1-independence free and all connected graphs are 2-independence free; (c) a graph with no edges and at least $k + 1$ vertices is not k-independence free for any $k \geq 2$; (d) if G is k-independence free and H is obtained by adding edges to G then H is also k-independence free; (e) a k-independence free graph is l-independence free for all $l \leq k$.

In general, a main difficulty in vertex-connectivity problems is that vertex cuts (and hence tight and dangerous sets) can cross each other in many different ways. In the case of an independence free graph G we can overcome these difficulties and prove the following results, including a complete characterisation of the case when there is no admissible split containing a specified edge in an extension of G.

Lemma 2. *Let $G + s$ be a (k, s)-connected extension of a k-independence free graph G and X, Y be fragments of G.*
(a) If X and Y are tight then either: $X \cup Y$ is tight, $X \cap Y \neq \emptyset$ and $\bar{d}(X \cap Y) = k$; or $X \cap Y^*$ and $Y \cap X^*$ are both tight and $d(s, X - Y^*) = 0 = d(s, Y - X^*)$.
(b) If X is a minimal tight set and Y is tight then either: $X \cup Y$ is tight, $d(s, X \cap Y) = 0$ and $n_G(X \cap Y) = k$; or $X \subseteq Y^*$.
(c) If X is a tight set and Y is a maximal dangerous set then either $X \subseteq Y$ or $d(s, X \cap Y) = 0$.
(d) If X is a tight set, Y is a dangerous set and $d(s, Y - X^*) + d(s, X - Y^*) \geq 2$ then $X \cap Y \neq \emptyset$ and $\bar{d}(X \cap Y) \leq k + 1$.

Lemma 3. *If $G + s$ is a k-critical extension of a k-independence free graph G then $d(s) = t(G)$. Furthermore there exists a unique minimal tight set in $G + s$ containing x for each $x \in N(s)$.*

Lemma 4. *Let $G + s$ be a k-critical extension of a k-independence free graph G and $x_1, x_2 \in N(s)$. Then the pair sx_1, sx_2 is not admissible for splitting in $G + s$ with respect to k if and only if there exists a dangerous set W in $G + s$ with $x_1, x_2 \in W$.*

Theorem 1. *Let $G + s$ be a k-critical extension of a k-independence free graph G and $x_0 \in N(s)$.*
(a) There is no admissible split in $G + s$ containing sx_0 if and only if either: $d(s) = b(G)$; or $d(s)$ is odd and there exist maximal dangerous sets W_1, W_2 in $G + s$ such that $N(s) \subseteq W_1 \cup W_2$, $x_0 \in W_1 \cap W_2$, $d(s, W_1 \cap W_2) = 1$, $d(s, W_1 \cap W_2^*) = (d(s) - 1)/2 = d(s, W_1^* \cap W_2)$, and $W_1 \cap W_2^*$ and $W_2 \cap W_1^*$ are tight.
(b) Furthermore if there is no admissible split containing sx_0 and $3 \neq d(s) \neq b(G)$ then there is an admissible split containing sx_1 for all $x_1 \in N(s) - x_0$.

Proof. Note that since $G + s$ is a k-critical extension, $d(s) \geq 2$. First we prove (a). Using Lemma 4, we may choose a family of dangerous sets $\mathcal{W} = \{W_1, W_2, \ldots, W_r\}$ in $G + s$ such that $x_0 \in \cap_{i=1}^r W_i$, $N(s) \subseteq \cup_{i=1}^r W_i$ and r is as small as possible. We may assume that each set in $\mathcal{W}$ is a maximal dangerous set in $G + s$. If $r = 1$ then $N(s) \subseteq W_1$ and $\bar{d}(W_1^*) = n_G(W_1^*) \leq n_G(W_1) \leq k + 1 - d(s, W_1) \leq k - 1$, since W_1 is dangerous. This contradicts the fact that $G + s$ is (k, s)-connected. Hence $r \geq 2$.

Claim. Let $W_i, W_j \in \mathcal{W}$. Then $W_i \cap W_j^* \neq \emptyset \neq W_j \cap W_i^*$ and $d(s, W_i - W_j^*) = 1 = d(s, W_j - W_i^*)$.

Proof. Suppose $W_i \cap W_j^* = \emptyset$. Since G is k-independence free, it follows that $W_i^* \cap W_j^* \neq \emptyset$ and hence $W_i \cup W_j$ is a fragment of G. The minimality of r now implies that $W_i \cup W_j$ is not dangerous, and hence $\bar{d}(W_i \cup W_j) \geq k + 2$. Applying (4) we obtain

$$2k + 2 \geq \bar{d}(W_i) + \bar{d}(W_j) \geq \bar{d}(W_i \cap W_j) + \bar{d}(W_i \cup W_j) \geq 2k + 2.$$

Hence equality holds throughout. Thus $\bar{d}(W_i \cap W_j) = k$ and, since $x_0 \in W_i \cap W_j$, $W_i \cap W_j$ is tight.

Choose $x_i \in N(s) \cap (W_i - W_j)$ and let X_i be the minimal tight set in $G + s$ containing x_i. Since $x_i \in N(s) \cap X_i \cap W_i$, it follows from Lemma 2(c) that $X_i \subseteq W_i$. Since G is k-independence free, $X_i \not\subseteq N(W_j)$. Thus $X_i \cap W_i \cap W_j \neq \emptyset$. Applying Lemma 2(b), we deduce that $X_i \cup (W_i \cap W_j)$ is tight. Now $X_i \cup (W_i \cap W_j)$ and W_j contradict Lemma 2(c) since $x_0 \in W_i \cap W_j$. Hence we must have $W_i \cap W_j^* \neq \emptyset \neq W_j \cap W_i^*$. The second part of the claim follows from (5) and the fact that $x_0 \in W_i \cap W_j$. $\qquad \square$

Suppose $r = 2$. Using the Claim, we have $d(s) = 1 + d(s, W_1 \cap W_2^*) + d(s, W_2 \cap W_1^*)$. Without loss of generality we may suppose that $d(s, W_1 \cap W_2^*) \leq d(s, W_2 \cap W_1^*)$. Then

$$\bar{d}(W_2^*) = d(s, W_1 \cap W_2^*) + n_G(W_2^*) \leq d(s, W_2 \cap W_1^*) + n_G(W_2) = \bar{d}(W_2) - 1 \leq k.$$

Thus equality must hold throughout. Hence $d(s, W_1 \cap W_2^*) = d(s, W_2 \cap W_1^*) = (d(s) - 1)/2$, $d(s)$ is odd, and $W_1 \cap W_2^*$ and $W_2 \cap W_1^*$ are tight.

Finally we suppose that $r \geq 3$. Choose $W_i, W_j, W_h \in \mathcal{W}$, $x_i \in (N(s) \cap W_i) - (W_j \cup W_h)$, $x_j \in (N(s) \cap W_j) - (W_i \cup W_h)$, and $x_h \in (N(s) \cap W_h) - (W_i \cup W_j)$. Then the Claim implies that $x_i \in W_i \cap W_j^* \cap W_h^*$. Applying (6), and using $d(s, W_i \cap W_j \cap W_h) \geq 1$, we get

$$3k + 3 \geq \bar{d}(W_i) + \bar{d}(W_j) + \bar{d}(W_h) \geq \bar{d}(W_i \cap W_j \cap W_h) + \bar{d}(W_i \cap W_j^* \cap W_h^*) +$$
$$+ \bar{d}(W_j \cap W_i^* \cap W_h^*) + \bar{d}(W_h \cap W_i^* \cap W_j^*) - |N(W_i) \cap N(W_j) \cap N(W_h)|$$
$$+ 2 \geq 4k - |N(W_i) \cap N(W_j) \cap N(W_h)| + 2 \geq 3k + 3. \qquad (7)$$

Thus equality must hold throughout. Hence $d(s, W_i \cap W_j \cap W_h) = 1$, and $W_i \cap W_j^* \cap W_h^*$ is tight. Furthermore, putting $S = N(W_i) \cap N(W_j) \cap N(W_h)$, we have $|S| = k - 1$ by (7). Hence $N(W_i \cap W_j \cap W_h) = S$. Thus $n_G(W_i) \geq k - 1$, and since W_i is dangerous, $d(s, W_i) = 2$ follows. Thus $d(s, W_i \cap W_j^* \cap W_h^*) = 1$ and $G - S$ has $r + 1 = d(s)$ components $C_0, C_1, ..., C_r$ where $C_0 = W_i \cap W_j \cap W_h$ and $C_i = W_i - C_0$ for $1 \leq i \leq r$. Thus S is a $d(s)$-shredder in G. Since the (k, s)-connectivity of $G + s$ implies that $b(G) \leq d(s)$, we have $b(G) = d(s)$.

(b) Using (a) we have $d(s)$ is odd and there exist maximal dangerous sets W_1, W_2 in $G + s$ such that $N(s) \subseteq W_1 \cup W_2$, $x_0 \in W_1 \cap W_2$, $d(s, W_1 \cap W_2) = 1$,

$d(s, W_1 \cap W_2^*) = d(s, W_1^* \cap W_2) = (d(s) - 1)/2 \geq 2$, and $W_1 \cap W_2^*$ and $W_1^* \cap W_2$ are tight. Suppose $x_1 \in N(s) \cap W_1$ and there is no admissible split containing sx_1. Then applying (a) to x_1 we find maximal dangerous sets W_3, W_4 with $x_1 \in W_3 \cap W_4$ and $d(s, W_3 \cap W_4) = 1$. Using Lemma 2(c) we have $W_1 \cap W_2^* \subseteq W_3$ and $W_1 \cap W_2^* \subseteq W_4$. Thus $W_1 \cap W_2^* \subseteq W_3 \cap W_4$ and $d(s, W_3 \cap W_4) \geq 2$. This contradicts the fact that $d(s, W_3 \cap W_4) = 1$.   □

We can use this result and the following lemma (which solves the case when $b(G)$ is large compared to $d(s)$) to determine $a_k(G)$ when G is k-independence free.

Lemma 5. *Let $G + s$ be a k-critical extension of a k-independence free graph G. If $d(s) \leq 2b(G) - 2$ then $a_k(G) = b(G) - 1$.*

Theorem 2. *If G is k-independence free then $a_k(G) = \max\{\lceil t(G)/2 \rceil, b(G) - 1\}$.*

Proof. Let $G + s$ be a k-critical extension of G. We shall proceed by induction on $d(s)$. By Lemma 3, $d(s) = t(G)$. If $d(s) \leq 3$ then $a_k(G) = \lceil t(G)/2 \rceil$ by Lemma 1. Hence we may suppose that $d(s) \geq 4$. If $d(s) \leq 2b(G) - 2$ then $a_k(G) = b(G) - 1$ by Lemma 5. Hence we may suppose that $d(s) \geq 2b(G) - 1$.

By Theorem 1, $G + s$ has an admissible pair su, sw. Let $G^* + s$ be the graph obtained from $G + s$ by splitting su, sw from s (where $G^* = G + uw$). Then G^* is k-independence free. Moreover $t(G^*) \geq t(G) - 2 = d_{G+s}(s) - 2 = d_{G^*+s}(s)$. Since the (k, s)-connectivity of $G^* + s$ implies that $t(G^*) \leq d_{G^*+s}(s)$, we have equality and hence $G^* + s$ is a k-critical extension of G^*. Furthermore

$$t(G^*) = d_{G^*+s}(s) = d_{G+s}(s) - 2 \geq 2b(G) - 3 \geq 2b(G^*) - 3.$$

Applying induction to G^* we deduce that $a_k(G^*) = \lceil t(G^*)/2 \rceil$. Thus $a_k(G) \leq a_k(G^*) + 1 = \lceil t(G)/2 \rceil$. By (1), equality must hold.   □

4   Graphs with Large Augmentation Number

Throughout this section we will be concerned with augmenting an l-connected graph G on at least $k+1$ vertices for which $a_k(G)$ is large compared to k. First we consider a k-critical extension $G+s$ of G for which $d(s)$ is large, and characterise when there is no admissible split containing a given edge at s.

Lemma 6. *(a) [7] If $d(s) \geq (k - l)(k - 1) + 4$ then $d(s) = t(G)$.*
(b) [8] If $l = k - 1$ and $d(s) \geq k + 1$ then $d(s) = t(G)$.

The proof of the next theorem is similar to that of Theorem 1. A weaker result with a similar proof was given in [7, Lemma 7].

Theorem 3. *Let sx_0 be a designated edge of a k-critical extension $G + s$ of G and suppose that there are $q \geq (k - l + 1)(k - 1) + \max\{4 - l, 1\}$ edges sy $(y \neq x)$ incident to s for which the pair sx_0, sy is not admissible. Then there exists a shredder K in G such that K has $q + 1$ leaves in $G + s$, and one of the leaves is the maximal tight set containing x_0.*

Next we show that if $b^*(G)$ is large compared to k and $b^*(G) - 1 \geq \lceil t(G)/2 \rceil$ then $a_k(G) = b^*(G) - 1$. We need several new observations on shredders. We assume that $G + s$ is a k-critical extension of G, and that, for some shredder K of G, we have $d(s) \leq 2b^*(K) - 2$.

Lemma 7. *Suppose $b^*(K) \geq 4k + 3(k - l) + 1$. Then (a) the number of components C of $G - K$ with $d(s, C) \geq 3$ is at most $b(K) - 2k - 1$, (b) $|N(s) \cap K| \leq 1$, and (c) if $d(s, x) = j \geq 1$ for some $x \in K$ then $k - d_G(x) = j$.*

We shall use the following construction to augment G with $b^*(G) - 1$ edges in the case when $d(s, K) = 0$, and hence $b(G) = b^*(G) =: b$. Let $C_1, ..., C_b$ be the components of $G - K$ and let $w_i = d_{G+s}(s, C_i)$, $1 \leq i \leq b$. Note that $w_i \geq 1$ by (3). Since $d(s, G - K) \leq 2b - 2$, there exists a tree F' on b vertices with degree sequence $d_1, ..., d_b$ such that $d_i \geq w_i$, $1 \leq i \leq b$. Let F be a forest on $N_{G+s}(s)$ with $d_F(v) = d_{G+s}(s, v)$ for every $v \in V(G)$ and such that $F/C_1/.../C_b = F'$ holds. Thus $|E(F)| = |E(F')| = b - 1$. We shall say that $G + F$ is a *forest augmentation* of G with respect to K and $G + s$.

Lemma 8. *Suppose $b^*(K) \geq 4k + 3(k - l) + 1$ and $d(s, K) = 0$. Let $G + F$ be a forest augmentation of G with respect to K and $G + s$. Then $G + F$ is k-connected.*

Our final step is to show how to augment G with $b^*(K) - 1$ edges when $d(s, K) \neq 0$. In this case, Lemma 7(b) implies that there is exactly one vertex $x \in K$ which is adjacent to s. We use the next lemma to split off all edges from s to x and hence reduce to the case when $d(s, K) = 0$.

Lemma 9. *Suppose $d(s, x) \geq 1$ for some $x \in K$ and $d(s) \geq k(k - l + 1) + 2$. Then there exists a sequence of $d(s, x)$ admissible spits at s which split off all edges from s to x.*

We can now prove our augmentation result for graphs with large shredders.

Theorem 4. *Suppose that G is l-connected, $b^*(G) \geq 4k + 4(k - l) + 1$, $t(G) \geq k(k - l + 1) + 2$ and $b^*(G) - 1 \geq \lceil t(G)/2 \rceil$. Then $a_k(G) = b^*(G) - 1$.*

We note that a stronger result holds when $l = k - 1$.

Theorem 5. *[8] Suppose G is a $(k - 1)$-connected graph such that $b(G) \geq k$ and $b(G) - 1 \geq \lceil t(G)/2 \rceil$. Then $a_k(G) = b(G) - 1$.*

4.1   Augmenting Connectivity by One

Throughout this subsection we assume that $G = (V, E)$ is a $(k-1)$-connected graph on at least $k+1$ vertices. We shall show that if $a_k(G)$ is large compared to k, then $a_k(G) = \max\{b(G) - 1, \lceil t(G)/2 \rceil\}$. Our proof uses Theorems 2 and 5. We shall show that if $a_k(G)$ is large, then we can add a set of new edges F so that we have $t(G + F) = t(G) - 2|F|$ and $G + F$ is k-independence free.

In order to do this we need to measure how close G is to being k-independence free. We use the following concepts. A set $X \subset V$ is *deficient* in G if $n(X) = k-1$ and $V - X - N(X) \neq \emptyset$. We call the (inclusionwise) minimal deficient sets in G the *cores* of G. A core B is *active* in G if there exists a $(k-1)$-cut K with $B \subseteq K$. Otherwise B is called *passive*. Let $\alpha(G)$ and $\pi(G)$ denote the numbers of active, respectively passive, cores of G. Since G is $(k-1)$-connected, the definition of k-independence implies that G is k-independence free if and only if $\alpha(G) = 0$. The following characterisation of active cores also follows easily from the above definitions.

Lemma 10. *Let B be a core in G. Then B is active if and only if $\kappa(G - B) = k - |B| - 1$.*

A set $S \subseteq V$ is a *deficient set cover* (or *$\mathcal{D}$-cover* for short) if $S \cap T \neq \emptyset$ for every deficient set T. Clearly, S covers every deficient set if and only if S covers every core. Note that S is a minimal $\mathcal{D}$-cover for G if and only if the extension $G + s$ obtained by joining s to each vertex of S is k-critical.

Lemma 11. *[8, p 16, Lemma 3.2] (a) Every minimal augmenting set for G induces a forest, (b) for every $\mathcal{D}$-cover S for G, there exists a minimal augmenting set F for G with $V(F) \subseteq S$, (c) if F is a minimal augmenting set for G, $e = xy \in F$, and $H = G + F - e$, then H has precisely two cores X, Y. Furthermore $X \cap Y = \emptyset$, $x \in X$, $y \in Y$ and, for any edge $e' = x'y'$ with $x' \in X, y' \in Y$, the graph $H + e'$ is k-connected.*

Based on these facts we can prove the following lemma.

Lemma 12. *Let S be a minimal $\mathcal{D}$-cover in a graph $H = (V, E)$ and let F be a minimal augmenting set with $V(F) \subseteq S$. Let $d_F(v) = 1$ and let $e = uv$ be the leaf of F incident with v. Let $u \in X$ and $v \in Y$ be the cores of $H + F - e$ and suppose that for a set F' of edges we have $\kappa(x, y, H + F') \geq k$ for some pair $x \in X, y \in Y$. Then $S - \{v\}$ is a $\mathcal{D}$-cover of $H + F'$.*

We need some further concepts and results from [8].

Lemma 13. *[8, Lemma 2.1, Claim I(a)] Suppose $t(G) \geq k$. Then the cores of G are pairwise disjoint and the number of cores of G is equal to $t(G)$. Furthermore, if $t(G) \geq k + 1$, then for each core X, there is a unique maximal deficient set $S_X \subseteq V$ with $X \subseteq S_X$ and $S_X \cap Y = \emptyset$ for every core Y of G with $X \neq Y$. In addition, for two different cores X, Y we have $S_X \cap S_Y = \emptyset$.*

The following lemma can be verified by using [8, Lemma 2.2].

Lemma 14. *Let K be a shredder in G with $b(K) \geq k$. Then (a) if $C = S_X$ for some component C of $G - K$ and for some core X then X is passive, (b) if some component D of $G - K$ contains precisely two cores X, Y and no edge of G joins S_X to S_Y then both X and Y are passive.*

A set F of new edges is *saturating* for G if $t(G + F) = t(G) - 2|F|$. Thus an edge $e = uv$ is *saturating* if $t(G + e) = t(G) - 2$. Motivation for considering saturating sets is provided by the following lemma, which follows from Theorem 5.

Lemma 15. *If F is a saturating set of edges for G with $b(G + F) - 1 = \lceil t(G + F)/2 \rceil \geq k - 1$ then $a_k(G) = \lceil t(G)/2 \rceil$.*

We say that two cores X, Y form a *saturating pair* if there is a saturating edge $e = xy$ with $x \in X, y \in Y$. For a core X let $\nu(X)$ be the number of cores Y $(Y \neq X)$ for which X, Y is not a saturating pair. The following lemma implies that an active core cannot belong to many non-saturating pairs.

Lemma 16. *Suppose $t(G) \geq k \geq 4$ and let X be an active core in G. Then $\nu(X) \leq 2k - 2$.*

We shall also need the following characterisation of saturating pairs.

Lemma 17. *[8, p.13-14] Let $t(G) \geq k + 2$ and suppose that two cores X, Y does not form a saturating pair. Then one of the following holds: (a) $X \subseteq N(S_Y)$, (b) $Y \subseteq N(S_X)$, (c) there exists a deficient set M with $S_X, S_Y \subset M$, which is disjoint from every core other than X, Y.*

Let $\bar{\mathcal{F}} = \{X \subset V : X$ is deficient in G, X contains no active core$\}$. Let $\mathcal{F}$ consist of the maximal members of $\bar{\mathcal{F}}$.

Lemma 18. *Suppose $t(G) \geq k$ and $\alpha(G) \geq k$. Then (a) $X \cap B = \emptyset$ for every $X \in \bar{\mathcal{F}}$ and for every active core B, (b) $X \cap Y = \emptyset$ for every pair $X, Y \in \mathcal{F}$.*

We shall use the following lemmas to find a saturating set F for G such that $G + F$ has many passive cores. Informally, the idea is to pick a properly chosen active core B and, by adding a set F of at most $2k - 2$ saturating edges between the active cores of G other than B, make $\kappa(G + F - B) \geq k - |B| =: r$. By Lemma 10, this will make B passive, and will not eliminate any of the passive cores of G. We shall increase the connectivity of $G - B$ by choosing a minimal r-deficient set cover S for $G - B$ of size at most $k - 1$ and then iteratively add one or two edges so that the new graph has an r-deficient set cover properly contained in S. Thus after at most $k - 1$ such steps (and adding at most $2k - 2$ edges) we shall make B passive.

Lemma 19. *Suppose $t(G) \geq k$. If $\pi(G) \leq 4(k - 1)$ and $\alpha(G) \geq 4(k - 1)^2 + 1$ then there exists an active core B with $B \cap \bigcup_{Y \in \mathcal{F}} (Y \cup N(Y)) = \emptyset$.*

Lemma 20. *Suppose $t(G) \geq k \geq 4$, $\pi(G) \leq 4(k-1)$ and $\alpha(G) \geq 8k^3 - 6k^2 - 3k+6$. Let B be an active core in G with $B \cap \bigcup_{Y \in \mathcal{F}} (Y \cup N(Y)) = \emptyset$, $H = G - B$, and $r = k - |B|$. Let S be a minimal r-deficient set cover of H with $S \subseteq N_G(B)$. Then there exists a saturating set of edges F for G such that $|F| \leq 2$, and either $\pi(G+F) > \pi(G)$, or $\pi(G+F) = \pi(G)$ and $G+F-B$ has an r-deficient set cover S' which is properly contained in S.*

Proof. Since B is active, $\kappa(H) = k-1-|B| = r-1$. Since B is deficient in G, we have $|S| \leq n_G(B) = k-1$. By Lemma 11 there exists a minimal r-augmenting set F^* for H such that F^* is a forest and $V(F^*) \subseteq S$. Let $d_{F^*}(v) = 1$ and let $e = uv$ be a leaf of F^*. By Lemma 11(c), there exist precisely two r-cores Z, W in $H + F^* - e$ with $u \in Z, v \in W$. Then Z, W are k-deficient in G. By the choice of B, Z contains an active k-core in G since otherwise Z would be a subset of some $Z' \in \mathcal{F}$ and $B \cap N_G(Z') \neq \emptyset$ would follow, contradicting the choice of B. This argument applies to W as well. Thus there exist active k-cores X, Y in G with $X \subseteq Z$ and $Y \subseteq W$.

If X and Y form a saturating pair in G then we may choose a saturating edge xy with $x \in X$ and $y \in Y$. Then $xy \notin E$ and hence $\kappa(x, y, G + xy) \geq k$ and $\kappa(x, y, H + xy) \geq r$ holds. Now $S - v$ is an r-deficient set cover in $H + xy$ by Lemma 12 and the lemma holds. Hence we may assume that X, Y is not a saturating pair in G. By Lemma 17 either

(i) there exists a k-deficient set M in G with $S_X \cup S_Y \subseteq M$ which is disjoint from every k-core other than X, Y, or

(ii) $Y \subseteq N(S_X)$ or $X \subseteq N(S_Y)$.

Choose $x \in X$ and $y \in Y$ arbitrarily and let $P_1, P_2, ..., P_{k-1}$ be $k-1$ openly disjoint xy-paths in G. Let $Q = \cup_{i=1}^{k-1} V(P_i)$. It is easy to see that if some edge of G joins S_X to S_Y, then one of the paths, say P_1, satisfies $V(P_1) \subseteq S_X \cup S_Y$. On the other hand, if no edge of G joins S_X to S_Y, then (ii) cannot hold. Hence (i) holds and, either one of the paths, say P_1, satisfies $V(P_1) \subseteq M$, or each of the $k-1$ paths intersects $N(M)$. In the latter case, since $n(M) = k-1$, we have $N(M) \subset Q$ and $Q \subset M \cup N(M)$ hold. We shall handle these two cases separately.

Case 1. No edge of G joins S_X to S_Y, (i) holds, and we have $N(M) \subset Q \subset M \cup N(M)$.

Let $C_1, C_2, ..., C_p$ be the components of $G - N(M)$. Using the properties of M (M intersects exactly two cores, M is the union of one or more components of $G - N(M)$, and $n(M) = k-1$) we can see that either, one component C_i contains S_X and S_Y and is disjoint from every core of G other than X, Y, or each of S_X and S_Y corresponds to the vertex set of a component of $G - N(M)$.

Since X and Y are active cores, Lemma 14 implies that $p \leq k-1$. Since $\alpha(G) \geq (k-2)(2k-1) + k + 3$, G has at least $(k-2)(2k-1) + 1$ active cores disjoint from B, X, Y, and $N(M)$. Thus some component C_j of $G - N(M)$ is disjoint from M and contains at least $2k$ active cores distinct from B. By Lemma 16, there exists a saturating edge xa_1 with $a_1 \in A_1$ for some active core

$A_1 \subset C_j$, $A_1 \neq B$. If $\pi(G+xa_1) \geq \pi(G)+1$ then we are done. Otherwise all the active cores in G other than X, A_1 remain active in $G + xa_1$. Applying Lemma 16 again, we may pick a saturating edge ya_2 with $a_2 \in A_2$ for some active core A_2 of $G + xa_1$, with $A_2 \subset C_j$, $A_2 \neq B$.

We have $\kappa(x, y, G + xa_1 + ya_2) \geq k$, since there is a path from x to y, using the edges xa_1, ya_2, and vertices of C_j only, and thus this path is disjoint from Q. Hence $\kappa(x, y, H + xa_1 + ya_2) \geq r$. Thus by Lemma 12, $S - v$ is an r-deficient set cover in $H + xa_1 + ya_2$.

Case 2. Either $V(P_1) \subseteq S_X \cup S_Y$ or (i) holds and $V(P_1) \subseteq M$.

Let us call a component D of $G - Q$ *essential* if D intersects an active core other than X, Y or B. Let $D_1, D_2, ..., D_r$ be the essential components of $G - Q$. We say that a component D_i is *attached to* the path P_j if $N(D_i) \cap V(P_j) \neq \emptyset$ holds. Let $T = S_X \cup S_Y$ if $V(P_1) \subseteq S_X \cup S_Y$ holds and let $T = M$ if $V(P_1) \subseteq M$. Then, T is disjoint from every active core other than X, Y.

We claim that at most $2k - 2$ essential components are attached to P_1. To see this focus on an essential component D which is attached to P_1 and let $w \in W \cap D$ for some active core $W \neq X, Y, B$ which has a vertex in D. There exists a path P_D from w to a vertex of P_1 whose inner vertices are in D. Since $w \notin T$ and $V(P_1) \subseteq T$, we have $D \cap N(T) \neq \emptyset$. The claim follows since the essential components are pairwise disjoint and $n(T) \leq 2k - 2$.

Suppose that one of the paths P_i intersects at least $4k - 2$ active cores in G other than X , Y or B. For every such active core A intersecting P_i choose a representative vertex $a \in A \cap P_i$. Since the cores are pairwise disjoint, the representatives are parwise distinct. Order the active cores intersecting P_i following the ordering of their representatives along the path P_i from x to y. By Lemma 16, we may choose a saturating edge xa_1 in G, where a_1 is among the $2k-1$ rightmost representatives and a_1 belongs to an active core A_1. If $\pi(G + xa_1) \geq \pi(G) + 1$ then we are done. Otherwise all the active cores of G other than X, A_1 remain active in $G + xa_1$. Again using Lemma 16, we may choose a saturating edge ya_2 in $G + xa_1$, where a_2 is among the $2k - 1$ leftmost representatives. By the choice of a_1 and a_2 there exist two openly disjoint paths from x to y in $G + xa_1 + ya_2$ using vertices of $V(P_i)$ only. Thus $\kappa(x, y, G + xa + yb) \geq k$. Hence, by Lemma 12, $S - v$ is an r-deficient set cover in $H + xa_1 + ya_2$.

Thus we may assume that for each path P_i intersects at most $4k - 3$ active cores in G other than X , Y or B. Hence there are at least $\alpha(G) - 3 - (k - 1)(4k-3) \geq (8k^3 - 6k^2 - 3k + 3) - (k-1)(4k-3) = 2k(4k^2 - 5k + 2)$ active cores other than B contained in $G - Q$. Note that since cores are minimal deficient sets, they induce connected subgraphs in G. Hence each core contained in $G - Q$ is contained in a component of $G - Q$. If some component of $G - Q$ contains at least $2k$ active cores of G other than B then the lemma follows as in Case 1. Hence we may assume that there are at least $4k^2 - 5k + 2$ essential components in $G - Q$ and each such component contains an active core distinct from X, Y, and B.

Since at most $2k - 2$ essential components are attached to P_1, we deduce that there are at least $4k^2 - 5k + 2 - (2k - 2) = (4k-3)(k-1)+1$ essential components

D_i with all their attachments on $P_2, P_3, \ldots, P_{k-1}$. Since G is $(k-1)$-connected, $n(D_i) \geq k-1$ and hence D_i has at least two attachments on at least one of the paths $P_2, P_3, \ldots, P_{k-1}$. Relabelling the components $D_1, \ldots, D_r$ and the paths $P_2, \ldots, P_{k-1}$ if necessary, we may assume that D_i has at least two attachments on P_{k-1} for $1 \leq i \leq 4k-2$.

Let z_i be the leftmost attachment of D_i on P_{k-1}. Without loss of generality we may assume that $z_1, z_2, \ldots, z_{4k-2}$ occur in this order on P_{k-1} as we pass from x to y. By Lemma 16, there exists a saturating edge ya_i where $a_i \in A_i$ for some active core $A_i \subseteq D_i$, where $A_i \neq B$ and $1 \leq i \leq 2k-1$. If $\pi(G+ya_i) \geq \pi(G)+1$ then we are done. Otherwise every active core in G other than Y, A_i remains active in $G + ya_i$. Using Lemma 16 again, there exists a saturating edge xa_j where $a_j \in A_j$ for some active core $A_j \subseteq D_j$, where $A_j \neq B$ and $2k \leq j \leq 4k-2$. Note that z_i is either to the left of z_j or $z_i = z_j$. Hence, using the fact that D_j has at least two attachments on P_k and by the choice of z_i, z_j, there exist two openly disjoint paths in $G + xa_i + ya_j$, using vertices from $V(P_k) \cup D_i \cup D_j$ only. Therefore $\kappa(x, y, G + xa_i + ya_j) \geq k$, and we are done as above. This completes the proof of the lemma.   □

By applying Lemma 20 recursively, we obtain the following.

Lemma 21. *Suppose $t(G) \geq k \geq 4$, $\pi(G) \leq 4(k-1)$ and $\alpha(G) \geq 8k^3 - 6k^2 + k - 2$. Then there exists a saturating set of edges F for G such that $|F| \leq 2k-4$ and $\pi(G+F) \geq \pi(G)+1$.*

Lemma 22. *Suppose $k \geq 4$ and $t(G) \geq 8k^3 + 10k^2 - 43k + 22$. Then there exists a saturating set of edges F for G such that $G + F$ is k-independence free and $t(G+F) \geq 2k-1$.*

Proof. If $\pi(G) \leq 4(k-1)$ then we may apply Lemma 21 recursively $4k-3-\pi(G)$ times to G to find a saturating set of edges F_1 for G such that $\pi(G+F_1) \geq 4k-3$. If $\pi(G) \geq 4k-3$ we set $F_1 = \emptyset$. Applying Lemma 16 to $G + F_1$, we can add saturating edges joining pairs of active cores until the number of active cores is at most $2k-2$. Thus there exists a saturating set of edges F_2 for $G + F_1$ such that $\alpha(G + F_1 + F_2) \leq 2k-2$ and $\pi(G + F_1 + F_2) \geq 4k-3$. Applying Lemma 16 to $G + F_1 + F_2$, we can add saturating edges joining pairs consisting of one active and one passive core until the number of active cores decreases to zero. Thus there exists a saturating set of edges F_3 for $G + F_1 + F_2$ such that $\alpha(G + F_1 + F_2 + F_3) = 0$ and $\pi(G + F_1 + F_2 + F_3) \geq 2k-1$.   □

The main theorem of this section is the following.

Theorem 6. *If $a_k(G) \geq 8k^3 + 10k^2 - 43k + 21$ then*

$$a_k(G) = \max\{\lceil t(G)/2 \rceil, b(G) - 1\}.$$

Proof. If $k \leq 3$ then the result follows from [2,13], even without the restriction on $a_k(G)$. Hence we may suppose that $k \geq 4$. We have $t(G) \geq a_k(G)+1 \geq 8k^3 +$

$10k^2 - 43k + 22$ by Lemmas 1 and 6. If $b(G) - 1 \geq \lceil t(G)/2 \rceil$ then $a_k(G) = b(G) - 1$ by Theorem 5 and we are done. Thus we may assume that $\lceil t(G)/2 \rceil > b(G) - 1$ holds. We shall show that $a_k(G) = \lceil t(G)/2 \rceil$. By Lemma 22, there exists a saturating set of edges F for G such that $G + F$ is k-independence free and $t(G + F) \geq 2k - 1$. Note that adding a saturating edge to a graph H reduces $\lceil t(H)/2 \rceil$ by exactly one and $b(H)$ by at most one. Thus, if $\lceil t(G + F)/2 \rceil \leq b(G + F) - 1$, then there exists $F' \subseteq F$ such that $\lceil t(G + F')/2 \rceil = b(G + F') - 1$ and the theorem follows by applying Lemma 15. Hence we may assume that $\lceil t(G + F)/2 \rceil \leq b(G + F) - 1$. Since $G + F$ is k-independence free, we can apply Theorem 2 to deduce that $a_k(G + F) = \lceil t(G + F)/2 \rceil$. Using (1) and the fact that $t(G) = t(G + F) + 2|F|$ we have $a_k(G) = \lceil t(G)/2 \rceil$, as required.   □

4.2   Augmenting Connectivity by at Least Two

Theorem 7. *Suppose G is an l-connected graph. If $a_k(G) \geq 3(k - l + 2)^3 (k+1)^3$ then $a_k(G) = \max\{b^*(G) - 1, \lceil t(G)/2 \rceil\}$.*

Outline of Proof: If $b^*(G) - 1 \geq \lceil t(G)/2 \rceil$ then the theorem follows from Theorem 4, so we may suppose $b^*(G) - 1 < \lceil t(G)/2 \rceil$ holds. Let $G + s$ be a (k, s)-critical extension of G. Construct a $(k - 1, s)$-critical extension $H + s$ of G from $G + s$ by deleting edges incident to s. We next perform a sequence of $(k - 1)$-admissible splittings in $H + s$ and obtain $H^* + s$. We then add all the edges of $(G + s) - (H + s)$ and obtain $G^* + s$. We will refer to the edges of $(G + s) - (H + s)$ as *new* edges. We first show that $G^* + s$ is a critical (k, s)-connected extension of G^* and that $G^* + s$ has at least $d_{G+s}(s)/(k - l + 1)$ new edges.

Using Lemma 3, we can either construct $H^* + s$ such that $d_{H^*+s}(s)$ is small or else there exists $K \subseteq V$ such that $|K| = k - 2$ and $H^* - K$ has $d_{H^*+s}(s)$ components. In the first case, we perform a sequence of admissible splits in $G^* + s$ such that, in the resulting graph $G' + s$, G' is $(k - 1)$-connected. We accomplish this by ensuring that $\kappa(x, y, G') \geq k - 1$ for every $x, y \in N_{H^*+s}(s)$, using a similar proof technique to that of Lemmas 20, 21 and 22. Since there are many new edges, this is feasible. We then apply Theorem 6. In the second case, we show directly that we can make G k-connected by adding $\lceil t(G)/2 \rceil$ edges. Our proof is similar to the forest augmentation of Lemma 8. Let $C_1, C_2, \ldots, C_r$ be the components of $H^* - K$. We add a set F of $\lceil d_{G^*+s}/2 \rceil$ edges to G^* between neighbours of s belonging to different components C_i, in such a way that $(G^* - K + F)/C_1/C_2 \ldots /C_r$ is 2-connected, and then show that $G^* + F$ is k-connected.   □

We note that Theorem 6 and Theorem 7 imply (partial) solutions to conjectures from [4, p.95], [5], [9, p.300], and [10].

5   The Algorithm

Let $G = (V, E)$ be an l-connected graph and let $k \geq 2$ be an integer. Based on the proofs of our min-max results we can develop an algorithm which delivers an optimal k-connected augmentation of G in $O(n^5) + O(f(k)n^3)$ time, where $n = |V|$ and $f(k)$ is an exponential function of k. If $a_k(G) \geq 3(k-l+2)^3(k+1)^3$ then the running time is simply $O(n^5)$. Here we give a sketch of the algorithm to indicate the main steps. Each of these steps can be implemented efficiently by using max-flow computations.

The algorithm first creates a k-critical extension $G+s$ of G. If $d(s)$ is small (or equivalently, if $a_k(G)$ is small), then the algorithm finds an optimal augmentation by checking all possible augmenting sets on the vertex set $N(s)$. It can be shown that there exists an optimal augmenting set on $N(s)$, and hence, since $a_k(G)$ and $|N(s)|$ are both bounded by a function of k, this procedure requires $f(k)$ max-flow computations. Thus this subroutine is polynomial for k fixed. If $d(s)$ is large then the algorithm follows the steps of our proofs and identifies an optimal solution in polynomial time, even if k is not fixed.

First suppose that G is $(k-1)$-connected (and $d(s)$ is large). Then the algorithm starts adding saturating edges and making the graph k-independence free. If $b(G') - 1 \geq \lceil t(G')/2 \rceil$ holds for the current graph G' after some iteration then an optimal solution is completed a forest augmentation and the algorithm terminates. Note that large shredders can be found in polynomial time, if exist. If G' becomes k-independence free then an optimal augmenting step is completed by an appropriate sequence of edge splittings.

Next suppose that G is not $(k-1)$-connected. In this case first we create a $(k-1)$-critical extension $H+s$ by deleting edges incident to s in $G+s$ and then start splitting off pairs of edges until either $d(s)$ becomes small or there are no more splits that preserve $(k-1)$-connectivity (and hence there is a large shredder K). After adding back all the edges of $(G+s) - (H+s)$ we proceed as follows. If $d(s)$ is small then we make the current graph $(k-1)$-connected by splittings and then continue as in the case when the input graph is $(k-1)$-connected. If there is a large shredder K then we complete the solution by adding an appropriate set of new edges for which $G' - K$ is 2-connected. Note that after every edge splitting step we check whether $b^*(G') - 1 \geq \lceil t(G')/2 \rceil$ holds and if yes, we complete the augmenting set by a forest augmentation and terminate.

References

1. G.R. Cai and Y.G. Sun, The minimum augmentation of any graph to a k-edge-connected graph, Networks 19 (1989) 151-172.
2. K.P. Eswaran and R.E. Tarjan, Augmentation problems, SIAM J. Computing, Vol. 5, No. 4, 653-665, 1976.
3. A. Frank, Augmenting graphs to meet edge-connectivity requirements, SIAM J. Discrete Mathematics, Vol.5, No 1., 22-53, 1992.
4. A. Frank and T. Jordán, Minimal edge-coverings of pairs of sets, J. Combinatorial Theory, Ser. B. 65, 73-110 (1995).
5. E. Győri, T. Jordán, How to make a graph four-connected, Mathematical Programming 84 (1999) 3, 555-563.
6. T.S. Hsu, On four-connecting a triconnected graph, Journal of Algorithms 35, 202-234, 2000.

7. B. Jackson, T. Jordán, A near optimal algorithm for vertex-connectivity augmentation, Proc. ISAAC 2000, (D.T. Lee and S.-H. Teng, eds) Springer Lecture Notes in Computer Science 1969, pp. 313-325, 2000.

8. T. Jordán, On the optimal vertex-connectivity augmentation, J. Combinatorial Theory, Ser. B. 63, 8-20,1995.

9. T. Jordán, A note on the vertex-connectivity augmentation problem, J. Combinatorial Theory, Ser. B. 71, 294-301, 1997.

10. T. Jordán, Extremal graphs in connectivity augmentation, J. Graph Theory 31: 179-193, 1999.

11. H. Nagamochi, Recent development of graph connectivity augmentation algorithms, IEICE Trans. Inf. and Syst., vol E83-D, no.3, March 2000.

12. T. Watanabe and A. Nakamura, Edge-connectivity augmentation problems, Computer and System Siences, Vol 35, No. 1, 96-144, 1987.

13. T. Watanabe and A. Nakamura, A minimum 3-connectivity augmentation of a graph, J. Computer and System Sciences, Vol. 46, No.1, 91-128, 1993.

The Throughput of Sequential Testing

Murali S. Kodialam

Bell Labs, Lucent Technologies
Holmdel, New Jersey 07733
muralik@lucent.com

Abstract. This paper addresses the problem of determining the maximum achievable throughput in a sequential testing system. The input to the system are n bit binary strings. There are n tests in the system, and if a string is subjected to test j then the test determines if bit j in that string is zero or one. The objective of the test system is to check if the sum of the bits in each incoming string is zero or not. The mean time taken to perform each test and the probability that bit j is one are known. The objective of this paper is to determine the maximum input rate that can be processed by this system. The problem of determining the maximum throughput is first formulated as a a quadratic programming problem over a polymatroid with some additional structure. The special structure of the polymatroid polyhedron is exploited to derive an $O(n^2)$ algorithm to solve the maximum throughput problem.

1   Introduction

This paper deals with the problem of determining the maximum throughput achievable in a sequential test system. This problem arises in the context of testing identical circuit boards with multiple sub-components. Each of these sub-components has to be tested individually and the circuit board is faulty if any one of sub-components is faulty. The tests on the sub-components can be performed in any order. The difficulty and the complexity of testing each sub-component is different and hence the time taken to perform each test is different. From historical data the probability that a particular sub-system is faulty is also known. Assume that there is a test system with as many tests as there are sub-systems on the circuit boards. At any given time at most one test can be conducted on any circuit board and any test can inspect at most one circuit board. In other words, as soon as it is determined that a particular sub-component on a circuit board is not functioning, the circuit board leaves the test system. The circuit boards that are currently not being examined by any test wait in a buffer.

The objective of the study was to determine the testing policy that maximizes the throughput of the system while keeping the number of circuit boards in the buffer bounded. In the case that each circuit board has only one sub-component, the maximum throughput of the system is bounded above by the processing rate for that test. The problem of determining the maximum throughput becomes

K. Aardal, B. Gerards (Eds.): IPCO 2001, LNCS 2081, pp. 280–292, 2001.
© Springer-Verlag Berlin Heidelberg 2001

non-trivial when each circuit board has more than one sub-component. One can either use some *static policy* where the order in which the components are inspected is determined apriori when a circuit board arrives into the system or a *dynamic policy*, where depending on the all the circuit boards that are currently waiting, a decision can be made on which sub-component is to be tested on which circuit board next. Therefore the space of policies is quite large. Since the space of dynamic policies subsumes the static policies, one can expect the maximum throughput that is achievable using a dynamic policy will be greater than the throughput achieved by static policies. In Section 3 of the paper, the surprising result that the maximum throughput can be achieved using a static policy where decisions are made independent of the state of the system is proved. In Section 4, an algorithm that determines the static testing policy that achieves this maximum throughput is outlined. It is shown that determining the maximum throughput as well as deriving the optimal testing policy are polynomial in the number of sub-components on each circuit board.

More abstractly, objects arriving into the system are assumed to have n binary attributes. One can view the input to the test system as a sequence of n bit binary strings. A zero in position j of a string represents that sub-component j is functioning and a one in position j represents that sub-component j is faulty. A string is defined to be a *zero string* if the sum of the n bits for the string is zero. The objective of the testing is to determine if each of the incoming strings is a zero string or not. The (mean) time taken to test whether bit j is zero or one as well as the probability that bit j is zero is known. There are n test stations in the testing system where the test station j determines if bit j is zero or one. At any given point in time, a test can be done on at most one string and and at most one test can be performed on a string. Any string that is not being tested waits in a storage buffer. The objective as stated earlier of this paper is to determine the maximum arrival rate for which there exists a testing policy which keeps the number of objects in the buffer bounded.

The approach that is followed in order to determine the maximum throughput of the system is to characterize the achievable space (this will be made clear in Section 2) over different testing policies. The polyhedral structure of the achievable space is then exploited to derive the maximum throughput. This line of research for solving dynamic and stochastic optimization problems was initiated by Coffman and Mitrani [3] and Gelenbe and Mitrani [7]. They formulated the optimal scheduling in a multi-class $M/G/1$ queue as a linear programming problem. Federgruen and Groenevelt [4] extended this work by showing that in certain cases, the achievable performance in multi-class system can be represented as a polymatroid. (Fujishige [6] is an excellent reference for polymatroids). More recently Bertsimas and Nino-Mora [2] showed that more general scheduling (multi-armed bandit) problems have a generalized polymatroid structure. The problem of maximizing the throughput of the test system is formulated as a min-max problem on the polymatroid. This is then transformed into a quadratic programming problem over a polymatroid polyhedron. Some additional structure of the polymatroid is exploited to get a simple algorithm to get the optimal throughput. In Section 4 of the paper, a static policy that achieves this optimal throughput is derived. All prior work in deriving the optimal policy to these problems have involved solving a linear programming problem [3] ,[10]. A combinatorial approach based on a constructive proof of Carathodory representation Theorem [1] is developed, again exploiting the special structure of the polymatroid polyhedron.

2   The Model

As stated in the introduction, the input to the testing system are n-bit binary strings $\{X^i\}_{i=1}^{\infty}$, i.e., $X^i = (X_1^i, X_2^i, \ldots, X_n^i) \in \{0,1\}^n$. Let λ represent the rate at which these strings enter the system. Let

$$\Pr[X_j^i = 1] = p_j = 1 - q_j \quad \forall i.$$

Test j determines whether bit j in X is zero or one in mean time $\frac{1}{\mu_j}$. An incoming string i is defined to be *zero-string* if $\sum_{j=1}^n X_j^i = 0$ and a *non-zero-string* otherwise. The objective of the test system is to determine if each arriving string is a zero-string. A string leaves the system as soon as it is determined that it is non-zero. Each string passes through each test at most once. All strings that are not currently being tested are kept in an infinite buffer. A testing policy is set of rules that one uses to load a string in the buffer to any empty test that the string has not already passed through. At any given point in time, a test can be done on at most one string and at most one test can be performed on a string. The objective is to determine the policy that maximizes the input rate λ^* at which strings can be processed by the system while keeping the number of strings in the buffer bounded.

3   Characterization of the Achievable Space

The output from test j is defined as the set of strings that leave the system because a one was detected in test j. Let γ_j represent the output rate from test j. The input to test j represented by $\lambda_j = \frac{\gamma_j}{p_j}$. Note that the value of γ_j and λ_j depends on the testing policy. Let $\mu = (\mu_1, \mu_2, \ldots, \mu_n)$ represent the vector of the processing rates at the different tests. Let $\mathcal{F}(\lambda, \mu)$ represent the space of $\gamma = (\gamma_1, \gamma_2, \ldots, \gamma_n)$ that is achievable over all testing policies. The objective of the problem can be restated as follows:

$$\lambda^* = \sup_{\lambda} \{\mathcal{F}(\lambda, \mu) \neq \emptyset\}.$$

For any $\lambda < \lambda^*$ there exists a resting policy that achieves a throughput of λ. A trivial upper bound on λ^* is $\sum_{i=1}^n \mu_i$. This is due to the fact that even if every string leaves the system after its first test, then the system cannot process strings at a rate greater than $\sum_{i=1}^n \mu_i$. In order to characterize $\mathcal{F}(\lambda, \mu)$ we first consider the case where $\mu_i = \infty \;\; \forall i$. In this case $\mathcal{F}(\lambda, \infty)$ represents the set of γ that can be achieved using some testing policy ignoring the finite processing rate for the tests. If the processing rate is finite, then the space of $\mathcal{F}(\lambda, \infty)$ is further restricted by the the stability conditions at each queue, i.e. at any test, the rate at which strings are input to the test must be less than the rate at which strings are processed by the test, i.e. $\lambda_j < \mu_j$ or equivalently $\gamma_j < p_j \mu_j$ for all j. Let $\mathcal{B}(\mu)$ represent these additional constraints.

$$\mathcal{B}(\mu) = \{\gamma_i < p_i \mu_i\}$$

Therefore,

$$\mathcal{F}(\lambda, \mu) = \mathcal{F}(\lambda, \infty) \cap \mathcal{B}(\mu).$$

Theorem 1 *Let $\mathcal{P}$ be the polyhedron represented by the following set of hyperplanes:*

$$\sum_{j \in S} \gamma_j \leq \lambda \left(1 - \prod_{j \in S} q_j \right) \quad S \subset \{1, 2, \ldots, n\} \tag{1}$$

$$\sum_{j \in N} \gamma_j = \lambda \left(1 - \prod_{j \in N} q_j \right) \tag{2}$$

where $N = \{1, 2, \ldots, n\}$. Then $\mathcal{F}(\lambda, \infty) \subseteq \mathcal{P}$.

Proof:
Note that γ_i will be the greatest when every string that enters the system gets bit i tested before any other bit. Under this scenario, note that

$$\gamma_i = \lambda(1 - q_i).$$

Therefore, under any testing policy

$$\gamma_i \leq \lambda(1 - q_i).$$

Now consider $\gamma_i + \gamma_j$ for some $i \neq j$. Note that $\gamma_i + \gamma_j$ will be greatest when bit i is first tested on all incoming strings and if bit i is zero bit j is tested before any other bit. Under this policy the value of $\gamma_i + \gamma_j = \lambda(1 - q_i q_j)$. Note that due to symmetry, the value of $\gamma_i + \gamma_j$ will be unchanged if bit j is tested first for all incoming strings and if bit j is zero then bit i is tested before any other bit. Therefore under any policy $\gamma_i + \gamma_j \leq \lambda(1 - q_i q_j)$. Inequalities of this form can be derived for any subset of tests. Note that the $\sum_{j \in N} \gamma_j$ is independent of the routing policy and this is shown as the equality constraint in the statement of the theorem.

It will now be shown that the conditions given in Theorem 1 are also sufficient. For proving this, a class, $\mathcal{C}$ of policies will be exhibited that will realize the entire polyhedron $\mathcal{P}$. Towards this end, it is first shown that $\mathcal{P}$ is a polymatroid polyhedron by proving that the right hand side of the inequalities in $\mathcal{P}$ is submodular.

Lemma 2 *Let*

$$f(S) = \lambda \left(1 - \prod_{j \in S} q_j \right) \quad S \subseteq N.$$

Then $f(S)$ is submodular.

Proof:
Note that f is submodular if and only if

$$f(S \cup \{j\}) - f(S) \geq f(S \cup \{j,k\}) - f(S \cup \{k\})$$
$$\text{for } j,k \in N, j \neq k, \text{ and } S \subseteq N \backslash \{j,k\}.$$

Pick an arbitrary $S \subseteq N$. Let $\Delta = \prod_{l \in S} q_l$. The above test can be rewritten as $j \neq k$ and $j,k \notin S$

$$\lambda(1 - q_j\Delta) - \lambda(1 - \Delta) \geq \lambda(1 - q_j q_k \Delta) - \lambda(1 - q_k\Delta).$$

The above statement is true since $q_k \leq 1$ and hence f is submodular.

The class $\mathcal{C}$ of policies that will realize the entire polyhedron $\mathcal{P}$ is defined as follows:

1. Associate with each permutation $\pi(i)$ of $\{1, 2, \ldots, n\}$, a weight $w_{\pi(i)}$ with the additional constraint that $\sum_{i=1}^{n!} w_{\pi(i)} = 1$.
2. Associate with each arriving string the permutation $\pi(i)$ with probability $w_{\pi(i)}$. This permutation represents the order in which bits will be tested in the string until a one is detected.
3. The infinite buffer is partitioned into n buffers one for each test. As soon as a string finishes a test and if the test determines that the corresponding bit is zero, then it is loaded on to the next test in its pre-determined sequence if the next test is free. Else it is loaded on to the buffer of the next test.
4. When a test completes processing a string, if its input buffer is non-empty, it picks up a string from its input buffer. If its input buffer is empty, then the test is idle.

In the next theorem, it is shown that the test policies in class $\mathcal{C}$ achieve all of $\mathcal{P}$. This is shown by proving that all the extreme points of the polyhedron are achievable by some policy in class $\mathcal{C}$.

Theorem 3 *Any $\gamma \in \mathcal{P}$ can be realized by a routing policy in class $\mathcal{C}$ and hence $\mathcal{F}(\lambda, \infty) = \mathcal{P}$.*

Proof:
It is shown in Welsh [11] that the set of exterme points of any polymatroid polyhedron can be represented as follows: Let π represent a permutation of $\{1, 2, \ldots, n\}$. Then an exteme point of the polyhedron corresponding to π is

$$(f(\{\pi(1)\}), f(\{\pi(1), \pi(2)\}) - f(\{\pi(1)\}), \ldots,$$
$$\ldots, f(\{\pi(1), \pi(2), \ldots, \pi(n)\}) - f(\{\pi(1), \pi(2), \ldots, \pi(n-1)\})).$$

In fact any vertex of the polyhedron can be represented in the above form with respect to some permutation π of $\{1, 2, \ldots, n\}$. Since the policies in the class $\mathcal{C}$ are all the permutation policies, these will achieve all the extreme points of $\mathcal{P}$. Since $\mathcal{P}$ is a convex polyhedron, the whole of $\mathcal{P}$ can be achieved with a weighted combination of these permutation policies.

The class of policies represented by $\mathcal{C}$ are static policies. By theorem 3, the entire space of $\mathcal{F}(\lambda, \infty)$ can be achieved by class $\mathcal{C}$. Note that $\mathcal{F}(\lambda, \infty)$ is a convex polyhedron in n-dimensions. The polyhedron is not full dimensional since there is one equality constraint. Therefore by the Caratheodory representation theorem any point in the polyhedron is representable as a convex combination of at most n extreme points of the polyhedron. There are an exponential $(2^n - 2)$ number of inequalities in the system representing $\mathcal{F}(\lambda, \infty)$. Since the polyhedron $\mathcal{F}(\lambda, \infty)$ is a polymatroid, a greedy algorithm can be used to determine an optimal solution to maximize or minimize any linear objective over this polyhedron. A slightly modified greedy algorithm can be used to minimize (maximize) separable convex (concave) objective functions over a polymatroid as shown in Federgruen and Groenevelt [5]. The stability problem is first formulated as a min-max optimization problem. This is then transformed into a quadratic optimization problem. This problem can be solved efficiently exploiting the structure of the polymatroid.

One of the key steps in the optimization procedure is checking for membership in the polyhedron. In general, the ellipsoid algorithm can be used to check for membership in a polymatroid polyhedron in polynomial time as shown by Grotschel, Lovasz and Schrijver [8]. However, in some special cases such as symmetric polymatroids [9] and generalized symmetric polymatroid [5], it is possible to devise combinatorial algorithms to check for membership. In the next lemma, it is shown that $\mathcal{P}$ is a variant of a generalized symmetric polymatroid and therefore a combinatorial algorithm can be devised to check for membership in $\mathcal{P}$.

Lemma 4 *Given a vector $\gamma \in \mathcal{F}(\lambda, \infty)$ renumber the components such that*

$$\frac{-\log q_1}{\gamma_1} \leq \frac{-\log q_2}{\gamma_2} \leq \cdots \leq \frac{-\log q_n}{\gamma_n}.$$

Then $\gamma \in \mathcal{F}(\lambda, \infty)$ if the following n inequalities hold:

$$\sum_{j=1}^{k} \gamma_j \leq \lambda \left(1 - \prod_{j=1}^{k} q_j \right) \quad k = 1, \ldots, n-1$$

$$\sum_{j=1}^{n} \gamma_j \leq \lambda \left(1 - \prod_{j=1}^{n} q_j \right)$$

and the complexity of checking for membership in $\mathcal{F}(\lambda, \infty)$ is $O(n \log n)$ steps.

Proof:

Note that after rearranging terms and taking logarithms equations (1) and (2) can be rewritten as

$$\sum_{j \in S} \log q_j \leq \log\left(1 - \frac{\sum_{j \in S} \gamma_j}{\lambda}\right) \quad S \subset \{1, 2, \ldots, n\}$$

$$\sum_{j \in N} \log q_j = \log\left(1 - \frac{\sum_{j \in N} \gamma_j}{\lambda}\right)$$

Let $g_1(S) = 1 - \frac{\sum_{j \in S} \gamma_j}{\lambda}$ and $g_2(S) = \sum_{j \in S} \log q_j$. Plot $g_1(S)$ versus $g_2(S)$. For a given value of γ if all the points in this graph lie below $\log(x)$ then the given γ is feasible. The following optimization problem is now solved for all values of $\theta > 0$.

$$S(\theta) = \text{Arg} \max_{S \subseteq N} -(1 - \frac{\sum_{j \in S} \gamma_j}{\lambda}) + \theta \sum_{j \in S} \log q_j$$

It is therefore enough to check the inequalities for all sets $S(\theta)$. This optimization problem is reformulated as

$$S(\theta) = \text{Arg} \max_{y} -(1 - \frac{\sum_j \gamma_j y_j}{\lambda}) + \theta \sum_j y_j \log q_j$$

$$y_j \in \{0, 1\} \quad j \in S.$$

This is a simple optimization problem and for a given value of θ, if

$$\frac{x_j}{\lambda} + \theta \log q_j > 0 \quad \text{imples } y_j = 1.$$

From the structure of the optimal solution notice that feasibility has to be checked for only n subsets and the result follows.

The problem of determining the maximum throughput is formulated as determining the largest λ^* such that the set $\mathcal{F}(\lambda^*, \infty) \cap \mathcal{B}(\mu)$ is not empty. Therefore for all $\lambda < \lambda^*$ there exits a feasible testing policy. Assume that λ is given. Note that since $\mathcal{B}(\mu)$ contains strict inequalities, the set $\mathcal{F}(\lambda, \mu)$ is not closed. An optimization problem is formulated to determine if $\mathcal{F}(\lambda, \infty) \cap \mathcal{B}(\mu)$ is empty.

Theorem 5 *Let*

$$z^* = \min_{\gamma \in \mathcal{F}(\lambda, \infty)} \max_j \frac{\gamma_j}{p_j \mu_j}$$

Then $\mathcal{F}(\lambda, \infty) \cap \mathcal{B}(\mu) \neq \emptyset$ *if and only if* $z^* < 1$.

The above theorem gives necessary and sufficient conditions for the existence of a testing policy for a given value of λ. In order to determine λ^*, first substitute $\frac{\gamma_j}{\lambda} = \phi_j$. The optimization problem in Theorem 5, can be restated as follows:

$$\lambda \min_{\phi \in \mathcal{F}(1, \infty)} \max_j \frac{\phi_j}{p_j \mu_j}$$

Ignore the term λ in the objective function and solve the optimization problem. If the optimal solution to this problem is w^*, then $\lambda^* = \frac{1}{w^*}$ is the upper bound on the achievable throughput. The algorithm MAX_LAMBDA(p, μ) shown in Figure 1, determines the value of λ^*. At this point the only quantity of interest

---

MAX_LAMBDA(p, μ)

INPUT : $(p_1, p_2, \ldots, p_n)$ and $(\mu_1, \mu_2, \ldots, \mu_n)$.
OUTPUT : The maximum throughput λ^* and vector γ^*.

1. Reorder the variables such that
$$\frac{-\log q_1}{p_1 \mu_1} \leq \frac{-\log q_2}{p_2 \mu_2} \leq \ldots \leq \frac{-\log q_n}{p_n \mu_n}.$$

2. Let $f(j) = \left(1 - \prod_{i \leq j} q_i\right)$ $g(j) = \sum_{i \leq j} \mu_i p_i$. $l = 1$

3. Let $\theta = \min_{l \leq j \leq n} \frac{f(j)}{g(j)}$ and $k = \text{Arg } \min_{l \leq j \leq n} \frac{f(j)}{g(j)}$.

4. If $k < n$ then let
$$f(j) = f(j) - f(k) \quad g(j) = g(j) - g(k) \quad \forall j > k.$$

Let $\gamma_j^* = \theta p_j \mu_j$ $l \leq j \leq k$ and set $l = k + 1$ and go to Step 2.

5. If $k = n$ then output $\lambda^* = \frac{1}{\theta}$ and γ^*.

Fig. 1. MAX_LAMBDA

is the value of λ^*. The vector γ^* that is also output by the algorithm is used in the next section to find the policy that achieves the maximum throughput.

Theorem 6 *The algorithm MAX_LAMBDA(p, μ) correctly computes the value of λ^* in $O(n^2)$ time.*

Proof:
(Outline) It can be shown [6], that

$$\text{Arg } \min_{\gamma \in \mathcal{F}(1, \infty)} \max_j \frac{\gamma_j}{p_j \mu_j} = \text{Arg } \min_{\gamma \in \mathcal{F}(1, \infty)} \sum_j \int_0^{\phi_i} \frac{x}{p_i \mu_i} dx.$$

Therefore the following optimization problem is solved:

$$\min_{\phi \in \mathcal{F}(1, \infty)} \sum_{j=1}^n \frac{\phi_j^2}{2 p_j \mu_j}.$$

Since the objective function is convex and separable, a (modified) greedy algorithm can be used to solve this problem. Due to the special structure of the membership test and the objective function, it can be shown that the potentially binding constraints always remain the same. This in turn leads an efficient algorithm. The running time is dominated by the finding the minimum in step 2 of the algorithm. For each iteration, finding the minimum is $O(n)$ and there are at most n iterations giving the the the $O(n^2)$ running time for the algorithm.

4   Determining the Representation

In the last section, the algorithm MAX_LAMBDA(p, μ) that determines the upper bound on the system throughput was outlined. In this section, the policy in class $\mathcal{C}$ that achieves this upper bound is derived. For any input rate that is less than λ^*, this policy can be used to guarantee that the system is stable. The vector γ^* output by the routine MAX_LAMBDA(p, μ) is used to derive this optimal policy. Since $\gamma^* \in \mathcal{F}(1, \infty)$, it can be represented as a convex combination of at most n extreme points of $\mathcal{F}(1, \infty)$. The objective of this section is to determine the appropriate extreme points along with their respective weights. A constructive proof of the representation theorem can be used to obtain the extreme points and their weights. It can be shown that (see for example [1]) that the representation for any point in convex polyhedron can be derived in $O(n^4)$. In this section an $O(n^3 \log n)$ algorithm is derived for generating the representation for a point $\gamma^* \in \mathcal{F}(1, \infty)$. Three routines that are used in deriving the representation are first defined. The first routine EXTREME_POINT(γ), takes as input a point $\gamma \in \mathcal{F}(1, \infty)$ and outputs an extreme point δ, that lies on the same face as γ, i.e., all the constraints that are binding at γ are also binding at δ. The algorithm EXTREME_POINT first renumbers the variables as given in Lemma 4. Then the vector δ is computed as follows:

$$\delta_1 = (1 - p_1)$$
$$\delta_k = \prod_{j=1}^{k-1} p_j (1 - p_k) \quad k = 2, 3, \ldots, n$$

A small modification can be made to the routine EXTREME_POINT(γ) to get the routine
NUM_BINDING_CONSTRAINTS(γ) that determines the number of binding constraints at a given point $\gamma \in \mathcal{F}(1, \infty)$. This is done as follows: Given a point $\gamma \in \mathcal{F}(1, \infty)$, the routine EXTREME_POINT(γ) is first executed. This routine outputs δ. Now the number of components where $\gamma_i = \delta_i$ is counted and is output by NUM_BINDING_CONSTRAINTS(γ). The third routine that is needed is EXIT_POINT(γ, d) which takes as input two n-vectors $\gamma \in \mathcal{F}(1, \infty)$ and a direction vector d. The routine outputs δ where

$$\delta = \sup_{\phi} \gamma + \phi d \in \mathcal{F}(1, \infty).$$

In other words, starting from the given point $\gamma \in \mathcal{F}(1, \infty)$ a ray is traced in the direction d. Let δ represent the point at which this ray exits from $\mathcal{F}(1, \infty)$. In the description of the algorithms below, the following two operations are defined on permutation $\pi = \{\pi(1), \pi(2), \ldots, \pi(n)\}$ of $\{1, 2, \ldots, n\}$. Let $T_j(\pi)$ to be the operation that finds a k such that $\pi(k) = j$ and outputs the the permutation $\{\pi(1), \pi(2), \ldots, \pi(k+1), \pi(k), \pi(k+2), \ldots \pi(n)\}$. Let $W_j(\pi)$ be the operation that determines a k such that $\pi(k) = j$ and outputs the set $\{\pi(1), \pi(2), \ldots, \pi(k)\}$. The description of EXIT_POINT(γ, d) is given in Figure 2.

EXIT_POINT(γ, d)

INPUT : $\gamma = (\gamma_1, \gamma_2, \ldots, \gamma_n) \in \mathcal{F}(1, \infty)$ and $d = (d_1, d_2, \ldots, d_n)$.
OUTPUT : $\alpha = (\alpha_1, \alpha_2, \ldots, \alpha_n) = \sup_\beta \gamma + \beta d \in \mathcal{F}(1, \infty)$.

1. The variables are ordered such that

$$\frac{-\log q_1}{\gamma_1} \leq \frac{-\log q_2}{\gamma_2} \leq \ldots \leq \frac{-\log q_n}{\gamma_n}.$$

2. Set $S_j = \{1, 2, \ldots, j\}$ $j = 1, 2, \ldots, n$. Let π_0 represent the identity permutation.
3. Compute

$$\theta_{ij} = \frac{\gamma_i \log q_j - \gamma_j \log q_i}{d_j \log q_i - d_i \log q_j} \qquad \forall i < j \leq n.$$

4. Let the p of these θ_{ij} values be positive. These positive values are sorted so that

$$\theta_{i_1 j_1} \leq \theta_{i_2 j_2} \leq \ldots \theta_{i_p j_p}.$$

5. Form the sets

$$S_{n+k} = V_{j_k} T_{i_k} T_{i_{k-1}} \ldots T_{i_1} \pi_0 \qquad k = 1, 2, \ldots, p$$

6. The value of β^* is computed as follows:

$$\beta^* = \min_{j=1,2,\ldots n+p} \frac{\left(1 - \prod_{k \in S_j} q_k\right) - \sum_{k \in S_j} \gamma_j}{\sum_{j \in S_j} d_k}.$$

7. Output $\gamma + \beta^* d$.

Fig. 2. EXIT_POINT

Theorem 7 *The algorithm EXIT_POINT(γ, d) determines the point on the polyhedron $\mathcal{F}(1, \infty)$ where a ray starting from γ in the direction d exits from $\mathcal{F}(1, \infty)$. The running time of the algorithm is $O(n^2 \log n)$.*

Proof:
(Outline) The sets $S_1, S_2, \ldots, S_n$ are the sets representing the potentially binding constraints at the start point γ. As the ray moves away from γ in the

direction d the potentially binding constraints can change. However, it can be shown that at most n^2 additional constraints have to be added to the problem. These additional constraints are represented by the sets $S_{n+1}, S_{n+2}, \ldots, S_{n+p}$. The running time of the algorithm is dominated by the sorting of potentially n^2 values in step 3 and hence the running time of $O(n^2 \log n)$.

Figure 4 gives a description of algorithm REPRESENTATION(γ) that is used to derive a policy that achieves the desired throughput vector γ.

REPRESENTATION(γ)

INPUT : $\gamma = (\gamma_1, \gamma_2, \ldots, \gamma_n) \in \mathcal{F}(1, \infty)$.
OUTPUT : Extreme points $Q_1, Q_2, \ldots, Q_k$ and weights $w_1, w_2, \ldots w_k$ such that

$$\gamma = \sum_{j=1}^{k} w_j Q_j.$$

1. Let $i = 1$, $P_1 = \gamma$, $B = \text{NUM\_BINDING\_CONST}(\gamma)$. If $B = n$ then set output i, P_1 and $w_1 = 1$. If $B < n$ then go to step 2.
2. $Q_i = \text{EXTREME\_POINT}(P_i)$.
3. $P_{i+1} = \text{EXIT\_POINT}(P_i, Q_i - P_i)$
4. Determine ϕ_i such that $P_i = \phi_i Q_i + (1 - \phi_i) P_{i+1}$.
5. Let $B = \text{NUM\_BINDING\_CONST}(P_i)$. If $B = n$ then go to step 6. If $B < n$ go to step 2.
6. $Q_1, Q_2, \ldots Q_k$ are the k extreme points. Compute

$$w_i = \prod_{j=1}^{i-1}(1 - \phi_j)\phi_i \qquad \text{for } i = 1, 2, \ldots, k-1.$$

$$w_k = \prod_{j=1}^{k-1}(1 - \phi_j).$$

Output the vectors Q and w.

Fig. 3. REPRESENTATION

Theorem 8 *The algorithm REPRESENTATION(γ) obtains $k \leq n$ extreme $Q_1, Q_2, \ldots, Q_k$ each corresponding to a different permutation of the testing vectors and a corresponding set of weights $w_1, w_2, \ldots, w_k$ such that*

$$\gamma = \sum_{j=1}^{k} w_j Q_j.$$

The running time complexity is $O(n^3 \log n)$.

Proof:
(Outline) In the first iteration of the algorithm, an extreme point on the same face as γ is determined. The algorithm traces a ray from this extreme point to γ and determines where this ray exits from the polyhedron using the EXIT_POINT routine. Note that the number of binding constraints at this point where the ray exits the polyhedron is at least one less than the number of binding constraints at γ. Note that γ is a convex combination of the extreme point determined and this exit point. This exit point is now taken as the new point and this process is repeated. This leads to the extreme points that can be used to represent γ. There can be at most $n-1$ iterations and the bottleneck operation in each iteration is the EXIT_POINT routine that takes time $O(n^2 \log n)$.

This leads to the main result of this paper.

Theorem 9 *Given any $\lambda < \lambda^*$, if the policy output by REPRESENTATION(γ^*) is used to test the strings then the number of strings in the buffer will be bounded.*

5   Conclusion

The problem of maximizing the throughput of a sequential testing system was posed as an optimization problem on a polymatroid. Polynomial time combinatorial algorithms were developed for getting the maximum throughput value as well as the policy that achieves this maximum. The techniques developed in this paper can be used to address other stable routing problems.

References

1. Bazaraa, M.S., Jarvis, J.J, and Sherali, H.D., *Linear Programming and Network Flows*, John Wiley and Sons, 1990.
2. Bertsimas, D., and Nino-Mora, J., "Conservation laws, extended polymatroids and multi-armed bandit problems: A unified approach to indexable systems," *Mathematics of Operations Research*
3. Coffman, E., and Mitrani, I., "A characterization of waiting time performance realizable by single server queues," *Operations Research 28*, 1980, pp. 810-821.
4. Federgruen, A., and Groenevelt, H., "M/G/c Queueing systems with multiple customer classes: Characterization and control of achievable performance under non-preemptive priority rules," *Management Science 34*, 1988, pp. 1121-1138.
5. Federgruen, A., and Groenevelt, H., "The greedy procedure for resource allocation problems: Necessary and sufficient conditions for optimality," *Operations Research 34*, 1988, pp. 909-918.
6. Fujishige, S., *Submodular functions and optimization*, North-Holland, 1991.
7. Gelenbe, E., and Mitrani, I., *Analysis and synthesis of computer systems*, Academic Press, 1980.
8. Grotschel, M., Lovasz, L., and Schrijver, A., "The ellipsoid method and its consequences in combinatorial optimization," *Cominatorica 1*, 1981, pp. 169-197.
9. Lawler, E., and Martel, C., "Computing maximal polymatroidal network flows," *Math. of Operations Research, 7*, 1982, pp. 334-348.

10. Ross, K., and Yao, D., "Optimal dynamic scheduling in Jackson networks," *IEEE Transaction on Automatic Control 34*, 1989, pp. 47-53.
11. Welsh, D., *Matroid Theory*, Academic Press, 1976.

An Explicit Exact SDP Relaxation for Nonlinear 0-1 Programs

Jean B. Lasserre

LAAS-CNRS, 7 Avenue du Colonel Roche,
31077 Toulouse Cédex, France
lasserre@laas.fr
http://www.laas.fr/~lasserre

Abstract. We consider the general nonlinear optimization problem in 0-1 variables and provide an explicit equivalent convex positive semidefinite program in $2^n - 1$ variables. The optimal values of both problems are identical. From every optimal solution of the former one easily find an optimal solution of the latter and conversely, from every solution of the latter one may construct an optimal solution of the former.

1 Introduction

This paper is concerned with the general nonlinear problem in 0-1 variables

$$\mathbb{P} \;\to\; p^* := \min_{x \in \{0,1\}^n} \{g_0(x) \,|\, g_k(x) \geq 0,\; k = 1, \ldots m\} \qquad (1)$$

where all the $g_k(x) : \mathbb{R}^n \to \mathbb{R}$ are real-valued polynomials of degree $2v_k - 1$ if odd or $2v_k$ if even. This general formulation encompasses 0-1 linear and nonlinear programs, among them the quadratic assignment problem. For the MAX-CUT problem, the discrete set $\{0,1\}^n$ is replaced by $\{-1,1\}^n$.

In our recent work [4,5,6], and for general optimization problems involving polynomials, we have provided a sequence $\{\mathbb{Q}_i\}$ of positive semidefinite (psd) programs (or SDP relaxations) with the property that $\inf \mathbb{Q}_i \uparrow p^*$ as $i \to \infty$, under a certain assumption on the semi-algebraic feasible set $\{g_k(x) \geq 0,\; k = 1, \ldots m\}$. The approach was based on recent results in algebraic geometry on the representation of polynomials, positive on a compact semi-algebraic set, a theory dual to the theory of *moments*. For general $0 - 1$ programs, that is, with the additional constraint $x \in \{0,1\}^n$, we have shown in Lasserre [5] that this assumption on the feasible set is automatically satisfied and the SDP relaxations $\{\mathbb{Q}_i\}$ simplify to a specific form with at most $2^n - 1$ variables, no matter how large is i. The approach followed in [4] and [5] is different in spirit from the *lift and project* iterative procedure of Lovász and Schrijver [7] for 0-1 programs (see also extensions in Kojima and Tunçel [3]), which requires that a weak separation oracle is available for the homogeneous cone associated with the constraints. In the *lift and project* procedure, the description of the convex hull of the feasible set is *implicit* (via successive projections) whereas we provide in this paper an

K. Aardal, B. Gerards (Eds.): IPCO 2001, LNCS 2081, pp. 293–303, 2001.

explicit description of the equivalent convex psd program, with a simple interpretation. Although different, our approach is closer in spirit to the successive LP relaxations in the RLT procedure of Sherali and Adams [10] for 0-1 linear programs in which each of the linear original constraints is multiplied by suitable polynomials of the form $\Pi_{i\in J_1} x_i \Pi_{j\in J_2}(1-x_j)$ and then linearized in a higher dimension space via several changes of variables to obtain an LP program. The last relaxation in the hierarchy of RLT produces the convex hull of the feasible set. This also extends to a special class of 0-1 polynomial programs (see Sherali and Adams [10]).

In the present paper, we show that in addition to the asymptotic convergence already proved in [4,5], the sequence of convex SDP relaxations $\{Q_i\}$ is in fact **finite**, that is, the optimal value p^* is also $\min Q_i$ for all $i \geq n + v$ with $v := \max_k v_k$. Moreover, every optimal solution y^* of Q_i is the (finite) vector of moments of some probability measure supported on optimal solutions of $\mathbb{P}$. Therefore, every 0-1 program is in fact **equivalent** to a continuous convex psd program in $2^n - 1$ variables for which an explicit form as well as a simple interpretation are available. The projection of the feasible set defined by the LMI constraints of this psd program onto the subspace spanned by the first n variables, provides the convex hull of the original feasible set. Note that the result holds for **arbitrary** 0-1 constrained programs, that is, with arbitrary polynomial criteria and constraints (no weak separation oracle is needed as in the lift and project procedure [7], and no special form of $\mathbb{P}$ is assumed as in [10]). This is because the theory of representation of polynomials positive on a compact semi-algebraic set and its dual theory of moments, make no assumption on the semi-algebraic set, except compactness (it can be nonconvex, disconnected). For illustration purposes, we provide the equivalent psd programs for quadratic 0-1 programs and MAX-CUT problems in $\mathbb{R}^3$.

For practical computational purposes, the preceding result is of little value for the number of variables grows exponentially with the size of the problem. Fortunately, in many cases, the optimal value is also the optimal value of some Q_i for $i \ll n$. For instance, on a sample of 50 randomly generated MAX-CUT problems in $\mathbb{R}^{10}$, the optimal value p^* was always obtained at the Q_2 relaxation (in which case Q_2 is a psd program with "only" 385 variables to compare with $2^{10} - 1 = 1023$).

2   Notation and Definitions

We adopt the notation in Lasserre [4] that for sake of clarity, we reproduce here. Given any two real-valued symmetric matrices A, B let $\langle A, B \rangle$ denote the usual scalar product trace(AB) and let $A \succeq B$ (resp. $A \succ B$) stand for $A - B$ positive semidefinite (resp. $A - B$ positive definite). Let

$$1, x_1, x_2, \ldots x_n, x_1^2, x_1 x_2, \ldots, x_1 x_n, x_2 x_3, \ldots, x_n^2, \ldots, x_1^r, \ldots, x_n^r, \qquad (2)$$

be a basis for the r-degree real-valued polynomials and let $s(r)$ be its dimension. Therefore, a r-degree polynomial $p(x) : \mathbb{R}^n \to \mathbb{R}$ is written

$$p(x) = \sum_\alpha p_\alpha x^\alpha, \qquad x \in \mathbb{R}^n,$$

where $x^\alpha = x_1^{\alpha_1} x_2^{\alpha_2} \ldots x_n^{\alpha_n}$. Denote by $p = \{p_\alpha\} \in \mathbb{R}^{s(r)}$ the coefficients of the polynomial $p(x)$ in the basis (2). Hence, the respective vectors of coefficients of the polynomials $g_i(x)$, $i = 0, 1, \ldots m$, in (1), are denoted $\{(g_i)_\alpha\} = g_i \in \mathbb{R}^{s(w_i)}$, $i = 0, 1, \ldots m$ if w_i is the degree of g_i. We next define the important notions of moment matrix and localizing matrix.

2.1   Moment Matrix

Given a $s(2r)$-sequence $(1, y_1, \ldots,)$, let $M_r(y)$ be the **moment** matrix of dimension $s(r)$ (denoted $M(r)$ in Curto and Fialkow [1]), with rows and columns labelled by (2). For instance, for illustration purposes, consider the 2-dimensional case. The moment matrix $M_r(y)$ is the block matrix $\{M_{i,j}(y)\}_{0 \le i,j \le 2r}$ defined by

$$M_{i,j}(y) = \begin{bmatrix} y_{i+j,0} & y_{i+j-1,1} & \cdots & y_{i,j} \\ y_{i+j-1,1} & y_{i+j-2,2} & \cdots & y_{i-1,j+1} \\ \cdots & \cdots & \cdots & \cdots \\ y_{j,i} & y_{i+j-1,1} & \cdots & y_{0,i+j} \end{bmatrix}. \tag{3}$$

To fix ideas, when $n = 2$ and $r = 2$, one obtains

$$M_2(y) = \begin{bmatrix} 1 & y_{10} & y_{01} & y_{20} & y_{11} & y_{02} \\ y_{10} & y_{20} & y_{11} & y_{30} & y_{21} & y_{12} \\ y_{01} & y_{11} & y_{02} & y_{21} & y_{12} & y_{03} \\ y_{20} & y_{30} & y_{21} & y_{40} & y_{31} & y_{22} \\ y_{11} & y_{21} & y_{12} & y_{31} & y_{22} & y_{13} \\ y_{02} & y_{12} & y_{03} & y_{22} & y_{13} & y_{04} \end{bmatrix}.$$

Another more intuitive way of constructing $M_r(y)$ is as follows. If $M_r(1, i) = y_\alpha$ and $M_r(y)(j, 1) = y_\beta$, then

$$M_r(y)(i, j) = y_{\alpha+\beta}, \quad \text{with } \alpha + \beta = (\alpha_1 + \beta_1, \cdots, \alpha_n + \beta_n). \tag{4}$$

$M_r(y)$ defines a bilinear form $\langle ., . \rangle_y$ on the space $\mathcal{A}_r$ of polynomials of degree at most r, by

$$\langle q(x), v(x) \rangle_y := \langle q, M_r(y)v \rangle, \quad q(x), v(x) \in \mathcal{A}_r,$$

and if y is a sequence of moments of some measure μ_y, then $M_r(y) \succeq 0$ because

$$\langle q, M_r(y)q \rangle = \int q(x)^2 \, \mu_y(dx) \ge 0, \quad \forall q \in \mathbb{R}^{s(r)}. \tag{5}$$

2.2   Localizing Matrix

If the entry (i, j) of the matrix $M_r(y)$ is y_β, let $\beta(i, j)$ denote the subscript β of y_β. Next, given a polynomial $\theta(x) : \mathbb{R}^n \to \mathbb{R}$ with coefficient vector θ, we define the matrix $M_r(\theta y)$ by

$$M_r(\theta y)(i, j) = \sum_\alpha \theta_\alpha y_{\{\beta(i,j)+\alpha\}}. \qquad (6)$$

For instance, with $x \mapsto \theta(x) = a - x_1^2 - x_2^2$, we obtain

$$M_1(\theta y) = \begin{bmatrix} a - y_{20} - y_{02}, & ay_{10} - y_{30} - y_{12}, ay_{01} - y_{21} - y_{03} \\ ay_{10} - y_{30} - y_{12}, ay_{20} - y_{40} - y_{22}, ay_{11} - y_{31} - y_{13} \\ ay_{01} - y_{21} - y_{03}, ay_{11} - y_{31} - y_{13}, ay_{02} - y_{22} - y_{04} \end{bmatrix}.$$

If y is a sequence of moments of some measure μ_y, then

$$\langle q, M_r(\theta y)q \rangle = \int \theta(x)q(x)^2 \, \mu_y(dx), \qquad \forall q \in \mathbb{R}^{s(r)}, \qquad (7)$$

so that $M_r(\theta y) \succeq 0$ whenever μ_y has its support contained in the set $\{\theta(x) \geq 0\}$. In Curto and Fialkow [1], $M_r(\theta y)$ is called a *localizing* matrix.

The theory of moments identifies those sequences $\{y_\alpha\}$ with $M_r(y) \succeq 0$, such that the y_α's are moments of a *representing* measure μ_y. The $\mathbb{K}$-moment problem identifies those y_α's whose representing measure μ_y has its support contained in the semi-algebraic set $\mathbb{K}$. In duality with the theory of moments is the theory of representation of positive polynomials (and polynomials positive on a semi-algebraic set $\mathbb{K}$), which dates back to Hilbert's 17th problem. For details and recent results, the interested reader is referred to Curto and Fialkow [1], Schmüdgen [9] and the many references therein.

Both sides (moments and positive polynomials) of this same theory is reflected in the primal and dual SDP relaxations proposed in Lasserre [4], and used below in the present context of nonlinear 0-1 programs.

3   Main Result

Consider the 0-1 optimization problem $\mathbb{P}$ in (1). Let the semi-algebraic set

$$\mathbb{K} := \{x \in \{0, 1\}^n \mid g_i(x) \geq 0, \, i = 1, \ldots m\}$$

be the feasible set. For the sake of simplicy in the proofs derived later on, we treat the 0-1 integrality constraints $x_i^2 = x_i$ as two opposite inequalities $g_{m+i}(x) = x_i^2 - x_i \geq 0$ and $g_{m+n+i}(x) = x_i - x_i^2 \geq 0$, and we redefine the set $\mathbb{K}$ to be

$$\mathbb{K} = \{g_i(x) \geq 0, \, i = 1, \ldots m + 2n\}. \qquad (8)$$

However, in view of the special form of the constraints $g_{m+k}(x) \geq 0$, $k = 1, \ldots 2n$, we will provide a simpler form of the LMI relaxations $\{Q_i\}$ below.

Depending on its parity, let $w_k := 2v_k$ or $w_k := 2v_k - 1$ be the degree of the polynomial $g_k(x)$, $k = 1, \ldots m + 2n$. When needed below, for $i \geq \max_k w_k$, the vectors $g_k \in \mathbb{R}^{s(w_k)}$ are extended to vectors of $\mathbb{R}^{s(i)}$ by completing with zeros. As we minimize g_0 we may and will assume that its constant term is zero, that is, $g_0(0) = 0$.

For $i \geq \max_k v_k$, consider the following family $\{\mathbb{Q}_i\}$ of convex psd programs

$$\mathbb{Q}_i \left\{ \begin{array}{c} \inf\limits_{y} \sum\limits_{\alpha} (g_0)_\alpha y_\alpha \\ M_i(y) \succeq 0 \\ M_{i-v_k}(g_k y) \succeq 0, \quad k = 1, \ldots m + 2n, \end{array} \right. \tag{9}$$

with respective dual problems

$$\mathbb{Q}_i^* \left\{ \begin{array}{c} \sup\limits_{X, Z_k \succeq 0} -X(1,1) - \sum\limits_{k=1}^{m+2n} g_k(0) Z_k(1,1) \\ \langle X, B_\alpha \rangle + \sum\limits_{k=1}^{m+2n} \langle Z_k, C_\alpha^k \rangle = (g_0)_\alpha, \; \forall \alpha \neq 0, \end{array} \right. \tag{10}$$

where we have written

$$M_i(y) = \sum_{\alpha} B_\alpha y_\alpha; \; M_{i-v_k}(g_k y) = \sum_{\alpha} C_\alpha^k y_\alpha, \; k = 1, \ldots m + 2n,$$

for appropriate real-valued symmetric matrices $B_\alpha, C_\alpha^k, k = 1, \ldots m + 2n$.

Interpretation of $\mathbb{Q}_i$. The LMI constraints of $\mathbb{Q}_i$ state necessary conditions for y to be the vector of moments up to order $2i$, of some probability measure μ_y with support contained in $\mathbb{K}$. This clearly implies that $\inf \mathbb{Q}_i \leq p^*$ because the vector of moments of the Dirac measure at a feasible point of $\mathbb{P}$, is feasible for $\mathbb{Q}_i$.

Interpretation of $\mathbb{Q}_i^*$. Let $X, Z_k \succeq 0$ be a feasible solution of $\mathbb{Q}_i^*$ with value ρ. Write

$$X = \sum_j t_j t_j' \quad \text{and} \quad Z_k = \sum_j t_{kj} t_{kj}', \; k = 1, \ldots m + 2n.$$

Then, from the feasibility of (X, Z_k) in $\mathbb{Q}_i^*$, it was shown in Lasserre [4,5] that

$$g_0(x) - \rho = \sum_j t_j(x)^2 + \sum_{k=1}^{m+2n} g_k(x) \left[\sum_j t_{kj}(x)^2 \right], \tag{11}$$

where the polynomials $\{t_j(x)\}$ and $\{t_{kj}(x)\}$ have respective coefficient vectors $\{t_j\}$ and $\{t_{kj}\}$, in the basis (2).

As $\rho \leq p^*$, $g_0(x) - \rho$ is nonnegative on $\mathbb{K}$ (strictly positive if $\rho < p^*$). One recognizes in (11) a decomposition into a weighted sum of squares of the polynomial $g_0(x) - p^*$, strictly positive on $\mathbb{K}$, as in the theory of representation of polynomials, strictly positive on a compact semi-algebraic set $\mathbb{K}$ (see e.g., Schmüdgen [9], Putinar [8], Jacobi and Prestel [2]). Indeed, when the set $\mathbb{K}$ has a certain property (satisfied here), the "linear" representation (11) holds (see Lasserre [4,5]). Hence, both programs $\mathbb{Q}_i$ and $\mathbb{Q}_i^*$ illustrate the duality between the theory of moments and the theory of positive polynomials. Among other results, it has been shown in Lasserre [5, Th. 3.3] that

$$\inf \mathbb{Q}_i \uparrow p^* \quad \text{as } i \to \infty. \tag{12}$$

From the construction of the localizing matrices in (6), and the form of the polynomials g_i for $i > m$, the constraints $M_{i-1}(g_{m+k}y) \succeq 0$ and $M_{i-1}(g_{m+n+k}y) \succeq 0$ for $k = 1, \ldots n$ (equivalently $M_{i-1}(g_{m+k}y) = 0$), simply state that the variable y_α with $\alpha = (\alpha_1, \ldots \alpha_n)$, can be replaced with the variable y_β with $\beta_i := 1$ whenever $\alpha_i \geq 1$. Therefore, a simpler form of $\mathbb{Q}_i$ is obtained as follows.

Ignore the constraints $M_{i-v_k}(g_k y) \succeq 0$ for $k = m+1, \ldots m+2n$, and make the above substitution $y_\alpha \to y_\beta$ in the matrices $M_i(y)$ and $M_{i-v_k}(g_k(y))$, $k = 1, \ldots m$. For instance, in $\mathbb{R}^2$ ($n = 2$), the matrix $M_2(y)$ now reads

$$M_2(y) = \begin{bmatrix} 1 & y_{10} & y_{01} & y_{10} & y_{11} & y_{01} \\ y_{10} & y_{10} & y_{11} & y_{10} & y_{11} & y_{11} \\ y_{01} & y_{11} & y_{01} & y_{11} & y_{11} & y_{01} \\ y_{10} & y_{10} & y_{11} & y_{10} & y_{11} & y_{11} \\ y_{11} & y_{11} & y_{11} & y_{11} & y_{11} & y_{11} \\ y_{01} & y_{11} & y_{01} & y_{11} & y_{11} & y_{01} \end{bmatrix}, \tag{13}$$

and only the variables y_{10}, y_{01}, y_{11} appear in all the relaxations $\mathbb{Q}_i$. Interpreted in terms of polynomials, the integrality constraints $x_i^2 = x_i$, $i = 1, \ldots n$, imply that a monomial $x_1^{\alpha_1} x_2^{\alpha_2} \cdots x_n^{\alpha_n}$ can be replaced by

$$x_1^{\beta_1} x_2^{\beta_2} \cdots x_n^{\beta_n} \quad \text{with } \beta_i = \begin{cases} 0 \text{ if } \alpha_i = 0 \\ 1 \text{ if } \alpha_i \geq 1 \end{cases} \tag{14}$$

Therefore, there are are no more than $2^n - 1$ variables y_β, the number of monomials as in (14), of degree at most n.

Remark 1. At any relaxation $\mathbb{Q}_i$, the matrix $M_i(y)$ can be reduced in size. When looking at the kth column $M_i(y)(., k)$, if $M_i(y)(1, k) = M_i(y)(p)$ for some $p < k$, then the whole column $M_i(y)(., k)$ as well as the corresponding line $M_i(y)(k, .)$ can be deleted. For instance, with $M_2(y)$ as in (13), the constraint $M_2(y) \succeq 0$ can be replaced by

$$\widehat{M_2}(y) := \begin{bmatrix} 1 & y_{10} & y_{01} & y_{11} \\ y_{10} & y_{10} & y_{11} & y_{11} \\ y_{01} & y_{11} & y_{01} & y_{11} \\ y_{11} & y_{11} & y_{11} & y_{11} \end{bmatrix} \succeq 0,$$

for the 4th and 6th columns of $M_2(y)$ are identical to the 2nd and 3rd columns, respectively. Thus, in the simplified form $\widehat{M_i}(y)$ of $M_i(y)$, one retains only the columns (and the rows) corresponding to the monomials in the basis (2) that are *distinct* after the simplification $x_i^2 = x_i$. The same simplification occurs for the matrices of the LMI constraints $M_{i-v_k}(g_k y) \succeq 0$, $k = 1, \dots m$.

Next, we begin with the following crucial result:

Proposition 1. *(a) All the relaxations $\mathbb{Q}_i$ involve at most $2^n - 1$ variables y_α.*
(b) For all the relaxations $\mathbb{Q}_i$ with $i > n$, one has

$$\operatorname{rank} M_i(y) = \operatorname{rank} M_n(y). \tag{15}$$

Proof. (a) is just a consequence of the comment preceding Proposition 1.
To get (b) observe that with $i > n$, one may write

$$M_i(y) = \begin{bmatrix} M_n(y) \mid B \\ - \quad - \\ B' \quad \mid C \end{bmatrix}$$

for appropriate matrices B, C, and we next prove that each column of B is identical to some column of $M_n(y)$. Indeed, remember from (4) how an element $M_i(y)(k, p)$ can be obtained. Let $y_\gamma = M_i(y)(k, 1)$ and $y_\alpha = M_i(y)(1, p)$. Then

$$M_i(y)(k, p) = y_\eta \quad \text{with } \eta_i = \gamma_i + \alpha_i, \ i = 1, \dots n. \tag{16}$$

Now, consider a column B_j of B, that is, some column $M_i(y)(., p)$ of $M_i(y)$, with first element $y_\alpha = B_j(1) = M_i(y)(1, p)$. Therefore, the element $B(k)$ (or, equivalently, the element $M_i(y)(k, p)$) is the variable y_η in (16). Note that α correspond to a monomial in the basis (2) of degree larger than n, say $\alpha_1 \cdots \alpha_n$. Associate to this column B_j the column $v := M_n(y)(., q)$ of $M_n(y)$ whose element $v(1) = M_n(y)(1, q) = y_\beta$ (for some q) with $\beta_i = 0$ if $\alpha_i = 0$ and 1 otherwise, for all $i = 1, \dots n$. Then, the element $v(k) = M_n(y)(k, q)$ is obtained as

$$v(k) = y_\delta \quad \text{with } \delta_i = \gamma_i + \beta_i, \ i = 1, \dots n.$$

But then, as for each entry y_α of $M_j(y)$, we can make the substitution $\alpha_i \leftrightarrow 1$ whenever $\alpha_i \geq 1$, it follows that the element $v(k)$ is identical to the element $B(k)$. In other words, each column of B is identical to some column of $M_n(y)$.

If we now write $M_j(y) = [A|D]$ with $A := \begin{bmatrix} M_n(y) \\ B' \end{bmatrix}$, and $D := \begin{bmatrix} B \\ C \end{bmatrix}$, then, with exactly same arguments, every column of D is also identical to some column of A, and consequently, (15) follows. $\square$

For instance, when $n = 2$ the reader can check that $M_3(y)$ has same rank as $M_2(y)$ in (13). We now can state our main result:

Theorem 1. *Let $\mathbb{P}$ be the problem defined in (1) and let $v := \max_{k=1,\dots m} v_k$. Then for every $i \geq n + v$:*
(a) $\mathbb{Q}_i$ is solvable with $p^* = \min \mathbb{Q}_i$, and to every optimal solution x^* of $\mathbb{P}$ corresponds the optimal solution

$$y^* := (x_1^*, \cdots, x_n^*, \cdots, (x_1^*)^{2i}, \cdots, (x_n^*)^{2i}) \tag{17}$$

of $\mathbb{Q}_i$.

(b) Every optimal solution y^* of $\mathbb{Q}_i$ is the (finite) vector of moments of a probability measure finitely supported on s optimal solutions of $\mathbb{P}$, with $s = \operatorname{rank} M_i(y) = \operatorname{rank} M_n(y)$.

Proof. Let y be an admissible solution of $\mathbb{Q}_{n+v}$. From Proposition 1, we have that $\operatorname{rank} M_i(y) = \operatorname{rank} M_n(y)$ for all $i > n$ (in particular for $i = n + v$). From a deep result of Curto and Fialkow [1, Th 1.6, p. 6], it follows that y is the vector of moments of some $\operatorname{rank} M_n(y)$ atomic measure μ_y. $M_{n+1}(y)$ is called a *flat positive extension* of $M_n(y)$ and it follows that $M_{n+1}(y)$ admits unique flat extension moment matrices $M_{n+2}(y)$, $M_{n+3}(y)$, etc. (see Curto and Fialkow [1, p. 3]). Moreover, from the constraints $M_{n+v-v_k}(g_k y) \succeq 0$, for all $k = 1, \ldots m + 2n$, it also follows that μ_y is supported on $\mathbb{K}$. Theorem 1.6 in Curto and Fialkow [1] is stated in dimension 2 for the complex plane, but is valid for n real or complex variables (see comments page 2 in [1]). In the notation of Theorem 1.6 in Curto and Fialkow [1], we have $M(n) \succeq 0$ and $M(n+1)$ has a flat positive extension $M(n+1)$ (hence unique flat positive extensions $M(n+k)$ for all $k \geq 1$), with $M_{g_k}(n+v_k) \succeq 0$, $k = 1, \ldots m + 2n$.

But then, as μ_y is supported on $\mathbb{K}$, it also follows that

$$\sum_\alpha (g_0)_\alpha y_\alpha = \int_{\mathbb{K}} g_0(x)\, \mu_y(dx) \geq p^*.$$

From this and $\inf \mathbb{Q}_i \leq p^*$, the first statement (a) in Theorem 1 follows for $i = n + v$. For $i > n + v$, the result follows from $p^* \geq \inf \mathbb{Q}_{i+1} \geq \inf \mathbb{Q}_i$, for all i. Finally, y^* in (17) is obviously admissible for $\mathbb{Q}_i$ with value p^*, and therefore, is an optimal solution of $\mathbb{Q}_i$.

(b) follows from same arguments as in (a). First observe that from (a), $\mathbb{Q}_i$ is solvable for all $i \geq n+v$. Now, let y^* be an optimal solution of $\mathbb{Q}_i$. From (a), y^* is the vector of moments of an atomic measure μ_{y^*} supported on $s (= \operatorname{rank} M_n(y))$ points $x^1, \cdots x^s \in \mathbb{K}$, that is, with $\delta_{\{.\}}$ the Dirac measure at ".",

$$\mu_{y^*} = \sum_{j=1}^s \gamma_j \delta_{x_j} \text{ with } \gamma_j \geq 0, \ \sum_{j=1}^s \gamma_j = 1.$$

Therefore, from $g_0(x^j) \geq p^*$ for all $j = 1, \ldots s$, and

$$p^* = \min \mathbb{Q}_i = \sum_\alpha (g_0)_\alpha y_\alpha^*$$

$$= \int_{\mathbb{K}} g_0(x)\, \mu_{y^*}(dx) = \sum_{j=1}^s \gamma_j g_0(x^j),$$

it follows that each point x^j must be an optimal solution of $\mathbb{P}$.   □

Hence, Theorem 1 shows that every 0-1 program is equivalent to a convex psd program with $2^n - 1$ variables. The projection of the feasible set of $\mathbb{Q}_{n+v}$ onto the space spanned by the n variables $y_{10...0}, \ldots, y_{0...01}$ is the convex hull of $\mathbb{K}$ in (8).

Interpretation . The interpretation of the relaxation $\mathbb{Q}_{n+v}$ is as follows. The unknown variables $\{y_\alpha\}$ should be interpreted as the moments of some probability measure μ, up to order $n + v$. The LMI constraints $M_{n+v-v_k}(g_k y) \succeq 0$ state that

$$\int g_k(x) q(x)^2 \, d\mu \geq 0, \qquad \text{for all polynomials } q \text{ of degree at most } n,$$

(as opposed to only $g_k(x) \geq 0$ in the original description). As the integral constraints $x_i^2 = x_i$ are included in the constraints, and due to the result of Curto and Fialkow, the above constraint (for $k = 1, \ldots m + 2n$) will ensure that the support of μ is concentrated on $\{0,1\}^n \cap [\cap_{k=1}^m \{g_k(x) \geq 0\}]$.

For illustration purposes, we provide below the explicit description of the equivalent convex psd programs for quadratic 0-1 programs and MAX-CUT programs in $\mathbb{R}^3$, respectively.

3.1   Quadratic 0-1 Programs

Consider the quadratic program $\min\{x'Ax \mid x \in \{0,1\}^n\}$ for some real-valued symmetric matrix $A \in \mathbb{R}^{n \times n}$. As the only constraints are the integral constraints $x \in \{0,1\}^n$, with $n = 3$, it is equivalent to the convex psd program

$$\inf A_{11}y_{100} + A_{22}y_{010} + A_{33}y_{001} + 2A_{12}y_{110} + 2A_{13}y_{101} + 2A_{23}y_{011}$$

$$\widehat{M}_3(y) = \begin{bmatrix} 1 & y_{100} & y_{010} & y_{001} & y_{110} & y_{101} & y_{011} & y_{111} \\ y_{100} & y_{100} & y_{110} & y_{101} & y_{110} & y_{101} & y_{111} & y_{111} \\ y_{010} & y_{110} & y_{010} & y_{011} & y_{110} & y_{111} & y_{011} & y_{111} \\ y_{001} & y_{101} & y_{011} & y_{001} & y_{111} & y_{101} & y_{011} & y_{111} \\ y_{110} & y_{110} & y_{110} & y_{111} & y_{110} & y_{111} & y_{111} & y_{111} \\ y_{101} & y_{101} & y_{111} & y_{101} & y_{111} & y_{101} & y_{111} & y_{111} \\ y_{011} & y_{111} & y_{011} & y_{011} & y_{111} & y_{111} & y_{011} & y_{111} \\ y_{111} & y_{111} & y_{111} & y_{111} & y_{111} & y_{111} & y_{111} & y_{111} \end{bmatrix} \succeq 0,$$

where $\widehat{M}_3(y)$ is the simplified form of $M_3(y)$ (see Remark 1).

3.2   MAX-CUT Problem

In this case, $g_0(x) := xAx$ for some real-valued symmetric matrix $A = \{A_{ij}\} \in \mathbb{R}^{n \times n}$ with null diagonal and the constraint set is $\{-1,1\}^n$. Now, we look for solutions in $\{-1,1\}^n$, which yields the following obvious modifications in the relaxations $\{\mathbb{Q}_i\}$. In the matrix $M_i(y)$, replace any entry y_α by y_β with $\beta_i = 1$ whenever α_i is odd, and $\beta_i = 0$ otherwise. As for 0-1 programs, the matrix $M_i(y)$

can be reduced in size to a simplified form $\widehat{M}_i(y)$ (by eliminating identical rows and columns). Hence, MAX-CUT in $\mathbb{R}^3$ is equivalent to the psd program

$$\inf A_{12}y_{110} + A_{13}y_{101} + A_{23}y_{011}$$

$$\widehat{M}_3(y) = \begin{bmatrix} 1 & y_{100} & y_{010} & y_{001} & y_{110} & y_{101} & y_{011} & y_{111} \\ y_{100} & 1 & y_{110} & y_{101} & y_{010} & y_{001} & y_{111} & y_{011} \\ y_{010} & y_{110} & 1 & y_{011} & y_{100} & y_{111} & y_{001} & y_{101} \\ y_{001} & y_{101} & y_{011} & 1 & y_{111} & y_{100} & y_{010} & y_{110} \\ y_{110} & y_{010} & y_{100} & y_{111} & 1 & y_{011} & y_{101} & y_{001} \\ y_{101} & y_{001} & y_{111} & y_{100} & y_{011} & 1 & y_{110} & y_{010} \\ y_{011} & y_{111} & y_{001} & y_{010} & y_{101} & y_{110} & 1 & y_{100} \\ y_{111} & y_{011} & y_{101} & y_{110} & y_{001} & y_{010} & y_{100} & 1 \end{bmatrix} \succeq 0.$$

Despite its theoretical interest, this result is of little value for computational purposes for the number of variables is exponential in the problem size. However, in many cases, low order SDP relaxations $\mathbb{Q}_i$, that is, with $i \ll n$, will provide the optimal value p^*. To illustrate the power of the SDP relaxations $\mathbb{Q}_i$, we have run a sample of 50 MAX-CUT problems in $\mathbb{R}^{10}$, with non diagonal entries randomly generated (uniformly) between 0 and 1 (in some examples, zeros and negative entries were allowed). In all cases, min $\mathbb{Q}_2$ provided the optimal value. The SDP relaxation $\mathbb{Q}_2$ has one LMI constraint $(\widehat{M}_i(y) \succeq 0)$ of dimension 56×56 and 385 variables y_β.

Remark 2. Consider the MAX-CUT problem with equal weights, that is,

$$p^* = \min_{x \in \{-1,1\}^n} \sum_{i<j} x_i x_j.$$

From

$$\left[\sum_{i<j} x_i x_j\right] + \frac{n}{2} = \frac{1}{2}\left[\sum_{i=1}^n x_i\right]^2 + \frac{1}{2}\left[\sum_{i=1}^n (1-x_i)^2\right],$$

we can prove that min $\mathbb{Q}_1 = -n/2$. Indeed, with $e_n \in \mathbb{R}^n$ being a vector of ones, let $X \in \mathbb{R}^{(n+1)\times(n+1)}, Z_k \in \mathbb{R}$, be defined as

$$Z_k = \frac{1}{2}, \; k = 1, \ldots n; \quad X := \frac{1}{2}\begin{bmatrix} 0 \\ e_n \end{bmatrix} \times [0, e_n], \tag{18}$$

so that $X \succeq 0$ and $-X(1,1) - \sum_{k=1}^n Z_k(1,1) = -n/2$. Thus, $(X, \{Z_k\})$ is admissible for $\mathbb{Q}_1^*$ with value $-n/2$.

Let $n = 2p$, and let $x \in \mathbb{R}^n$ be defined as $x_i = 1, \, i = 1, \ldots p; \, x_i = -1, \, i = p+1, \ldots n$. Let y be the vector of first and second moments of the Dirac at x (which, from $M_1(y) \succeq 0$, implies that y is admissible for $\mathbb{Q}_1$). Its value, i.e., the sum of all y_α's corresponding to $x_i x_j, \forall i < j$, is $-n/2$. Moreover,

$$\langle M_1(y), X \rangle = \frac{1}{2}\langle \begin{bmatrix} 0 \\ e_n \end{bmatrix}, X \begin{bmatrix} 0 \\ e_n \end{bmatrix}\rangle = \left[\sum_{i=1}^n x_i\right]^2 = 0,$$

so that y and $(X, \{Z_k\})$ are optimal solutions of $\mathbb{Q}_1$ and $\mathbb{Q}_1^*$ respectively. It also follows that $-n/2 = p^*$, as x is feasible for $\mathbb{P}$.

Next, consider the case where n is odd. Let y be defined with the correspondances

$$x_i \to y_\alpha := 0; \; i = 1, 2, \ldots n; x_i x_j \to y_\alpha := -1/(n-1), \; \forall i < j.$$

With $X \in \mathbb{R}^{(n+1) \times (n+1)}$, $Z_k \in \mathbb{R}$ as in (18), one may check that $M_1(y) \succeq 0$ and $\langle M_1(y), X \rangle = 0$. The sum of all y_α corresponding to $x_i x_j$, $i < j$ is $-n(n-1)/2 \times (n-1)^{-1} = -n/2$. Hence, again, y and $(X, \{Z\}_k)$ are optimal solutions of $\mathbb{Q}_1$ and $\mathbb{Q}_1^*$ respectively. But y is not a moment vector and $-n/2$ is only a (strict) lower bound on p^*.

4 Conclusion

We have provided an explicit equivalent continuous convex psd program for arbitrary constrained nonlinear 0-1 programs, the last in a finite hierarchy of SDP relaxations. For practical computation, in many cases, an SDP relaxation of low order is exact, i.e., provides the optimal value p^*, as confirmed on the MAX-CUT problem as well as on a (small) sample of nonconvex continuous problems in Lasserre [4]. However, for large or even moderate size problems, the resulting SDP relaxations $\mathbb{Q}_i$ might still be too large for the present status of SDP software packages. Therefore, we hope that this work will help stimulate further effort in designing efficient solving procedures for large SDP programs.

References

1. Curto, R.E. , Fialkow, L.A.: The truncated complex K-moment problem. Trans. Amer. Math. Soc. **352** (2000) 2825–2855.
2. Jacobi, T., Prestel, A.: On special representations of strictly positive polynomials. Technical-report, Konstanz University, Germany (2000).
3. Kojima, M., Tunçel, L.: Cones of matrices and successive convex relaxations of non convex sets. SIAM J. Optim. **10** (2000) 750–778.
4. Lasserre, J.B.: Global optimization with polynomials and the problem of moments. SIAM J. Optim. **11** (2001) 796–817.
5. Lasserre, J.B.: Optimality conditions and LMI relaxations for 0-1 programs. Technical report #00099, LAAS-CNRS Toulouse, France (2000).
6. Lasserre, J.B.: New LMI relaxations for the general nonconvex quadratic problem. In: Advances in Convex Analysis and Global Optimization, Eds: N. Hadjisavvas and P.M. Pardalos, Kluwer Academic Publishers, 2001.
7. Lovász, L., Schrijver, A.: Cones of matrices and set-functions and 0-1 optimization. SIAM J. Optim. **1** (1991) 166–190.
8. Putinar, M.: Positive polynomials on compact semi-algebraic sets. Ind. Univ. Math. J. **42** (1993) 969–984.
9. Schmüdgen, K.: The K-moment problem for compact semi-algebraic sets. Math. Ann. **289** (1991) 203–206.
10. Sherali, H.D., Adams, W.P.: A hierarchy of relaxations between the continuous and convex hull representations for zero-one programmimg problems. SIAM J. Disc. Math. **3** (1990) 411–430.

Pruning by Isomorphism in Branch-and-Cut

François Margot

Department of Mathematics, University of Kentucky
Lexington, KY 40506-0027
fmargot@ms.uky.edu

Abstract. The paper presents a Branch-and-Cut for solving (0, 1) integer linear programs having a large symmetry group. The group is used for pruning the enumeration tree and for generating cuts. The cuts are non standard, cutting integer feasible solutions but leaving unchanged the optimal value of the problem. Pruning and cut generation are performed by backtracking procedures using a Schreier-Sims table for representing the group. Applications to the generation of covering designs and error correcting codes are presented.

1 Introduction

Let Π^n be the set of all permutations of the ground set $I^n = \{1, \ldots, n\}$. A permutation in Π^n is represented by an n-vector π, with $\pi[i]$ being the image of i under π. If v is an n-vector and $\pi \in \Pi^n$, let $w = \pi(v)$ denote the vector w obtained by permuting the coordinates of v according to π, i.e.

$$w[\pi[i]] = v[i] \text{ for all } i \in I^n.$$

We consider an ILP problem of the form

$$
\begin{aligned}
\min \quad & c^T \cdot x \\
\text{s.t.} \quad & Ax \geq b, \\
& x \in \{0, 1\}^n \ ,
\end{aligned}
\tag{1}
$$

where A is an $m \times n$ matrix. For a permutation π of the n variables such that $\pi(c) = c$ and a permutation σ of the m rows of A such that $\sigma(b) = b$, let $A(\pi, \sigma)$ be the matrix obtained from A by permuting its columns according to π and its rows according to σ. Let

$$G = \{\pi \mid \text{there exists } \sigma \text{ s.t. } A(\pi, \sigma) = A\} \ .$$

Clearly, G is a permutation group of I^n. Moreover, for $\pi \in G$, a point $\bar{x}$ is feasible (resp. optimal) for the linear relaxation of the ILP (1) if and only if $\pi(\bar{x})$ is feasible (resp. optimal) for that ILP. Hence, G is a symmetry group of the feasible (and of the optimal) set of the ILP.

ILPs with large symmetry groups are natural when formulating classical problems in combinatorics, a.o. problems looking for a family of subsets of a

K. Aardal, B. Gerards (Eds.): IPCO 2001, LNCS 2081, pp. 304–317, 2001.

given set E with specified properties. In most cases, the elements in E are undistinguishable and G is a group with order at least $|E|!$. The problem of scheduling jobs on p parallel identical machines also yields ILPs with a natural symmetry group with at least $p!$ elements. For relatively modest size problems, it turns out that the corresponding ILPs become very difficult (if not impossible) to solve by traditional Branch-and-Cut techniques. The trouble comes from the fact that many subproblems in the enumeration tree will be isomorphic, forcing a wasteful duplication of effort.

In this paper, we assume that an ILP together with its symmetry group G is given. We show how to use G in order to prune efficiently isomorphic subproblems and to help the search by generating isomorphism cuts (cutting integer feasible solutions, but leaving the value of the optimal solution unchanged). This isomorphism pruning is compatible with standard cut generation techniques (Gomory cuts, Lift-and-Project cuts, or specially designed cuts for the problem at hand). The price to pay for the pruning is that the branching variable can no longer be chosen arbitrarily.

While isomorphism rejection in backtracking searches has been used in many applications [1], [3], [4], [7], [8], [10], [11], [12], [13], [15], [17], [18], it is not commonly used in a Branch-and-Cut (B&C) context. In most instances, the symmetry group G is not assumed to be known and the backtracking search has the additional task to produce it. The originality of the proposed approach resides essentially in (i) the possibility of generating isomorphism cuts (that will be shown to be efficient for the covering design problem), and (ii) the development of algorithms for computing orbits and stabilizers of sets under a group, taking advantage of the type of stabilizers and points in the queries needed by the B&C.

Section 2 describes the pruning algorithm and Section 3 presents basic data structures and algorithms for group operations. Section 4 describes the restrictions that can be put on queries for orbits and stabilizers generated during the B&C. Section 5 introduces the isomorphism cuts. Finally, Section 6 presents results on two applications, covering designs and error correcting codes.

We close this section with two basic definitions and a notation:

Let $S \subseteq I^n$. To simplify the notation, we make no difference between a set S and its characteristic vector. The *orbit* of S under G is

$$orb(S, G) = \{S' \subseteq I^n \mid S' = g(S) \text{ for } g \in G\} .$$

The *stabilizer* of S in G is the subgroup of G given by:

$$stab(S, G) = \{g \in G \mid g(S) = S\} .$$

For $1 \le a \le b \le n$, we write $v[a..b]$ the entries $\{v[a], v[a+1], \ldots, v[b]\}$ of v as an unordered set.

2   Isomorphism Test and Pruning

The proposed B&C will branch by fixing the value of one variable x_j to 0 or 1. Since the ILP (1) has a large automorphism group G, it is very likely that several nodes in the enumeration tree will correspond to isomorphic problems. Obviously, solving one of these isomorphic problems and pruning the others would result in huge savings. One important goal is to do so without having to keep in memory the list of all non isomorphic subproblems encountered since the start of the algorithm. One way to achieve this is to define, for each isomorphism class of subproblems, one particular subproblem (called *representative* of the class) that will be solved. Given a subproblem, we then just need to be able to decide if it is a representative or not and, in the latter case, we can prune the corresponding node of the B&C. Some care must be taken to ensure that the representative subproblems form a subtree of the B&C tree including the root. The general approach of isomorphism free generation of combinatorial structures based on representatives was studied by Read [17]. A general theory for isomorphism free generation, developed by McKay, can be found in [15].

Let a be a node of the B&C enumeration tree. Let F_1^a (resp. F_0^a) be the set of indices of variables fixed to 1 (resp. to 0) at a. Let F^a be the set of indices of variables that are not fixed to 0 or 1 at a, variables also called *free* at a. Let b be an other node and let F_1^b, F_0^b and F^b be the corresponding set of indices of fixed and free variables at b. The subproblems associated with nodes a and b of the B&C are isomorphic if and only if there exist a permutation $g \in G$, such that $g(F_i^a) = F_i^b$ for $i = 0, 1$.

Unfortunately, using this isomorphism test to identify subproblems that can be pruned during the B&C would require the storage of a maximal set of non isomorphic subproblems generated so far in the enumeration. Moreover, the computation needed to find if g exists is not trivial and would be required for many pairs of subproblems. Using the definition of a representative, we can use a slight relaxation of the isomorphism test that turns out to be practical. The price to pay for the simplification is that we will no longer be free to branch on any variable of the ILP: At node a, the branching variable will have to be x_f where f is the minimum index in F^a (even if the value of x_f in the current solution of the LP relaxation is 0 or 1). The variable x_f is called the *branching variable* at a. This branching strategy is called *minimum index branching* (MIB). The enumeration tree (containing only nodes that are not pruned) generated by a Branch-and-Bound $\mathcal{B}$ using the LP relaxation of (1) to prune only infeasible subproblems is called the *full enumeration tree* of $\mathcal{B}$.

A set $S \subseteq I^n$ is a *representative* if S is lexicographically minimum among the sets in its orbit under G. The following property is crucial for the validity of the pruning:

Lemma 1. *Let $S \subseteq I^n$ be a representative under G. Let $S' := S - v$ with $v = \max \{w \in S\}$. Then S' is also a representative.*

Proof. If S' is not a representative, then there exists $g \in G$ such that $g(S')$ is lexicographically smaller than S'. Then $g(S)$ is lexicographically smaller than S, a contradiction.   □

Consider the following *isomorphism pruning* (IP) to be applied on nodes of the enumeration tree of a *B&C*: If F_1^a is not a representative, then prune node a.

Lemma 2. *Let τ be the full enumeration tree of a B&C $\mathcal{B}$ using MIB. Let S be the nodes in τ that are not pruned by IP. Then*

(i) S *induces a subtree of τ containing the root of τ;*
(ii) *The B&C $\mathcal{B}'$ obtained by adding IP to $\mathcal{B}$ returns the same optimal value as $\mathcal{B}$.*

Proof. (i): Let $a \in S$ and let $b \in \tau$ on the path between the root and a in τ. Then F_1^a is a representative and, by the choice of branching strategy, F_1^b is the set of the $|F_1^b|$ smallest entries in F_1^a. By Lemma 1, F_1^b is a representative, i.e. $b \in S$.

(ii): Let a be a node of τ for which F_1^a is an optimal solution to ILP (1). Then the representative of the orbit of F_1^a under G is a set F^*, and thus there is a node $b \in S$ with $F_1^b = F^*$. By (i), the full enumeration tree of $\mathcal{B}'$ is the subtree induced by S in τ, implying that $\mathcal{B}'$ will process node b at some point, yielding the same optimal value as the one returned by $\mathcal{B}$. $\square$

When solving a subproblem a, it is sometimes possible to identify variables that may be set to 0 without affecting the optimal solution returned by a B&C using MIB and IP. Consider the following operations:

(i) Let b be the father of a in the enumeration tree and let x_f the branching variable at b. If a is the son of b where x_f is set to 0 then set to 0 all free variables in $orb(f, stab(F_1^a, G))$.
(ii) Let $f = \min \{r \in F^a\}$. If $F_1^a \cup f$ is not a representative, then fix to 0 all free variables in $orb(f, stab(F_1^a, G))$.

Applying these operations (repeatedly for (ii) if possible, i.e. until no free variable exists or until $F_1^a \cup f$ is a representative) is called a 0-*fixing*. The output of the 0-fixing is the value f in (ii) for which $F_1^a \cup f$ is a representative, or $n+1$ is no such f exists.

Remark 1. Trivially, the variables set to 0 during a 0-fixing at node a have all an index larger than the maximum index M in F_1^a, since all variables in F^a have index larger than M. $\square$

Lemma 3. *Consider a B&C $\mathcal{B}$ using MIB and IP and let $\mathcal{B}'$ be the B&C obtained by adding 0-fixing in $\mathcal{B}$. Then the optimal values returned by $\mathcal{B}$ and $\mathcal{B}'$ are equal.*

Proof. Let a be a node of the full enumeration tree τ of $\mathcal{B}$ for which F_1^a is an optimal solution to ILP (1). Then F_1^a is a representative. Assume that no node b in the full enumeration tree τ' of $\mathcal{B}'$ has $F_1^b = F_1^a$. Hence there exists a node $c \in \tau'$ such that F_1^c contains the $|F_1^c|$ smallest indices in F_1^a and, during the

0-fixing at c, one of the variables in $F_1^a - F_1^c$ is fixed to 0. Assuming that c is chosen as close as possible to the root, we then have $j \in orb(f, stab(F_1^c, G))$ for some $j \in F_1^a - F_1^c$ and $f \in F_0^c$ with

$$\max\{r \in F_1^c\} < f < m := \min\{r \in (F_1^a - F_1^c)\} \le j .$$

The first inequality comes from Remark 1 and the second one from the fact that $F_1^c \cup m$ is a representative: If m is fixed to 0 during the 0-fixing at c, then it is from a $f < m$ and if m is not fixed to 0, then all f considered during the 0-fixing are smaller than m.

Thus there exists $g \in stab(F_1^c, G)$ such that $g[j] = f$. Then $g(F_1^c \cup j) = F_1^c \cup f$ which is lexicographically smaller than $F_1^c \cup m$, proving that F_1^a is not a representative as $F_1^c \cup j \subseteq F_1^a$, a contradiction.   □

It remains to show how to compute $orb(f, stab(F_1^a, G))$ and how to test if a set is a representative or not. This will be covered in Section 4. In the remainder of the paper, the B&C is assumed to use MIB, IP and 0-fixing. The operations performed at node a in the enumeration tree are thus:

$r := $ 0-fixing(a);
Repeat until a criterion is met
    solve the LP relaxation;
    generate cuts;
If $r < n + 1$ then create two sons of a by fixing x_r to 0 or 1;

3   Group Representation and Basic Algorithms

Essentially two options are available to represent a permutation group G: The explicit representation or a representation by generators. The explicit representation simply store in a list each permutation in G. A representation by generators store only a subset $\{g_1, \ldots, g_k\}$ of the permutations in G, with the property that any permutation in G can be written as a product of permutations in the subset. If $|G|$ is small, the explicit representation might work well, but in most cases of interest a representation by generators is required. The operations of interest listed above are also, usually, faster with the representation by generators.

We use the *Schreier-Sims* representation of G (also called *strong generators*) [1], [2], [3], [4], [10], [11], [12]. Let

$$G_0 = G$$
$$G_1 = \{g \in G_0 \mid g[1] = 1\}$$
$$G_2 = \{g \in G_1 \mid g[2] = 2\}$$
$$\ldots$$
$$G_n = \{g \in G_{n-1} \mid g[n] = n\} .$$

G_1 is simply the stabilizer of 1 in G and G_i is the stabilizer of i in G_{i-1}. It follows that $G_0, G_1, \ldots, G_n$ are nested subgroups of G.

For $k = 1, \ldots, n$, let $orb(k, G_{k-1}) = \{j_1, \ldots, j_p\}$ be the orbit of k under G_{k-1}. Then for each $1 \leq i \leq p$, let h_{k,j_i} be a permutation in G_{k-1} sending k on j_i, i.e. $h_{k,j_i}[k] = j_i$. Let $U_k = \{h_{k,j_1}, \ldots, h_{k,j_p}\}$. Note that U_k is never empty as $orb(k, G_{k-1})$ always contains k.

Arrange the permutations in the sets U_k, $k = 1, \ldots, n$ in an $n \times n$ table T, with

$$T_{k,j} = \begin{cases} h_{k,j} & \text{if } j \in orb(k, G_{k-1}), \\ \emptyset & \text{otherwise.} \end{cases}$$

The table T is called the Schreier-Sims representation of G. This table is not uniquely defined, as there is usually a choice for the permutations included in the sets U_k. However, the general shape of the table (i.e. which entries are empty or not) is fixed.

Remark 2. It is more efficient to implement the table as a vector of ordered lists instead of a 2-dimensional table, as most entries in the table are usually empty. However, algorithms are simpler to describe and understand for the 2-dimensional table. The actual implementation uses a vector of ordered lists. □

Remark 3. The most interesting property of this representation of G is that each $g \in G$ can be uniquely written as

$$g = g_1 \cdot g_2 \cdots \cdots g_n \tag{2}$$

with $g_i \in U_i$ for $i = 1, \ldots n$. Hence the permutations in the table form a set of generators of G. It is called a strong set of generators, since the equation (2) shows that $g \in G$ can be expressed as a product of at most n permutations in the sets.

Given a permutation $g \in G$, it is easy to find the n permutations $g_1, \ldots, g_n$ of equation (2): the permutations $g_2, \ldots, g_n$ all stabilize point 1, forcing g_1 to be $T[1, g[1]]$. Then, as $g_3, \ldots, g_n$ all stabilize point 2, we must have $(g_1 \cdot g_2)[2] = g[2]$, i.e. $g_2[2] = (g_1^{-1} \cdot g)[2]$ and thus $g_2 = T[2, (g_1^{-1} \cdot g)[2]]$. A similar reasoning yields $g_3, \ldots, g_n$. □

It is possible to make a small generalization of the presentation by ordering the points of the ground set in an arbitrary order β, called the *base* of the table. In that case, the subgroups $G(\beta)_k$ for $k = 1, \ldots, n$ are defined as the stabilizer of $\beta[k]$ in $G(\beta)_{k-1}$, with $G(\beta)_0 = G$. The corresponding table is denoted by $T(\beta)$. Row k of $T(\beta)$ corresponds to the element k, $U(\beta)_k$ is the set of non empty entries in row k of $T(\beta)$ and $J(\beta)_k$ denotes the set of indices $\{j \in I^n \mid T(\beta)[k,j] \neq \emptyset\}$, also called the *basic orbit* of k in T, following the terminology of [12]. When the base β is fixed, we sometimes drop the qualifier (β) in these symbols, but from now on each table T is defined with respect to a base.

Remark 4. For any $k \in \{1, \ldots, n\}$, replacing rows $\beta[1], \ldots, \beta[k-1]$ in $T(\beta)$ by identity rows yields a Schreier-Sims representation of $G(\beta)_{k-1}$. Hence the permutations on rows $\beta[k], \ldots, \beta[n]$ of $T(\beta)$ form a set of generators of $G(\beta)_{k-1}$. □

Two natural questions arise: How can we create the table $T(\beta)$, knowing the group G either explicitly or by a family of generators, and how can we change the base β of the representation. Algorithms for performing these operations can be found in [1], [3], [4], [10], [11], [12]. The implemented algorithm to create the table is closest to [10], with worst case complexity in $O(n^6)$. The algorithm for changing the base is essentially the algorithm in [4], blended with basic algorithms of [10], taking advantage of the fact that the base changes that arise during the course of the B&C are of a particular kind: Suppose that $T(\beta)$ is the table at a node of the enumeration tree. As we will see in Section 4, the base β' for any of its sons can be obtained through a few applications of the following operation (called *downing of a point v*): Assume that $v = \beta[k]$ and let $k \leq r \leq n$. Let β' be the permutation obtained from β by moving the entry v to position r of β', keeping the other entries in the same order as in β. The worst case complexity of this modified base change algorithm is in $O(n^4)$ for downing a single point.

An algorithm with worst case complexity of $O(n^6)$ or even $O(n^4)$ might seem impractical for values of $n \geq 100$. It turns out that these bounds are very pessimistic and that the amount of time spent in performing these group operations during the B&C stays well below 5% of the total cpu time in typical applications.

4   Orbits, Stabilizers, and Representatives

We are interested in performing the following operations that were mentioned in Section 2: Computing the orbit of a point in the stabilizer of a set and deciding if a set is lexicographically minimum in its orbit under G.

Standard algorithms for computing orbits under a group G' [2], [10] usually return a partition of the ground set into orbits. Here, we only need the orbit of a particular point v and it seems a waste of effort to compute all the disjoint orbits and keep only the one we are interested in. If G' was given by a Schreier-Sims table, it would be of course possible to make a change of basis so that v becomes the first entry of the basis, since, by definition, non empty entries in row v of the table will be the orbit of v under $G'_0 = G'$. In our particular case, however, G' is the stabilizer of a set in G and building the table for G' would be quite expensive.

We thus devised a backtracking algorithm for computing the orbit of a single point in the stabilizer of a set in G. It takes advantage of the fact that we might assume that the basis β of the group at node a of the enumeration tree has the following structure: Variables set to 1 at a (i.e. F_1^a) come first in β, then the free variables (F^a), and then the variables set to 0 at a (F_0^a).

The data structure associated with group G at node a of the B&C is the following:

table: T base: β
integer: *fixed_one*   vector: *part_zero*   .

The table T is just a Schreier-Sims representation of the group with base β. The variable $fixed\_one$ gives the number of variables in F_1^a and

$$F_1^a = \beta[1..fixed\_one] \qquad \text{with} \qquad \beta[1] < \cdots < \beta[fixed\_one] \ .$$

The vector $part\_zero$ is used to store information about variables fixed to 0. For $i = 1, \ldots, fixed\_one$, $\beta[part\_zero[i]..n]$ are the variables that have been fixed to 0 before $\beta[i]$ was set to 1. For $i = fixed\_one + 1$, $\beta[part\_zero[i]..n] = F_0^a$, i.e. all the variables currently fixed to 0 at a. The remaining variables (the free ones) appear in β in increasing order of their index, after variables in F_1^a and before variables in F_0^a. Note that this structure of β is easy to maintain throughout the B&C: When the 0-fixing is performed (or a variable is set to 0 by branching), free variables in a set U become fixed to 0. To update the table, simply use the procedure for changing the base by downing a point, moving one by one the variables in U. When a variable is fixed to 1 by branching, it is always the free variable with smallest index, and the basis (and thus the table) remains the same.

The backtracking procedure computes the orbit of $\beta[k]$ in the stabilizer of the points in $\beta[1..k-1]$. Due to the particular structure of the base β, this is exactly the operation of computing $orb(f, stab(F_1^a, G))$ with $f = \min\{r \in F^a\}$ needed in Section 2 if we use $k = |F_1^a| + 1$.

The procedure is a recursive procedure orb_in_stab() (described below) called from the following initializing procedure:

```
orbit_in_stabilizer(n, T, β, k)

/* Returns the orbit of β[k] in stab(β[1..(k − 1)], G) where G
is the group represented by T with base β */

    Jₖ = basic orbit of β[k] in T;
    ident = identity permutation in Πⁿ;
    remain := β[1..k − 1];
    orbit := Jₖ;
    orb_in_stab(n, T, β, k, Jₖ, ident, remain, orbit, 1);
    return(orbit);
```

The parameters of the call to orb_in_stab() have the following interpretation: $perm$ is a permutation in G sending $\beta[1..ind-1]$ on a subset $B \subseteq \beta[1..k-1]$; $remain$ is the set $perm^{-1}(\beta[1..k-1] - B)$; J_k is the basic orbit of $\beta[k]$ in T; $orbit$ is the set of points currently known in the orbit of $\beta[k]$ in $stab(\beta[1..(k-1)], G)$; ($orbit$ is passed by reference during the recursive calls;) ind refers to the point $\beta[ind]$ being treated during the current call.

```
orb_in_stab(n, T, β, k, J_k, perm, remain, orbit, ind)

    For each i ∈ remain do
      h := T[β[ind], i];
      If h ≠ ∅ then
        loc_remain := remain − i;
        loc_remain := h⁻¹(loc_remain);
        loc_perm := perm · h;
        If ind < k − 1 then
          orb_in_stab(n, T, β, k, J_k, loc_perm, loc_remain, orbit, ind + 1);
        else
          For each j ∈ J_k do orbit := orbit ∪ perm[j];
```

Proposition 1. *The algorithm orbit_in_stabilizer() is correct.*

Proof. Let $S = \beta[1..(k-1)]$. If $k = 1$, $stab(\emptyset, G) = G$ and the orbit of $\beta[1]$ in G is J_1, as returned by the algorithm. Otherwise, we have $k \geq 2$. By Remark 2, $stab(S, G)$ is generated by all permutations g such that $g(S) = S$ with

$$g = g_1 \cdot \cdots \cdot g_{k-1} \cdot g_k \cdot h$$

and $g_i \in U_{\beta[i]}$ for $i = 1, \ldots k$, $h \in G_k$. Since $h[\beta[k]] = \beta[k]$,

$$orb(\beta[k], stab(S, G)) =$$
$$\{v \in I^n \mid v = (g_1 \cdot \cdots \cdot g_k)[\beta[k]], g_i \in U_{\beta[i]} \text{ for } i = 1, \ldots k, g(S) = S\} \, .$$

Assume that $g_i = T[\beta[i], j_i]$ for $i = 1, \ldots, k-1$. The condition $g(S) = S$ implies $j_1 \in S$. Moreover, if $k \geq 3$ then $(g_1 \cdot g_2)[2] \in S - j_1$ and thus $g_2[2] \in g_1^{-1}(S - j_1)$. In general, for index $2 \leq ind \leq k - 1$, we have

$$g_{ind}[ind] \in (g_{ind-1}^{-1}(\ldots (g_2^{-1}((g_1^{-1}(S - j_1)) - j_2)) - \cdots - j_{ind-1})) \, . \tag{3}$$

Note that the set in (3) is exactly the parameter *remain* of the call to orb_in_stab() with value *ind* as last parameter. That procedure simply selects an index in this set, update *perm* and *remain* and calls itself recursively with $ind + 1$ until $ind = k - 1$ or no permutation h is found. In the former case, $g_1 \cdot \cdots \cdot g_{k-1}(S) = perm(S) = S$ and it adds $perm[j]$ for all $j \in J_k$. This amounts to compute $(perm \cdot g_k)[\beta[k]]$ for all $g_k \in U_{\beta[k]}$. In the later case, the algorithm backtracks to $ind - 1$, since no permutation in G stabilizes S with the current choice of permutations $g_1, \ldots, g_{ind-1}$. Since at each level in the recursion, all possible choices for g_{ind} are explored, the algorithm indeed returns the desired orbit. □

Remark 5. As observed in the justification above, the set in (3) is the current set *remain*. A weaker statement about this set is that $remain \subseteq perm^{-1}(S)$, as

$$perm^{-1} = g_{ind-1}^{-1} \cdots g_2^{-1} \cdot g_1^{-1} .$$

□

Let us now turn to the question of deciding if set $S = \beta[1..k]$ is the lexicographically minimum set in $orb(S, G)$. Note that for $k = |F_1^a| + 1$, this is exactly the same question as deciding if $F_1^a \cup f$ is a representative, with $f = \min \{r \in F^a\}$ mentioned in Section 2. We assume that β has the structure stated at the beginning of this section.

```
first_in_orbit(n, T, k)

/* Returns ''true'' if and only if β[1..k] is
lexicographically minimum in orb(β[1..k], G) */

    ident := identity permutation in Πⁿ;
    remain := β[1..k];
    is_first := true;
    f_in_orb(n, T, k, ident, remain, 1, is_first);
    return(is_first);
```

The parameters *perm*, *remain* and *ind* in the call to f_in_orb() are similar to the same parameters in the call of orb_in_stab(). The parameter *is_first* is passed by reference and is used to stop the procedure as soon as it is known that $\beta[1..k]$ is not lexicographically minimum in $orb(\beta[1..k], G)$.

```
f_in_orb(n, T, k, perm, remain, ind, is_first)

    If is_first = false then return;
    For each i ∈ remain do
        h := T[β[ind], i];
        If h ≠ ∅ then
            loc_remain := remain − i;
            loc_remain := h⁻¹(loc_remain);
            loc_perm := perm · h;
            For each j ∈ loc_remain do
                If β⁻¹[j] ≥ part_zero[ind + 1] then
                    is_first := false;
                    return;
            If ind < k then
                f_in_orb(n, T, k, loc_perm, loc_remain, ind + 1, is_first);
```

Proposition 2. *The algorithm first_in_orbit() is correct.*

Proof. Suppose that the condition

$$\beta^{-1}[j] \geq part\_zero[ind + 1]$$

in procedure f_in_orb() is satisfied. This condition means that there exists a point j in *loc_remain* that has been fixed to 0 before fixing $\beta[t]$ to 1 and (if $t \geq 2$) after fixing $\beta[t - 1]$ to 1, for some $t \leq ind + 1$. Let

$$S := perm(\beta[1..ind]) \subseteq \beta[1..k] \qquad \text{i.e.} \qquad perm^{-1}(S) = \beta[1..ind] .$$

Moreover, as pointed out in Remark 5 (the algorithms are similar, so this remark holds here too), *loc_remain* $\subseteq perm^{-1}(\beta[1..k])$ and since it is disjoint from S, we have

$$j = perm^{-1}[\beta[s]] \qquad \text{for some } s \in \{ind + 1, \ldots, k\} .$$

Since j was fixed to 0 before setting $\beta[t]$ to 1, we have, for some $w < \beta[t]$,

$$j \in orb(w, stab(\beta[1..t - 1], G)) .$$

Hence there exists a permutation

$$p \in stab(\beta[1..t - 1], G) \qquad \text{with} \qquad p(j) = w .$$

Let $T := perm(\beta[1..t - 1]) \subseteq S$. As $p(\beta[1..t - 1]) = \beta[1..t - 1]$, we have

$$(p \cdot perm^{-1})(T) = \beta[1..t - 1] \qquad \text{and} \qquad (p \cdot perm^{-1})[\beta[s]] = w < \beta[t] .$$

Thus $(p \cdot perm^{-1})(\beta[1..t - 1] \cup \beta[s]) = \beta[1..t - 1] \cup w$ is lexicographically smaller than $\beta[1..t]$. It follows that when the algorithm returns "false", the set $\beta[1..k]$ is indeed not lexicographically minimal in its orbit under G.

Suppose now that the set $\beta[1..k]$ is not lexicographically minimal in its orbit under G. Hence, there is a smallest index $1 \leq t \leq k$ such that $\beta[1..t - 1]$ is lexicographically minimal in its orbit under G, but $\beta[1..t]$ is not. Let $p \in G$ such that $p(\beta[1..t])$ is lexicographically smaller than $\beta[1..t]$. By Remark 3, we can write

$$p = h_1 \cdots \cdots h_n$$

with $h_i \in U_{\beta[i]}$ for $i = 1, \ldots, n$. Observe that the choice of t implies that

$$p(\beta[1..t]) = \beta[1..t - 1] \cup w \qquad \text{for some } w < \beta[t]$$

and w fixed to 0 before setting $\beta[t]$ to 1. We have

$$p^{-1}(\beta[1..t-1] \cup w) = \beta[1..t] \quad \text{and thus} \quad p^{-1}[w] = \beta[s] \text{ for some } s \in \{1, \ldots, t\}.$$

During the recursive calls to f_in_orb(), a permutation *perm* will occur with $perm[\beta[i]] = p^{-1}[\beta[i]]$ for $i = 1, \ldots, t - 1$, namely

$$perm = h_1 \cdots \cdots h_{t-1}$$

with $h_i = T[\beta[i], p^{-1}[\beta[i]]]$ for $i = 1, \ldots, t - 1$. Let $z := perm^{-1}[\beta[s]]$. Observe that

$$(perm^{-1} \cdot p^{-1})[w] = z \qquad \text{and} \qquad perm^{-1} \cdot p^{-1} \in stab(\beta[1..t - 1], G) \ .$$

Hence $z \in orb(w, stab(\beta[1..t - 1], G))$ and z was fixed to 0 with w (or earlier). It follows that loc_remain contains z and that $\beta^{-1}[z] \geq part\_zero[t]$, implying that the algorithm will return "false".   □

Crude bounds on the worst case complexities for these two backtracking procedures are $O(n \cdot k!)$ and $O(n \cdot (k + 1)!)$, respectively, but they turn out to be orders of magnitude faster on average, making them practical. (Values of k in the range of 20 to 40 with $n \geq 200$ appear routinely in applications and are handled efficiently).

Remark 6. For clarity, the algorithms orb_in_stab() and first_in_orbit() were presented separately, but it is possible to take advantage of their similarities to merge them into one single recursive procedure.   □

5   Isomorphism Inequalities

Let a be a node of the enumeration tree and H^a be the set of variables that are not fixed to 0 at node a. Suppose that there exists $J \subseteq H^a$ such that the representative J^* of the orbit of J under $stab(F_1^a, G)$ is lexicographically smaller than F_1^a. Then if a node b in the descendants of a with $J \subseteq F_1^b$ exists, this node will be pruned by IP. Hence, the *isomorphism inequality*

$$\sum_{j \in J} x_j \leq |J| - 1 \tag{4}$$

is valid in the subtree rooted at a. Moreover, if the whole restricted enumeration tree is explored by a depth-first search, always selecting first the son d where the branching variable is set to 1, then the sets F_1^d are enumerated in lexicographic order, starting with the smallest one. It follows that if an inequality (4) is generated at a, it is valid for the rest of the enumeration, i.e. it can be considered global.

The (exact) separation algorithm for the isomorphism inequalities is similar to the backtracking procedure for testing if a set is lexicographically minimal in its orbit under a stabilizer. Crude estimates for its worst case complexity is in $O(n \cdot |H^a|!)$ but, in practice, it is able to handle efficiently instances with $|H^a| \geq 100$ and $n \geq 200$. The algorithm can also be turned to an heuristic separation algorithm by working with a subset $H \subseteq H^a$ instead of H^a, if needed.

6   Applications

We use the software ABACUS (version 2.2) developed by Thienel [9] as generic implementation of all B&C steps (isomorphism pruning excepted) and the LP solver is CPLEX6.6. We briefly describe preliminary results obtained on two applications: covering designs and error correcting codes.

Let V be a set of elements of cardinality v and let k and t be integers such that $v \geq k \geq t \geq 0$. Let $\mathcal{K}$ be the set of all k-subsets of V and $\mathcal{T}$ be the set of all t-subsets of V. A (v, k, t)-*covering design* is a collection $\mathcal{C}$ of sets in $\mathcal{K}$ such that each $t \in \mathcal{T}$ is contained in at least one set of $\mathcal{C}$. A (v, k, t)-covering design $\mathcal{C}$ is *minimum* if the cardinality of $\mathcal{C}$ is as small as possible.

Covering designs have a long history and have applications in statistics, coding theory and combinatorics, among others. Numerous theorems give the value of a minimum covering design under certain assumptions on the parameters (see the survey [16]). Yet, for particular values of the parameters, only lower and upper bounds are available. A case point is the $(10, 5, 4)$-covering design, for which a lower bound of 50 and an upper bound of 51 are known [6].

Running the described B&C algorithm for the $(10, 5, 4)$-covering design problem, pruning nodes as soon as their associated LP relaxation has value strictly larger than 50, we obtain a proof that no solution better than the best known solution of 51 exists (see [14] for the ILP formulation and details). The ILP has 252 variables, 384 inequalities and the symmetry group G has order $10! = 3'628'800$. The average number of non empty entries in the Schreier-Sims table over all nodes of B&C is about 550. There are only 313 nodes in the enumeration tree and the cpu time (in seconds) is distributed as follows (the machine used is an HP-J5000 running HP-UX10.20 with two 440MHz PA-8500 RISC CPUs): Total cpu time: 75.99, LP cpu time: 66.13, Pool separation for inactive inequalities: 0.97, Separation for isomorphism inequalities: 4.72, Representative test algorithm: 2.34.

Although the separation for isomorphism inequalities might seem time consuming, this should be balanced with the fact that not using these inequalities makes the B&C enumeration tree grow from 313 nodes to well above 400. (These numbers and running times are better than those in [14] where no 0-fixing and a less general isomorphism pruning were used). It is worth noting that proving that this ILP has no solution with value 50 is not doable by the B&C of CPLEX6.6.

An error correcting binary code with distance d and word length w is a collection $\mathcal{C}$ of binary w-vectors such that the Hamming distance between any pair of vectors in $\mathcal{C}$ is at least d [5]. The maximum number of vectors in $\mathcal{C}$ is denoted by $A(w, d)$. Here also, for small values of w and d, only bounds on $A(w, d)$ are known. For example, $72 \leq A(10, 3) \leq 79$. A simple set partitioning problem with one variable per binary w-vector with at least three 1's yields an ILP with a group of order $w!$. This ILP for finding $A(8, 3)$ is difficult for the B&C of CPLEX6.6 as it needs about half a million nodes and 4 hours CPU to prove optimality of a given solution. The isomorphism pruning algorithm described in this paper, however, does it in 95 nodes and 13 seconds CPU. The ILP has 219 variables, 347 inequalities and the symmetry group G has order $8! = 40'320$. The average number of non empty entries in the Schreier-Sims table over all nodes of the B&C is about 315. The cpu time (in seconds) is distributed as follows: Total cpu time: 13.46, LP cpu time: 11.80, Pool separation for inactive inequalities: 0.04, Separation for isomorphism inequalities: 0.04, Representative test algorithm: 0.79.

References

1. Butler G. "Computing in Permutation and Matrix Groups II: Backtrack Algorithm", *Mathematics of Computation* 39 (1982), 671-680.
2. Butler G., *Fundamental Algorithms for Permutation Groups, Lecture Notes in Computer Science* 559, Springer (1991).
3. Butler G., Cannon J.J., "Computing in Permutation and Matrix Groups I: Normal Closure, Commutator Subgroups, Series", *Mathematics of Computation* 39 (1982), 663-670.
4. Butler G., Lam W.H., "A General Backtrack Algorithm for the Isomorphism Problem of Combinatorial Objects", *J. Symbolic Computation* 1 (1985), 363-381.
5. Conway J.H., Sloane N.J.A., *Sphere Packings, Lattices and Groups*, Springer (1993).
6. Etzion T., Wei V., Zhang Z., "Bounds on the Sizes of Constant Weight Covering Codes", *Designs, Codes and Cryptography* 5 (1995), 217–239.
7. Gibbons P.B., "Computational Methods in Design Theory", in: *The CRC Handbook of Combinatorial Designs*, Colbourn C.J., Dinitz J.H (eds.), CRC Press (1996), 718–740.
8. Ivanov A.V., "Constructive Enumeration of Incidence Systems", *Annals of Discrete Mathematics* 26 (1985), 227–246.
9. Jünger M., Thienel S., "Introduction to ABACUS – A Branch-And-CUt System", *Operations Research Letters* 22 (1998), 83–95.
10. Kreher D.L., Stinson D.R., *Combinatorial Algorithms, Generation, Enumeration, and Search*, CRC Press (1999).
11. Leon J.S., "On an Algorithm for Finding a Base and a Strong Generating Set for a Group Given by Generating Permutations", *Mathematics of Computation* 35 (1980), 941-974.
12. Leon J.S., "Computing Automorphism Groups of Combinatorial Objects", in *Computational Group Theory*, Atkinson M.D. (ed.), Academic Press (1984), 321–335.
13. Luetolf C., Margot F., "A Catalog of Minimally Nonideal Matrices", *Mathematical Methods of Operations Research* 47 (1998), 221–241.
14. Margot F., "Small Covering Designs by Branch-and-Cut", Research report 2000-27, Department of Mathematics, University of Kentucky.
15. McKay D., "Isomorph-free Exhaustive Generation", *Journal of Algorithms* 26 (1998), 306–324.
16. Mills W.H., Mullin R.C., "Coverings and Packings", in: *Contemporary Design Theory: A collection of Surveys*, Dinitz H., Stinson D.R. (eds.), Wiley (1992), 371–399.
17. Read R.C., "Every One a Winner or How to Avoid Isomorphism Search When Cataloguing Combinatorial Configurations", *Annals of Discrete Mathematics* 2 (1978), 107-120.
18. Seah E., Stinson D.R., "An Enumeration of Non-isomorphic One-factorizations and Howell Designs for the Graph K_{10} minus a One-factor", *Ars Combinatorica* 21 (1986), 145–161.

Facets, Algorithms, and Polyhedral Characterizations for a Multi-item Production Planning Model with Setup Times*

Andrew J. Miller[1], George L. Nemhauser[2], and Martin W.P. Savelsbergh[2]

[1] CORE, 34 Voie du Roman Pays, 1348 Louvain-la-Neuve, Belgium,
miller@core.ucl.ac.be
[2] Georgia Institute of Technology, School of Industrial and Systems Engineering,
Atlanta, GA 30332-0205, USA, {george.nemhauser,mwps}@isye.gatech.edu

Abstract. We present and study a mixed integer programming model that arises as a substructure in many industrial applications. This model provides a relaxation of various capacitated production planning problems, more general fixed charge network flow problems, and other structured mixed integer programs. We analyze the polyhedral structure of the convex hull of this model; among other results, we present valid inequalities that induce facets of the convex hull in the general case, which is $\mathcal{NP}$–hard. We then present an extended formulation for a polynomially solvable case for which the LP always gives an integral solution. Projecting from this extended formulation, we show that the inequalities presented for the general model suffice to solve this polynomially solvable case by linear programming in the original variable space.

Keywords: Mixed integer programming, production planning, polyhedral combinatorics, capacitated lot-sizing, setup times, fixed charge network flow

1 Introduction

One of the most successful techniques that has been used to solve integer programming (IP) and mixed integer programming (MIP) problems is the application of polyhedral results for structured relaxations of these problems. Pioneering work in this area includes Crowder, Johnson, and Padberg [1983], who showed that valid inequalities for knapsack sets defined by single rows can be used to solve IP's, and Padberg, Van Roy, and Wolsey [1985], who showed that valid inequalities for single–node fixed charge (SNFC) flow models can be used to solve structured MIP's.

---

* This paper presents research results of the Belgian Program on Interuniversity Poles of Attraction initiated by the Belgian State, Prime Minister's Office, Science Policy Programming. The scientific responsibility is assumed by the authors. This research was also supported by NSF Grant No. DMI-9700285 and by Philips Electronics North America.

K. Aardal, B. Gerards (Eds.): IPCO 2001, LNCS 2081, pp. 318–332, 2001.

In this paper we introduce a model that occurs as a relaxation of many structured MIP problems. We derived this model from a single period of the multi–item capacitated lot–sizing problem with setup times (MCL), studied by Trigiero, Thomas, and McClain [1989], among others. In this relaxation the decision variables considered include those that represent inventory carried over from the preceding period; therefore we call this relaxation PI, for *preceding inventory*. PI can be formulated

$$\min \; \sum_{i=1}^{P} p^i x^i + \sum_{i=1}^{P} q^i y^i + \sum_{i=1}^{P} h^i s^i \tag{1}$$

subject to

$$x^i + s^i \geq d^i, i = 1, ..., P, \tag{2}$$

$$\sum_{i=1}^{P} x^i + \sum_{i=1}^{P} t^i y^i \leq c, \tag{3}$$

$$x^i \leq (c - t^i) y^i, i = 1, ..., P, \tag{4}$$

$$x^i, s^i \geq 0, i = 1, ..., P, \tag{5}$$

$$y^i \in \{0, 1\}, i = 1, ..., P. \tag{6}$$

PI can be viewed as a single–node, multi–item fixed charge flow model with two inbound arcs (see Figure 1). The first of these arcs has a capacity c jointly imposed on P items. A fixed capacity usage $t^i \geq 0$ is incurred if any amount of item i is shipped along this arc. (If capacity is measured in time units, then t^i can be considered a *setup time* for item i.) The variable x^i is the flow along this arc, with unit cost p^i; y^i is the setup variable for i, and the setup cost is q^i. The second arc can also be used by each item. This arc is uncapacitated, and there are no fixed costs associated with flow on it. The variables s^i represent flow on this second arc; the unit cost for shipping i along this arc is h^i. The total amount of flow shipped along these two arcs must be at least $d^i, i = 1, ..., P$. Let $\mathcal{P} = \{1, ..., P\}$; if $h^i < 0$ for any $i \in \mathcal{P}$, PI has unbounded optimum. We will therefore assume that $h^i \geq 0, i = 1, ..., P$. Also, we assume that $q^i > 0, i = 1, ..., P$, which ensures that in every optimal solution, $y^i = 1$ if and only if $x^i > 0$. Thus, a setup in PI occurs if and only if there is flow or production, which is also generally true in industrial applications.

In addition to MCL, PI provides a relaxation for other production planning problems, as well as for other structured MIP's, such as fixed charge network flow problems. When $t^i = 0, i = 1, ..., P$, PI resembles SNFC, though even in this case PI has a different, more complicated structure than SNFC, due to the demand constraints (2) and the slack/inventory variables s^i. The possibility of having strictly positive setup times in PI allows it to provide a tighter relaxation of many structured MIP's. The results for PI presented here are valid for the case when setup times are 0 for all items, as well as for the case when they are strictly positive for some or all items. In both cases the results are new, to the best of our knowledge.

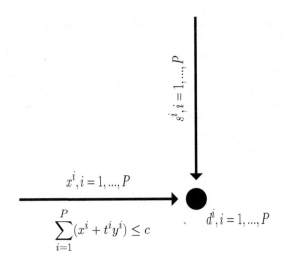

Fig. 1. Single–node multi–item flow model with two arcs

PI is $\mathcal{NP}$–hard, but the special case of PI in which setup times and demand are constant for all items, called PIC, is polynomially solvable. PI remains $\mathcal{NP}$–hard if either setup times or demand, but *not* both, are constant for all items.

We now summarize the contents of this paper. In Section 2, we give basic polyhedral results that will be used in the development of later sections. In Section 3, we define new, facet–defining valid inequalities for PI. In Section 4, we discuss polynomial reductions from other $\mathcal{NP}$–hard problems to PI, and we define a polynomial algorithm for the special case PIC. In Section 5, we give an integral extended formulation for PIC that is related to the algorithm defined in Section 4. By considering the projection of this formulation onto the (x, y, s) space, we show that the valid inequalities that we define for the general case of PI suffice to solve the special case PIC by linear programming (LP). We define combinatorial separation algorithms for these inequalities for PIC in Section 6.

Throughout, many proofs are omitted or summarized. Detailed proofs for many of the results in Sections 2 and 3 can be found in Miller et al. [2000a]; detailed proofs for many of the results in Sections 4 and 5 can be found in Miller et al. [2001]. Even more detail can be found in Miller [1999].

Finally, we remark here that computational experience using the inequalities presented for PI in solving MCL confirms the expectation that they should be effective in solving MIP problems that have substructures of the form PI (Miller et al. [2000b]).

2   Basic Polyhedral Results

Denote the set of points defined by (2)–(6) as X^{PI}. If $c > t^i, i = 1, ..., P$, conv(X^{PI}) is full–dimensional; under similarly mild conditions, constraints (2),

(3), and (4), as well as the bounds $x^i \geq 0, i = 1, .., P$, and $y^i \leq 1, i = 1, ..., P$, yield facets of the convex hull. One family of nontrivial facet–inducing inequalities is given by

Proposition 1. *If $c > t^i + d^i$, the valid inequalities*

$$s^i + d^i y^i \geq d^i \tag{7}$$

induce facets of conv(X^{PI}), $i = 1, ..., P$.

These inequalities correspond to the (l, S) inequalities for the uncapacitated lot–sizing problem, introduced by Barany, Van Roy, and Wolsey [1984]. Moreover, they imply the bounds $s^i \geq 0, i = 1, ..., P$, which therefore never induce facets of conv(X^{PI}).

We now characterize the extreme points and rays of conv(X^{PI}). Given an extreme point $(\bar{x}, \bar{y}, \bar{s})$ of conv(X^{PI}), let $Q = \{i \in \mathcal{P} : \bar{y}^i = 1\}$. Also, let $Q_u = \{i \in Q : \bar{x}^i = d^i, \bar{s}^i = 0\}$, let $Q_l = \{i \in Q : \bar{x}^i = 0, \bar{s}^i = d^i\}$, and let $Q_r = \{i \in Q : \bar{x}^i > 0, \bar{x}^i \neq d^i\}$.

Proposition 2. *In every extreme point $(\bar{x}, \bar{y}, \bar{s})$ of conv(X^{PI}), $Q = Q_u \cup Q_l \cup Q_r$, and $\bar{x}^i = 0, \bar{y}^i = 0, \bar{s}^i = d^i, i \in \mathcal{P} \setminus Q$. Moreover, either (i) $Q_r = \emptyset$; or (ii) $|Q_r| = 1$ and $\bar{x}^r = c - \sum_{i \in Q_u}(t^i + d^i) - \sum_{i \in Q_l} t^i - t^r, \bar{y}^r = 1, \bar{s}^r = (d^r - \bar{x}^r)^+$, where $Q_r = \{r\}$.*

It is clear that the extreme rays of conv(X^{PI}) are $s^i = 1, i = 1, ..., P$. Thus, we can characterize conv(X^{PI}) by its extreme points and rays. We now discuss which extreme points are possible optimal solutions of X^{PI}.

Definition 1. *The point $(\bar{x}, \bar{y}, \bar{s})$ is a **dominant solution** of X^{PI} if there exists some cost vector (p, q, h), such that $q^i > 0, h^i > 0, i = 1, ..., P$, for which optimizing (1) over X^{PI} yields $(\bar{x}, \bar{y}, \bar{s})$ as the **unique** optimal solution.*

This following result is crucial for the development in Section 5.

Proposition 3. *The point (x, y, s) is a dominant solution of X^{PI} for PI if and only if (i) (x, y, s) is an extreme point of conv(X^{PI}); and (ii) $y^i = 1$ if and only if $x^i > 0, i = 1, ..., P$.*

This proposition says that a setup for an item does not take place unless production of that item also occurs. It also implies that $Q_l = \emptyset$ in every dominant solution.

3   New Valid Inequalities for PI

Define a *reverse cover* of PI to be a set $S \neq \emptyset$ such that $\mu = c - \sum_{i \in S}(t^i + d^i) > 0$. Note that a reverse cover S does not have to be maximal in any sense, merely non-empty. Similarly, define a *cover* of PI to be a set S such that $\lambda = \sum_{i \in S}(t^i + d^i) - c \geq 0$.

We present inequalities derived from reverse covers first because they are simpler. Given $(x, y, s) \in X^{PI}$, let S be a reverse cover, and let i' be some element not in S. If $x^{i'} = (c - t^{i'})y^{i'}$, either (i) $y^{i'} = 1$, and no capacity is left for $i \in S$, thus implying that $y^i = 0, i \in S$, and that $\sum_{i \in S} s^i \geq \sum_{i \in S} d^i$; or (ii) $y^{i'} = 0$. Such reasoning yields

$$\sum_{i \in S} s^i \geq (\sum_{i \in S}(t^i + d^i))y^{i'} - \sum_{i \in S} t^i(1 - y^i) - ((c - t^{i'})y^{i'} - x^{i'}),$$

which is generalized in the following proposition.

Proposition 4. (Reverse Cover Inequalities) *Let S be a reverse cover of PI, let $T = \mathcal{P} \setminus S$, and let (T', T'') be any partition of T. Then*

$$\sum_{i \in S} s^i \geq (\sum_{i \in S}(t^i + d^i)) \sum_{i \in T'} y^i - \sum_{i \in S} t^i(1 - y^i) - \sum_{i \in T'}((c - t^i)y^i - x^i) \quad (8)$$

is valid for X^{PI}.

Note that T' can be seen as a candidate set for the choice of the element $i' \notin S$ mentioned in the paragraph motivating Proposition 4. Under general conditions, reverse cover inequalities define facets of $\text{conv}(X^{PI})$.

Proposition 5. *If $S \neq \emptyset$, $T' \neq \emptyset$, and $t^i < \mu, i \in T$, (8) induces a facet of $\text{conv}(X^{PI})$.*

This proposition can be proven by exhibiting $3P - |T|$ points that lie in the hyperplane defined by 8, and that are linearly independent of each other and of the $|T|$ extreme rays $s^i = 1, i \in T$.

Now consider a given $(x, y, s) \in X^{PI}$, and a cover S of PI. If $y^i = 1, i \in S$, then $\sum_{i \in S} s^i \geq \sum_{i \in S} d^i - \sum_{i \in S} x^i \geq \sum_{i \in S} d^i - (c - \sum_{i \in S} t^i) = \sum_{i \in S}(t^i + d^i) - c \geq \lambda$. Moreover, if $y^{i'} = 0$ for exactly one $i' \in S$, then both (i) $\sum_{i \in S} s^i \geq \lambda - t^i$ and (ii) $\sum_{i \in S} s^i \geq s^{i'} \geq d^{i'}$ must hold. From such reasoning we can derive

Proposition 6. (Basic Cover Inequalities) *Given a cover S of PI, the inequality*

$$\sum_{i \in S} s^i \geq \lambda + \sum_{i \in S} \max\{-t^i, d^i - \lambda\}(1 - y^i) \quad (9)$$

is valid for PI.

Under general conditions, these inequalities also induce facets of $\text{conv}(X^{PI})$.

Proposition 7. *Given an inequality of the form (9), order the $i \in S$ such that $t^{[1]} + d^{[1]} \geq \ldots \geq t^{[|S|]} + d^{[|S|]}$, let $\mu^1 = t^{[1]} + d^{[1]} - \lambda$, and define $T = \mathcal{P} \setminus S$. If $\lambda > 0$, $t^{[2]} + d^{[2]} > \lambda$, and $t^i < \mu^1, i \in T$, then (9) induces a facet of $\text{conv}(X^{PI})$.*

This proposition can be proven by exhibiting $2P + |S|$ points that lie in the hyperplane defined by (9), and that are linearly independent of each other and of the $P - |S|$ extreme rays $s^i = 1, i \in \mathcal{P} \setminus S$.

We can expand this class by incorporating items not in S into the inequality. Order the $i \in S$ and define μ^1 as in Proposition 7. Let $k' = |\{i \in S : t^i + d^i > \lambda\}|$, and define the lifting function

$$
f(t^0, d^0) = \begin{cases}
-d^0 & \text{if} \quad t^0 + d^0 \leq \mu^1; \\
t^0 + (j-1)\lambda - [\mu^1 + \sum_{i=2}^{j}(t^{[i]} + d^{[i]})] & \text{if} \quad \mu^1 + \sum_{i=2}^{j}(t^{[i]} + d^{[i]}) \\
& \qquad \leq t^0 + d^0 \leq \\
& \qquad \mu^1 + \lambda + \sum_{i=2}^{j}(t^{[i]} + d^{[i]}), \\
& \qquad j \leq k' - 1; \\
j\lambda - d^0 & \text{if } \mu^1 + \lambda + \sum_{i=2}^{j}(t^{[i]} + d^{[i]}) \\
& \qquad \leq t^0 + d^0 \leq \\
& \qquad \mu^1 + \sum_{i=2}^{j+1}(t^{[i]} + d^{[i]}), \\
& \qquad j \leq k' - 1; \\
t^0 + (k'-1)\lambda - [\mu^1 + \sum_{i=2}^{k'}(t^{[i]} + d^{[i]})] & \text{if} \quad t^0 + d^0 \geq \\
& \qquad \mu^1 + \sum_{i=2}^{k'}(t^{[i]} + d^{[i]}).
\end{cases}
$$

We define $f(\cdot, \cdot)$ so that the function $\phi(t' + d') = f(t', d') + d'$ is superadditive. It can be checked that

$$
\phi(u) = \begin{cases}
0 & \text{if} \quad u \leq \mu^1; \\
(j-1)\lambda + (u - [\mu^1 + \sum_{i=2}^{j}(t^{[i]} + d^{[i]})]) & \text{if} \quad \mu^1 + \sum_{i=2}^{j}(t^{[i]} + d^{[i]}) \\
& \qquad \leq u \leq \\
& \qquad \mu^1 + \lambda + \sum_{i=2}^{j}(t^{[i]} + d^{[i]}), \\
& \qquad 1 \leq j \leq k' - 1; \\
j\lambda & \text{if } \mu^1 + \lambda + \sum_{i=2}^{j}(t^{[i]} + d^{[i]}) \\
& \qquad \leq u \leq \\
& \qquad \mu^1 + \sum_{i=2}^{j+1}(t^{[i]} + d^{[i]}), \\
& \qquad 1 \leq j \leq k' - 1; \\
(k'-1)\lambda + (u - [\mu^1 + \sum_{i=2}^{k'}(t^{[i]} + d^{[i]})]) & \text{if } u \geq \mu^1 + \sum_{i=2}^{k'}(t^{[i]} + d^{[i]}).
\end{cases}
$$

Figure 2 shows an example of $\phi(\cdot)$ in which $k' = 3$. The superadditivity of $\phi(\cdot)$ is crucial to the proof of the next proposition.

Proposition 8. (Lifted Cover Inequalities) *Given a cover S of PI, order the $i \in S$ and define μ^1 as in Proposition 7. Let (T, U) be any partition of $\mathcal{P} \setminus S$, let (T', T'') be any partition of T, and define $D' = \max\{t^{[2]} + d^{[2]}, \max_{i \in U}\{t^i + d^i\}\}$. If $\mu^1 \geq 0$ and $|S| \geq 2$, then*

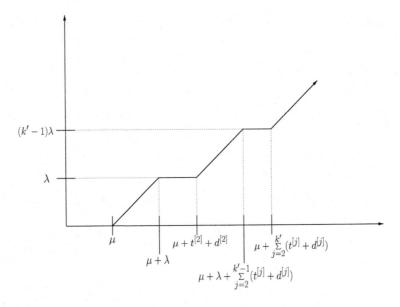

Fig. 2. A graphical example of the function $\phi(\cdot)$ with $k = 3$

$$\sum_{i \in S \cup U} s^i \geq \lambda + \sum_{i \in S} \max\{-t^i, d^i - \lambda\}(1 - y^i) +$$
$$\sum_{i \in U} f(t^i, d^i)y^i + \sum_{i \in U} d^i +$$
$$\frac{\sum_{i \in S} \min\{t^i + d^i, \lambda\} + \sum_{i \in U} \phi(t^i + d^i) - \lambda}{(|S| + |U| - 1)D'} \sum_{i \in T'} (x^i - (\mu^1 - t^i)y^i) \qquad (10)$$

is valid for X^{PI}.

Note that in lifting an item $i \in U$ into (10), we lift two variables, both s^i and y^i. Also note that the constant term on the right hand side of (10) changes with every $i \in U$ that we lift into it. Note further that, given $i' \in T'$, when $x^{i'}$ is much larger than $(\mu^1 - t^{i'})y^{i'}$ and uses up so much capacity that $y^i = 0$ for several $i \in S$, incorporating i' into (10) provides a valid lower bound on $\sum_{i \in S \cup U} s^i$ that is much tighter than the same inequality in which i' is not present.

Under general conditions, inequalities of the form (10) induce facets of the convex hull. For example, we have

Proposition 9. *Given an inequality of the form (10), let* $\lambda > 0$, $t^{[2]} + d^{[2]} > \lambda$, $T' = \emptyset$, *and* $t^i < \mu^1, i \in T''$. *Also, let* k' *be defined as in Proposition 8; for each* $i \in U$, *assume that* $t^i + d^i = \sum_{j=1}^{k}(t^{[j]} + d^{[j]})$, *for some* $k : 1 \leq k \leq k' - 1$. *Then* (10) *induces a facet of* $conv(X^{PI})$.

There are facets induced by inequalities of the form (10) with $T' \neq \emptyset$, $U \neq \emptyset$. For example, we have

Proposition 10. *Given an inequality of the form (10), let $\lambda > 0$, $t^{[2]} + d^{[2]} > \lambda$, and $t^i < \mu^1, i \in T$. Assume also that $t^i + d^i = t^{[2]} + d^{[2]}, i \in \{S \setminus [1]\} \cup U$. Then (10) induces a facet of $conv(X^{PI})$.*

Finally, we remark here that the extension of (9) to (10) is critical for Theorem 1 to hold.

4   Complexity of PI and PIC

Proposition 11. *If either setup times or demand, but not both, are constant, then PI is $\mathcal{NP}$–hard.*

For the case in which setup times are all 0 and $h^i = h > p = p^i = 0, i = 1, ..., P$, there is a reduction from the continuous 0–1 knapsack problem (see Miller [1999], Marchand and Wolsey [1999]). For the case in which $d^i = 1, i = 1, ..., P$, and $h^i > 0, p = p^i = 0, i = 1, ..., P$, there is a reduction from the 0–1 knapsack problem (see Miller [1999]).

Note that this result implies that the general case of PI is $\mathcal{NP}$-hard. Since PI is $\mathcal{NP}$–hard, we cannot expect to find an explicit description of the convex hull in general, or expect to solve PI by optimizing (1) over a specific set of linear inequalities. Now we consider the case when demand and setup times are constant for all i, i.e., PIC.

Proposition 12. *PIC is polynomially solvable.*

Proof: We describe an $\mathcal{O}(P^2)$ algorithm for PIC. To initialize the algorithm, we sort the items so that $dp^{[1]} + q^{[1]} - dh^{[1]} \leq ... \leq dp^{[P]} + q^{[P]} - dh^{[P]}$. Then, for each $i' \in \mathcal{P}$, we find an extreme point solution with $Q_r = \{i'\}$ that has the minimum objective function value among all extreme points with $Q_r = \{i'\}$, where Q_r is defined as in Proposition 3.

To do this, for each $i' \in \mathcal{P}$, we first set $y^{i'} = 1$, $x^{i'} = c - t$, and $s^{i'} = 0$, and we also set $y^i = 0$, $x^i = 0$, $s^i = d$, $i \in \mathcal{P} \setminus i'$. Then we put $i \in \mathcal{P} \setminus i'$ greedily into $Q_u = \{i : y^i = 1, x^i = d\}$, proceeding in the order in which we have sorted the elements, decreasing $x^{i'}$ by $t + d$ for each element we put into Q_u. We continue until doing so no longer decreases the value of the objective function (1), or until we do not have enough capacity left to include any more items in Q_u.

We then find the extreme point solution with $Q_r = \emptyset$ that has the minimum objective function value among all such extreme points, using a similar greedy procedure. Because of Proposition 3, the solution having the minimum value among these $P + 1$ points must be an optimal solution. $\square$

5   An Extended Formulation for PIC

Recall that in PIC $t^i = t \geq 0, d^i = d \geq 0, i = 1, ..., P$. Let $M = \lfloor \frac{c}{t+d} \rfloor$; to avoid uninteresting cases, we assume throughout that $c > M(t + d)$. Let $\lambda =$

$(M+1)(t+d) - c$; observe that $\lambda > 0$, and that $\lambda = \sum_{i \in S}(t+d) - c$ for any minimal cover S. Now for PIC cover inequalities take the form

$$\sum_{i \in S^*} s^i \geq (|S^*| - |M|)\lambda + \sum_{i \in S^*} (d - \lambda)(1 - y^i) + \frac{\lambda}{t+d} \sum_{i \in T'} (x^i - (d - \lambda)y^i),$$

(11)

for any S^* such that $|S^*| \geq M + 1$. (Note that S^* corresponds $S \cup U$, rather than to S, in (10)). Reverse cover inequalities take the form

$$\sum_{i \in S} s^i \geq |S|(t+d) \sum_{i \in T'} y^i - \sum_{i \in S} t(1 - y^i) - \sum_{i \in T'} ((c-t)y^i - x^i), \quad (12)$$

for any S such that $|S| \leq M$.

Given an instance of PIC, let $\tilde{X}^{PIC}$ be the set of dominant solutions of X^{PI}. In defining new variables to reformulate PIC, we say that capacity is met if (3) is tight, and we say that i is produced at demand if exactly d units of i are produced. The variables are

$$\Delta_m = \begin{cases} 1 \text{ if } m \text{ items are produced at demand} \\ 0 \text{ otherwise} \end{cases}$$

$$\delta_m^i = \begin{cases} 1 \text{ if } m \text{ items are produced at demand, and } i \text{ is produced at demand} \\ 0 \text{ otherwise} \end{cases}$$

$$\rho_m^i = \begin{cases} 1 \text{ if } m \text{ items are produced at demand, and } i \text{ is produced but } not \text{ at} \\ \quad \text{demand} \\ 0 \text{ otherwise} \end{cases}$$

$$\alpha_m^i = \begin{cases} 1 \text{ if } m \text{ items are produced at demand, and } i \text{ is not produced at all} \\ 0 \text{ otherwise} \end{cases}$$

In addition, we define the continuous variables $\gamma^i, i = 1, ..., P$ to be the amount of inventory of item i that is carried in excess of what is needed to satisfy demand. We call the set of points defined by the following set of constraints $X^{PIC}_{(\Delta, \delta, \rho, \alpha, \gamma)}$:

$$x^i = d \sum_{m=0}^{M} \delta_m^i + \sum_{m=0}^{M} (d - \lambda + (M - m)(t+d))\rho_m^i, i = 1, ..., P \quad (13)$$

$$y^i = \sum_{m=0}^{M} (\delta_m^i + \rho_m^i), i = 1, ..., P \quad (14)$$

$$s^i = d(\sum_{m=0}^{M} \alpha_m^i) + \lambda \rho_M^i + \gamma^i, i = 1, ..., P \quad (15)$$

$$\sum_{m=0}^{M} \Delta_m = 1 \quad (16)$$

$$\delta_m^i + \rho_m^i + \alpha_m^i = \Delta_m, i = 1, ..., P, m = 0, ..., M \tag{17}$$

$$\sum_{i=1}^{P} \delta_m^i = m\Delta_m, m = 0, ..., M \tag{18}$$

$$\sum_{i=1}^{P} \rho_m^i \leq \Delta_m, m = 0, ..., M \tag{19}$$

$$\gamma^i \geq 0, i = 1, ..., P \tag{20}$$

$$\Delta \in \mathbb{B}^{M+1}, \delta \in \mathbb{B}^{P(M+1)}, \rho \in \mathbb{B}^{P(M+1)}, \alpha \in \mathbb{B}^{P(M+1)}. \tag{21}$$

Also, let $X_{(\Delta,\delta,\rho,\alpha)}^{PIC} = \{(x, y, s, \Delta, \delta, \rho, \alpha) \in X_{(\Delta,\delta,\rho,\alpha,\gamma)}^{PIC} : \gamma^i = 0, i = 1, ..., P\}$, and define $Proj_{(x,y,s)}X$ to be the projection of a set of points X onto the space of the (x, y, s) variables.

Proposition 13. $Proj_{(x,y,s)}X_{(\Delta,\delta,\rho,\alpha)}^{PIC} = \tilde{X}^{PIC}$.

This follows from Proposition 3 and from the variable definitions. In fact, the algorithm in Proposition 12 can be seen as identifying the solution to $X_{(\Delta,\delta,\rho,\alpha)}^{PIC}$ that minimizes (1) by searching among the $\mathcal{O}(P^2)$ possible choices for values of ρ_m^i.

   To see this, let $(\tilde{x}, \tilde{y}, \tilde{s})$ be the optimal solution to PIC identified by the algorithm; clearly $(\tilde{x}, \tilde{y}, \tilde{s}) \in \tilde{X}^{PIC}$. If $Q_r = \{i'\}$ for this solution, this corresponds to setting $\rho_{m'}^{i'} = 1$, where m' is the cardinality of Q_u for $(\tilde{x}, \tilde{y}, \tilde{s})$. Moreover, we can set $\Delta_{m'} = 1$, $\delta_{m'}^i = 1, i \in Q_u$, and $\delta_{m'}^i = 0, i \notin Q_u \cup \{i'\}$. If $Q_r = \emptyset$ for $(\tilde{x}, \tilde{y}, \tilde{s})$, this corresponds to setting $\rho_m^i = 0, i = 1, ..., P, m = 0, ..., M$. As before, letting m' be the cardinality of Q_u for $(\tilde{x}, \tilde{y}, \tilde{s})$, we can also set $\Delta_{m'} = 1$, $\delta_{m'}^i = 1, i \in Q_u$, and $\delta_{m'}^i = 0, i \notin Q_u$. Given the above specifications for the values of the variables ρ_m^i, $\Delta_{m'}$, and $\delta_{m'}^i$, the rest of the variables are uniquely determined by the constraints of $X_{(\Delta,\delta,\rho,\alpha)}^{PIC}$. Thus, the feasible points of $X_{(\Delta,\delta,\rho,\alpha)}^{PIC}$ correspond to the points in $\tilde{X}^{PIC}$.

Corollary 1. *The projection onto the (x, y, s) space of the dominant solutions of $X_{(\Delta,\delta,\rho,\alpha,\gamma)}^{PIC}$ is $\tilde{X}^{PIC}$.*

This follows from Proposition 13 and from the fact that $\gamma^i = 0, i = 1, ..., P$, in every dominant solution of $X_{(\Delta,\delta,\rho,\alpha,\gamma)}^{PIC}$. Let $X_{LP(\Delta,\delta,\rho,\alpha,\gamma)}^{PIC}$ be defined by replacing (21) in $X_{(\Delta,\delta,\rho,\alpha,\gamma)}^{PIC}$ with

$$\delta_m^i, \rho_m^i, \alpha_m^i \geq 0, i = 1, ..., P, m = 0, ..., M. \tag{22}$$

Note that these bounds imply $0 \leq \Delta_m \leq 1, m = 0, ..., M$.

Proposition 14. *The polyhedron associated with (16)–(19), (22) is integral.*

The proof proceeds by considering the $M + 1$ submatrices defined by the constraints (17)–(19) for $m = 0, ..., M$. If Δ_m is fixed, each of these submatrices can be shown to be a network flow matrix. Then, using (16), every fractional solution of (16)–(19), (22) can be shown to be convex combination of $M + 1$ solutions to the network flow subproblems indexed by $m = 0, ..., M$. Thus no fractional solution can be an extreme point.

Corollary 2. *The polyhedron associated with (16)–(20), (22) is integral.*

Proposition 15. $Proj_{(x,y,s)}X^{PIC}_{LP(\Delta,\delta,\rho,\alpha)} = conv(\tilde{X}^{PIC})$.

Proof: From Proposition 14, $X^{PIC}_{LP(\Delta,\delta,\rho,\alpha)} = conv(X^{PIC}_{(\Delta,\delta,\rho,\alpha)})$. From Proposition 13, the projection of $conv(X^{PIC}_{(\Delta,\delta,\rho,\alpha)})$ onto the (x,y,s) space is $conv(\tilde{X}^{PIC})$. $\square$

Corollary 3. *The projection of the dominant solutions of $X^{PIC}_{LP(\Delta,\delta,\rho,\alpha,\gamma)}$ onto the (x,y,s) space is $\tilde{X}^{PIC}$, and thus optimizing (1) over $X^{PIC}_{LP(\Delta,\delta,\rho,\alpha,\gamma)}$ solves PIC.*

Thus we have shown how to solve PIC by LP. Moreover, since (13)–(20), (22) has $\mathcal{O}(P^2)$ variables and constraints, we have shown how to solve PIC by LP in polynomial time. We still seek to show that the set of inequalities that we have defined for PI suffices to solve PIC by LP in the (x,y,s) space. To do this, let $X^{PIC'}_{LP(\Delta,\delta,\rho,\gamma)}$ to be the polyhedron defined by (14), (16), (18)–(20), (22), (2)–(5), (7), (11), (12), the bounds $y^i \leq 1, i = 1, ..., P$, and the constraints

$$\sum_{i=1}^{P^*} x^i \leq \sum_{i=1}^{P}(d\sum_{m=0}^{M}\delta^i_m + \sum_{m=0}^{M}(d - \lambda + (M-m)(t+d))\rho^i_m) \quad (23)$$

$$\sum_{i=1}^{P}(s^i + dy^i - d) \leq \lambda\sum_{i=1}^{P}\rho^i_M + \sum_{i=1}^{P}\gamma^i, i = 1, ..., P. \quad (24)$$

$$x^i + s^i - d \geq \sum_{m=0}^{M-1}(-\lambda + (M-m)(t+d))\rho^i_m + \gamma^i, i = 1, ..., P \quad (25)$$

$$\delta^i_m + \rho^i_m \leq \Delta_m, i = 1, ..., P, m = 0, ..., M. \quad (26)$$

Define $\bar{X}^{PIC} = Proj_{(x,y,s)}X^{PIC'}_{LP(\Delta,\delta,\rho,\gamma)}$. By extensive algebraic manipulation we can show that

Proposition 16. $\bar{X}^{PIC} = Proj_{(x,y,s)}X^{PIC}_{LP(\Delta,\delta,\rho,\alpha,\gamma)}$.

Let $\hat{X}^{PIC}$ to be the polyhedron defined by (2)–(5), (7), (11), (12), and the bounds $y^i \leq 1, i = 1, ..., P$. We see immediately that $\bar{X}^{PIC} \subset \hat{X}^{PIC}$. Moreover, we have

Proposition 17. *Given any $(\bar{x}, \bar{y}, \bar{s}) \in \hat{X}^{PIC}$, $(\bar{x}, \bar{y}, \bar{s}) \in \bar{X}^{PIC}$ if and only if*
$$\eta + \sum_{i=1}^{P}\omega^i\bar{y}^i + \sigma\sum_{i=1}^{P}(\bar{s}^i + d\bar{y}^i - d) - \sum_{i=1}^{P}\pi^i(\bar{x}^i + \bar{s}^i - d) + \sum_{i=1}^{P}\beta x^i \leq 0$$
holds for all extreme rays of the dual cone

$$(\delta_m^i) \quad \omega^i + d\beta + \tau_m - \nu_m^i \le 0, i = 1, ..., P, m = 0, ..., M \tag{27}$$

$$(\rho_m^i) \quad \omega^i + (d - \lambda + (M - m)(t + d))\beta - (-\lambda + (M - m)(t + d))\pi^i \tag{28}$$

$$-\kappa_m - \nu_m^i \le 0, i = 1, ..., P, m = 0, ..., M - 1 \tag{29}$$

$$(\rho_M^i) \quad \omega^i + (d - \lambda)\beta + \lambda\sigma - \kappa_M - \nu_M^i \le 0, i = 1, ..., P \tag{30}$$

$$(\gamma^i) \quad \sigma - \pi^i \le 0, i = 1, ..., P \tag{31}$$

$$(\Delta_m) \quad \eta = m\tau_m - \kappa_m - \sum_{i=1}^{P} \nu_m^i, m = 0, ..., M \tag{32}$$

$$\beta \ge 0, \sigma \ge 0; \pi^i \ge 0, i = 1, ..., P; \kappa_m \ge 0, m = 0, ..., M;$$

$$\nu_m^i \ge 0, i = 1, ..., P, m = 0, ..., M. \tag{33}$$

To prove this proposition, for a fixed $(\bar{x}, \bar{y}, \bar{s})$ we take the dual of (14), (16), (18)–(20), (22)–(26) and apply Farkas' lemma. The next lemma is crucial to the main result of this section.

Lemma 1. *Let $(\bar{x}, \bar{y}, \bar{s})$ be a dominant solution of $\hat{X}^{PIC}$. Then $\eta + \sum_{i=1}^{P} \omega^i \bar{y}^i + \sigma \sum_{i=1}^{P} (\bar{s}^i + d\bar{y}^i - d) - \sum_{i=1}^{P} \pi^i(\bar{x}^i + \bar{s}^i - d) + \sum_{i=1}^{P} \beta x^i \le 0$ holds for all extreme rays of the dual cone (27)–(33).*

The proof of Lemma 1 follows from a detailed analysis of the rays in question. This analysis proceeds by cases, which are defined by values of β and σ.

Proposition 18. *$\hat{X}^{PIC}$ and of $\bar{X}^{PIC}$ have the same dominant solutions.*

The proof proceeds by considering any dominant solution $(\bar{x}, \bar{y}, \bar{s})$ of $\hat{X}^{PIC}$, and using Lemma 1 to show that $(\bar{x}, \bar{y}, \bar{s}) \in \bar{X}^{PIC}$. It is straightforward that if $(\bar{x}, \bar{y}, \bar{s})$ is a dominant solution of $\hat{X}^{PIC}$ and $(\bar{x}, \bar{y}, \bar{s}) \in \bar{X}^{PIC}$, then $(\bar{x}, \bar{y}, \bar{s})$ is also a dominant solution of $\bar{X}^{PIC}$; thus every dominant solution of $\hat{X}^{PIC}$ is also a dominant solution of $\bar{X}^{PIC}$. We can use this fact, and the fact that $\bar{X}^{PIC} \subseteq \hat{X}^{PIC}$, to show that every dominant solution of $\bar{X}^{PIC}$ is also a dominant solution of $\hat{X}^{PIC}$.

Theorem 1. *$\tilde{X}^{PIC}$ consists of the dominant solutions to the polyhedron defined by (2)–(5), (7), (11), (12), and the bounds $y^i \le 1, i = 1, ..., P$. Thus, given that $q^i > 0, h^i \ge 0, i = 1, ..., P$, optimizing (1) over this set of linear inequalities solves PIC.*

This follows from Corollary 3 and Propositions 16 and 18. Thus the set of facet–defining inequalities we presented for the general case of PI suffice to solve the special case PIC by LP in the original space of variables.

6   Separation for Cover and Reverse Cover Inequalities for PIC

While separation for inequalities of the forms (2)–(5), (7), is trivial, there are an exponential number of both cover and reverse cover inequalities. As a consequence of our results in the previous section, separation for PIC can be performed in polynomial time by solving an LP in a higher–dimensional space. However, it would be preferable to have fast combinatorial algorithms to separate for cover and reverse cover inequalities for PIC. In this section we describe such separation algorithms.

Proposition 19. (Separation for Cover Inequalities) *Given a point* $(\bar{x}, \bar{y}, \bar{s})$, *there exists a violated inequality of the form (11) if and only if the optimal solution to the IP*

$$\max \sum_{i=1}^{P}((d-\lambda)(1-\bar{y}^i) + \lambda - \bar{s}^i)\chi^i + \frac{\lambda}{t+d}\sum_{i=1}^{P}(\bar{x}^i - (d-\lambda)\bar{y}^i)\xi^i$$

(34)

$$\text{subject to } \chi^i + \xi^i \leq 1, i = 1, ..., P$$

(35)

$$\sum_{i=1}^{P}\chi^i \geq M+1$$

(36)

$$\chi^i, \xi^i \in \{0,1\}, i = 1, ..., P$$

(37)

is strictly greater than $M\lambda$. *Moreover, if the optimal solution is strictly greater than* $M\lambda$, *then a most violated inequality of the form (11) is obtained by taking* $S^* = \{i \in \mathcal{P} : \chi^i = 1\}$, $T' = \{i \in \mathcal{P} : \xi^i = 1\}$, *and* $T'' = \{i \in \mathcal{P} : \chi^i = \xi^i = 0\}$.

The IP (34)–(37) can be solved by finding a maximum weight perfect matching on a complete bipartite graph G. In G the vertex set V is partitioned into (A, B), with $|A| = |B| = P$. The nodes in A represent items in $\mathcal{P}$, while the nodes in B are used to represent positions in S^*, T' or T''.

We partition B further into B' and B'' such $|B'| = M + 1$. For each $i \in A$ and $j \in B'$, we let the edge weight w_{ij} be

$$w_{ij} = (d-\lambda)(1-\bar{y}^i) + \lambda - \bar{s}^i;$$

that is, w_{ij} is the value added to the objective function (34) by setting $\chi^i = 1$. For each $i \in A$ and $j \in B''$, we let the edge weight w_{ij} be

$$w_{ij} = \max\{(d-\lambda)(1-\bar{y}^i) + \lambda - \bar{s}^i, \frac{\lambda}{t+d}(\bar{x}^i - (d-\lambda)\bar{y}^i, 0\};$$

that is, w_{ij} is the value added to (34) by setting to 1 the choice between χ^i or ξ^i that increases the value of (34) the most, or by setting $\chi^i = \xi^i = 0$ if both $(d-\lambda)(1-\bar{y}^i) + \lambda - \bar{s}^i$ and $\frac{\lambda}{t+d}(\bar{x}^i - (d-\lambda)\bar{y}^i)$ are nonpositive.

A maximum weight perfect matching in G yields the optimal solution to (34)–(37). A most violated cover inequality is then defined as prescribed in Proposition 19.

It is well known that finding a maximum weight matching in a bipartite graph can be accomplished by using a maximum flow algorithm (see e.g. Cook et al. [1998]). For a general graph with unit capacities, it is possible to solve the maximum flow problem in $\mathcal{O}(E^{\frac{3}{2}})$, where E is the number of edges (see e.g. Even and Tarjan [1975]). Since G has P^2 edges, we have

Proposition 20. *The separation problem for cover inequalities for PIC can be solved in* $\mathcal{O}(P^3)$ *time.*

Separation for reverse cover inequalities can be accomplished in a manner similar to cover inequality separation.

Proposition 21. (Separation for Reverse Cover Inequalities) *Given a point* $(\bar{x}, \bar{y}, \bar{s})$*, there is a violated inequality of the form (12) if and only if, for some* m*,* $1 \leq m \leq M$*, the value of the optimal solution to the IP*

$$\max \ \sum_{i=1}^{P}(t\bar{y}^i - \bar{s}^i)\chi^i + \sum_{i=1}^{P}((m(t+d) - c + t)\bar{y}^i + \bar{x}^i)\xi^i \qquad (38)$$

$$\text{subject to } \chi^i + \xi^i \leq 1 \qquad (39)$$

$$\sum_{i=1}^{P}\chi^i = m \qquad (40)$$

$$\chi^i, \xi^i \in \{0, 1\}, \qquad (41)$$

is strictly greater than tm*. Moreover, if the optimal solution is strictly greater than* tm *for some such* m*, then a most violated inequality of the form (12) with* $|S| = m$ *is defined by putting* i *into* S *if* $\chi^i = 1$*, putting* i *into* T' *if* $\xi^i = 1$*, and putting* i *into* T'' *otherwise.*

The separation problem for all reverse cover inequalities amounts to solving M IP's of the form (38)–(41), which are identical except for the value of m. Each of these IP's can be solved by finding a maximum weight perfect matching in a complete bipartite graph similar to the graph described in the discussion on cover inequality separation. Therefore, for each $m = 1, ..., M$, the separation problem for reverse cover inequalities in which $|S| = m$ can be solved in $\mathcal{O}(P^3)$ time. This fact yields

Proposition 22. *The separation problem for reverse cover inequalities for PIC can be solved in* $\mathcal{O}(P^4)$ *time.*

Given the separation algorithms that we have described in this section, we can now define an ellipsoid algorithm for PIC that runs entirely in the space of (x, y, s) variables. In addition, these separation algorithms can clearly be applied to problems for which PIC is a substructure. Moreover, while they are not directly applicable to the general case of PI, it may be possible to use the results of this section to help to develop separation heuristics for PI, and for problems for which PI provides a relaxation.

References

Barany et al., 1984. Barany, I., Roy, T. V., and Wolsey, L. (1984). Uncapacitated lot–sizing: the convex hull of solutions. *Mathematical Programming Study*, 22:32–43.

Cook et al., 1998. Cook, W., Cunningham, W., Pulleyblank, W., and Schrijver, A. (1998). *Combinatorial Optimization*. Wiley, New York.

Crowder et al., 1983. Crowder, H., Johnson, E., and Padberg, M. (1983). Solving large scale zero–one linear programming problems. *Operations Research*, 31:803–834.

Even and Tarjan, 1975. Even, S. and Tarjan, R. (1975). Network flow and testing graph connectivty. *SIAM Journal on Computing*, 4:507–518.

Miller, 1999. Miller, A. (1999). *Polyhedral Approaches to Capacitated Lot–Sizing Problems*. PhD thesis, Georgia Institute of Technology.

Miller et al., 2000a. Miller, A., Nemhauser, G., and Savelsbergh, M. (2000a). On the polyhedral structure of a multi–item production planning model with setup times. CORE DP 2000/52, Université Catholique de Louvain, Louvain-la-Neuve.

Miller et al., 2000b. Miller, A., Nemhauser, G., and Savelsbergh, M. (2000b). Solving multi–item capacitated lot–sizing problems with setup times by branch–and–cut. CORE DP 2000/39, Université Catholique de Louvain, Louvain-la-Neuve.

Miller et al., 2001. Miller, A., Nemhauser, G., and Savelsbergh, M. (2001). A multi–item production planning model with setup times: Algorithms, reformulations, and polyhedral characterizations for a special case. CORE DP 2001/06, Université Catholique de Louvain, Louvain-la-Neuve.

Padberg et al., 1985. Padberg, M., Roy, T. V., and Wolsey, L. (1985). Valid linear inequalities for fixed charge problems. *Operations Research*, 33:842–861.

Trigiero et al., 1989. Trigiero, W., Thomas, L., and McClain, J. (1989). Capacitated lot–sizing with setup times. *Management Science*, 35:353–366.

Fences Are Futile: On Relaxations for the Linear Ordering Problem

Alantha Newman[1] and Santosh Vempala[2]

[1] Laboratory for Computer Science, Massachusetts Institute of Technology,
Cambridge, MA 02139
alantha@theory.lcs.mit.edu
[2] Department of Mathematics, Massachusetts Institute of Technology,
Cambridge, MA 02139
vempala@math.mit.edu

Abstract. We study polyhedral relaxations for the linear ordering problem. The integrality gap for the standard linear programming relaxation is 2. Our main result is that the integrality gap remains 2 even when the standard relaxations are augmented with k-*fence* constraints for any k, and with k-*Möbius ladder* constraints for k up to 7; when augmented with k-Möbius ladder constraints for general k, the gap is at least $\frac{33}{17} \approx 1.94$. Our proof is non-constructive–we obtain an extremal example via the probabilistic method. Finally, we show that no relaxation that is solvable in polynomial time can have an integrality gap less than $\frac{66}{65}$ unless P=NP.

1   Introduction

Given a complete weighted directed graph, the *linear ordering* problem is to find a linear ordering of the vertices that maximizes the weight of the forward edges (edge (i,j) is a *forward* edge if i precedes j in the ordering). This problem is equivalent to finding a maximum acyclic subgraph of a given graph.

The linear ordering problem is NP-hard [8], motivating the question of polynomial time approximation algorithms. It is in fact easy to find a solution with weight at least half the optimum: take *any* linear ordering of the vertices; partition the edges into two sets, those going forward in the ordering and those going backward. Both sets are acyclic; one of these sets has weight at least half the total weight of all the edges in the graph (and hence at least half the optimum). This simple algorithm gives the best-known polynomial-time computable approximation factor for the problem (namely $\frac{1}{2}$).

In this paper, we study the quality of polyhedral relaxations for this optimization problem [4,5]. The quality of a relaxation can be measured by the *integrality gap*, the maximum possible ratio between the linear programming optimum and the true integral optimum. A well-known linear programming relaxation for the problem is based on the simple idea of requiring that from every directed cycle C, a solution contains at most $|C|-1$ edges. The corresponding linear constraints are exponential in number (one for each cycle), but can be solved in polynomial

K. Aardal, B. Gerards (Eds.): IPCO 2001, LNCS 2081, pp. 333–347, 2001.

time via an efficient separation oracle. This and another well-known relaxation are described in Section 2. How good are the relaxations? The integrality gap for both of these standard relaxations turns out to be at least $2 - \epsilon$ for any $\epsilon > 0$. Thus, the estimate they provide on the optimum is no better (in the worst-case) than the trivial upper bound of the total edge weight.

A natural next step is to strengthen these standard relaxations by adding constraints. To this end, a promising set of constraints are the *k-fence* constraints [4,5]. Although these constraints are NP-complete to separate in general [9], they can be separated in polynomial time for any fixed k. Another set of constraints that have been proposed are the *k-Möbius* ladder constraints [4,5]. These are known to be separable in polynomial time [1,12]. In Section 3, we present our main result: the integrality gap is 2 even with k-fence constraints for any k and with k-Möbius ladder constraints for $k \leq 7$; when augmented with k-Möbius ladder constraints for arbitrary k, the gap is at least $\frac{33}{17} \approx 1.94$. Our proofs of the integrality gap start with a probabilistic construction which is molded to have the desired structure (thus we demonstrate the existence of extremal graphs without explicitly describing them).

Finally, we establish a concrete lower bound on approximability: it is NP-hard to approximate the optimum to within a factor better than $\frac{65}{66}$, i.e. no polynomial-time solvable relaxation can have an integrality gap less than $\frac{66}{65}$. The reduction, described in Section 4, is from the problem of finding a maximum satisfiable subset of a given set of linear equations modulo 2.

2   Standard LP Relaxations

In this section, we describe two standard linear programming relaxations, prove that they have the same optimal value for any graph, and show that both relaxations can be arbitrarily close to twice the value of the optimum in the worst case.

2.1   LP$_1$

The maximum acyclic subgraph problem can be viewed as maximizing the number of edges subject to a constraint for every cycle. Grötschel, Jünger, and Reinelt refer to these constraints as *dicycle inequalities* [4]. We will call them *cycle constraints*. The constraints specify that the sum of the edge variables on any cycle C is at most $|C| - 1$.

maximize $\sum_{(i,j) \in E} w_{ij} x_{ij}$
subject to:
$$\sum_{ij \in C} x_{ij} \leq |C| - 1 \quad \forall C$$
$$x_{ij} \in \{0, 1\} \quad \forall ij \in E$$

The solutions to this integer program are acyclic subgraphs. It is NP-hard to solve this integer program. However, we can relax the requirement that the x_{ij}

are in $\{0, 1\}$ and replace it with the requirement that $0 \leq x_{ij} \leq 1$. We refer to this linear programming relaxation as LP_1. We can solve LP_1 in polynomial time using the Ellipsoid Algorithm [6] via the following polynomial-time separation oracle. Given an assignment for the variables x_{ij}, we consider the graph with each edge (i, j) assigned a weight of $1 - x_{ij}$. In this graph, we find the minimum weight cycle. If there is any cycle with weight less than 1, then the corresponding cycle of length C actually has weight more than $C-1$, which highlights a violated constraint.

2.2   LP$_2$

Another integer program is based on the linear ordering problem. It has a variable for every pair of vertices $i, j \in V$. In this program, there are only constraints for 2- and 3-cycles. This set of constraints is discussed by Grötschel, Jünger, and Reinelt [5].

maximize $\sum_{ij} w_{ij} x_{ij}$
subject to:

$$x_{ij} + x_{ji} = 1 \quad \forall i, j \in V$$
$$x_{ij} + x_{jk} + x_{ki} \leq 2 \quad \forall i, j, k \in V$$
$$x_{ij} \in \{0, 1\} \quad \forall i, j \in V$$

Solutions for this integer program correspond to linear orderings. Again, it is NP-hard to solve this integer program. We refer to the corresponding relaxation as LP_2. Although LP_2 only contains constraints for 2- and 3-cycles, we can show that a valid solution for LP_2 does not violate *any* cycle constraints.

Lemma 1. *A solution for LP_2 does not violate any cycle constraints.*

Proof. We will prove by induction on k that a solution for LP_2 does not violate any k-cycle constraints. Clearly, a valid solution for LP_2 does not violate any 2- or 3-cycle constraints. Assume all k-cycle constraints are satisfied. Then we will show that all $(k+1)$-cycle constraints are satisfied. Consider a cycle C of length $k+1$. Choose two non-adjacent vertices i, j in C. Now consider the following two edge-disjoint cycles: C_1 is composed of edge (i, j) and the path from j to i in C and C_2 is composed of edge (j, i) and the path from i to j in C. By induction, we have $\sum_{e \in C_1} x_e + \sum_{e \in C_2} x_e \leq |C_1| - 1 + |C_2| - 1 \leq |C|$ and $x_{ij} + x_{ji} = 1$. Thus, $\sum_{e \in C} x_e = \sum_{e \in C_1} x_e + \sum_{e \in C_2} x_e - x_{ij} - x_{ji} \leq |C| - 1$.   □

A maximum acyclic subgraph has the same weight as a maximum linear ordering, i.e. the optimal integral solutions for the two integer programs above are equal. We now prove that the optimal solutions for the two linear programming relaxations are equal. For some graph $G = (V, E)$ with edge weights $w = \{w_{ij}\}$, let $OPT(LP_1)$ denote an optimal solution for LP_1 and $|OPT(LP_1)|$ denote its objective value. Define $OPT(LP_2)$ similarly with $w_{ij} = 0$ for all $(i, j) \notin E$.

Theorem 1. $|OPT(LP_1)| = |OPT(LP_2)|$.

Proof. First, we will show that $|OPT(LP_1)| \geq |OPT(LP_2)|$, i.e. given an optimal solution for LP_2, we can find a solution for LP_1 with the same value. We simply let the solution for LP_1 be the subset of $\{x_{ij}\}$ such that $(i,j) \in E$. By Lemma 1, this solution does not violate any cycle constraints and is therefore a valid solution for LP_1.

Second, we will show that $|OPT(LP_2)| \geq |OPT(LP_1)|$, i.e. given an optimal solution for LP_1, we can construct a solution for LP_2 with the same objective value. Assign all edges in E value x_{ij} where x_{ij} is taken from the given solution for LP_1. Since this is a valid solution for LP_1, no cycle constraints have been violated thus far. Now consider an arbitrary order for the pairs (j,i) such that $(i,j) \in E$ and $(j,i) \notin E$ and assign $x_{ji} = 1 - x_{ij}$ in that order. Let (j,i) be the first edge causing a violated cycle constraint. Then there is some path p_{ij} of length ℓ from i to j such that the total value of the edges in p_{ij} is more than $\ell - x_{ji}$. Since $(i,j) \in E$, and the solution for the edges in E is optimal, it must be the case that there is some path p_{ji} of length ℓ' such that the value of the edges in p_{ji} equals to $\ell' - x_{ij}$, i.e. a cycle constraint for some cycle containing edge (i,j) must be tight, otherwise we could increase the value of x_{ij}. Thus, p_{ij} and p_{ji} form a cycle of length $\ell + \ell'$ of value more than $\ell + \ell' - (x_{ij} + x_{ji}) = \ell + \ell' - 1$. So it is a contradiction that this edge is the first to cause a violated cycle constraint. Also note that all 2-cycles in E have total value exactly 1. Otherwise, for some $x_{ij} + x_{ji} < 1$, we can find a cycle composed of the paths p_{ji} and p_{ij} which violates a cycle constraint. Simply let p_{ji} be the path in the cycle with x_{ij} for which a cycle constraint is tight, and define p_{ij} similarly.

Let $\bar{G} = (V, \bar{E})$ be the graph with edge set $\bar{E} = \{(i,j)\}$ such that $(i,j) \in E$ or $(j,i) \in E$. By the argument above, all 2-cycles in $\bar{G}$ have value exactly 1 and no cycle constraints are violated. Now we will assign values x_{ij} to all edges (i,j) such that $i, j \in V$ and neither (i,j) nor (j,i) are in E. We define the *shortest path* between i and j to be the path with the least total value, where x_{ij} is the value of an edge. Let α_{ij} be the length of the shortest directed path from i to j in $\bar{G}$. Define α_{ji} similarly. For any i, j the shortest paths from i to j and from j to i form a cycle in the current graph. Therefore, $\alpha_{ij} + \alpha_{ji} \geq 1$. Without loss of generality, assume $\alpha_{ij} \leq \alpha_{ji}$. Then let $x_{ij} = \min\{\frac{1}{2}, \alpha_{ij}\}$ and $x_{ji} = 1 - x_{ij}$. Thus, every cycle that includes edge (i,j) or edge (j,i) will have value at least 1, which implies that every cycle C will have value at most $|C| - 1$. This implies that all 3-cycle constraints are satisfied and all 2-cycles have value exactly 1.   □

2.3   Integrality Gap

The *integrality gap* of a linear program is the worst case ratio between the value of an optimal fractional solution and the value of an optimal integral solution over all weight functions $w = \{w_{ij}\}$. Formally, the integrality gap is defined as,

$$\max_{w>0} \frac{|OPT(LP)|}{|OPT(IP)|}$$

In this section, we show that the integrality gap for both LP_1 and LP_2 is $2 - \epsilon$ for any $\epsilon > 0$. As the basis of the construction, we use the fact that there exists a class of undirected graphs with girth g and $\Theta(n^{1+1/g})$ edges. This result is due to Erdös and Sachs [2]. Graphs from this class have been used to prove integrality gaps for the maximum cut problem [11]. Based on these graphs, we define $G(n)$ to be a family of graphs with the following properties. A graph $G \in G(n)$ has n vertices, girth $g = \Theta(\frac{\log n}{\log \log n})$, and $n^{1+1/g}$ edges. Then we have the following lemma.

Lemma 2. *For any $\epsilon > 0$, there exists an $n \geq f(\epsilon)$ such that at least one directed orientation of $G - (V, E) \in G(n)$ has the following property: for any ordering of the vertices, the number of forward edges is at most $(1 + \epsilon)|E|/2$.*

A proof of Lemma 2 can be found in [10]. We now define $G(\epsilon)$ to be the family of *directed* graphs on $n \geq f(\epsilon)$ vertices whose underlying undirected graphs belong to $G(n)$ and which have maximum acyclic subgraphs of size at most $(1 + \epsilon)|E|/2$. All edges in $G \in G(\epsilon)$ have weight $w_{ij} = 1$. We use $G(\epsilon)$ to prove the following theorem.

Theorem 2. *The integrality gap of LP_1 is at least $2 - \epsilon$ for any $\epsilon > 0$.*

Proof. For a graph $G = (V, E) \in G(\epsilon)$, we assign $x_{ij} = 1 - 1/g$ for every edge in G, where g is the girth of G. This is a feasible solution for LP_1 since there are no cycles of length less than g. Thus, the optimal solution of LP_1 has size at least $|E|(1 - 1/g)$. The ratio of the optimal fractional solution to the optimal integral solution is at least $(1 - 1/g)/(\frac{1}{2}(1 + \epsilon))$. Since $g = \Theta(\frac{\log \log n}{\log n})$, then for any $\epsilon' > 0$, we can choose ϵ and n so that $2(1 - 1/g)/(1 + \epsilon) \geq 2 - \epsilon'$. □

Theorem 3 follows from Theorem 2 and Theorem 1.

Theorem 3. *The integrality gap of LP_2 is at least $2 - \epsilon$ for any $\epsilon > 0$.*

3   Augmented LP Relaxations

In the previous section, we saw that a rather non-trivial LP has an integrality gap arbitrarily close to 2, thus providing an upper bound that is no better than the total weight of the edges in the worst case. How can we get a better upper bound? One way would be to add new constraints to this LP. Some well-known constraints for this problem are the so-called *fence* constraints and *Möbius ladder* constraints presented by Grötschel, Jünger, and Reinelt [4,5]. In this section, we will show that that if we augment LP_2 with k-fence constraints for any k and with k-Möbius ladder constraints for $k = 3$, then the integrality gap remains 2. This is also true for 5, 7-ladders, but the proofs are omitted here. For $k \geq 9$, the integrality gap of LP_2 augmented with k-Möbius ladder constraints is at least $\frac{33}{17}$. (Note that the integrality gap of LP_1 augmented with these constraints trivially remains 2: a graph belonging to $G(\epsilon)$ has girth greater than 4 for sufficiently small ϵ and therefore does not contain any k-fences or k-Möbius ladders.)

Throughout this section, it will be convenient to have the following definitions. An edge (i, j) is the *complementary* edge to edge (j, i). The *value* of a set S of edges is defined to be $x(S) = \sum_{(i,j) \in S} x_{ij}$. We define the *shortest* path from i to j in a graph G as a path from i to j with the least total value.

3.1   The Bad Example Graph

We will now describe the *bad example* graph–the graph which we use to prove our lower bound on the integrality gap of the augmented LP relaxation. We use the family of graphs $G(\epsilon)$ defined in Section 2.3. We begin with a graph $G = (V, E) \in G(\epsilon)$. (We used this graph to prove Theorem 2. However, now we need to an assign a value to every edge in the complete graph.) For every $(i, j) \notin E$, assign $w_{ij} = 0$. For every edge $(i, j) \in E$, we assign x_{ij} value $1 - 1/g$ and x_{ji} value $1/g$, where g is the girth of G. For all i, j such that neither (i, j) nor (j, i) are in E, we assign a value to x_{ij} using the rule given in the proof of Theorem 1. We will restate this rule here for the sake of convenience. Define $\bar{G}$ to be the graph consisting of the edges in E and their complementary edges. Let α_{ij} be the shortest path from i to j in $\bar{G}$. Define α_{ji} similarly. Without loss of generality, assume $\alpha_{ij} \leq \alpha_{ji}$. Then assign x_{ij} and x_{ji} using the following rule.

Edge Assignment Rule: $x_{ij} = \min\{\frac{1}{2}, \alpha_{ij}\}, \quad x_{ji} = 1 - x_{ij}$

The following corollary holds for the complete directed graph $\tilde{G}$ in which every edge has been assigned a value.

Corollary 1. *If the shortest path from i to j in $\bar{G}$ is $\alpha < \frac{1}{2}$, then the value of the shortest path between i and j in $\tilde{G}$ is α. If the value of the shortest path from i to j in $\bar{G}$ is at least $\frac{1}{2}$, then the value of the shortest path from i to j in $\tilde{G}$ is at least $\frac{1}{2}$.*

Recall that the optimal objective value of LP_2 for $\tilde{G}$ is at least $|E|(1 - 1/g)$. Therefore, if we can show that the edges of $\tilde{G}$ also satisfy other specified constraints, then we can show that the integrality gap of LP_2 augmented with these constraints remains the same as LP_2.

3.2   Möbius Ladders

A Möbius ladder for an odd integer k (a k-ladder) is defined to be a set of $2k$ vertices $\{a_1, b_1, \cdots, a_k, b_k\}$ and $3k$ edges such that each vertex a_i has a directed edge to b_{i+1} and $b_{i-1}, 1 \leq i \leq k$. (We define $b_0 = b_k$ and $b_{k+1} = b_1$.) There is also an edge from b_i to a_i. A 5-ladder is shown in Figure 1. (A 3-ladder is isomorphic to a 3-fence, which will be defined later on.)

An acyclic subgraph of a 5-Möbius ladder includes at most 12 of the 15 edges. However, there is a fractional solution of $12\frac{1}{2}$ that satisfies LP_2: each edge from b_i to a_i is assigned a value of $\frac{1}{2}$ and all other edges are assigned value 1. In general, an acyclic subgraph of a k-ladder includes at most $3k - (\frac{k+1}{2})$ of the edges. However, we can always find a fractional solution with value $2\frac{1}{2}k$ that satisfies every cycle constraint. So we add the following constraint to LP_2 for every subset of edges that forms a k-ladder.

$$\sum_{(i,j) \in k-ladder} x_{ij} \leq 3k - (\frac{k+1}{2}) \tag{1}$$

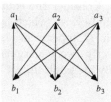 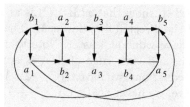

Fig. 1. A 3-ladder (or 3-fence) and a 5-ladder.

Recall that LP$_2$ yields an assignment for the complete graph. Therefore, in order to show that the total value of the edges in any k-ladder is at most $3k - \frac{k+1}{2}$, it suffices to show that the total value of the edges in any k-ladder is at least $\frac{k+1}{2}$. This is because the set of edges complementary to a k-ladder also form a k-ladder and the sum of the values of both k-ladders is exactly $3k$.

Now we will show that constraint (1) is satisfied for all 3-ladders in $\tilde{G}$, which we defined in Section 3.1. Let M be the set of $3k$ edges in a k-ladder and let C be the subset of edges (a_i, b_{i+1}) and (a_i, b_{i-1}) for $i \in \{1, \ldots k\}$. The edges in C make up an undirected cycle of value $2k$. For example, in the 5-ladder shown in Figure 1, C is the set of all edges expect for the 5 vertical edges. Furthermore, define T to be the set of edges in $\bar{G}$ (each graph $\tilde{G}$ has a corresponding $\bar{G}$, which is also defined in Section 3.1) that belong to the shortest paths for each pair of vertices i, j such that edge $(i, j) \in C$. If there are multiple shortest paths for some pair, only one of these paths is included in T. The edges in T will play a key role in our proof.

We say that T contains an *undirected* cycle if the underlying undirected graph contains a cycle. There are two cases to consider: T can either be a tree or contain an undirected or directed cycle. In the case that T contains a cycle, we will show that constraint (1) is satisfied, and when T is a tree, we will show that constraint (1) is satisfied for $k = 3$. Constraint (1) is also satisfied for $k = 5, 7$, but the proofs are omitted. However, this constraint is not necessarily satisfied when $k = 9$. In other words, our edge assignment rules could violate some k-ladder constraints for $k \geq 9$.

Case 1: T Contains a Cycle. We consider two subcases based on the values of the $2k$ paths in T from a_i to b_{i+1} and from a_i to b_{i-1} for $i \in \{1, \ldots k\}$. For each of these cases, we will show that the total value of the edges in M is at least $\frac{k+1}{2}$. In the proof, we will use the following lemma.

Lemma 3. *If the value of C is at least 1, then the total value of the edges in M is at least $\frac{k+1}{2}$.*

Proof. If the total value of C is $1 + \epsilon$ for some $\epsilon \geq 0$, then we will show that the total value of the other k edges in M is at least $\frac{k-1-\epsilon}{2}$. This would imply that the total value of the edges in M is $(1 + \epsilon) + (\frac{k-1-\epsilon}{2}) = \frac{k+1+\epsilon}{2}$ as required. We just need to show that $x(M \setminus C)$ is at least $\frac{k-1-\epsilon}{2}$. Consider the equations for each of the directed 4-cycles in M. There are k such 4-cycles, which by Lemma 1 each have value at least 1. Note that each edge in C appears in exactly one equation, and each edge of $M \setminus C$ appears in exactly two equations. Adding these k equations, we have: $x(C) + 2x(M \setminus C) \geq k$. By assumption, $x(C) = 1 + \epsilon$. Therefore, we have: $x(M \setminus C) \geq \frac{k-1-\epsilon}{2}$.   □

Lemma 4. *If T contains a cycle, then the total value of M is at least $\frac{k+1}{2}$.*

Proof. Note that if two or more of the $2k$ paths in C have value at least $\frac{1}{2}$, then by Corollary 1 and Lemma 3, the total value of M is at least $\frac{k+1}{2}$. Now we will consider two remaining cases.

(i) *All $2k$ paths in T have value strictly less than $\frac{1}{2}$.* By Claim 1, all edges in C have value equal to the value of their respective paths in T. Since some subset of the edges in T form a cycle, the total value of the edges in T is at least 1. Since every edge in T, by definition, belongs to at least one of the $2k$ shortest paths, the total value of C is at least the total value of T, which is at least 1. By Lemma 3, the total value of M is at least 2.

(ii) *Only one of the $2k$ paths in T has value at least $\frac{1}{2}$.* We can assume the total value of the other $2k - 1$ edges in C is less than $\frac{1}{2}$. In this case, note that T does not contain a directed cycle. If it did, the total value of the edges in C would be at least 1. Thus, if we removed the edge with value at least $\frac{1}{2}$, the remaining edges would sum to at least $\frac{1}{2}$.

Without loss of generality, assume that edge (a_1, b_2) in C is the edge with value at least $\frac{1}{2}$. Consider the 4-cycle $\{a_1, b_2, a_2, b_1\}$ in M that contains this edge. Since T contains a path from a_2 to b_1 and a path from a_1 to b_2, it cannot contain a path from b_1 to a_1 and from b_2 to a_2, since then it would contain a directed cycle. Without loss of generality, assume T does not contain a path from b_1 to a_1. Let T' be the set of edges in G that correspond to the shortest paths for all edges in C except (a_1, b_2). Then T' contains a directed or undirected path from a_1 to b_1. Since the total value of the edges in T' is less than $\frac{1}{2}$, there is a directed or undirected path from a_1 to b_1 in G with value less than $\frac{1}{2}$. Thus the shortest path in G from b_1 to a_1 must have value at least $\frac{1}{2}$. Consider the $\frac{k-1}{2}$ edge-disjoint 4-cycles in M that remain when we remove edges with either endpoint in the set $\{a_1, b_1\}$. Then, since edges (b_1, a_1) and (a_1, b_2) each have value at least $\frac{1}{2}$ and the 4-cycle has value at least $\frac{k-1}{2}$, then the total value of the edges in M is at least $\frac{k+1}{2}$.   □

Corollary 2. *For a 3-fence in which the corresponding T contains a cycle, either $x(C)$ is at least 1, or C contains an edge of value at least $\frac{1}{2}$.*

Case 2: T is a Tree. We will consider two subcases based on the total value of the edges in T. For these subcases, we will use the following two lemmas.

Lemma 5. *If T is a tree, then every edge in T is included in the shortest paths corresponding to at least two edges in C.*

Proof. Assume there is an edge e in T that belongs to only one shortest path corresponding to an edge in C. Consider the set of edges that corresponds to the $2k-1$ shortest paths corresponding to the other $2k-1$ edges in C. By assumption, this set of edges does not include edge e. This set forms a connected graph, since the $2k-1$ corresponding edges form a connected graph. It also contains all $2k$ vertices in M. Thus, if we add edge e to this graph, it will contain a cycle, implying that T contains a cycle, which is a contradiction. □

Lemma 6. *If T is a tree corresponding to a 3-ladder, then for some $i \in \{1,2,3\}$, T contains a directed path from a_i to b_i and from a_{i+1} to b_{i+1}.*

Proof. First we will show there is a path from a_i to b_i for some i. Let $p(j,k)$ denote the set of edges in the directed path from j to k in T. Since T is a tree, if $p(j,k)$ exists, it is unique. Assume there is not a path from a_i to b_i for any i. We will show that this leads to a contradiction. Consider the paths $p(a_1,b_2)$ and $p(a_1,b_3)$. The first case is that, without loss of generality, b_2 is in $p(a_1,b_3)$. Since there must be a directed path from a_3 to b_2 and there is a directed path from b_2 to b_3, there is a directed path from a_3 to b_3. The second case is that b_2 is not in $p(a_1,b_3)$ and b_3 is not in $p(a_1,b_2)$. Let v be the vertex that belongs to both $p(a_1,b_2)$ and $p(a_1,b_3)$ and is farthest from a_1, as shown in Figure 2.

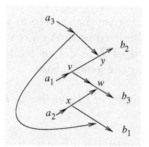

Fig. 2.

There must also be a path from a_2 to b_3 but not from a_2 to b_2. Thus, there must be a path from a_2 to some vertex $w \neq v$ in the path $p(v,b_3)$. Similarly, since there is no path from a_1 to b_1, there must be a path from some vertex $x \neq w$ in $p(a_2,w)$ to b_1. There is also a path from a_3 to a vertex $y \neq v$ in $p(v,b_2)$, since there is a path from a_3 to b_2 but not from a_3 to b_3. Thus, T contains a simple undirected path from a_3 to b_1. But T also contains a directed path from a_3 to b_1. So T contains a cycle, which is a contradiction.

Now, without loss of generality, assume there is path from a_1 to b_1 in T and assume all the vertices on this path are numbered in increasing order. Consider the vertex v_1 and v_2 where paths from a_2 and to b_2, respectively, intersect the path from a_1 to b_1. If v_1 is less than v_2, then $p(a_1,b_1)$ and $p(a_2,b_2)$ intersect and we are done. So assume v_1 is greater than v_2. But then the path from a_3 must intersect $p(a_1, b_1)$ at a point before v_2 and the path to b_3 must intersect $p(a_1, b_1)$ at a point after v_1, so the paths $p(a_1, b_1)$ and $p(a_3, b_3)$ intersect. $\square$

Lemma 7. *The total value of a 3-ladder is at least 2.*

Proof. By Lemma 4, this is true for the case when T contains a cycle. If T is a tree such that the total value of the edges in T is at least $\frac{1}{2}$, then by Lemma 5 every edge in T belongs to the corresponding path for at least two edges in C. Thus, the total value of C is at least 1. By Lemma 3 the total value of M is at least $\frac{k+1}{2}$.

If the total value of the edges in T is less than $\frac{1}{2}$, then by Lemma 6 and without loss of generality, assume T contains a directed path from a_1 to b_1 and from a_2 to b_2. Let s be the value of the directed path from a_1 to b_1 in T, i.e. $s = \sum_{ij \in p(a_1, b_1)} x_{ij}$. Note that $s < \frac{1}{2}$ since the total value of the edges in T is less than $\frac{1}{2}$. The value of $x_{b_1 a_1}$ is at least $1 - s$ and each edge in the path from a_1 to b_1 is on a path from a_1 to b_2 or from a_2 to b_1, so the sum of $x_{a_1 b_2}$ and $x_{a_2 b_1}$ is at least s. Therefore, the total value of the edges $(b_1, a_1), (a_1, b_2)$, and (a_2, b_1) is at least 1. The 4 edges in M that have neither endpoint in the set $\{a_1, b_1\}$ make up a directed 4-cycle, so they have total value at least 1. Thus any 3-ladder has value at least 2.

$\square$

For $k = 5, 7$, it is also the case that the total value of the edges in M is at least $\frac{k+1}{2}$. These proofs are omitted here. For $k = 9$, it is not necessarily the case that the total sum of the edges in a Möbius ladder is at least 5. We conclude this section with a bound for the general case.

Theorem 4. *The integrality gap of LP_2 augmented by Möbius ladder constraints is at least $\frac{33}{17} - \epsilon$ for any $\epsilon > 0$.*

Proof. When k is at least 9, then T is a tree with at least 18 unique vertices and at least 17 edges. If every edge original edge in G is assigned value $\frac{33}{34}$ and every complementary edge is assigned value $\frac{1}{34}$, then the total value of the edges in T is at least $\frac{1}{2}$. Thus, the objective value of the LP relaxation augmented with constraint (1) for all k will be at least $33|E|/34$. Since the optimal integral value is arbitrarily close to $|E|/2$ for sufficiently large n, the integrality gap of LP_2 extended by Möbius ladder constraints is at least $\frac{33}{17} - \epsilon$ for any $\epsilon > 0$. $\square$

3.3 Fence Constraints

A *k-fence* is obtained by directing the edges of a complete undirected bipartite graph on $2k$ vertices as follows: the $2k$ vertices are divided into two sets, A and B, of k vertices each. Each vertex from set A is paired with a vertex from set B and each of these pairs is connected with a *up* edge. All other edges are directed *down*. A 3-fence is shown in Figure 1.

An acyclic subgraph of a 3-fence includes at most 7 of the 9 edges. However, there is a fractional solution of $7\frac{1}{2}$ that satisfies LP$_2$: each down edge is assigned a value of 1 and each up edge is assigned $\frac{1}{2}$. In general, an acyclic subgraph of a k-fence includes at most $k^2 - k + 1$ of the edges. However, we can always find a fractional solution with value $k^2 - \frac{k}{2}$ that satisfies every cycle constraint. Thus, if we add fence constraints to the LP relaxation, we may get a fractional solution that is a better approximation of the integral solution. Fence constraints state that the total value of the edges in any k-fence cannot exceed $k^2 - k + 1$.

$$\sum_{(i,j)\in k-fence} x_{ij} \leq k^2 - k + 1 \tag{2}$$

We will show that despite the fact that LP$_2$ is strengthened by adding constraint (2), the integrality gap of this augmented LP relaxation remains 2. To show this, we will again use the graph $\tilde{G}$ discussed in Section 3.1. Since a 3-ladder is also a 3-fence, we know that the total value of any 3-fence is at most 7 in this solution. Thus, we can show that the total value of any k-fence is at most $k^2 - k + 1$.

Lemma 8. *The total value of any k-fence is at most $k^2 - k + 1$.*

Proof. The set of complementary edges of a k-fence also form a k-fence and hence (2) is equivalent to the condition that the total value of edges in a k-fence is *at least* $k - 1$. In Lemma 7, we showed that the lemma is true for $k = 3$, which will be the base case for our inductive proof. We will assume that the total value of any $(k - 1)$-fence is at least $k - 2$ and show that the total value of any k-fence is at least $k - 1$.

A k-fence contains $\binom{k}{3}$ distinct 3-ladders (or 3-fences) as subgraphs. For some 3-ladder contained in a k-fence, if the corresponding T is a tree (T and C corresponding to a 3-ladder are defined in Section 3.2) with value less than $\frac{1}{2}$, then by Lemma 6 for some $i \in \{1, 2, 3\}$, there is a directed path from a_i to b_i and from a_{i+1} to b_{i+1}. Thus, the total value of the edges (b_i, a_i), (a_i, b_{i+1}), and (a_{i+1}, b_i) is at least 1. When we remove all edges that have one endpoint in $\{a_i, b_i\}$ from the k-fence, we are left with a $(k - 1)$-fence. By induction, this fence has value at least $k - 2$. Thus, the total value of the k-fence is at least $k - 1$.

If it is the case that for none of the 3-ladders contained in the k-fence, the corresponding T is a tree with value less than $\frac{1}{2}$, then we will show that the total value of the edges directed from A to B in the k-fence is at least 1. Consider a particular 3-ladder that is a subgraph of the k-fence. If the corresponding T is a tree with value at least $\frac{1}{2}$, then this is true by Lemma 5. If T contains a cycle, then by Corollary 2, the set of edges C corresponding to this 3-ladder has value at least 1 or contains an edge e with at least $\frac{1}{2}$. In the latter case, consider another 3-ladder subgraph that does not contain edge e. Using Corollary 2 again, one of the edges in C corresponding to *this* 3-ladder must also contain an edge with value at least $\frac{1}{2}$. Thus, the total value of the edges directed from A to B is at least 1.

Let Y be the set of edges from b_i to a_i and let X be the set of edges from a_i to b_j. If we consider the k possible $(k-1)$-fences that are subgraphs of the k-fence, we see that each edge in Y is used in $k-1$ of these $(k-1)$-fences, and each edge in X is used in $k-2$ of these $(k-1)$-fences. By the induction hypothesis, each $(k-1)$-fence has value at least $k-2$ by induction. Combining these k equations, we have: $(k-1)x(Y) + (k-2)x(X) \geq k(k-2)$. We know that $x(X)$ is at least 1. The minimum value of $x(X) + x(Y)$ that satisfies the equation is $x(X) + x(Y) = k-1$, i.e. $x(X) = 1, x(Y) = k-2$. Thus, the total value of the edges in a k-fence is at least $k-1$, which implies that it is also at most $k^2 - k + 1$.   □

Theorem 5. *The integrality gap of LP_2 augmented with fence constraints is $2 - \epsilon$ for any $\epsilon > 0$.*

4   Lower Bounds on Approximation

In this section, we describe a reduction from the problem of finding a maximum satisfiable subset of a given set of linear equations modulo 2 with three variables per equation to the maximum acyclic subgraph problem. For this problem, Håstad proved the following tight bound.

Theorem 6 (Håstad [7]). *For every $\epsilon > 0$, it is NP-hard to tell if a given set of linear equations modulo 2 with three variables is satisfiable or at most $m(\frac{1}{2}+\epsilon)$ of its clauses are satisfiable.*

We use Theorem 6 and the reduction described below to obtain the following lower bound on the maximum acyclic subgraph problem (and hence the linear ordering problem).

Theorem 7. *It is NP-hard to approximate the maximum acyclic subgraph to within $\frac{65}{66} + \epsilon$ for any $\epsilon > 0$.*

Given a set of m linear equations on n variables, we construct a graph G using the following rules: (we assume all equations have the right hand side zero by negating one literal if necessary.)

1. For each variable $x \in F$, we create two vertices and two edges. The vertices are x_0 and x_1 and the edges are (x_0, x_1) and (x_1, x_0). These vertices and edges will form the *variable* gadget for x.
2. For each clause $C_j \in F$, we construct a *clause* gadget. Each clause has the form $x + y + z \equiv 0$ where x, y and z are literals. For a literal x in the clause we create a 4-cycle $\{x_2, x_3, x_4, x_5\}$. We label edge (x_5, x_2) as $x = 1$ and edge (x_3, x_4) as $x = 0$. We also do this for the literals y and z. Then we add the following 12 edges as shown in Figure 3: (z_2, x_5), (z_2, y_3), (z_4, y_3), (z_4, x_3), (x_2, z_3), (x_2, y_5), (x_4, z_5), (x_4, y_5), (y_2, z_3), (y_2, z_5), (y_4, x_3), (y_4, x_5).
3. Each clause gadget is linked to the appropriate variable gadgets as follows.

· For a literal x, we connect the corresponding 4-cycle in the clause gadget to the variable gadget by adding edges $(x_2, x_1), (x_1, x_3), (x_0, x_5), (x_4, x_0)$.
· For a literal $\overline{x}$, we connect the corresponding 4-cycle in the clause gadget to the variable gadget by adding edges $(x_2, x_0), (x_0, x_3), (x_1, x_5), (x_4, x_1)$.

The resulting graph G has $36m + 2n$ edges, 36 edges for each clause gadget and two edges for each variable gadget. In order to relate variable assignments to acyclic subgraphs of G, we say that removing edge (x_1, x_0) (labeled $x = 1$ in Figure 3) corresponds to setting the variable x to true, and removing edge (x_0, x_1) (labeled $x = 0$ in Figure 3) corresponds to setting variable x to false. Throughout the proof, we will refer to edges labeled $x = 0$ or $x = 1$ in a clause gadget for a literal x (i.e edges (x_5, x_2) and (x_3, x_4)) and the edges in the variable gadgets as *labeled* edges. The proof of Theorem 7 uses the lemmas below. A feedback arc set of a graph is defined as a set of edges whose removal results in an acyclic graph.

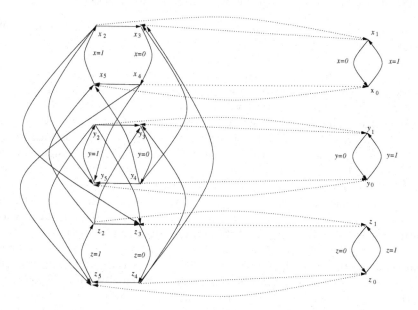

Fig. 3. The clause and variable gadgets for $x + y + z \equiv 0$.

Lemma 9. *A minimal feedback arc set is acyclic.*

Proof. For any acyclic graph, there is an ordering of the vertices such that all edges in the graph are forward edges, i.e. for an edge (i, j), i precedes j in the ordering. Given a feedback arc set, consider such an ordering for the acyclic graph that is obtained by deleting the feedback arc set. If any edges in the feedback arc set are forward edges, then the feedback arc set is not minimal, since such edges can be added to the acyclic graph without creating any cycles. Thus, the feedback arc set must consist only of backward edges and hence is itself acyclic.   □

Lemma 10. *There is a minimum feedback arc set of G consisting of only labeled edges.*

Proof. Note that every cycle in G contains labeled edges. This is because for every non-labeled edge (i, j) in G, i is a vertex such that the only incoming edge is a labeled edge, and j is a vertex such that the only outgoing edges is a labeled edge. Thus, given a feedback arc set F containing a non-labeled edge in a clause gadget, we can find another feedback arc set F' with $|F'| \leq |F|$ by replacing each non-labeled edge with either of its adjacent labeled edges.   □

Lemma 11. *The minimum feedback arc set for the graph G contains $n + 3m + u$ edges, where u is the minimum number of unsatisfied equations.*

Proof. By Lemma 9, exactly one edge from every variable gadget is in the minimum feedback arc set. In addition, we show the following things:

(i) For a clause, $x + y + z = 0$, and an assignment that satisfies this clause, we need to remove only three edges from the corresponding clause gadget so that the subgraph consisting of the clause gadget and its three corresponding variable gadgets is acyclic. By Lemma 10, we only need to remove labeled edges from the clause gadget. There are four satisfying assignments for the variables x, y, z in this clause. They are: $\{\{0,0,0\}, \{0,1,1\}, \{1,1,0\}, \{1,0,1\}\}$. For each of these four assignments, we remove the three edges with the opposite assignment from the clause gadget. For example, for the assignment $x = y = z = 0$, we remove the edges labeled $x = 1, y = 1, z = 1$.

(ii) For a clause, $x + y + z = 0$, and any assignment that does not satisfy this clause, we need to remove four edges from the corresponding clause gadget so that the subgraph consisting of the clause gadget and its three corresponding variable gadgets is acyclic. There are four assignments to the variables x, y, z that do not satisfy this clause. They are: $\{\{1,1,1\}, \{0,0,1\}, \{1,0,0\}, \{0,1,0\}\}$. For each of these four assignments, if we remove the labeled edges corresponding to the opposite assignment, then the clause gadget will still contain a cycle. For example, for the assignment $x = y = z = 1$, we remove edges labeled $x = 0, y = 0, z = 0$. However, the edges labeled $x = 1, y = 1, z = 1$ remain and form a cycle. So we must remove one more of these edges for the resulting subgraph of the clause gadget to be acyclic.

(iii) For each variable gadget, if we remove one of the edges in the corresponding 2-cycle and the corresponding edge from the clause gadgets representing clauses that contain this variable, then the resulting graph does not contain any cycle composed of edges from multiple clause gadgets. For the clause $x + y + z = 0$, consider the edge (x_2, x_1). If a cycle contains this edge, it must also contain the only incoming edge to vertex x_2, which is the edge labeled $x = 1$. If these edges are contained in a cycle with edges from another clause gadget, then at vertex

x_1, we can move to another gadget. However, we arrive at a vertex such that the only out edge corresponds to the edge that remains iff x has been set to 0, which is not the case if the edge labeled $x = 1$ was present. So there cannot be any cycles that use edges from more than one clause gadget.

It follows from (i),(ii), and (iii) that the minimum feedback arc set has size $n + 3m + u$. □

Corollary 3. *The Maximum Acyclic Subgraph for G is of size $n + 33s + 32u$ where s and u represent the number of satisfied and unsatisfied clauses, respectively, for an assignment that satisfies the maximum number of clauses.*

Proof of Theorem 7. By Corollary 3 and by Theorem 6 it is NP-hard to distinguish between a graph that has a maximum acyclic subgraph of size $n + 33(\frac{1}{2} + \epsilon)m + 32(\frac{1}{2} - \epsilon)m$ and a graph that has a maximum acyclic subgraph of size $n + 33m$. If we could approximate the maximum acyclic subgraph to within $\frac{2n+65}{2n+66} + \epsilon$, then we could distinguish between these two cases. Therefore it is NP-hard to approximate to approximate the maximum acyclic subgraph to within $\frac{2n+65}{2n+66} + \epsilon$. We can make n arbitrarily small compared to m by creating another set of linear equations in which each original equation appears k times for some k so that we have km clauses and only n variables. The ratio $\frac{2n+65}{2n+66}$ is arbitrarily close to $\frac{65}{66}$ as k becomes large. Therefore, it is NP-hard to approximate the maximum acyclic subgraph to within $\frac{65}{66} + \epsilon$ for any $\epsilon > 0$. □

References

1. Alberto Caprara and Matteo Fischetti, 1/2-Chvatal-Gomory cuts, *Mathematical Programming*, 74A, 1996, 221–235.
2. P. Erdös and H. Sachs, Regulare Graphe gegebener Taillenweite mit minimaler Knotenzahl, Wiss. Z. Univ. Halle-Wittenberg, *Math. Nat.* 12, 1963, 251–258.
3. M. X. Goemans and L. A. Hall, The Strongest Facets of the Acyclic Subgraph Polytope are Unknown, *Proceedings of IPCO 1996*, 415–429.
4. M. Grötschel, M. Jünger, and G. Reinelt, On the Maximum Acyclic Subgraph Polytope, *Mathematical Programming*, 33, 1985, 28–42.
5. M. Grötschel, M. Jünger, and G. Reinelt, Facets of the Linear Ordering Polytope, *Mathematical Programming*, 33, 1985, 43–60.
6. M. Grötschel, L. Lovász and A. Schrijver, *Geometric Algorithms and Combinatorial Optimization*, Springer-Verlag, New York, 1988.
7. Johan Håstad, Some Optimal Inapproximability Results, *Proceedings of STOC*, 1997.
8. Richard M. Karp, Reducibility Among Combinatorial Problems, *Complexity of Computer Computations*, Plenum Press, 1972.
9. Rudolf Mueller, On the Partial Order Polytope of a Digraph, *Mathematical Programming*, 73, 1996, 31–49.
10. Alantha Newman, Approximating the Maximum Acyclic Subgraph, M.S. Thesis, MIT, June 2000.
11. S. Poljak, Polyhedral and Eigenvalue Approximations of the Max-Cut Problem, *Sets, Graphs and Numbers, Colloq. Math. Soc. Janos Bolyai* 60, 1992, 569–581.
12. Andreas Schulz and Rudolf Mueller, The Interval Order Polytope of a Digraph, *Proceedings of IPCO 1995*, 50–64.

Generating Cuts from Multiple-Term Disjunctions

Michael Perregaard and Egon Balas

Carnegie Mellon University, Pittsburgh PA, USA

Abstract. The traditional approach towards generating lift-and-project cuts involves solving a cut generating linear program (CGLP) that grows prohibitively large if a multiple-term disjunction is used instead of the classical dichotomy. In this paper we present an alternative approach towards generating lift-and-project cuts that solves the cut generation problem in the original space (i.e. does not introduce any extra variables). We present computational results that show our method to be superior to the classical approach. As soon as the number of terms in the disjunction grows larger than 4, the benefit of not solving the large CGLP directly is clear, and the gap grows larger as more terms are included in the disjunction.

1 Introduction

Disjunctive cuts were first introduced in the early seventies [1,2]. In the late nineties a subclass named lift-and-project cuts [3] were implemented in a branch-and-cut framework and proved to be particularly fruitful. These cuts, as implemented and tested in the MIPO solver [4] (see also [7]), are derived from a two-term disjunction, usually of the form $(x_k \leq 0) \vee (x_k \geq 1)$ for some 0-1 constrained variable x_k.

The lift-and-project approach formulates and solves a block-angular higher dimensional Cut Generating Linear Program (CGLP). Each block of CGLP corresponds to the constraints of one term in the underlying disjunction. In the standard case there are two blocks, but if a cut is generated by the lift-and-project approach from imposing the 0-1 condition on e.g. three 0-1 variables, there are 8 ways to assign 0-1 values to the variables and hence the underlying disjunction will have 8 terms. In principle, the dimension of the CGLP grows as 2^k, where k is the number of 0-1 variables on which the 0-1 condition is imposed. Thus, as the complexity of the disjunctions used for cut generation increases, the time to solve a CGLP quickly becomes prohibitive.

In a recent paper by Balas and Perregaard [6] it was shown how lift-and-project cuts from a two-term disjunction can be solved directly in the simplex tableau of the LP relaxation, thereby bypassing the whole higher-dimensional linear program. This result, however, only applies to cuts from two-term disjunctions. The focus of this paper is on cuts derived from stronger disjunctions, typically involving at least 3-4 integer variables.

K. Aardal, B. Gerards (Eds.): IPCO 2001, LNCS 2081, pp. 348–360, 2001.
© Springer-Verlag Berlin Heidelberg 2001

We cannot get around the exponential growth in generating optimal disjunctive cuts as we impose the 0-1 condition on more variables, because the problem in itself is $\mathcal{NP}$-hard. However, as we claim to show in this paper, there is a more efficient way to generate cuts from stronger disjunctions, than by solving the corresponding CGLP in the standard fashion.

Consider a mixed integer program of the form

$$
\begin{aligned}
\min \ & cx \\
\text{s.t.} \ & Ax \geq b \\
& x_j \in \mathbb{Z} \quad \text{for } j \in I
\end{aligned}
\tag{MIP}
$$

where A is an $m \times n$ matrix and $Ax \geq b$ subsumes the constraints $x \geq 0$ and $x_j \leq 1$, $j \in I$. The LP-relaxation of MIP is

$$
\begin{aligned}
\min \ & cx \\
\text{s.t.} \ & Ax \geq b
\end{aligned}
\tag{LP}
$$

To generate one or more cuts we will consider a disjunctive relaxation of MIP of the form

$$
\begin{aligned}
\min \ & cx \\
\text{s.t.} \ & Ax \geq b \\
& \bigvee_{q \in Q} D^q x \geq d^q
\end{aligned}
\tag{DP}
$$

where each term of the disjunction imposes integrality on several x_j, $j \in I$, and no feasible solution to MIP is excluded. We will use P_{IP}, P_{LP} and P_{DP} to denote the feasible regions of MIP, LP and DP, respectively, and we have $P_{IP} \subset P_{DP} \subset P_{LP}$. We will further let x^{IP}, x^{LP} and x^{DP} denote an optimal solution to each of the three programs.

Our objective is to determine one or more cutting planes of the form $\alpha x \geq \beta$ which are valid for P_{DP} but cut off parts of P_{LP}, and in particular the point x^{LP}.

An inequality $\alpha x \geq \beta$ is valid for P_{DP} (see [1,2]) if and only if there exist non-negative multipliers u^q and v^q for each term $q \in Q$ such that

$$
\begin{aligned}
\alpha &= u^q A + v^q D^q \\
\beta &= u^q b + v^q d^q \qquad \forall q \in Q \\
& u^q \geq 0, v^q \geq 0
\end{aligned}
\tag{1}
$$

In order to generate cuts, we truncate the higher-dimensional cone defined by (1) by introducing a *normalization* constraint of the form $\alpha y = 1$, where $y \in \mathbb{R}^n$ is the constant vector introduced in [5] (to be specified later).

To obtain a cut that is maximally violated by the point x^{LP}, we minimize an objective function $\alpha x^{LP} - \beta$ over the above set of constraints. The result is a lift-and-project Cut Generating Linear Program:

$$
\begin{aligned}
\min \ & \alpha x^{LP} - \beta \\
\text{s.t.} \ & \alpha = u^q A + v^q D^q \\
& \beta = u^q b + v^q d^q \quad \forall q \in Q \\
& u^q \geq 0, v^q \geq 0 \\
& \alpha y = 1
\end{aligned}
\tag{CGLP$_Q$}
$$

If we replace the generic disjunction $\bigvee_{q \in Q} D^q x \geq d^q$ with a simple two-term disjunction $(x_k \leq 0) \vee (x_k \geq 1)$ we recover the CGLP used in [3,4] (although with the normalization constraint introduced in [5]).

2   An Iterative Approach to Cut Generating

An alternative formulation of the cut generating linear program is to define it as optimization over the reverse polar cone of P_{DP}, which after normalization yields the problem

$$
\begin{aligned}
\min\ & \alpha x^{LP} - \beta \\
\text{s.t.}\ & \alpha x \geq \beta \qquad \forall x \in P_{DP} \\
& \alpha y = 1
\end{aligned}
\tag{2}
$$

In principle we have here a constraint in α and β for each $x \in P_{DP}$, an infinite set. However, $\mathrm{conv}(P_{DP})$ is defined by its set of extreme points and extreme rays, so only constraints of (2) associated with these are necessary. Thus (2) is equivalent to

$$
\begin{aligned}
\min\ & \alpha x^{LP} - \beta \\
\text{s.t.}\ & \alpha x^i \geq \beta \qquad \forall x^i \in S \\
& \alpha x^i \geq 0 \qquad \forall x^i \in R \\
& \alpha y = 1
\end{aligned}
\tag{3}
$$

where S and R are the sets of extreme points and (directions of) extreme rays of $\mathrm{conv}(P_{DP})$. Although we do not know S and R a priori, we can solve (3) iteratively and generate the extreme points and rays as needed. In other words, we propose is solve (3) by row generation.

Suppose we have $S_1 \subset S$ and $R_1 \subset R$. If we solve

$$
\begin{aligned}
\min\ & \alpha x^{LP} - \beta \\
\text{s.t.}\ & \alpha x^i - \beta \geq 0 \quad \forall x^i \in S_1 \\
& \alpha x^i \geq 0 \qquad \forall x^i \in R_1 \\
& \alpha y = 1
\end{aligned}
\tag{4}
$$

we obtain an inequality $\alpha^1 x \geq \beta^1$ valid for all extreme points in S_1 and all extreme rays in R_1, but not necessarily for those in S and R. To check if $\alpha^1 x \geq \beta^1$ is valid for all of $\mathrm{conv}(P_{DP})$, we can solve a linear program of the form

$$
\begin{aligned}
\min\ & \alpha^1 x \\
\text{s.t.}\ & x \in \mathrm{conv}(P_{DP})
\end{aligned}
\tag{5}
$$

If (5) is bounded and x^1 is an optimal, extreme solution with $\alpha^1 x^1 < \beta^1$, then x^1 is a point from $S \setminus S_1$ that violates the current inequality $\alpha^1 x \geq \beta^1$. Hence we can replace S_1 by the larger set $S_2 = S_1 \cup \{x^1\}$. If (5) is unbounded then we can find an extreme ray of (5) with a direction vector x^1 such that $\alpha^1 x^1 < 0$, which we can add to R_1 to obtain a larger set $R_2 = R_1 \cup \{x^1\}$. We then repeat this with the new sets S_2 and R_2 to obtain a new inequality $\alpha^2 x \geq \beta^2$, and keep repeating until in some iteration k we obtain an optimal solution x^k to (5) which satisfies $\alpha^k x \geq \beta^k$ for the last solution (α^k, β^k) to (4). This solution demonstrates that we have found a valid inequality.

Since the x^i we obtain from solving (5) are *extreme* points or rays of P_{DP}, the finiteness of S and R guarantees that the process will terminate. The procedure is outlined in Figure 1. In the following we will refer to the problem in Step 2 as the *master problem* and to the problem in Step 3 as the *separation problem.* The procedure described here is isomorphic to applying Benders' decomposition to $(CGLP)_Q$.

Step 1 Let $k = 1$, $R_1 \subset R$ and $S_1 \subset S$ with $S_1 \neq \emptyset$

Step 2 Let (α^k, β^k) be an optimal solution to the master problem:

$$\min \alpha x^{LP} - \beta$$
$$\text{s.t.} \quad \alpha x^i - \beta \geq 0 \ \forall x^i \in S_k$$
$$\alpha x^i \geq 0 \qquad \forall x^i \in R_k$$
$$\alpha y = 1$$

Step 3 Solve the separation problem:

$$\min \alpha^k x$$
$$\text{s.t.} \quad Ax \geq b$$
$$\bigvee_{q \in Q} D^q x \geq d^q$$

If the problem is *bounded*, let x^k be an optimal solution. If $\alpha^k x^k \geq \beta^k$ then go to Step 4. Otherwise, set $S_{k+1} = S_k \cup \{x^k\}$ and $R_{k+1} = R_k$.

If the problem is *unbounded*, let x^k be the direction vector of an extreme ray satisfying $\alpha^k x^k < 0$. Set $R_{k+1} = R_k \cup \{x^k\}$ and $S_{k+1} = S_k$.

Set $k \leftarrow k + 1$ and repeat from Step 2.

Step 4 The inequality $\alpha^k x \geq \beta^k$ is a valid inequality for P_{DP}. Stop.

Fig. 1. Iterative procedure for generating a valid inequality: version 1.

So far we have not considered the possibility that the master problem in Step 2 of Figure 1 could be unbounded. This is where we need a certain property of the normalization $\alpha y = 1$. If we choose $y = x^{LP} - x^*$, where $x^* \in \text{conv}(S_1)$, then the master problem will always be bounded (see [5]).

The iterative procedure of Figure 1 can be modified in its Step 3 as follows. Instead of adding to the master problem the inequality corresponding to the extreme point x^k that minimizes $\alpha^k x$, i.e. violates the inequality $\alpha^k x \geq \beta^k$ by a maximum amount, we add *all* the violating extreme points or rays encountered in solving the separation problem. We call this *version 2*. Since the separation problem is usually solved by solving an LP over each term of the disjunction, version 2 of the iterative procedure does not require more time to solve the separation problem. On the other hand, it builds up faster the master problem, but it also creates one with a larger number of constraints. On balance, version 2 seems better (see the section on computational results).

3   Generating Adjacent Extreme Points

In this section we consider one approach towards reducing the number of extreme points we need to consider in the separation problem.

A reasonable constraint to impose on the cut is to require it to be tight at x^{DP}, an optimal solution to (DP). Suppose we impose this restriction, i.e. add to (4) the equation $\alpha x^{DP} - \beta = 0$, and (α^k, β^k) is the solution to the master problem at iteration k. Then either $\alpha^k x \geq \beta^k$ is a valid inequality for P_{DP} or there exists a vertex *adjacent* to x^{DP} (or possibly an extreme ray incident with x^{DP}) in conv(P_{DP}) which violates $\alpha^k x \geq \beta^k$. This is an immediate consequence of the convexity of conv(P_{DP}).

It follows from this observation that when searching for a violating extreme point (or ray) in the separation problem of Figure 1 we only need to consider extreme points adjacent to x^{DP} (or extreme rays incident with x^{DP}) in conv(P_{DP}).

We now turn to the problem of identifying the extreme points adjacent to x^{DP}. Consider the disjunctive cone $C_{x^{DP}}$ defined by

$$C_{x^{DP}} = \{(x', x_0') \in \mathbb{R}^n \times \mathbb{R}_+ \mid Ax' + (Ax^{DP} - b)x_0' \geq 0 \\ \bigvee_{q \in Q} D^q x' + (D^q x^{DP} - d^q)x_0' \geq 0 \}$$

This cone is obtained from P_{DP} by first translating P_{DP} by $-x^{DP}$ such that x^{DP} is translated into the origin, and then homogenizing the translated polyhedron. The following Theorem gives the desired property. Let cone($C_{x^{DP}}$) be the conical hull (positive hull) of $C_{x^{DP}}$.

Theorem 1. *Let C be the projection of cone($C_{x^{DP}}$) onto the x-space. Then the extreme rays of the convex cone C are in one-to-one correspondence with the edges of conv(P_{DP}) incident with x^{DP}.*

Proof. We consider the mapping $x \to x' = x - x^{DP}$ from points $x \in$ conv(P_{DP}) to rays $x' \in C$. It should be clear that $x' \in C$, since we have $(x - x^{DP}, 1) \in$ cone($C_{x^{DP}}$).

Now, take two distinct points $x^1, x^2 \in$ conv(P_{DP}) on any edge of conv(P_{DP}). These points are mapped into the vectors $(x^1 - x^{DP})$ and $(x^2 - x^{DP})$. If and only if the two points lie on an edge of conv(P_{DP}) incident with x^{DP} will the vectors $(x^1 - x^{DP})$ and $(x^2 - x^{DP})$ describe the same ray in C. Thus only edges of conv(P_{DP}) correspond to edges of C.

The converse is also true. Let x' be an extreme ray of C. Then there exists x_0' such that (x', x_0') is a ray of cone($C_{x^{DP}}$). In particular we can scale the point such that $(x', x_0') \in C_{x^{DP}}$. Consider the line segment $[x^{DP}, x^{DP} + x']$. We first show that there exists $\gamma > 0$ such that $\tilde{x} = x^{DP} + \gamma x' \in \text{conv}(P_{DP})$. If we plug $\tilde{x}$ into (DP) we obtain $A(x^{DP} + \gamma x') \geq b$ and $D^q(x^{DP} + \gamma x') \geq d^q$ for each $q \in Q$. If $x_0' = 0$ we have $Ax' \geq 0$ and $D^q x' \geq 0$ for some $q \in Q$. In other words, x' is a ray of P_{DP}, and in particular $\tilde{x} \in \text{conv}(P_{DP})$ for any $\gamma > 0$. If $x_0' > 0$ and we choose $\gamma \leq \frac{1}{x_0'}$ then $\tilde{x}$ satisfies (DP) since $(x', x_0') \in C_{x^{DP}}$.

Next we show that $\tilde{x}$ lies on an edge of $\text{conv}(P_{DP})$ incident with x^{DP}. For a contradiction, suppose that $\tilde{x}$ is *not* on such an edge. Then there exists $x^1, x^2 \in \text{conv}(P_{DP})$ not on the line through x^{DP} and $\tilde{x}$ (i.e., affinely independent from x^{DP} and $\tilde{x}$), such that $\tilde{x}$ is a strict convex combination of x^1 and x^2. Hence there exists $\lambda \in \mathbb{R}$ with $0 < \lambda < 1$ such that $x = \lambda x^1 + (1 - \lambda)x^2$. Now, consider the rays $(\tilde{x}', \tilde{x}_0') = (\tilde{x} - x^{DP}, 1)$, $(x^{1'}, x_0^{1'}) = (x^1 - x^{DP}, 1)$ and $(x^{2'}, x_0^{2'}) = (x^2 - x^{DP}, 1)$. Since $\tilde{x}, x^1, x^2 \in \text{conv}(P_{DP})$ it follows that $(\tilde{x}', \tilde{x}_0'), (x^{1'}, x_0^{1'}), (x^{2'}, x_0^{2'}) \in \text{cone}(C_{x^{DP}})$. Hence $(\tilde{x} - x^{DP}), (x^1 - x^{DP})$ and $x^2 - x^{DP}$ are rays of C. We have that $(\tilde{x} - x^{DP}) = \gamma x'$, so $x' = \frac{\lambda}{\gamma}(x^1 - x^{DP}) + \frac{1-\lambda}{\gamma}(x^2 - x^{DP})$. We are thus able to write x' as a positive combination of linearly independent vectors, contradicting that x' is an *extreme* vector of C. This shows that the extreme ray x' is the map of the edge of $\text{conv}(P_{DP})$ incident with x^{DP} containing $\tilde{x}$.   □

Since any vertex adjacent to x^{DP} in $\text{conv}(P_{DP})$ by definition shares an edge incident with x^{DP}, the immediate result of this theorem is that we only need to consider the extreme rays of C. The relationship between rays (x', x_0') of $C_{x^{DP}}$ and points or rays x of P_{DP} is

$$x = \begin{cases} \frac{x'}{x_0'} + x^{DP} & \text{if } x_0' > 0 \\ x' & \text{if } x_0' = 0 \end{cases}$$

Let $\alpha^k x \geq \beta^k$ be the current iterate from our procedure. To check if there is a violating point adjacent to x^{DP}, we first need to translate and homogenize the inequality, in accordance with what was done to obtain $C_{x^{DP}}$. The translation results in the inequality $\alpha^k x' \geq \beta^k - \alpha^k x^{DP}$, but since we imposed on (4) the constraint $\alpha x^{DP} = \beta$, the righthand side becomes zero, and the coefficient for x_0' after homogenizing will also be zero; hence we obtain the inequality $\alpha^k x' \geq 0$. We can thus state

Proposition 1. *Let $x \in P_{DP}$. $\alpha^k x < \beta^k$ if and only if $\alpha^k x' < 0$ for $(x', x_0') = (x - x^{DP}, 1) \in C_{x^{DP}}$.*

To obtain a violating vertex of $\text{conv}(P_{DP})$ adjacent to x^{DP} we solve the following disjunctive program:

$$\begin{aligned}
\min \ & \tilde{\alpha}x' \\
\text{s.t. } & Ax' + (Ax^{DP} - b)x_0' \geq 0 \\
& \bigvee_{q \in Q} D^q x' + (D^q x^{DP} - d^q)x_0' \geq 0 \\
& \alpha^k x' = -1 \\
& x_0' \geq 0
\end{aligned} \qquad (6)$$

We impose the equation $\alpha^k x' = -1$ both to truncate the cone $C_{x^{DP}}$ and to restrict the feasible set to those solutions that satisfy the condition of Proposition 1. Any solution to problem (6) corresponds to a violating point in P_{DP}, but to obtain a violating *extreme* point, we minimize an objective over this set. If we choose $\tilde{\alpha}$ such that $\tilde{\alpha}x' \geq 0$ is valid for $C_{x^{DP}}$ then the problem (6) will be bounded.

To guarantee that the solution we obtain corresponds to a vertex of P_{DP} adjacent to x^{DP}, we must first project out x_0', according to Theorem 1. This can be done by e.g. applying the Fourier-Motzkin projection method. The size of the resulting set of constraints will depend on the number of constraints present in the system $D^q x' \geq d^q$, but since in most cases of interest $D^q x' \geq d^2$ can be replaced with a single constraint, the cost of projecting out x_0' is typically not high.

4   How to Generate a Facet of conv(P_{DP}) in n Iterations

For the iterative procedure in Figure 1, we do not have a bound on the number of iterations required to obtain a valid inequality, except the trivial bound which is the total number extreme points and extreme rays of conv(P_{DP}). In this section we present a method which will find a facet-defining inequality for conv(P_{DP}) in a number of iterations that only depends on the dimension of the problem. When we do this we can no longer guarantee that the resulting inequality will be optimal in (2).

The basic idea is to start with an inequality that is known to be valid and supporting for P_{DP}. Finding such an inequality should not pose a problem. Using x^{DP} we can easily give such an inequality: $cx \geq cx^{DP}$. Then through a sequence of rotations we turn this inequality into a facet-defining inequality for P_{DP}. Each rotation will be chosen such that the new cut will be tight at one more vertex of P_{DP} than the previous cut.

An illustration is provided in Figure 2. This figure presents two polyhedra, P_1 and P_2, whose union is P_{DP}. Our initial plane is H_1 which supports P_{DP} only at the point x_1. The first rotation we perform is around the axis a_1 through x_1 which rotates H_1 into H_2. Now H_2 is a plane touching P_{DP} at the two points x_1 and x_2. Finally, we rotate H_2 around the axis a_2 through the points x_1 and x_2. This brings us to the final plane H_3, which is tight at x_1, x_2 and x_3, a maximum independent set on a facet of conv(P_{DP}).

The idea of hyperplane rotation is implemented by performing a linear transformation, in which the current inequality $\alpha^k x \geq \beta^k$ is combined with some target inequality $\tilde{\alpha}x \geq \tilde{\beta}$. Thus, we want to find a maximal γ such that $(\alpha^k + \gamma\tilde{\alpha})x \geq (\beta^k + \gamma\tilde{\beta})$ is a valid inequality for P_{DP}. Suppose S_k is the set of extreme points of conv(P_{DP}) for which $\alpha^k x \geq \beta^k$ is tight. If we choose $(\tilde{\alpha}, \tilde{\beta})$ such that the inequality $\tilde{\alpha}x \geq \tilde{\beta}$ is also tight at S_k then the resulting inequality must be tight at S_k. If we further ensure that $\tilde{\alpha}x \geq \tilde{\beta}$ is *invalid* for P_{DP} then there is a finite maximal γ for which $(\alpha^k + \gamma\tilde{\alpha})x \geq (\beta^k + \gamma\tilde{\beta})$ is valid for P_{DP}.

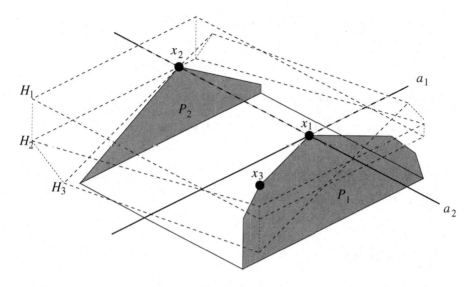

Fig. 2. Example showing how an initial supporting hyperplane H_1 is rotated through H_2 into a facet-defining hyperplane H_3.

It can be shown (which we do not do here) that the maximum value γ^* of γ is given by the optimal objective value of the disjunctive program

$$\gamma^* = \min \alpha^k x - \beta^k x_0$$
$$\text{s.t.} \quad Ax - bx_0 \geq 0$$
$$\bigvee_{q \in Q} D^q x - d^q x_0 \geq 0 \qquad (7)$$
$$x_0 \geq 0$$
$$\tilde{\alpha} x - \tilde{\beta} x_0 = -1$$

The optimal solution (x', x_0') we obtain from solving (7) defines a point $x = \frac{x'}{x_0'} \in P_{DP}$ (or a ray of P_{DP} if $x_0' = 0$) affinely independent of S_k, for which the new inequality is tight.

We are now able to present an outline of a procedure that finds a facet-defining inequality of $\text{conv}(P_{DP})$ in n iterations. This procedure is given in Figure 3. For simplicity we have assumed that P_{DP} is bounded and thus omitted the possibility of extreme rays.

We will not specify how to choose $(\tilde{\alpha}, \tilde{\beta})$ since there are many technical details involved in this. Suffice it to say that the inequality $\tilde{\alpha} x \geq \tilde{\beta}$ should be chosen as one that is "deeper" than $\alpha^k x \geq \beta^k$ with respect to x^{LP}, the point we want to cut off. If we do this and if the initial inequality $\alpha^1 x \geq \beta^1$ already cuts off x^{LP^1} then the above procedure produces a facet-defining inequality that also cuts off x^{LP}.

There are other technical details pertaining to this procedure which we have no space to cover here, e.g. how to choose the initial inequality $\alpha^1 x \geq \beta^1$, how to decide when a cut $\tilde{\alpha} x \geq \tilde{\beta}$ is "deeper" with respect to x^{LP}.

Step 1 Let $\alpha^1 x \geq \beta^1$ be a valid inequality for P_{DP} tight for $x^1 \in P_{DP}$. Set $S_1 = \{x^1\}$ and $k = 1$.

Step 2 Choose a target inequality $\tilde{\alpha} x \geq \tilde{\beta}$ tight for S_k and not valid for P_{DP}.

Step 3 Solve (7) to obtain γ^* and a point x^k.

Step 4 Set $(\alpha^{k+1}, \beta^{k+1}) = (\alpha^k, \beta^k) + \gamma^*(\tilde{\alpha}, \tilde{\beta})$ and $S_{k+1} = S_k \cup \{x^k\}$. Increment $k \leftarrow k + 1$.

Step 5 If $k = n$ stop, otherwise repeat from Step 2.

Fig. 3. Procedure to obtain a facet-defining inequality for P_{DP} in n iterations.

5   Cut Lifting

An important ingredient of the lift-and-project method is that of working in a subspace [3,4]. If a variable is at its lower or upper bound in the optimal solution x^{LP} to the LP-relaxation, it can be ignored for the purpose of cut generation. Thus cuts are generated in a subspace and are then *lifted* to a cut that is valid for the full space by computing the coefficients of the missing variables. These coefficients are computed using the multipliers $\{u^q\}_{q \in Q}$ that satisfy the constraints of $(\text{CGLP})_Q$ for the subspace cut coefficients (see [3,4] for details). The procedures featured in Figures 1 and 3 do not cover the cut lifting aspect and thus do not specify how to compute these multipliers; but once we have determined the cut $\alpha x \geq \beta$, we can fix the value of α and β in (CGLP). This will decouple the constraints and leave $|Q|$ independent linear equality problems from which the multipliers associated with (α, β) are easy to calculate. One potential problem with working in a subspace is the choice of the latter: if we restrict the space too much, the feasible region may become empty. To avoid this, we require that each term of the disjunction be non-empty in the subspace. In our testing to be discussed in the next section, we used the smallest subspace that contains the nonzero components of the optimal solution from each of the separate linear programs of the disjunction. This was easy to implement, since the method we used to solve the disjunctive programs of Step 3 in the procedures of Figures 1 and 3 was to solve a linear program over each term of the disjunction and retain the best solution found.

6   Computational Testing

To test the ideas presented in this paper experimentally, we need specific disjunctive relaxations of (MIP). We are mainly interested in comparing the effect on the various methods of an increase in the number of terms in the disjunction. There are many ways to create a disjunction involving multiple 0-1 variables. The simplest one is to assign all possible values to a fixed number, k, of 0-1 variables, thus creating a disjunction with 2^k terms. However, a little thinking and experimenting shows that this way is not the best. Instead, we use a partial branch and bound procedure with no pruning except for infeasibility, to generate a search tree with a fixed number, k, of leaves. The union of subproblems corresponding to these k leaves is then guaranteed to contain the feasible solutions to (MIP). Therefore the disjunction whose terms correspond to the leaves of this partial search tree is a valid one, although the number of variables whose values are fixed at 0 or 1 in the different terms of the disjunction need not be the same.

The branch-and-bound procedure used here is a simple one whose only purpose is to provide us with a disjunction of a certain size. As a branching rule, we branch on an integer constrained variable whose fractional part is closest to $\frac{1}{2}$. For node selection we use the best-first rule. This search strategy will quickly grow a sufficiently large set of leaf nodes for our disjunction. Further, by using best-first search we also ensure a strong disjunction with respect to the objective function.

For each problem instance we generate a round of up to 50 cuts, each from a disjunction coming from a search tree initiated by first branching on a different 0-1 variable fractional in the LP solution. The cuts themselves are generated by five different methods, each using the same disjunctions:

1. By using the simplex method to solve $(CGLP)_Q$ in the higher dimensional space;
2. By using the iterative procedure of Figure 1, version 1;
3. By using the iterative procedure of Figure 1, version 2;
4. By using the iterative procedure that generates only extreme points adjacent to x^{DP};
5. By using the n-step procedure of Figure 3 to find a facet defining inequality.

These procedures have been implemented in C on a SUN Ultra 60 with a 360 MHz Ultra SPARC-II processor and 512 MB of memory. To solve the linear programs that arise, CPLEX version 6.60 was used. The test set for our experiments consisted of a set of 14 pure or mixed 0-1 problems from the MIPLIB library of mixed integer programs (http://www.caam.rice.edu/~bixby/miplib/miplib.html).

The main purpose of our computational testing was to compare the proposed procedures with each other and with the standard procedure of solving $(CGLP)_Q$ from the point of view of their sensitivity to the number of terms in the disjunctions from which the cuts are generated. A first comparison, shown in Table 1, features the total time required to generate up to 50 cuts for each of the 14 test problems, (a) by solving the higher dimensional $(CGLP)_Q$ as a standard linear program (using CPLEX), and (b) by using the iterative procedure of Figure 1, version 2 (which adds to the master problem all the violators found in Step 3). These numbers are compared for disjunctions with 2, 4, 8 and 16 terms, with the outcome that the times in column (b) are worse than those in column (a) for disjunctions with 2 terms ($|Q| = 2$), roughly equal or slightly worse for $|Q| = 4$, considerably better for $|Q| = 8$, and vastly better for $|Q| = 16$. For the method featured in column (b), the total computing time grows roughly linearly with

$|Q|$: in about half of the 14 problems, the growth is slightly less than linear, and in the other half it is slightly more than linear. The numbers in column (a) grow much faster, which is understandable in light of the fact that the number of variables *and* constraints of (CGLP)$_Q$ increases with $|Q|$.

Table 1. Total time for up to 50 cuts

	$|Q| = 2$		$|Q| = 4$		$|Q| = 8$		$|Q| = 16$	
	(a)	(b)	(a)	(b)	(a)	(b)	(a)	(b)
BM21	0.07	0.21	0.32	0.40	1.91	0.85	7.85	1.52
EGOUT	0.08	0.16	0.23	0.25	0.82	0.46	5.76	0.82
FXCH.3	0.52	0.80	1.53	1.50	9.86	2.54	65.19	4.01
LSEU	0.07	0.14	0.23	0.27	0.97	0.56	6.33	1.47
MISC05	1.13	1.82	5.15	5.07	52.25	13.72	658.88	33.00
MOD008	0.09	0.13	0.19	0.33	0.60	0.67	2.69	1.54
P0033	0.07	0.13	0.19	0.24	0.78	0.57	2.89	1.14
P0201	1.40	2.09	8.12	4.80	88.26	10.67	609.17	26.45
P0282	0.85	1.90	2.02	4.51	8.94	17.75	86.24	44.36
P0548	2.77	11.85	7.94	9.00	37.75	20.51	276.18	44.25
STEIN45	36.42	148.38	99.71	157.86	280.71	159.04	1082.78	222.32
UTRANS.2	0.50	0.81	1.55	1.50	10.26	3.53	111.06	13.65
UTRANS.3	0.78	1.26	2.71	2.29	26.47	4.93	173.20	9.31
VPM1	0.45	0.79	1.29	1.62	11.38	2.97	81.29	6.82

(a) Solving (CGLP)
(b) Using the Iterative Method of Figure 1, version 2

Figure 4 shows 5 graphs featuring the behavior of the 5 procedures listed above, as a function of $|Q|$, the number of terms in the disjunction. On the horizontal axis we represent $|Q|$, on the vertical axis the total time needed to generate up to 50 cuts, normalized by setting to 1 the time needed by procedure 1 (solving (CGLP)$_Q$ directly) for the case $|Q| = 8$.

Graph 3 of Figure 4 corroborates what we said above concerning version 2 of the iterative procedure, whose performance is featured in the columns (b) of Table 1: namely, total time grows roughly linearly with $|Q|$. Also, graph 1 illustrates the much faster growth of the total time needed for solving (CGLP)$_Q$ directly, featured in the columns (a) of Table 1.

7   Conclusions

We have described several methods for generating cuts for pure or mixed 0-1 programs (MIP) from more complex disjunctions than the standard dichotomy $(x_j \leq 0) \vee (x_j \geq 1)$.

These methods solve the cut generating linear program iteratively, in the space of the original MIP. For the classical dichotomy, these procedures are

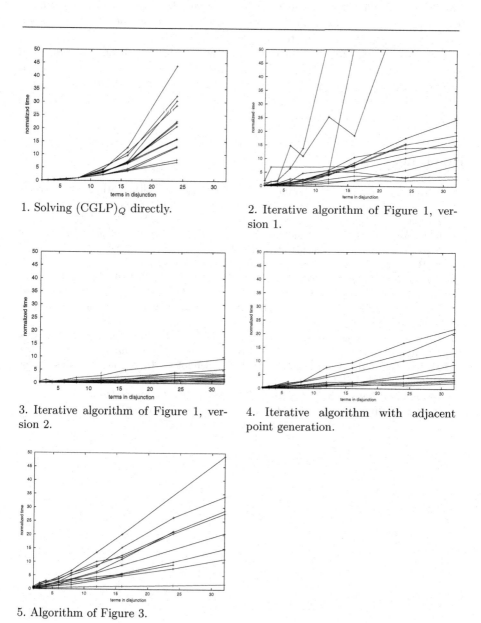

1. Solving $(CGLP)_Q$ directly.

2. Iterative algorithm of Figure 1, version 1.

3. Iterative algorithm of Figure 1, version 2.

4. Iterative algorithm with adjacent point generation.

5. Algorithm of Figure 3.

Fig. 4. Scaled running time versus number of terms in the disjunction

inferior to the standard lift-and-project method which solves a linear program in a higher dimensional space; but for generating cuts from disjunctions with more than 4 terms, i.e. involving 3 or more variables, at least one of the proposed methods is definitely superior to the standard one.

This opens up for further research problems like: modifying the procedures to generate multiple cuts from the more complex disjunctions studied here, identifying those multiple-term disjunctions most likely to provide stronger cuts, analyzing the behavior of these cuts as compared to the ones generated from the standard two-term disjunction.

References

1. E. Balas, *Disjunctive Programming*, Annals of Discrete Mathematics, 5 (1979) 3-51.
2. E. Balas, *Disjunctive Programming: Properties of the Convex Hull of Feasible Points*, Invited paper in Discrete Applied Mathematics, 89 (1998) 1-44.
3. E. Balas, S. Ceria, G. Cornuéjols, *A Lift-and-Project Cutting Plane Algorithm for Mixed 0-1 Programs*, Mathematical Programming, 58 (1993) 295-324.
4. E. Balas, S. Ceria, G. Cornuéjols, *Mixed 0-1 Programming by Lift-and-Project in a Branch-and-Cut Framework*, Management Science, 42 (1996) 1229-1246.
5. E. Balas, M. Perregaard, *Lift and Project for Mixed 0-1 Programming: Recent Progress*, MSRR No. 627, September 1999, Graduate School of Industrial Administration, Carnegie Mellon University.
6. E. Balas, M. Perregaard, *A Precise Correspondence Between Lift-and-Project Cuts, Simple Disjunctive Cuts, and Mixed Integer Gomory Cuts for 0-1 Programming*, MSRR No. 631, August 2000, Graduate School of Industrial Administration, Carnegie Mellon University
7. S. Ceria, G. Pataki, *Solving Integer and Disjunctive Programs by Lift-and-Project*, in R.E. Bixby, E.A. Boyd and R.Z. Rios-Mercado (editors), IPCO VI, Lecture Notes in Computer Science 1412, Springer, (1998) 271-283.

A $(2 + \varepsilon)$-Approximation Algorithm for Generalized Preemptive Open Shop Problem with Minsum Objective

Maurice Queyranne[*1] and Maxim Sviridenko[2]

[1] Faculty of Commerce and Business Administration, University of British Columbia, Vancouver, B.C., Canada V6T 1Z2,
`Maurice.Queyranne@commerce.ubc.ca`,
[2] IBM T.J. Watson Research Center, Yorktown, P.O. Box 218, NY 10598, USA,
`sviri@us.ibm.com`,
`http://www.research.ibm.com/people/s/sviri/sviridenko.html`

Abstract. In this paper we consider generalized version of the classical preemptive open shop problem with sum of weighted job completion times objective. The main result is a $(2 + \varepsilon)$-approximation algorithm for this problem. In the last section we also discuss the possibility of improving our algorithm.

1 Introduction

The generalized open shop is formulated as follows: There are m machines $\mathcal{M} = \{M_1, \ldots, M_m\}$, on which n jobs $\mathcal{J} = \{J_1, \ldots, J_n\}$ are to be scheduled for processing. Each job J_j consists of μ operations $O_{1j}, \ldots, O_{\mu j}$. Each operation O_{hj} has specified processing requirement p_{hj} and release dates r_{ihj} before which it cannot be processed on machine M_i. Machine M_i is capable of processing operation O_{hj} at a specified speed s_{ihj}.

Jobs are to be scheduled subject to the usual ground rules. A given machine can process at most one operation at any time. At most one operation of a given job can be processed by at most one machine at any time. The processing of any operation may be interrupted at any time and resumed at the same time on a different machine or at a later time on any machine. There is no penalty for such interruption or *preemption*. Let C_{hj} be a completion time of operation O_{hj}. Our objective will be to find a schedule with minimum value of the objective function $\sum_S w_S C_S$ where $C_S = \max_{O_{hj} \in S} C_{hj}$. This objective function includes few interesting special cases: operation completion times ($S = \{O_{hj}\}$), job completion times ($S = \{O_{1j}, \ldots, O_{\mu j}\}$), stage completion times for the instances of the classical open shop problem and makespan. Our general objective function also includes all variants of weighted sums of above special cases. We will consider instances with polynomial number of different sets S and $w_S > 0$ for all sets S.

* Research supported by a research grant from NSERC (the Natural Sciences and Research Council of Canada) to the first author.

K. Aardal, B. Gerards (Eds.): IPCO 2001, LNCS 2081, pp. 361–369, 2001.
© Springer-Verlag Berlin Heidelberg 2001

Special cases of the preemptive generalized open shop problem are extensively studied in the literature (we use standard three-field notations [8] for these problems):

- Most non-trivial special cases of the problem are strongly NP-hard [8].
- Gonzalez and Sahni's polynomial algorithm [4] for $O|pmtn|C_{max}$ is one of the classical results in the field. It can be extended to preemptive generalized open shop problem with makespan criteria by applying linear programming [9].
- Another well-known polynomially solvable case is $Q|r_j, pmtn|\sum C_j$ by Labetoulle et al. [7] and it is a famous open question to find a polynomial time algorithm for $R|pmtn|\sum C_j$ [8].
- Chakrabarti et al. [1] obtain a $(2.89 + \varepsilon)$-approximation algorithm for $Om|r_j, pmtn|\sum w_j C_j$,
- Queyranne and Sviridenko [10,11] present a 3-approximation algorithm for $O|r_j, pmtn|\sum w_j C_j$ and a 2e-approximation algorithm for $O(P)|r_j, pmtn|\sum w_S C_S$,
- Skutella [13] obtain a 3-approximation algorithm for $R|r_{ij}, pmtn|\sum w_j C_j$ and a 2-approximation algorithm for $R|pmtn|\sum w_j C_j$,

In this paper for any $\varepsilon > 0$ we present a polynomial time $(2 + \varepsilon)$-approximation algorithm for preemptive generalized open shop problem with general minsum criteria. The performance guarantee of our algorithm is better then performance guarantees of all approximation algorithms known for special cases of the problem (except Skutella's 2-approximation algorithm for $R|pmtn|\sum w_j C_j$).

2   An Interval-Indexed Formulation

Let $\mathcal{O}$ be the set of all operations and U be the set of different sets S. We define new release dates $r'_{ihj} = r_{ihj} + \delta$ where δ is some small constant. The optimal value of modified instance is at most $1 + \delta$ times of the optimal value of original instance. For a given $\varepsilon > 0$, L is a smallest integer such that $\delta(1+\varepsilon)^L \geq \sum_{O_{hj} \in \mathcal{O}} p_{hj} + \max r'_{ihj}$. Consequently, L is polynomially bounded in the input size of the considered scheduling problem. Let $\gamma_k = \delta(1 + \varepsilon)^k$ for $k = 0, \ldots, L$ and let $A = \{\gamma_k | k = 0 \ldots, L\} \cup \{r'_{ihj} | M_i \in \mathcal{M}, O_{hj} \in \mathcal{O}\}$ be the union of the set of different release dates with the set of points γ_k. Assume that the elements in A are indexed so that $a_1 < a_2 < \ldots < a_{|A|}$. We partition the time interval from δ to $\delta(1+\varepsilon)^L$ into $|A| - 1$ intervals $I_t = (a_t, a_{t+1}], t = 1, \ldots, |A| - 1$. Let Δ_t be the length of the tth interval, i.e. $\Delta_t = a_{t+1} - a_t$. For each machine M_i, operation O_{hj} and interval I_t we introduce variables y_{ihjt} with the following interpretation: $y_{ihjt}\Delta_t$ is the time operation O_{hj} is processed on machine M_i within interval I_t, or, equivalently: $s_{ihj}y_{ihjt}\Delta_t/p_{hj}$ is the fraction of operation O_{hj} that is being processed on machine M_i within I_t. For each set $S \in U$ and interval I_t we introduce variables x_{St} with the following intuitive interpretation: x_{St} is the fraction of set S that is being processed within I_t. Consider the following linear program:

$$\min \sum_{S \in U} w_S C_S, \tag{1}$$

$$\sum_{M_i \in \mathcal{M}} \sum_{t=1}^{|A|-1} s_{ihj} y_{ihjt} \Delta_t = p_{hj}, \quad O_{hj} \in \mathcal{O}, \tag{2}$$

$$\sum_{O_{hj} \in \mathcal{O}} y_{ihjt} \leq 1, \quad M_i \in \mathcal{M}, t = 1, \ldots, |A| - 1, \tag{3}$$

$$\sum_{M_i \in \mathcal{M}} \sum_{h=1}^{\mu} y_{ihjt} \leq 1, \quad J_j \in \mathcal{J}, t = 1, \ldots, |A| - 1, \tag{4}$$

$$\sum_{t=1}^{|A|-1} x_{St} = 1, \quad S \in U, \tag{5}$$

$$\sum_{\tau=1}^{t} x_{S\tau} \leq \sum_{M_i \in \mathcal{M}} \sum_{\tau=1}^{t} \frac{s_{ihj} y_{ihj\tau} \Delta_\tau}{p_{hj}}, \quad t = 1, \ldots, |A| - 1, S \in U, O_{hj} \in S, \tag{6}$$

$$C_S = \sum_{t=1}^{|A|-1} a_t x_{St}, \quad S \in U, \tag{7}$$

$$y_{ihjt} = 0, \quad M_i \in \mathcal{M}, O_{hj} \in \mathcal{O}, a_t < r'_{ihj}, \tag{8}$$

$$y_{ihjt} \geq 0, x_{St} \geq 0, \quad M_i \in \mathcal{M}, O_{hj} \in \mathcal{O}, S \in U, t = 1, \ldots, |A| - 1. \tag{9}$$

Consider a feasible schedule for the generalized preemptive open shop problem and assign values to variables y_{ihjt} as defined above. Variables x_{St} are defined by the following formulas

$$x_{S1} = \min_{O_{hj} \in S} \sum_{M_i \in \mathcal{M}} \frac{s_{ihj} y_{ihj1} \Delta_1}{p_{hj}}$$

and

$$x_{St} = \left(\min_{O_{hj} \in S} \sum_{M_i \in \mathcal{M}} \sum_{\tau=1}^{t} \frac{s_{ihj} y_{ihj\tau} \Delta_\tau}{p_{hj}} \right) - \sum_{\tau=1}^{t-1} x_{S\tau}.$$

This solution is clearly feasible for linear program (1)-(9): constraints (2), (5), (6), (8) and (9) are satisfied by definition of variables y_{ihjt} and x_{St}, constraints (3) are satisfied since the total amount of processing on machine M_i of parts of operations that is processed within the interval I_t cannot exceed its size, constraints (4) are satisfied since the total amount of job J_j that is processed within interval I_t cannot exceed its size, too. Finally, if C_S is the completion time of set S in a feasible schedule then

$$\sum_{t=1}^{|A|-1} a_t x_{St} \leq C_S \sum_{t=1}^{|A|-1} x_{St} = C_S$$

where the equality follows from the constraints (5) and the inequality follows from the fact that if $x_{St} > 0$ then $y_{ihjt} > 0$ for some operation $O_{hj} \in S$ and some machine M_i and therefore $a_t \leq C_{hj} \leq C_S$. Therefore, the linear program (1)-(9) is a relaxation of the preemptive generalized open shop problem with general minsum objective.

Notice that we cannot obtain a polynomial time approximation algorithm with performance guarantee better than 2 for the generalized preemptive open shop problem using the linear relaxation (1)-(9). Indeed, consider an instance with m machines and one job $J_1 = \{O_{1j}, \ldots, O_{mj}\}$. Operation O_{ij} has unit processing requirement (i.e. $p_{ij} = 1$), unit speed on machine M_i (i.e. $s_{ij} = 1$) and zero speeds on another machines (i.e. $s_{kj} = 0$ for $k \neq i$) that is we have an instance of the classical open shop problem. Each operation has a zero release date on each machine (i.e. $r_{ihj} = 0$). Set U consists of one set $S = \{O_{1j}, \ldots, O_{mj}\}$ that is objective function is a job completion time. Assume that $\delta(1 + \varepsilon)^L = m + \delta$ (we can always decrease δ and ε to enforce this equality). Then the optimum completion time of job J_1 is m whereas the optimum solution of linear program is defined by $y_{ihjt} = x_{St} = \Delta_t/m$ for all $O_{hj}, M_i \in \mathcal{M}, t = 1, \ldots, |A| - 1$ and has LP completion time equal to

$$\sum_t \frac{\Delta_t}{m} a_t = \frac{1}{m} \sum_t \left(\delta(1 + \varepsilon)^{t+1} - \delta(1 + \varepsilon)^t \right) \cdot \delta(1 + \varepsilon)^t = \frac{\delta^2 \varepsilon}{m} \sum_t (1 + \varepsilon)^{2t} =$$

$$= \frac{\delta^2 \varepsilon}{m} \cdot \frac{(1 + \varepsilon)^{2L} - 1}{(1 + \varepsilon)^2 - 1} \approx \frac{\varepsilon}{m} \cdot \frac{m^2}{2\varepsilon} = \frac{m}{2}.$$

The approximate equality holds for sufficiently small ε, δ and sufficiently big m.

3   Randomized Approximation Algorithm

The following algorithm takes an optimum LP solution and then constructs a feasible schedule by using the polynomial time algorithm of Gonzalez and Sahni [4] for $O|pmtn|C_{max}$ and the algorithm SLOW-MOTION of Schultz and Skutella [12]. Let y be an optimum solution of the linear program (1)-(9). This solution defines an instance of classical preemptive open shop problem in each interval I_t. Each job J_j^t in the instance corresponding to the interval I_t consists of m operations $O_{1j}^t, \ldots, O_{mj}^t$. Operation O_{ij}^t must be processed on machine M_i and has processing time $p_{ij}^t = \sum_{h=1}^{\mu} y_{ihjt}\Delta_t$, i.e. O_{ij}^t is the sum of all pieces of job J_j scheduled by LP to interval t on machine M_i. It follows from the constraints (3) and (4) that the maximum job length and the maximum machine load in the constructed instance are at most length of the interval I_t, i.e. $\max_{J_j^t} \sum_{i=1}^m p_{ij}^t \leq \Delta_t$ and $\max_{M_i} \sum_{j=1}^n p_{ij}^t \leq \Delta_t$. Using the classical algorithm of Gonzalez and Sahni for $O|pmtn|C_{max}$ we obtain a preemptive schedule in each interval I_t with makespan at most Δ_t. The union of the schedules for these intervals yields a feasible solution for the original preemptive generalized open shop problem.

The next step of our algorithm is to apply algorithm SLOW-MOTION of Schultz and Skutella [12] to constructed preemptive feasible schedule $\bar{S}$.

Algorithm SLOW-MOTION

1. Consider the given preemptive schedule $\bar{S}$ and process each operation β times slower (β will be defined later). In other words, we map every point t in time onto βt. This defines a new schedule $\bar{S}'$.

2. Let τ_{hj} be the earliest point in time when operation O_{hj} has been processed for p_{hj} units in schedule $\bar{S}'$, i.e.

$$\sum_{i=1}^{m} s_{ihj} Y_{ihj}(\tau_{hj}, S') = p_{hj}$$

where $Y_{ihj}(\tau_{hj}, \bar{S}')$ is the total processing time of operation O_{hj} in the schedule $\bar{S}'$ on machine M_i in the time interval $[0, \tau_{hj}]$. Convert $\bar{S}'$ to a feasible schedule $\bar{S}''$ by leaving the machine idle whenever it processed operation O_{hj} after τ_{hj}.

Let $C_{hj}(\alpha)$ for $0 < \alpha \leq 1$ be the earliest point in time when an α-fraction of operation O_{hj} has been completed in schedule $\bar{S}$, i.e.

$$\sum_{i=1}^{m} s_{ihj} Y_{ihj}(C_{hj}(\alpha), \bar{S}) = \alpha p_{hj}.$$

By analogy, we define $C_S(\alpha)$ for $0 < \alpha \leq 1$ be the starting point a_t of the earliest interval when an α-fraction of set S has been completed, i.e.

$$\sum_{\tau=1}^{t} x_{S\tau} \geq \alpha \quad \text{and} \quad \sum_{\tau=1}^{t-1} x_{S\tau} < \alpha.$$

We will use the following three lemmas in the analysis of the algorithm.

Lemma 1 (Goemans [3]). *Let C_S^{LP} be the LP completion time of set S defined by (7) then $\int_0^1 C_S(\alpha) d\alpha = C_S^{LP}$.*

Proof. Denote by α_t the fraction of set S that is completed at time a_t for $t = 1, \ldots, |A|$, i.e. $\alpha_t = \sum_{\tau=1}^{t-1} x_{St}$. Clearly, that $\alpha_1 = 0 \leq \alpha_2 \leq \ldots \leq \alpha_{|A|} = 1$ and $C_S(\alpha) = a_t$ for $\alpha \in [\alpha_t, \alpha_{t+1})$. Therefore,

$$\int_0^1 C_S(\alpha) d\alpha = \sum_{t=1}^{|A|-1} \int_{\alpha_t}^{\alpha_{t+1}} C_S(\alpha) d\alpha = \sum_{t=1}^{|A|-1} (\alpha_{t+1} - \alpha_t) a_t = \sum_{t=1}^{|A|-1} x_{St} a_t = C_S^{LP}.$$

Lemma 2 (Schultz and Skutella [12]). *Algorithm SLOW-MOTION with input $\beta \geq 1$ computes a feasible preemptive schedule. The completion time of operation O_{hj} equals $\beta C_{hj}(\frac{1}{\beta})$.*

Proof. Recall that τ_{hj} is the completion time of operation O_{hj} in the schedule $\bar{S}''$ and that exactly $1/\beta$ fraction of operation O_{hj} in schedule $\bar{S}'$ is finished to the point τ_{hj}. Therefore, $\tau_{hj} = \beta C_{hj}(1/\beta)$ by construction (see step 1 of the algorithm SLOW-MOTION).

Lemma 3.
$$\max_{O_{hj} \in S} C_{hj}(\alpha) \le (1+\epsilon)C_S(\alpha)$$

Proof. We claim that this lemma follows from constraints (6). Indeed, if $C_{hj}(\alpha) > a_t$ for some $t = 1, \ldots, |A| - 1$ and operation $O_{hj} \in S$ then

$$\sum_{\tau=1}^{t-1} x_{S\tau} \le \sum_{M_i \in \mathcal{M}} \sum_{\tau=1}^{t-1} \frac{s_{ihj} y_{ihj\tau} \Delta_\tau}{p_{hj}} < \alpha$$

by definition of the $C_{hj}(\alpha)$ and (6). Hence, $C_S(\alpha) \ge a_t$ and if $\max_{O_{hj} \in S} C_{hj}(\alpha) \in [a_t, a_{t+1}]$ then $a_{t+1} \le (1+\varepsilon)a_t \le (1+\varepsilon)C_S(\alpha)$.

Finally we can prove a theorem about expected performance of the algorithm SLOW-MOTION with β chosen randomly.

Theorem 1. *If we apply the algorithm SLOW-MOTION to preemptive schedule $\bar{S}$ with parameter β such that $1/\beta$ is randomly drawn from $(0,1]$ with density function $f(x) = 2x$ then the expected heuristic completion time C_S^H of set S in the resulting schedule is bounded above by $2(1+\varepsilon)C_S^{LP}$.*

Proof. We prove theorem by straightforward applications of Lemmas 1, 2 and 3

$$E(C_S^H) = E\left(\max_{O_{hj} \in S} C_{hj}^H\right) \overset{\text{Lemma 2}}{=} \int_0^1 \left(\max_{O_{hj} \in S} \frac{C_{hj}(x)}{x}\right) f(x)dx =$$

$$2\int_0^1 \max_{O_{hj} \in S} C_{hj}(x)dx \overset{\text{Lemma 3}}{\le} 2(1+\varepsilon)\int_0^1 C_S(x)dx \overset{\text{Lemma 1}}{=} 2(1+\varepsilon)C_S^{LP}.$$

Since our bound is a set-by-set bound, we obtained a randomized approximation algorithm with performance guarantee $2(1+\varepsilon)(1+\delta)$ for the generalized open shop problem with minsum objective.

4   Derandomization

In this section we will use the following notations:

$(b_{hj}^k, b_{hj}^{k+1}]$, $k \in K_{hj}$ are the time intervals where operation O_{hj} is processed nonpreemptively in the schedule $\bar{S}$,

$\alpha_{hj}(x)$ is the fraction of operation O_{hj} that is finished in the schedule $\bar{S}$ at time x,

$X_{hj}(\alpha)$ is the latest moment in time such that exactly an α-fraction of operation O_{hj} is finished before $X_{hj}(\alpha)$ in the schedule $\bar{S}$, notice that operation O_{hj} is not processed during time interval $(C_{hj}(\alpha), X_{hj}(\alpha)]$,

$F(\alpha) = \sum_{S \in U} \frac{w_S \max_{O_{hj} \in S} C_{hj}(\alpha)}{\alpha}$ is the value of the heuristic schedule $\bar{S}''$ obtained by the algorithm SLOW-MOTION with parameter $\beta = 1/\alpha$.

Note that $|K_{hj}| = O(m^2|A|)$ since the total number of preemptions in each interval I_t in schedule $\bar{S}$ is at most $O(m^2)$ [4].

We now show that $\min_{\alpha \in (0,1]} F(\alpha) = \min_{\alpha \in D} F(\alpha)$ for some set D containing a polynomial number of points from the interval $(0, 1]$. Therefore, we can find in polynomial time the point $\alpha_{min} \in (0, 1]$ such that $F(\alpha_{min}) \leq 2(1 + \varepsilon)(1 + \delta) \sum_{S \in U} w_S C_S^{LP}$ by enumerating all $\alpha \in D$.

Consider the set $Q = \{\alpha_{hj}(b_{hj}^k)|k \in K_{hj}, O_{hj} \in \mathcal{O}\}$. Assume that the elements in Q are indexed so that $q_1 = 0 < q_2 < \ldots < q_{|Q|} = 1$. Since for each $O_{hj} \in \mathcal{O}$ there are no preemptions in intervals $(X_{hj}(q_s), C_{hj}(q_{s+1})]$, $s = 1, \ldots, |Q| - 1$ we have that

$$C_{hj}(\alpha) = X_{hj}(q_s) + \frac{\alpha - q_s}{q_{s+1} - q_s}(C_{hj}(q_{s+1}) - X_{hj}(q_s))$$

for $\alpha \in (q_s, q_{s+1}]$, i.e. all functions $C_{hj}(\alpha)$ are linear in intervals $(q_s, q_{s+1}]$. Therefore, functions $f_S(\alpha) = \max_{O_{hj} \in S} C_{hj}(\alpha)$ are continuous piecewise linear convex functions in intervals $(q_s, q_{s+1}]$ for all $S \in U$ and have at most $|S|$ breakpoints between linear pieces. Hence we have at most $|\mathcal{O}| \cdot |U|$ points $d_1^s, \ldots, d_{p(s)}^s = q_{s+1}$ (notice that $d_1^s \neq q_s$) from the interval $(q_s, q_{s+1}]$ such that all functions $f_S(\alpha)$ are linear in intervals $(q_s, d_1^s], [d_i^s, d_{i+1}^s]$, $i = 1, \ldots, p(s) - 1, s = 1, \ldots, |Q| - 1$. Therefore, we proved that the function $F(\alpha)$ is monotone in these intervals as a sum of linear functions divided by α. Let

$$D = \bigcup_{s=1}^{|Q|-1} \{d_1^s, \ldots, d_{p(s)}^s\}$$

be the set of all breakpoints of the function $F(\alpha)$.

Lemma 4.

$$\min_{\alpha \in D} F(\alpha) = \min_{\alpha \in (0,1]} F(\alpha)$$

Proof. We now prove that the function $F(\alpha)$ is nonincreasing in interval $(q_1, d_1^1]$. Indeed,

$$C_{hj}(\alpha) = X_{hj}(0) + \frac{\alpha}{q_2}(C_{hj}(q_2) - X_{hj}(0))$$

for $\alpha \in (q_1, q_2]$. Therefore, $F(\alpha) = K_1 + K_2/\alpha$ for $\alpha \in (q_1, d_1^1]$ where K_1 and K_2 are some nonnegative constants. Since

$$C_{hj}(q_s) \leq X_{hj}(q_s) = \lim_{y \to +0} C_{hj}(q_s + y)$$

we obtain the similar inequality for the function $F(\alpha)$

$$F(q_s) \leq \lim_{y \to +0} F(q_s + y). \qquad (10)$$

For any point $\alpha \in (q_s, d_1^s]$, $s = 2, \ldots, |Q| - 1$, the inequality $F(\alpha) \geq \min\{F(q_s), F(d_1^s)\}$ follows from monotonicity of $F(\alpha)$ in interval $(q_s, d_s^1]$ and the inequality (10) (we gave the separate proof for the first interval since $F(0)$ is undefined).

Therefore, we can restrict our attention to the union of closed intervals $\cup_{s=1}^{|Q|-1}[d_1^s, d_{p(s)}^s]$. Since the function $F(\alpha)$ is continuous in each interval $[d_1^s, d_{p(s)}^s]$ there exists a point $\alpha_{min} \in \cup_{s=1}^{|Q|-1}[d_1^s, d_{p(s)}^s]$ which minimizes the function $F(\alpha)$. Assume that $\alpha_{min} \in [d_k^s, d_{k+1}^s]$. Since $F(\alpha)$ is monotone in this interval we obtain that $F(\alpha_{min}) \geq \min\{F(d_k^s), F(d_{k+1}^s)\}$ and therefore $F(\alpha)$ is minimized on the point belonging to the set D.

5   Conclusion

In this paper we obtained a $(2+\varepsilon)$-approximation algorithm for the preemptive generalized open shop problem with minsum criteria. An obvious open question is to build an algorithm with better performance guarantee. It would be also interesting to remove ε from our performance guarantee, the problem is that we don't know how to define operation completion times in time-indexing formulations of scheduling problems with preemptions, probably we need another type of mathematical programming relaxations. Another open question is to obtain non-approximability results for the classical minsum open shop problem with preemptions, i.e. to prove that $O|pmtn| \sum C_j$ does not admit a PTAS. Probably technique developed by Hoogeveen, Schuurman and Woeginger [6] for nonpreemptive scheduling problems with minsum objective can be applied to resolve this question.

Our last note is that we believe that 2 is the best possible performance guarantee of approximation algorithm for the problem studied in this paper even in very restrictive case: $1|pmtn| \sum w_S C_S$. We explain reason for this conjecture below. First of all, by the corollary of Theorem 1.1 in [14] there always exists an optimal non-preemptive schedule for $1|pmtn| \sum w_S C_S$. So, we can restrict our attention to the non-preemptive version of this problem $1|| \sum w_S C_S$. We now claim that this problem includes the following special case of the problem $1|prec| \sum w_j C_j$. In this special case we have a set of jobs such that either $p_j > 0$ and $w_j = 0$ or $p_j = 0$ and $w_j > 0$, there are also specific precedence constraints in this problem. If $J_j \prec J_k$ then J_j is a job of the first type, i.e. $p_j > 0$ and $w_j = 0$, and J_k is a job of the second type, i.e. $p_k = 0$ and $w_k > 0$. It is well-known that the problem $1|prec| \sum w_j C_j$ has a 2-approximation algorithm [5]. Moreover, many researchers believe that there is no better approximation algorithms even in the special case defined above. The main reason for such believe is that all known linear programming relaxations of this problem have integrality gap of 2, in particular the strongest known LP for such kind of problems is the linear ordering relaxation of Potts. Chekuri and Motwani [2] recently proved the integrality gap of 2 for this LP relaxation for above special case of $1|prec| \sum w_j C_j$. This special case can be reduced to $1|| \sum w_S C_S$ as follows. For any job of the first type we define a job with the same processing time in the new instance, for any job J_j of the second type we define a set S_j containing all predecessors of J_j and with weight $w_{S_j} = w_j$.

References

1. S. Chakrabarti, C. Phillips, A. Schulz, D. Shmoys, C. Stein and J. Wein, Improved Scheduling Algorithms for Minsum Criteria , In Proceedings of ICALP96, LNCS 1099, pp. 646–657.
2. C. Chekuri and R. Motwani, Precedence constrained scheduling to minimize sum of weighted completion times on a single machine, Discrete Appl. Math. 98 (1999), 29–38.
3. M. X. Goemans, Improved Approximation Algorithms for Scheduling with Release Dates, *In Proceedings of the 8th ACM-SIAM Symposium on Discrete Algorithms* (1997), 591–598.
4. T. Gonzalez and S. Sahni, Open Shop Scheduling to Minimize Finish Time, *Journal of the ACM* **23** (1976) 665–679.
5. L. Hall, A. Schulz, D. Shmoys and Wein, Scheduling to minimize average completion time: off-line and on-line approximation algorithms, Math. Oper. Res. 22 (1997), 513–544.
6. H. Hoogeveen, P. Schuurman and G. Woeginger, Non-approximability results for scheduling problems with minsum criteria, In Proceedings of IPCO98, LNCS v. 1412, pp. 353–366.
7. J. Labetoulle, E. Lawler, J. K. Lenstra and A.H. G. Rinnooy Kan, Preemptive scheduling of uniform machines subject to release dates, Progress in combinatorial optimization, pp. 245–261, Academic Press, Toronto, Ont., 1984.
8. E.L. Lawler, J.K. Lenstra, A.H.G. Rinnooy Kan, and D.B. Shmoys, "Sequencing and Scheduling: Algorithms and Complexity." In: S.C. Graves, A.H.G. Rinnooy Kan, and P.H. Zipkin, eds., *Logistics of Production and Inventory*, Handbooks in Operations Research and Management Science 4, North–Holland, Amsterdam, The Netherlands (1993) 445–522.
9. E. L. Lawler, M. G. Luby and V. V. Vazirani, Scheduling Open Shops with Parallel Machines, *Operations Research Letters* **1** (1982) 161–164.
10. M. Queyranne and M. Sviridenko, "Approximation Algorithms for Shop Scheduling Problems with Minsum Objective", submitted for publication.
11. M. Queyranne and M. Sviridenko, New and Improved Algorithms for Minsum Shop Scheduling, Proceedings of SODA00, pp.871–878.
12. A.S. Schulz and M. Skutella, "Random-Based Scheduling: New Approximations and LP Lower Bounds." In: J. Rolim, ed., *Randomization and Approximation Techniques in Computer Science* (Random'97 Proceedings, 1997) Lecture Notes in Computer Science **1269**, Springer, 119–133.
13. M. Skutella, Convex quadratic and semidefinite programming relaxations in scheduling, to appear in JACM.
14. V. Tanaev, V. Gordon, and Y. Shafransky, Scheduling theory. Single-stage systems, Mathematics and its Applications, v. 284, Kluwer Academic Publishers Group, Dordrecht, 1994.

Performance Guarantees of Local Search for Multiprocessor Scheduling

Petra Schuurman and Tjark Vredeveld*

Department of Mathematics and Computing Science
Technische Universiteit Eindhoven
P.O.Box 513, 5600 MB Eindhoven, The Netherlands
petra,tjark@win.tue.nl

Abstract. This paper deals with the worst-case performance of local search algorithms for makespan minimization on parallel machines. We analyze the quality of the local optima obtained by iterative improvements over the jump, the swap, and the newly defined push neighborhood.

1   Introduction

We analyze the quality of local optima from a worst-case perspective. We consider multiprocessor scheduling problems. We are given a set of n jobs, $J_1, \ldots, J_n$, each of which has to be processed without preemption on one of m machines, $M_1, \ldots, M_m$. A machine can process at most one job at a time, and all jobs and machines are available at time 0. The objective we consider is *makespan minimization*, that is, we want the last job to complete as early as possible. The time, p_{ij}, it takes for a job J_j to be fully processed on a machine M_i depends on the machine environment:

- Identical parallel machines, denoted by P: a job has the same processing time on all machines, i.e., $p_{ij} = p_j$, where p_j is a positive integer.
- Uniform parallel machines, denoted by Q: the machines have positive integral speeds, $s_1, \ldots, s_m$, and each job has a given positive integral processing requirement p_j; the processing time is $p_{ij} = p_j/s_i$.
- Unrelated parallel machines, denoted by R: the time it takes to process job J_j on machine M_i is dependent on the machine as well as the job; p_{ij} is a positive integer.

In the case that the number of machines is part of the input, Graham et al. [7] denote these problems by $P\|C_{\max}$, $Q\|C_{\max}$, and $R\|C_{\max}$. If the number of machines is a constant m, then they denote the problems by $Pm\|C_{\max}$, etc.

Even the simplest case, $P2\|C_{\max}$, is NP-hard, see Garey and Johnson [5]. Therefore, we search for approximate solutions. If an algorithm is guaranteed to

* Supported by the project "High performance methods for mathematical optimization" of the Netherlands Organization for Scientific Research (NWO)

K. Aardal, B. Gerards (Eds.): IPCO 2001, LNCS 2081, pp. 370–382, 2001.
© Springer-Verlag Berlin Heidelberg 2001

deliver a solution that has value at most ρ times the optimal solution value, we call this algorithm a ρ-approximation algorithm. The value ρ is called *worst-case performance guarantee*. For $P\|C_{\max}$ and $Q\|C_{\max}$, Hochbaum and Shmoys [8,9] develop so called polynomial-time approximation schemes, that is, for each $\epsilon > 0$, there exists a polynomial-time $(1 + \epsilon)$-approximation algorithm. For $R\|C_{\max}$, Lenstra, Shmoys, and Tardos [12] present a polynomial-time 2-approximation algorithm; they also prove that there does not exists a polynomial-time $(\frac{3}{2} - \epsilon)$-algorithm for $\epsilon > 0$, unless P=NP.

A way to find approximate solutions is through *local search*. Local search methods iteratively search through the set of feasible solutions. Starting from an initial solution, a local search procedure moves from a feasible solution to a neighboring solution until some stopping criteria are met. The choice of a suitable neighborhood function has an important influence on the performance of local search. In this paper, we analyze the performance of local search for the jump, the swap, and the newly defined push neighborhood from a worst-case perspective.

The simplest form of local search is *iterative improvement*, also called local improvement or, in the case of minimization problems, descent algorithms. This method iteratively chooses a better solution in the neighborhood of the current solution; it stops when no better solution is found; we say that the current solution is a *local optimum*.

Previous results on the worst-case analysis of local search include the following. Korupolu, Plaxton, and Rajarman [11] show that iterative improvement for the metric uncapacitated facility location problem in polynomial time yields a solution with value no more than $5 + \epsilon$ times the optimal solution value, for $\epsilon > 0$. They also show that for the metric uncapacitated k-median problem local search finds a solution using at most $(3+5/\epsilon)k$ facilities and having costs at most $1 + \epsilon$ times the optimal costs, for $\epsilon > 0$. On the max-cut problem it is easy to see that a local optimal solution with respect to the flip-neighborhood has value at least half the optimal cut value. Using the same neighborhood, Feige, Karpinski, and Langberg [4], show that local search improves the performance guarantee of the Goemans-Williamson approximation algorithm [6]. To the best of our knowledge, the only papers in the area of scheduling on worst-case analysis of local search are those by Brucker, Hurink, and Werner [1,2]. The authors show, among others, that for $P2\|C_{\max}$ and $P\|C_{\max}$ a local optimum with respect to the jump neighborhood can be found by iterative improvement in $\mathcal{O}(n^2)$ iterations. It can easily be seen that the performance bound of this local optimum is $2 - \frac{2}{m+1}$.

2 Neighborhoods

Before discussing the neighborhoods, we first describe our representation of a schedule. As the sequence in which the jobs are processed does not influence the makespan of a schedule for a given assignment of the jobs to the machines, we represent a schedule by such an assignment. This is equivalent to a partitioning of the set of jobs into m disjoint subsets $\mathcal{J}_1, \ldots, \mathcal{J}_m$, where $\mathcal{J}_i$ is the set of jobs scheduled on M_i. The *load* of a machine is the total processing time of its jobs. A *critical machine* is a machine with maximum load.

The first neighborhood that we consider is the *jump* neighborhood, also known as the *move* neighborhood: we move a job J_j to a machine M_i on which job J_j is not scheduled (see Figure 1). We say that we are in a *jump optimal solution*, if no jump decreases the makespan or the number of critical machines without increasing the makespan.

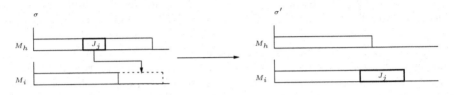

Fig. 1. Jump

In the *swap* neighborhood, we select two jobs, J_j and J_k, scheduled on different machines. The neighbor is formed by interchanging the machine allocations of the jobs (see Figure 2). If all jobs are scheduled on the same machine, then the swap neighborhood is empty; therefore, we define the swap neighborhood as one that consists of all possible jumps and all possible swaps. A *swap optimal solution* is a solution in which no swap or jump decreases the makespan or the number of critical machines without increasing the makespan.

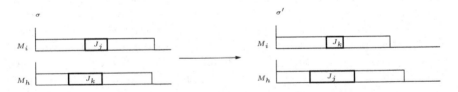

Fig. 2. Swap

As we will see in the next section, the jump and swap neighborhoods have no constant performance guarantee for $Q\|C_{\max}$. Therefore, we introduce a *push* neighborhood, for which any local optimum is within a factor $2 - \frac{2}{m+1}$ of optimal for $Q\|C_{\max}$. The push neighborhood is a form of variable-depth search, introduced by Kernighan and Lin for graph partitioning [10] and the traveling salesman problem [13]. A push is a sequence of jumps.

Starting with a schedule $\sigma = (\mathcal{J}_1, \ldots, \mathcal{J}_m)$ having makespan $C_{\max}(\sigma)$, a push is initiated by selecting a job J_k on a critical machine and a machine M_i to move it to. We say that J_k fits on M_i if $\sum_{J_j \in \mathcal{J}_i : p_{ij} \geq p_{ik}} p_{ij} + p_{ik} < C_{\max}(\sigma)$. If J_k does not fit on any machine, then it cannot be pushed. If, after moving J_k to M_i, the load of this machine is at least as large as the original makespan, that is, if $\sum_{J_j \in \mathcal{J}_i} p_{ij} + p_{ik} \geq C_{\max}(\sigma)$, then we iteratively remove the smallest job from M_i until the load of this machine is less than $C_{\max}(\sigma)$. The removed jobs are gathered in a queue. We now have a queue of pending jobs and a partial schedule that has lower makespan or fewer critical machines. If this queue is non-empty, then the largest job in the queue is removed and moved to some machine on which it fits, in the same way as the first job was pushed. Thus, if necessary, we allow some smaller jobs to be removed. If the largest job in the queue does not fit on any machine, then we say that the push is unsuccessful. We repeat the procedure of moving the largest job in the queue to a machine until the queue is empty or we have determined that the push is unsuccessful. When pushing all jobs on the critical machines is unsuccessful, we are in a *push optimal solution*.

We illustrate a push in the following example.

Example 1. Consider an instance for $P3\|C_{\max}$, with $n = 8$ jobs. The processing times are $p_1 = 8$, $p_2 = p_3 = p_4 = 6$, $p_5 = p_6 = 5$, $p_7 = 3$, and $p_8 = 2$. The starting schedule is $\sigma = (\mathcal{J}_1, \mathcal{J}_2, \mathcal{J}_3)$, with $\mathcal{J}_1 = \{J_2, J_5, J_6\}$, $\mathcal{J}_2 = \{J_1, J_7, J_8\}$, and $\mathcal{J}_3 = \{J_3, J_4\}$, having makespan $C_{\max}(\sigma) = 16$. This schedule is depicted in Figure 3a. In the first step of the push, we select job J_6, with $p_6 = 5$, to be pushed onto machine M_2. When moving J_6 to M_2, jobs J_8 and J_7 have to be removed from M_2 (Figure 3b). At this point, we have a partial schedule $\sigma' = (\mathcal{J}_1', \mathcal{J}_2', \mathcal{J}_3')$, $\mathcal{J}_1' = \{J_2, J_5\}$, $\mathcal{J}_2' = \{J_1, J_6\}$, $\mathcal{J}_3' = \{J_3, J_4\}$ and a queue of pending jobs containing J_7 and J_8. In the next step, we remove J_7 from the queue and move it to M_1 and then we move J_8 to M_3. After moving J_8, the queue is empty and we have a new schedule $\bar{\sigma} = (\bar{\mathcal{J}}_1, \bar{\mathcal{J}}_2, \bar{\mathcal{J}}_3)$, with $\bar{\mathcal{J}}_1 = \{J_2, J_5, J_7\}$, $\bar{\mathcal{J}}_2 = \{J_1, J_6\}$, and $\bar{\mathcal{J}}_3 = \{J_3, J_4, J_8\}$, which has makespan $C_{\max}(\bar{\sigma}) = 14$ (Figure 3c).

A push optimal schedule is given in Example 2. As the schedule is not swap optimal, this example shows that a push optimal solution is not necessarily swap optimal. Of course, as a push is a sequence of one or more jumps, a push optimal solution is jump optimal.

Example 2. In Figure 4, we give an example for $P2\|C_{\max}$. There are $n = 7$ jobs, $J_1, \ldots, J_7$, with processing times $p_1 = 9$, $p_2 = 8$, $p_3 = 6$, $p_4 = 5$, $p_5 = 4$, and $p_6 = p_7 = 3$. The schedule $\sigma = (\mathcal{J}_1, \mathcal{J}_2)$, with $\mathcal{J}_1 = \{J_1, J_3, J_4\}$ and $\mathcal{J}_2 = \{J_2, J_5, J_6, J_7\}$, is push optimal. When trying to push J_1, jobs J_7, J_6, and J_5 are moved to the queue of pending jobs. Then jobs J_5 and J_6 are moved to machine M_1, resulting in a partial schedule with a load of 18 for M_1 and of 17 for M_2. The queue of pending jobs consists of job J_7 of length $p_7 = 3$. As $18+3 \geq 20 = C_{\max}(\sigma)$ and $17+3 \geq 20$, J_7 does not fit on both machines. Hence, pushing J_1 is unsuccessful. In the same manner, we see that pushing J_3 or J_4 is also unsuccessful. The schedule can be improved by swapping e.g. J_1 and J_2.

By our way of defining a push, we know that when moving a job J_k, only smaller jobs than J_k can be removed from the machine. Hence, during one push, at most n jobs need to be moved and one move can straightforwardly be done

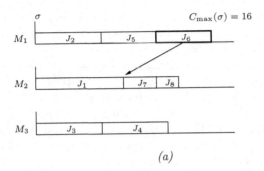

(a)

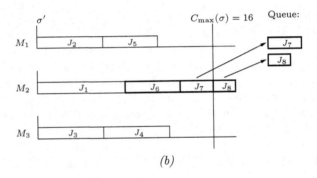

(b)

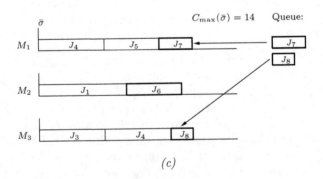

(c)

Fig. 3. Push

Fig. 4. Push optimal schedule

in $\mathcal{O}(n)$ time. Thus, a push requires $\mathcal{O}(n^2)$ elementary operations. Using appropriate data structures and selecting, in a greedy manner, the machine to move a job to, a push neighbor can be found in $\mathcal{O}(n \log n)$ time.

Note that, as we always take the largest job from the queue, the push neighborhood is only defined for scheduling problems where the largest job is defined unambiguously. Therefore, in the case of unrelated parallel machines, a push is not well defined.

3   Performance Guarantees

In this section, we establish performance guarantees for the various local optima and scheduling problems. These are given in Table 1. "UB $= \rho$" denotes that ρ is a performance guarantee and "LB $= \rho$" denotes that the performance guarantee cannot be less than ρ; "ρ" denotes that both UB $= \rho$ and LB $= \rho$.

The guarantee of jump optimal solutions for $Q\|C_{\max}$ can easily be proven by adapting the first part of the proof of Cho and Sahni [3] for list schedules on uniform parallel machines. In the following we focus on the guarantees for the push optimal solutions, as this neighborhood is the most effective one.

Table 1. Performance guarantees: $C_{\max}^{LS}/C_{\max}^{*}$

	jump	swap	push
$P2\|C_{\max}$	$\frac{4}{3}$	$\frac{4}{3}$	$\frac{8}{7}$
$P\|C_{\max}$	$2 - \frac{2}{m+1}$	$2 - \frac{2}{m+1}$	UB $= 2 - \frac{2}{m+1}$ LB $= \frac{4m}{3m+1}$
$Q2\|C_{\max}$	$\frac{1+\sqrt{5}}{2}$	$\frac{1+\sqrt{5}}{2}$	$\frac{\sqrt{17}+1}{4}$
$Q\|C_{\max}$	$\frac{1+\sqrt{4m-3}}{2}$	$\frac{1+\sqrt{4m-3}}{2}$	UB $= 2 - \frac{2}{m+1}$ LB $= \frac{3}{2} - \epsilon$
$R2\|C_{\max}$	LB $= \frac{p_{\max}}{C_{\max}^{*}}$	LB $= n - 1$	undefined
$R\|C_{\max}$	LB $= \frac{p_{\max}}{C_{\max}^{*}}$	LB $= \frac{p_{\max}-1}{C_{\max}^{*}}$	undefined

Theorem 1. *A push optimal solution for $P2\|C_{\max}$ has value at most $8/7$ times the optimal solution value.*

Proof. Suppose, to the contrary, that there exists a push optimal schedule with makespan $C_{\max}^{\mathrm{P}} > \frac{8}{7}C_{\max}^*$, where $C_{\max}^*$ is the value of an optimal schedule. Let $L_i = \sum_{J_j \in \mathcal{J}_i} p_j$ be the load of machine M_i ($i = 1, 2$). W.l.o.g. we assume that $L_1 \geq L_2$, thus $C_{\max}^{\mathrm{P}} = L_1$.

For the difference in loads of the two machines, we know that

$$L_1 - L_2 = L_1 - \left(\sum_j p_j - L_1\right) \geq 2L_1 - 2C_{\max}^* > \frac{1}{4}L_1.$$

The first inequality is due to the lower bound $C_{\max}^* \geq \frac{1}{2}\sum_j p_j$ and the second inequality is due to the assumption that $L_1 > \frac{8}{7}C_{\max}^*$. Let J_1 be the smallest job on M_1. Then, by push optimality, we know that $p_1 \geq L_1 - L_2 > \frac{1}{4}L_1$. Hence, there are at most three jobs on M_1. Let $\mathcal{J}_1 = \{J_1, J_2, J_3\}$, $\mathcal{J}_2 = \{J_4, \ldots, J_n\}$, and $p_1 \leq p_2 \leq p_3$ and $p_4 \geq \ldots \geq p_n$. Let J_k be the job such that the smaller of J_k and J_1 is the largest job that has to be removed, when pushing J_1 onto M_2, i.e.,

$$\begin{aligned} p_4 + \ldots + p_{k-1} + p_1 &< L_1, \\ p_4 + \ldots + p_k + p_1 &\geq L_1. \end{aligned} \tag{1}$$

As $L_1 - L_2 > \frac{1}{4}L_1 \geq \frac{3}{4}p_1$, we know that $L_2 + p_1 < L_1 + \frac{1}{4}p_1$ and

$$\frac{1}{4}p_1 > L_2 + p_1 - L_1 \overset{(1)}{\geq} L_2 - (p_4 + \ldots + p_k) = p_{k+1} + \ldots + p_n. \tag{2}$$

Because of push optimality, we know that $p_1 \leq p_k + \ldots + p_n$ and thus

$$p_k \geq p_1 - (p_{k+1} + \ldots + p_n) \overset{(2)}{>} \frac{3}{4}p_1. \tag{3}$$

If $p_k < p_1$, then by pushing J_1 onto M_2, J_k is moved to M_1 and jobs $J_{k+1}, \ldots, J_n$ are distributed among M_1 and M_2 yielding a schedule with makespan

$$C_{\max}' \leq \max(L_1 - p_1 + p_k, L_2 + p_1 - p_k) \overset{(3)}{<} \max(L_1, L_1 - \frac{1}{2}p_1) = L_1.$$

Thus the schedule is not push optimal and it must be the case that $p_1 \leq p_k$.

Suppose M_2 processes at least three large jobs, i.e., at least as large as J_k. Then $L_2 \geq 3p_1$. By push optimality, we know that $p_1 \geq L_1 - L_2 > \frac{1}{4}L_1$ and thus $L_2 \geq 3p_1 > \frac{3}{4}L_1$. However, as $L_1 - L_2 > \frac{1}{4}L_1$, it must be that $L_2 < \frac{3}{4}L_1$. Therefore, M_2 can process at most two large jobs.

If M_2 processes only one large job, then the current schedule is optimal, as the sub-schedule for J_1, J_2, J_3, and J_4 is optimal, because by (1) we know that $p_1 + p_4 \geq p_1 + p_2 + p_3$. In the case that M_2 processes two large jobs, i.e., $p_4 \geq p_5 = p_k$, we consider two sub-cases: $p_3 \leq p_5$ and $p_3 > p_5$. If $p_3 \leq p_5$, then the sub-schedule for $J_1, \ldots, J_5$ is optimal and $C^*_{\max} \geq C^*_{\max}[1,5] = L_1 = C^P_{\max}$, where $C^*_{\max}[1,5]$ denotes the makespan of an optimal sub-schedule for $J_1, \ldots, J_5$.

If $p_3 > p_5$, then an optimal schedule on $J_1, \ldots, J_5$ has value $C^*_{\max}[1,5] \geq p_1 + p_2 + p_5$. As $L_2 < 3p_1$, we know that $p_4 < 2p_1$ and by push optimality we know that $p_5 + \sum_{j \geq 6} p_j \geq p_3$. Hence,

$$\frac{C^P_{\max}}{C^*_{\max}} \leq \frac{p_1 + p_2 + p_3}{p_1 + p_2 + p_5} \leq \frac{p_1 + p_2 + p_5 + \ldots + p_n}{p_1 + p_2 + p_5} \overset{(2)}{\leq} 1 + \frac{1/4 p_1}{3 p_1} = \frac{13}{12} < \frac{8}{7}.$$

This contradicts the assumption that $C^P_{\max} > \frac{8}{7} C^*_{\max}$.

In the case that $\mathcal{J}_1 = \{J_1, J_2\}$ and $\mathcal{J}_2 = \{J_3, \ldots, J_n\}$, we can prove in a similar way that $C^*_{\max} = C^P_{\max}$. If M_1 only processes J_1, then it is trivial to see that $C^P_{\max} = C^*_{\max}$.

Hence, $C^P_{\max} \leq \frac{8}{7} C^*_{\max}$.   □

To see that the bound in Theorem 1 is tight, consider the following example.

Example 3. We have $n = 6$ jobs, $J_1, \ldots, J_6$, two of which have processing times $p_1 = p_2 = 3$ and the others have processing times $p_3 = p_4 = p_5 = p_6 = 2$. In an optimal schedule each machine processes one job with $p_j = 3$ and two jobs with $p_j = 2$ and $C^*_{\max} = 7$. The schedule in which machine M_1 processes all jobs of size 2 and machine M_2 processes all jobs of size 3 is a push optimal schedule and has makespan $C^P_{\max} = 8$ (see Figure 5).

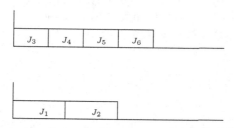

Fig. 5. Push optimal schedule for $P2\|C_{\max}$ with $C^P_{\max} = \frac{8}{7} C^*_{\max}$

Theorem 2. *A push optimal schedule for $Q\|C_{\max}$ has makespan at most $2 - \frac{2}{m+1}$ times the optimal solution value.*

Proof. In $Q\|C_{\max}$, we are given n jobs with processing requirements $p_1, \ldots, p_n$ and m machines with speeds $s_1 \geq \ldots \geq s_m$. If we consider an unsuccessful push, then there is a partial schedule, $(\mathcal{J}'_1, \ldots, \mathcal{J}'_m)$, and a queue of pending jobs. The largest job in this queue does not fit on any machine. Let job J_k be the largest job in the queue and let $L'_i = \sum_{J_j \in \mathcal{J}'_i : p_j \geq p_k} \frac{p_j}{s_i}$ be the total processing time of the large jobs, that is, at least as large as J_k, on machine M_i. Because of push optimality, we know that for all $i = 1, \ldots, m$ we have

$$L_i' + \frac{p_k}{s_i} \geq C_{\max}^P. \tag{4}$$

Let M_h be the slowest machine on which job J_k has a processing time that is not larger than the optimal makespan, that is, $h = \max\{i : \frac{p_k}{s_i} \leq C_{\max}^*\}$. Thus $C_{\max}^* \geq \frac{p_k}{s_h}$. Another lower bound on the optimal makespan is then

$$C_{\max}^* \geq \frac{\sum_{j : p_j \geq p_k} p_j}{\sum_{i=1}^h s_i}. \tag{5}$$

By push optimality (4), we know that

$$\sum_{i=1}^h s_i C_{\max}^P \leq \sum_{i=1}^h s_i L_i' + h p_k \leq \sum_{j : p_j \geq p_k} p_j + (h-1) p_k. \tag{6}$$

If $s_1 \geq 2 s_h$, then $\sum_{i=1}^h s_i \geq (h+1) s_h$, and rearranging the terms in (6) yields

$$C_{\max}^P \leq C_{\max}^* + \frac{h-1}{h+1} \frac{p_k}{s_h} \leq C_{\max}^* + \frac{h-1}{h+1} C_{\max}^* \leq \frac{2m}{m+1} C_{\max}^*.$$

The first inequality is due to the lower bound (5) and the second inequality is because of the lower bound $C_{\max}^* \geq \frac{p_k}{s_h}$.

If $s_1 \leq 2 s_h$, then $\sum_{i=1} h s_i \geq \frac{h+1}{2} s_1$. We may assume that $C_{\max}^* \geq \frac{2 p_k}{s_1}$, as otherwise there are at most h large jobs and the push optimal schedule is optimal. Rearranging the terms in (6) yields

$$C_{\max}^P \leq C_{\max}^* + \frac{h-1}{(h+1)/2} \frac{p_k}{s_1} \leq \frac{2m}{m+1} C_{\max}^*.$$

$\square$

In the case of two uniform parallel machines, we establish a better performance guarantee. To do so, we need two lemmata, in which we consider instances with only three jobs and where the machines do not have the same speed. We prove in Theorem 3 that a smallest worst-case instance for push on two uniform parallel machines has exactly three jobs.

Lemma 1. *Consider an instance for $Q2\|C_{\max}$ with three jobs in which the machines do not have the same speed, and assume w.l.o.g that $p_1 \geq p_2 \geq p_3$ and $s_1 > s_2$. If in an optimal schedule J_1 is processed on M_1, then in any push optimal schedule a job of size p_1 is scheduled on M_1 and this push optimal schedule is globally optimal.*

Proof. Suppose to the contrary that there is a push optimal schedule in which J_1 is processed on M_2. Then J_2 is scheduled on M_1, as otherwise a push is possible.

If $p_1 = p_2$, then the first part of the lemma is proven. Consider the case that $p_1 > p_2$. If M_2 is the critical machine, then J_1 can be pushed, and the schedule is not push optimal. Therefore, M_1 is the critical machine, and it processes J_2 as well as J_3. As in the optimal schedule J_1 is processed on machine M_1, the optimal makespan has value $C^*_{\max} \geq \min\{\frac{p_1+p_3}{s_1}, \frac{p_2+p_3}{s_2}\}$. As $\frac{p_1+p_3}{s_1} > \frac{p_2+p_3}{s_1} = C^P_{\max}$ and $\frac{p_2+p_3}{s_2} > \frac{p_2+p_3}{s_1} = C^P_{\max}$, we have that $C^*_{\max} > C^P_{\max}$, which is a contradiction. Therefore, in a push optimal schedule, J_1 must be processed by M_1, whenever J_1 is scheduled on M_1 in an optimal schedule.

By enumerating over all possible schedules with J_1 scheduled on M_1 for the push optimal schedule as well as the optimal schedule, it is easy to see that whenever a schedule of this form is push optimal, it is globally optimal. □

Lemma 2. *Consider an instance for $Q2\|C_{\max}$ with three jobs and different speeds for the machines and assume w.l.o.g. that $p_1 \geq p_2 \geq p_3$ and $s_1 > s_2$. If $C^P_{\max} > C^*_{\max}$, then in the optimal schedule, M_1 processes J_2 and J_3, and J_1 is scheduled on M_2. In a push optimal schedule with $C^P_{\max} > C^*_{\max}$, the machine allocation of the jobs is reversed, that is, J_1 is scheduled on M_1, and M_2 processes J_2 and J_3. This push optimal schedule has makespan $C^P_{\max} = \frac{p_2+p_3}{s_2}$.*

Proof. By Lemma 1, we know that whenever $C^P_{\max} > C^*_{\max}$, in the optimal schedule J_1 is scheduled on M_2. As $\frac{p_1+p_2}{s_2} \geq \frac{p_1+p_3}{s_2} > \frac{p_2+p_3}{s_1}$, J_2 as well as J_3 are processed by M_1 in the optimal schedule.

If, in a push optimal schedule, J_1 is processed by M_2, then this schedule must be globally optimal. Hence, for each push optimal schedule with $C^P_{\max} > C^*_{\max}$, J_1 is scheduled on M_1, and the critical machine is M_2 as otherwise J_1 can be pushed. If J_2 or J_3 are also scheduled on M_1, then M_2 cannot be critical and the schedule is not push optimal. Therefore, in a push optimal schedule with makespan $C^P_{\max} > C^*_{\max}$ J_1 is processed on M_1 and M_2 processes J_2 and J_3, and as M_2 is the critical machine $C^P_{\max} = \frac{p_2+p_3}{s_2}$. □

Theorem 3. *A push optimal schedule for $Q2\|C_{\max}$ has performance guarantee $\frac{\sqrt{17}+1}{4}$.*

Proof. Consider a push optimal schedule with $C^P_{\max} > \frac{5}{4}C^*_{\max}$. We may assume that such a schedule exists, as otherwise $C^P_{\max}/C^*_{\max} \leq \frac{5}{4} < \frac{\sqrt{17}+1}{4}$. Pushing the smallest job onto the critical machine leads to an unsuccessful push. Hence, there is a largest job in the queue of pending jobs that cannot be pushed onto both machines. Let this job be J_k. Note that this job is at most as large as the smallest job on the critical machine. Because of push optimality, we now have

$$\sum_{J_j \in \mathcal{J}_i : p_j \geq p_k} \frac{p_j}{s_i} + \frac{p_k}{s_i} \geq C^P_{\max}, \qquad i = 1, 2.$$

Thus,

$$(s_1 + s_2)C^P_{\max} \leq \sum_{j: p_j \geq p_k} p_j + p_k \leq (s_1 + s_2)C^*_{\max} + p_k.$$

By the assumption that $C_{\max}^P > \frac{5}{4}C_{\max}^*$, we have that $\sum_j p_j \leq (s_1 + s_2)C_{\max}^* < 4p_k$. Hence, there are at most three large jobs, that is, at least as large as J_k.

If we remove all jobs that are smaller than J_k from the push optimal schedule, then we still have a push optimal schedule and the makespan has not changed, as all the small jobs are scheduled on the non-critical machine. As the optimal makespan of the instance with only the large jobs is at most equal to the optimal makespan of the original instance, the smallest worst-case instance consists of only those, at most three, large jobs.

Any push optimal schedule on an instance with at most two jobs is an optimal schedule, and therefore the worst-case instance for the ratio $C_{\max}^P/C_{\max}^*$ consists of three jobs. Note that if $s_1 = s_2$, then we actually have two identical parallel machines and by Theorem 1 we know that $C_{\max}^P/C_{\max}^* \leq \frac{8}{7}$.

Consider a worst-case instance, and assume w.l.o.g. that $p_1 \geq p_2 \geq p_3$ and that $s_1 > s_2$. By Lemma 2, we know that in this worst-case push optimal schedule J_1 is scheduled on M_1 and M_2 processes J_2 and J_3. We also know that $C_{\max}^* = \max\{\frac{p_1}{s_2}, \frac{p_2+p_3}{s_1}\}$ and $C_{\max}^P = \frac{p_2+p_3}{s_2}$.

By push optimality, we know that $\frac{p_1+p_3}{s_1} \geq \frac{p_2+p_3}{s_2}$, and thus $C_{\max}^P/C_{\max}^*$ is bounded by

$$C_{\max}^P/C_{\max}^* = \min\{\frac{s_1}{s_2}, \frac{p_2+p_3}{p_1}\} \leq \min\{\frac{p_1+p_3}{p_2+p_3}, \frac{p_2+p_3}{p_1}\}.$$

This minimum is maximal when $\frac{p_1+p_3}{p_2+p_3} = \frac{p_2+p_3}{p_1}$. Then $(p_2+p_3)^2 = p_1^2 + p_1 p_3$ and thus

$$C_{\max}^P/C_{\max}^* \leq \frac{\sqrt{p_1^2 + p_1 p_3}}{p_1} = \sqrt{1 + \frac{p_3}{p_1}}. \qquad (7)$$

As $p_2 \geq p_3$, we know that $p_1^2 + p_1 p_3 = (p_2+p_3)^2 \geq 4p_3^2$ and thus $p_1 \geq \frac{\sqrt{17}-1}{2}p_3$. Using this bound in inequality (7) yields

$$C_{\max}^P/C_{\max}^* \leq \sqrt{1 + \frac{2}{\sqrt{17}-1}} = \frac{\sqrt{17}+1}{4}.$$

$\square$

In the following example, we have an instance for $Q2||C_{\max}$ and a push optimal schedule for which $C_{\max}^P = \frac{\sqrt{17}+1}{4}C_{\max}^*$.

Example 4. Consider the following instance with three jobs: $p_1 = \frac{\sqrt{17}-1}{2}$, $p_2 = p_3 = 1$, and $s_1 = \frac{\sqrt{17}+1}{4}$ and $s_2 = 1$. In the optimal schedule M_1 processes J_2 and J_3 and J_1 are scheduled on M_2. The optimal makespan is $C_{\max}^* = \frac{\sqrt{17}-1}{2}$. The schedule in which J_1 is processed by M_1 and J_2 and J_3 are scheduled on M_2 is a push optimal schedule with makespan $C_{\max}^P = 2$. This schedule is depicted in Figure 6; $C_{\max}^P/C_{\max}^* = 2/(\frac{\sqrt{17}-1}{2}) = \frac{4}{\sqrt{17}-1} = \frac{\sqrt{17}+1}{4}$.

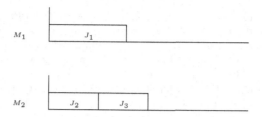

Fig. 6. Push optimal schedule for $Q2\|C_{\max}$

4   Concluding Remarks

In this paper, we have investigated the quality of local optima with respect to three neighborhoods. We have seen that the neighborhood based on variable-depth search gives the best guarantees on the quality of a local optimum for identical and uniform parallel machines. It would be interesting to see whether we can extend the push neighborhood to the unrelated parallel machines environment.

It is also interesting to know the number of iterations that iterative improvement performs to find these local optima. Brucker, Hurink, and Werner [1,2] showed that iterative improvement needs $\mathcal{O}(n^2)$ iterations to find a jump optimal solution for $P2\|C_{\max}$ and $P\|C_{\max}$. The result for $P2\|C_{\max}$ can easily be improved to $\mathcal{O}(n)$ iterations if we always do the best jump. We can show that to find a jump optimal solution for $Q2\|C_{\max}$, iterative improvement performs $\mathcal{O}(n^2)$ iterations and even only $\mathcal{O}(n)$ iterations if always the best jump is done. For $Q\|C_{\max}$ iterative improvement needs $\mathcal{O}(n^2 m)$ iterations. It is still an open question how many iterations iterative improvement needs to find a swap or a push optimal solution. We conjecture that a push optimal solution cannot be found in polynomial time via an iterative improvement procedure.

Acknowledgments. The authors thank Jan Karel Lenstra for his useful comments.

References

1. P. Brucker, J. Hurink, and F. Werner. Improving local search heuristics for some scheduling problems I. *Discrete Applied Mathematics*, 65:97–122, 1996.
2. P. Brucker, J. Hurink, and F. Werner. Improving local search heuristics for some scheduling problems II. *Discrete Applied Mathematics*, 72:47–69, 1997.
3. Y. Cho and S. Sahni. Bounds for list schedules on uniform processors. *SIAM Journal on Computing*, 9:511–522, 1980.
4. U. Feige, M. Karpinski, and M. Langberg. Improved approximation of MAX-CUT on graphs of bounded degree. Technical Report 85215 CS, Institut für Informatik, Universität Bonn, 2000.

5. M.R. Garey and D.S. Johnson. *Computers and Intractability: A Guide to the Theory of NP-Completeness*. Freeman, San Francisco, 1979.

6. M.X. Goemans and D.P. Williamson. Improved approximation algorithms for maximum cut and satisfiability problems using semidefinite programming. *Journal of the ACM*, 42:1115–1145, 1995.

7. R.L. Graham, E.L. Lawler, J.K. Lenstra, and A.H.G. Rinnooy Kan. Optimization and approximation in deterministic sequencing and scheduling: a survey. *Annals of Discrete Mathematics*, 5:287–326, 1979.

8. D.S. Hochbaum and D.B. Shmoys. Using dual approximation algorithms for scheduling problems: theoretical and practical results. *Journal of the ACM*, 34:144–162, 1987.

9. D.S. Hochbaum and D.B. Shmoys. A polynomial approximation scheme for machine scheduling on uniform processors: using the dual approximation approach. *SIAM Journal on Computing*, 17:539–551, 1988.

10. B.W. Kernighan and S. Lin. An efficient heuristic procedure for partitioning graphs. *The Bell System Technical Journal*, 49:291–307, 1970.

11. M.R. Korupolu, C.G. Plaxton, and R. Rajaraman. Analysis of a local search heuristic for facility location problems. Technical Report 98-30, DIMACS, 1998.

12. J.K. Lenstra, D.B. Shmoys, and É. Tardos. Approximation algorithms for scheduling unrelated parallel machines. *Mathematical Programming*, 46:259–271, 1990.

13. S. Lin and B.W. Kernighan. An effective heuristic for the traveling salesman problem. *Operations Research*, 21:489–516, 1973.

Connected Joins in Graphs

András Sebő and Eric Tannier

Laboratoire Leibniz-Imag, 46, Avenue Félix Viallet, 38000 Grenoble, France
{Andras.Sebo, Eric.Tannier}@imag.fr

Abstract. A join in a graph is a set F of edges such that for every circuit C, $|C \cap F| \leq |C \setminus F|$. We study the problem of finding a connected join covering a given subset of vertices of the graph, that is a Steiner tree which is a join at the same time. This turns out to contain the question of finding a T-join of minimum cardinality (or weight) which is, in addition, connected. This last problem is mentioned to be open in a survey of Frank [7], and is motivated by its link to integral packings of T-cuts: if a minimum T-join F is connected, then there exists an integral packing of T-cuts of cardinality $|F|$.

The problems we deal with are closely related to some known NP-complete problems: deciding the existence of a connected T-join; finding the minimum cardinality of a connected T-join; the Steiner tree problem; subgraph isomorphism. We also explore some of these connections.

1 Introduction

Graphs $G = (V, E)$ are always undirected, $V = V(G)$ denotes the vertex-set of G and $E = E(G)$ its edge-set. Paths, trees and circuits will be considered as edge-sets.

A set $F \subseteq E(G)$ is said to be a *join* if for every circuit C in $E(G)$, $|C \cap F| \leq |C \setminus F|$. We say that $F \subseteq E(G)$ *covers* $v \in V(G)$, if $\deg_F(v) > 0$ ($\deg_F(v)$ denotes the *degree* of v in F, that is, the number of edges of F incident to v), and covers $S \subseteq V(G)$ if it covers every $v \in S$. We will denote by $V(F)$ the set of vertices covered by F. We will say that F is *connected* if the graph $(V(F), F)$ is connected, and generally F will also denote the subgraph $(V(F), F)$. The main result is a polynomial algorithm an related theorems for the following problem.
Connected join
INSTANCE. A graph G, a subset $S \subseteq V(G)$.
QUESTION. Is there a connected join covering S?

The connected join problem has interesting connections to well-known objects of combinatorial optimization, one of them is T-joins, and a 'dual' one is integral packings of T-cuts. For basic facts and context related to T-joins, see Frank [7]. If T is an even cardinality subset of vertices of G, $F \subseteq E(G)$ is called a T-*join* if for all $v \in V$, $\deg_F(v)$ is odd exactly when v is in T. A T-join of minimum cardinality is a join according to a result of Guan [10], [6]. Conversely, a join F is a minimum cardinality T-join, where T is defined as the set of vertices incident to an odd number of edges of F. (Indeed, F is not a join if and only if there is

K. Aardal, B. Gerards (Eds.): IPCO 2001, LNCS 2081, pp. 383–395, 2001.
© Springer-Verlag Berlin Heidelberg 2001

a circuit C in G such that $|C \cap F| > |C \setminus F|$, that is, if and only if $F \Delta C$ is a T-join of smaller cardinality). A T-*cut* is a set of edges connecting two subsets of $V(G)$ both intersecting T in an odd number of vertices. A consequence of the characterization of connected joins is the solution the problem of deciding the existence of a connected minimum weight T-join. This problem is an (the unique) open problem stated in Frank's survey paper [7].

Another interesting related problem is minimizing the cardinality (or weight) of a Steiner tree. A Steiner tree is a tree covering a given set of vertices; to require that it is a join looks like a natural substitute for minimizing its cardinality. The main result of this paper tells that *in spite of the NP-completeness of Steiner tree minimization, the 'substitute' is polynomially solvable.*

We need to introduce some basic notions about metric spaces, an important tool in our solution. Let X be any finite set. A function $d : X \times X \longrightarrow \mathbb{R}_+$ is a *metric* if it is symmetric ($d(x,y) = d(y,x)$ for all $x, y \in X$), $d(x,y) = 0$ implies $x = y$ for all $x, y \in X$, and it satisfies the triangle inequality : $d(x,y) \leq d(x,z) + d(z,y)$ for all $x, y, z \in X$. (X, d) is a *metric space*. The *restriction* of d to $S \subseteq X$, denoted by $d|_S$, is the metric on $S \times S$ such that $d|_S(x,y) = d(x,y)$ for all x, y in S. The restriction of a function f to a subset S of its domain will also be denoted by $f|_S$.

If X is the set of vertices of a connected graph G, let $d_G(x,y)$ be the minimum number of edges in a path between x and y, $x, y \in V(G)$. It is easy to see that $(V(G), d_G)$ is a metric space. With an abuse of notation, this metric space $(V(G), d_G)$ will simply be denoted G.

An *isometry* from a metric space (X, d_X) into another (Y, d_Y) is a mapping $f : X \to Y$, such that for all $u, v \in X$, $d_X(u,v) = d_Y(f(u), f(v))$. Given a metric space (X, d), we say that a graph H *realizes* d if there is an isometry from (X, d) into H. (An alternative terminology in the literature: (X, d) is 'embeddable' in H). Note that f is an injection and therefore $|X| = |f(X)| \leq |V(H)|$ and the strict inequality may hold. If there is an isometry between two graphs G and H with $|V(G)| = |V(H)|$, then this isometry is in fact a bijection and the two graphs are *isomorphic*.

A metric d is said to be a *tree metric* if there exists a tree realizing it. (This definition slightly differs from the one used in the literature: all the weights in the realization must be equal to 1. This corresponds to 'tree realizable' metrics in terms of [2]. For instance the metric m which is 1 on every pair of a three element set is not a tree metric, but $2m$ is a tree metric in our sense.) For tree metrics, $d(x,y) + d(x,z) - d(y,z)$ is even for all x, y, z in X : this is twice the length of the path joining x to the y, z-path in any tree that realizes d. We note this length $d_{y,z}^x = \frac{1}{2}(d(x,y) + d(x,z) - d(y,z))$.

Metric spaces and embeddability are treated in details in Deza and Laurent's book [5].

The following observation, is not yet a good characterizaton, just an equivalent reformulation of the connected join problem. However, it provides a key to the solution.

Proposition 1. *Let G be a graph, $S \subseteq V(G)$, and F a subset of edges of G covering S. Then F is a connected join if and only if it is a tree, and $d_F = d_G|_{V(F)}$.*

Proof. The necessity is straightforward: a connected join F covering S is indeed a tree, so we only have to show that the path between any two covered vertices is a minimum path in G. Indeed, if not, then let $a, b \in V(F)$, and let P_1 be an a, b-path in F, and P_2 a strictly smaller a, b-path in G. Then $C = P_1 \triangle P_2$ is the disjoint union of circuits, and $|C \cap F| = |C \cap P_1| > |C \cap P_2| = |C \setminus F|$, contradicting that F is a join.

To prove the sufficiency, let F be a tree in G such that $d_G|_{V(F)} = d_F$. We have to prove that F is a join, that is, F is a minimum T-join, where $T = \{v \in V(G), \deg_F(v) \text{ is odd}\}$. Let J be a minimum T-join in G, with $|F \cap J|$ maximum.

Claim: $J \subseteq F$.

Indeed, suppose not: then there exists vertices $a, b \in V(F)$ so that both F and J contain an a, b-path, P and R respectively, so that the only two common vertices of these paths are a and b. (If $V(J) \subseteq V(F)$, then any $R = \{ab\}$, $ab \in J \setminus F$ will do, otherwise consider a vertex $c \in V(J) \setminus V(F)$. Since $c \notin T$, its degree is even, and one can walk in both directions on even vertices of J until reaching vertices a, b of F. In this way R is determined, and P can be defined as the a, b-path of F.)

Now $C = P \cup R$ is a circuit. By the assumption $d_F = d_G|_{V(F)}$, we have $|C \cap F| = |R| \leq |P| = |C \setminus F|$. It follows that $J \triangle C$ is also a minimum T-join and has bigger intersection with F than J, contradicting the definition of F, and finishing the proof of the claim.

So $J \subseteq F$. As F and J are both T-joins, $F \triangle J = F \setminus J$ is the union of disjoint circuits, which is necessarily empty since F is a tree. This implies $J = F$.   □

We will treat the condition of Proposition 1 in two parts. In order to test whether there exists a join in G covering $S \subseteq V(G)$, we will first test the weaker condition :

(1) *The restriction $d_G|_S$ of d_G to S is a tree metric.*

In other words, there exists a tree A, and an isometry f_S from $(S, d_G|_S)$ into A. In order to check the condition of Proposition 1, choose A to be *inclusionwise minimal*, that is, all paths in A are subpaths of paths between vertices of $f_S(S)$. It is easy to see that (1) is not sufficient for the existence of a connected join: C_6 with every second vertex in S satisfies (1), and $K_{1,3}$ is the realization but there is no connected join covering S in C_6. This shows that the following condition (2) is essential. It is easy to see that (1) and (2) together are already equivalent to the condition of Proposition 1 (see the proof of Theorem 1).

(2) *There is a subgraph F of G and an isometry f from A into F, with $f(f_S(x)) = x$ for all x in S.*

It is easy and well-known to decide the condition (1) in polynomial time (and, more generally to decide if any metric is a tree metric, see section 2). In section 3, it is shown that (2) can also be tested in polynomial time: we construct the subgraph F whose existence is stated in (2), or certify its nonexistence, supposing that a tree satisfying (1) is already known. Since the tree realizing $d_G|_S$ in (1) *turns out to be unique*, applying the proposition to $S := T$, the solution of Frank's connected T-join problem follows (section 4). Then, in section 5, we show a way of constructing Korach's maximum integral packing of T-cuts whenever a connected minimum T-join exists. Finally, in section 6 the relation of the results to some well-known problems of combinatorial optimization is shown.

2   Constructing the Shape of a Join

In this section we provide a polynomial algorithm that either finds a tree realizing a metric d given on a finite set X, or certifies that such a tree does not exist. This problem has many applications in various fields, such as phylogeny [16], data mining, or psychology [4], and has been solved long ago [1], [2], [11] in a sharper sense; since we make only a very simple use of tree metrics, and for the sake of completeness, we include the simple treatment we worked out for our restricted goals.

Start with a copy $\{f(x), x \in X\}$ of the set X, and construct a path of length $d(a,b)$ between $f(a)$ and $f(b)$ by adding new vertices and edges. This path is a tree realizing $d|_{\{a,b\}}$, and $f|_{\{a,b\}}$ is an isometry.

Suppose now that we have a tree A realizing the restriction of d to some proper subset Y of X, and $f|_Y$ is the corresponding isometry. Let $c \in X \setminus Y$. Choose $a, b \in Y$ so that $d_{a,b}^c$ is minimum, and let d be the vertex of $V(A)$ at distance $d_{b,c}^a$ from $f(a)$ on the $f(a), f(b)$-path. Construct a path P from $f(c)$ to d, of length $d_{a,b}^c$.

Then either $A \cup P$ realizes $d|_{Y \cup \{c\}}$, and then $f|_{Y \cup \{c\}}$ is an isometry or there is no tree realizing $d|_{Y \cup \{c\}}$.

Note that an inclusionwise minimal tree realizing d, whenever it exists, is *uniquely determined* by this procedure. The unicity up to isomorphism is actually straightforward to prove from scratch: vertices of degree 1 of the tree can be characterized with the help of distances, and after deleting them we can continue by induction. A variant of this statement will be important in the sequel:

Proposition 2. *Let G be a graph and $S \subseteq V(G)$. An inclusionwise minimal connected join covering S is a subtree of G that realizes $d_G|_S$ and is uniquely determined up to isomorphisms leaving S fixed.*

The proof is easy, along any of the lines explained above (by the unicity of the construction, or from scratch), we leave it to the reader.

Note that tree metrics have been characterized with a certain 'four-point-condition', that ensures the existence of a realizing weighted tree (Buneman [2]):

$$\text{for all } a,b,c,d \in X, d(a,b) + d(c,d) \leq max \begin{cases} d(a,c) + d(b,d) \\ d(a,d) + d(b,c) \end{cases}$$

However, the realizing tree may have noninteger weights. As mentioned before, we also need the following condition : for all a, b, c in X, $d(a,b) + d(a,c) - d(b,c)$ is an even integer. Together with the four-point-condition, this is sufficient for a metric to be a tree metric, as it can be shown with the simple proof above (this is irrelevant for the main results of this paper).

3   Finding a Connected Join

The previous section allows us to decide whether the condition (1) holds or not. In this section we show how to test condition (2) in polynomial time, that is, we construct a subgraph of G covering S, isomorphic to A.

Suppose that condition (1) holds, and let A be the inclusionwise minimal tree that realizes the restriction of d_G to S. Let f_S be the isometry from $(S, d_G|_S)$ into A.

Let us recursively construct a set P_u^i for every $u \in V(A)$ and $i = 0, \ldots, n$, where $n = |V(G)|$. (P_u^i is the set of vertices of G that are not excluded to be in the image of u after step i).

For $u \in f_S(S)$, define $P_u^i := \{f_S^{-1}(u)\}$ for all $i = 0, \ldots, n$. Let now $u \in V(A)$ be arbitrary:

- $P_u^0 := \{v \in V(G) : d_G(v, s) = d_A(u, f_S(s)) \text{ for all } s \in S\}$, and
- for some $i \in 0 \ldots n - 1$, if P_u^i has already been defined for all $u \in V(A)$, then we check for all $v \in P_u^i$, and all $x \in V(A)$ such that $ux \in A$, whether there is an edge from v to some vertex of P_x^i. If not, we define $P_u^{i+1} := P_u^i \setminus \{v\}$, and $P_{u'}^{i+1} := P_{u'}^i$ for all $u' \neq u$.

Clearly, the sets P_u^i ($u \in V(A)$) are pairwise disjoint, and after $k \leq n$ steps, either some P_u^k is empty, or for all $u \in V(A)$ all the vertices of P_u^k are adjacent to some vertex of P_x^k for every $x \in V(A)$ such that $ux \in A$. We then define $P_u := P_u^k$ the *representing set* of $u \in V(A)$ in G. If one of the representing sets is empty, then all are empty, and there is no connected join. If none of the representing sets is empty, then a connected join can be constructed 'greedily', as it is shown in the proof of the following good characterisation theorem for connected joins covering S:

Theorem 1. *Let G be a graph, and $S \subseteq V(G)$. There exists a connected join F covering S if and only if the distances in G between pairs of vertices of S form a tree metric, and the representing sets of the vertices of the realizing tree are nonempty.*

Proof. The necessity is obvious. Conversely, let A be the tree realizing the restriction of d_G to S, P_u the representing set of $u \in A$, and let us construct a tree F which is a subgraph of G covering S isomorphic to A, so that $d_G|_{V(F)} = d_F$. Then by Proposition 1 it follows that F is a connected join.

Suppose that the representing sets are nonempty. We construct the isomorphism f from A into an appropriate subgraph of G by choosing one $f(u) \in P_u$ for all $u \in V(A)$ as follows:

Start with an arbitrary $u \in V(A)$, and let $f(u) \in P_u$ be arbitrary.

- Choose one vertex $f(v)$ from every P_v such that uv is an edge of A; add to F the edges $f(u)f(v)$ for all $uv \in E(A)$. (The property of the representing sets guarantees that these edges exist in G for every $uv \in E(A)$.)

Then we can repeat the same procedure for any vertex $u \in V(A)$ such that $f(u) \in P_u$ has already been defined, but $0 < \deg_F(v_u) < \deg_A(u)$:

- Choose a vertex $f(v) \in P_v$ for all $v \in V(A)$ such that $uv \in E(A)$, and $f(v)$ has not yet been defined; add to F all edges $f(u)f(v)$, $uv \in E(A)$.

Since A is connected, after at most $|V(A)|$ steps, one vertex v_u in every representing set P_u is chosen, and $\deg_F(f(u)) = \deg_A(u)$. (The procedure is similar to usual 'in width label and scan' procedures.)

The result is a tree F, and clearly, f is an isomorphism between A and F. Then for all $x, y \in S$,

$$d_G(x,y) = d_A(f_S(x), f_S(y)) = d_F(f(f_S(x)), f(f_S(y))) = d_F(x,y).$$

Every path in F is a subpath of a path between vertices in S, that is, a subpath of a minimum path in G, hence is a minimum path in G. As a consequence, $d_G(x,y) = d_F(x,y)$ for all $x, y \in V(F)$, as claimed. □

4   Finding a Connected Minimum T-Join

The following problem was mentioned to be open by Frank [7] :

Connected Minimum T-join

INSTANCE. A graph G, a subset $T \subseteq V(G)$ of even cardinality.

QUESTION. Is there a T-join of minimum cardinality, which in addition is connected?

As an immediate corollary of Theorem 1, this problem can be solved in polynomial time:

Corollary 1. *A graph $G = (V, E)$, with a subset T of vertices has no connected minimum T-join if and only if it has no connected join covering T, or it has an inclusionwise minimal connected join F which is not a T-join.*

Proof. The necessity of the condition is obvious. To prove the sufficiency, suppose that F is an inclusionwise minimal connected join, which is not a T-join. We have to prove that there is no connected T-join in G at all !

Indeed, by Proposition 1 $d_F = d_G|_{V(F)}$, moreover, since F is inclusionwise minimal, by Proposition 2, F is uniquely determined up to isomorphisms leaving T fixed. Therefore, in any connected inclusionwise minimal join F' (and note that connected T-joins are like this), the degrees in the vertices of T are the same as in F, and the number of vertices of odd degree in $V(F') \setminus T$ is also the same as the number of such vertices in $V(F) \setminus T$. So F' is a T-join if and only if F is a T-join. □

Note that this corollary can also be proved independently of the theorem: the connected minimum T-join problem can be reduced to finding a connected join, and this reduction does not have to rely on the solution of the problem, it is actually much easier than solving either of the two subproblems.

5   Connected T-Joins and Integral Packings of T-Cuts

In this section we develop an important motivation for the connected join problem, which is the problem of finding integral packings of T-cuts.

We call $\nu(G, T)$ the maximum cardinality of a family of pairwise disjoint T-cuts, and $\tau(G, T)$ the minimum cardinality of a T-join. It is easy to see that for all G and T, $\nu(G, T) \leq \tau(G, T)$ (a T-cut and a T-join always have an edge in commun). The characterization of the equality is the subject of extensive studies in the literature because of its links to integral multiflows. Middendorf and Pfeiffer [15] proved that deciding if equality holds is an NP-complete problem. However, in [12], Korach and Penn proved that for a minimum cardinality T-join F with k connected components, $\nu(G, T) - k + 1 \leq \nu(G, T) \leq |F|$, and therefore, if a minimum T-join is connected, then $\nu(G, T) = \tau(G, T)$.

We will present in this section a simple polynomial algorithm that computes a packing of disjoint T-cuts of cardinality $\tau(G, T)$, or provides a certificate that a connected minimum T-join does not exist.

Such algorithms can be derived from any proof of Korach and Penn's above mentioned theorem [12], [8], [18]. The one we present here in the spirit of the present work is short, elementary, and maybe shows a somewhat simpler way of dealing with vertices of degree 1. The skeleton of the proof is similar to the proof in [17] of $\nu(G, T) = \tau(G, T)$ for bipartite graphs.

First we need one more definition. For a graph G, and $T \subseteq V(G)$, the T-contraction of $xy \in E(G)$ is the operation of contracting xy in the graph, and redefining T with $T' := T \setminus \{x, y\} \cup \{v_{xy}\}$ if exactly one of x and y are in T, and $T' := T \setminus \{x, y\}$ otherwise. In this paper we will apply this operation uniquely to the T-contraction of (all edges of) the star of a vertex, that is to identifying a vertex with all its neighbors, (and redefining T depending on the number of vertices of T in the identified set).

Lemma 1 can be viewed as one iteration of a polynomial algorithm finding :
- either a packing of disjoint T-cuts of cardinality $\tau(G, T)$
- or a certificate that a minimum connected join does not exist (when one of the conditions in lemma 1 is violated).

Lemma 1. *Let G be a graph, T a subset of its vertices. Let a and b be elements of T such that $d_G(a, b)$ is the maximum over all distances between pairs of vertices of T. If there exists a connected minimum T-join, then:*

(i)   *a (and also b) has degree 1 in any connected minimum T-join in G; in particular a and b are in T.*

(ii)   *there exists at least one neighbor a' of a, such that $d_G(a', x) = d_G(a, x) - 1$ for all $x \in T$, and at most one of such neighbors is in T.*

(iii) *$\delta(a)$ is a T-cut and, if G', T' are obtained by T-contracting $\delta(a)$, then G' has a connected minimum T'-join of cardinality $\tau(G, T) - 1$.*

Proof. Let F be a connected minimum T-join, and let P be the a, b-path of F. By Proposition 1 P is also a shortest path in G, and by definition, the longest among all shortest paths between vertices of T.

Proof of (i): If indirectly, F contains an edge incident to a which is not in P, then starting on that edge from a until reaching a vertex c, $deg_F(c) = 1$, the obtained a, c-path $R \subseteq F$ is vertex-disjoint of P, because F is a tree. But then $P \cup R$ is also a path, and is also a shortest path of G, by Proposition 1. Because of $deg_F(c) = 1$ we have $c \in T$, and therefore $P \cup R$ is a path between two vertices of T, and longer than P, a contradiction with the choice of a, b.

Proof of (ii): Let a' be the neighbor of a on P. Again, by Proposition 1, $d_G(a', b) = d_F(a', b) = d_F(a, b) - 1$, and if a had another neighbor in G such that $a'' \in T$, then a'' is also a vertex of F, and $d_G(a, a'') = d_F(a, a'') = 1$, that is $aa'' \in F$, contradicting (i).

Proof of (iii): Let us prove that $F \setminus \{aa'\}$ is a connected minimum T'-join in G'. By (i), it is connected. By Proposition 1 it is sufficient to prove that any path $R \subseteq P$ between two vertices $x, y \in V(F)$, $deg_F(x) = deg_F(y) = 1$, is also a shortest path in G'. Since we know that in R is a shortest path in G, a shorter x, y-path R' of G' must contain the new vertex a^* of G'. But $d_{G'}(x, a^*) \geq d_G(x, a) - 1 = d_G(x, a')$ by (ii), and therefore

$$|R| = d_G(x, y) \leq d_G(x, a') + d_G(a', y) \leq d_{G'}(x, a^*) + d_{G'}(a^*, y) = |R'|,$$

a contradiction. □

Clearly, the sets $\delta(a)$ found by successive application of the lemma are disjoint T-cuts, and each intersects any minimum T-join in one element. Hence, if there exists a connected T-join, we arrive at a packing of T-cuts of cardinality $\tau(G, T)$, as desired. Note that this algorithm neither implies nor is implied by the algorithm developed in the preceding sections for deciding the existence of a connected minimum T-join; indeed, in case a packing of disjoint T-cuts of cardinality $\tau(G, T)$ is found, no connected T-join is exhibited. We do not see how to decide the existence of a connected T-join in this way.

6   Concluding Remarks

6.1   Complexity

The complexity of the algorithm based on our results finds a connected join covering a subset S of vertices of a graph on n vertices (or certifies its nonexistence), in time at most $O(n^3)$.

Indeed, in the first step (constructing A from $d_G|_S$), at each step of a recurrence, we have to compute $d_{a,b}^c$ and $d_{c,b}^a$ for all a, b in $Y \subseteq S$; it takes $O(|Y|^2)$ operations. Then we construct a path (in $O(|Y|)$ time) and check whether the tree realizes $d_G|_Y$, that is we compute $d(c, a)$ for all $a \in Y$. The entire procedure takes $O(\sum_{i=1...n} i^2) = O(n^3)$ time. Note that it is possible to determine if a metric satisfies the four-point-condition (which is equivalent to our first step) in time $O(n^2 \log n)$ with the help of ultrametrics as it is proved in [1], that could improve our computation time.

The second step (embedding A in the graph G) has also a $O(n^3)$ time complexity: constructing a P_u^0 requires the computation of all the distances between

u and every $x \in S$, then for all $u \in V(A)$, it takes at most $O(n^2)$ operations. Then computing P_u^i from P_u^{i-1} may need $O(nm)$ checks for a single $u \in V(A)$, and this operation is achieved at most n times. In conclusion, the entire process takes at most a time which is a function $O(n^3)$.

6.2 Weighted Case

As we mentioned in the introduction, the results of this paper, including poly-nomial algorithms, work for the weighted case too. As far as the results are concerned they can be immediately derived by subdivivision of the edges of a weighted graph G, w (subdivide every edge $e \in E(G)$ $w(e)$ times, or contract it if $w(e) = 0$, and apply Theorem 1). But this provides only an algorithm that depends linearly on the weights (instead of being bounded by a polynomial of the logarithm, or not depending on the numbers at all). Anyway it is possible to achieve a strongly polynomial algorithm for the weighted case. The problem can be reformulated as follows (a join is then defined as a set F of edges such that for all circuit C, $w(C \cap F) \le w(C \setminus F)$):

Weighted Connected Join

INSTANCE. A graph G, an nonnegative integer weight function $w : E(G) \longrightarrow \mathbb{N}$, a subset S of vertices, an integer k.

QUESTION. Is there a connected join F covering S such that $w(F) \le k$?

In particular, if $|S|$ is even, the question could concern the existence of a S-join of minimum weight which is connected. We reformulate Theorem 1 (and Corollary 1) and the principles of the construction in the following way.

Let the distances between pairs of vertices in G be the minimum weights of paths between the vertices. Then it is possible to restate the two steps of the algorithm.

First, construct a weighted tree realizing a metric d on a set X.

Start with a, b in X. Construct the images of a and b by an isometry f, and an edge between $f(a)$ and $f(b)$ with a weight $w(ab) = d(a, b)$.

Then suppose a weighted tree A, w realizes the restriction of d to $Y \subset X$. Choose $c \in X \setminus Y$ such that $d_{b,c}^a$ is minimum among all a, b in Y. Subdivide the edge e incident to $f(a)$ (it is unique, because of the choice of a), in two edges e_1, e_2, of respective weight $d_{b,c}^a$ and $w(e) - d_{b,c}^a$. Then join $f(c)$ and the new vertex incident to both e_1 and e_2 by an edge of weight $d_{a,b}^c$. Check that the result is a tree realizing $d_{Y \cup \{c\}}$ and start again. Remark that in A, there is no vertex of degree two, and therefore a minimum connected join in G is homeomorphic to A (if it is incusionwise minimal).

Then, construct the representing sets P_u of every vertex $u \in V(A)$: first, $x \in V(G)$ is in P_u if $d_{G,w}(x, s) = d_A(x, s)$ for all $t \in S$. Check for every $x \in P_u$, and $v \in V(A)$ such that uv is an edge of $E(A)$, that there is a path of weight $w(uv)$ between x and some $y \in P_v$. If not, delete x from P_u, and start again. Finally Theorem 1 can be reformulated for the weighted case as follows.

Theorem 2. *Let G be a graph, $w : E(G) \longrightarrow \mathbb{N}$, $S \subseteq V(G)$. There exists a minimum weight connected join F covering S if and only if the distances in G*

with respect to w form a tree metric, and the representing sets of the vertices of the realizing tree are nonempty.

6.3   Related Problems

There are several problems which are closely related to the connected join problem, which have been proved to be NP-complete. We provide some simple NP-completeness proofs and show their relation to our problem.

Connected T-join. This is a NP-complete variant of the connected minimum T-join problem : we don't require the T-join to have minimum cardinality.
Connected T-join
INSTANCE. A graph G, a subset $T \subseteq V(G)$, $|T|$ even.
QUESTION. Is there a connected T-join?

   This problem is mentioned by Frank [7], and NP-competeness is said to be solved by a proof of Pulleyblank, with a reduction of the Hamilonian circuit problem, for 3-regular graphs, probably similar to the following :

Proof. Let G be a 3-regular graph. Let us construct G' in the following way : let f be a bijection between $V(G)$ and a new set U of the same cardinality. Let $V(G') = V(G) \cup U$, and $E(G') = E(G) \cup \{xf(x)$, for all $x \in V(G)\}$. Let $T = V(G')$. There exists a Hamiltonian circuit in G if and only if there exists a connected T-join in G'. Indeed, let $F \in E(G')$ be a connected T-join, then all degrees of vertices of $V(G)$ in F are equal to 3 (they are odd, less than 4 because G is 3-regular, and greater than 2 because F is connected). Then $F \cap E(G)$ is a connected subgraph of G such that every vertex has degree 2, in consequence a Hamiltonian circuit. The 'only if' part of the proof follows in the opposite way.

   Note that in consequence, it is NP-hard to find a connected T-join, with a minimum number of edges.

Steiner Tree. Let G be a graph and $w : E(G) \longrightarrow \mathbb{N}$ a weight function, and T a subset of $V(G)$. We call a *Steiner tree* for T a tree in G covering T. The *optimal Steiner tree* is a Steiner tree of minimum weight. The following NP-complete problem has been much studied.
Steiner Tree
INSTANCE. A graph G, and $T \subseteq V(G)$, a weight function $w : E(G) \longrightarrow \mathbb{N}$, an integer k.
QUESTION. Is there a tree F in G covering T, such that $w(F) \leq k$?

   For NP-completeness proofs, exact algorithms, polynomial heuristics, see [13], [20]. The connection to joins is that a connected join turns out to be an optimal Steiner tree.

Proposition 3. *Let G be a graph, $T \subseteq E(G)$ and $w : E(G) \longrightarrow \mathbb{N}$. If F is a w-minimum connected join covering T, then F is an optimal Steiner tree for the terminal set T.*

Proof. If $|T| = 2$, then the statement is obvious. Suppose now $|T| > 2$. Let A be a Steiner tree covering T, and F a connected join covering T, such that $w(F) > w(A)$. Choose A and F so that $|T|$ is minimum.

For any $t \in T$, call A_t (resp. F_t) the subset of edges of A (resp. F) not belonging to any path of A (resp. F) between two vertices of $T \setminus \{t\}$. For all $t \in T$, $F \setminus F_t$ is a minimum weight connected join covering $T \setminus \{t\}$, and $A \setminus A_t$ is a Steiner tree for $T \setminus \{t\}$. Then $w(F \setminus F_t) \leq w(A \setminus A_t)$, by the minimality of the example. Consequently, $w(F_t) > w(A_t)$ for all $t \in T$.

Now it is easy to see that there exists two vertices t_1, t_2 such that the t_1, t_2-path in A is exactly $A_{t_1} \cup A_{t_2}$ (For example, fix $x \in V(A)$, and let t_1, t_2 maximize $d_A(x, t_1) + d_A(x, t_2) - d_A(t_1, t_2)$). Then

$$d_A(t_1, t_2) = w(A_{t_1}) + w(A_{t_2}) < w(F_{t_1}) + w(F_{t_2}) \leq d_F(t_1, t_2)$$

This contradicts (the trivial part of) Proposition 1.   □

Approximations, λ-joins and tree λ-spanners. The links to the Steiner tree problem lead us to the study of a relaxation of the connected join problem. It is possible for example to relax the property of Proposition 1: a λ-*tree* covering S ($\lambda \in \mathbb{R}_+$) is a tree $F \subseteq E(G)$ covering S, with the property that $d_F(x, y) \leq \lambda \times d_G(x, y)$ for all x, y in $V(F)$.

In fact, a λ-tree F is a Steiner tree, which is a most λ times bigger than the optimal Steiner tree. (Indeed, if w is the weight function on the edges of G, define $w'(e) := 1/\lambda w(e)$ for all $e \in F$, and $w'(e) = w(e)$ otherwise. F is a connected join according to w', and therefore $1/\lambda w(F) = w'(F) \leq w'(S) \leq w(S)$, where S is an optimal Steiner tree.)

We could not generalize our arguments to λ-trees, and this is not surprising in the view of the work of Cai and Corneil's on tree λ-spanners [3]: a *tree λ-spanner* in a weighted graph G, is a spanning tree F (that means 'covering $V(G)$') such that $d_F(x, y) \leq \lambda \times d_G(x, y)$ for all $x, y \in V(G)$.

Note that a tree λ-spanner is a connected λ-tree covering $S = V(G)$, a special case of our problem. In consequence, as a corollary of our work, it is polynomially solved when $\lambda = 1$ (it was also solved in [3] in a simpler way); but when $\lambda > 1$, it is NP-complete to determine whether a graph contains a tree λ-spanner, and that proves NP-completeness of the λ-tree. Note that tree λ-spanners become tractable for $\lambda \leq 2$ if the graph is unweighted, according to [3].

Isometric Subgraphs. We put the algorithm of Section 3 to the more general context of subgraph isomorphisms, showing also the limits of our method. It is well-known, that it is NP-complete to decide whether a graph G given as input contains a subgraph isomorphic to another given graph H, even if H is a tree [9]. Isomorphisms are exactly the isometries of the distance functions of unweighted graphs: indeed, xy is an edge of a graph, if and only if the distance of x and y in the graph is 1. The problem we have been studying in this paper (and mainly in Section 3) has the specificity, that the distances in the isomorphic subgraph must be the same as in the whole graph.

Subgraph Isometry
INSTANCE. Two graphs G, and A.
QUESTION. Is there an isometry from A into G?

In other words, does G contain a subgraph H isomorphic to A so that the distances in H are the same as in G. We will call H an A-*isometric* subgraph of G.

If A is a clique of size k, then an A-isometric subgraph of G is still a clique of size k: the subgraph isometry problem is in consequence NP-complete. However, the problem is not the same as subgraph isomorphism: indeed, suppose for instance that A is a path of length $|V(G)|$. The problem of deciding whether G contains a subgraph isomorphic to A is the Hamiltonian path problem, whereas the A-isometric subgraph problem is trivial in this case: G has such a subgraph if and only if G *is* a path of length n.

Another specificity of the problems we had to solve here is that the images of some vertices of A are fixed in advance, moreover, that the distances from these prefixed vertices uniquely determine every vertex of the graph. Indeed, the polynomial running time of the algorithm in Section 3 is based on the pairwise disjointness of the representing sets.

We say that $R \subseteq V(A)$ *determines* A if for all $x \in V(A)$, there exists $u, v \in R$, such that $d_A(u, x) \neq d_A(v, x)$.

Determined Subgraph Isometry
INSTANCE: Two graphs G and A, a set $R \subseteq V(A)$ that determines A and an isometry f from $(R, d_A|_R)$ into G.
QUESTION: Is there an extension of f, which is an isometry from A into G?

In other words, is there an isomorphism $\bar{f}$ between A and a subgraph H of G so that $\bar{f}(r) = f(r)$ for all $r \in R$, and $d_H(x, y) = d_G(x, y)$ for all $x, y \in V(H)$.

The algorithm in Section 3 can now be copied to provide a polynomial algorithm for this more general problem. The only fact to notice is that since R determines A, the representing sets are disjoint again.

In other words, the main point of this work expressed in Section 3 is working because the vertices of degree 1 of a tree determine it (and at least all the vertices of degree 1 are fixed in advance). In fact all the vertices which degree is at least 2 and some of the vertices of degree 1 can be deleted from the determining set (one from among each vertex in an equivalence class of vertices having degree one and the same neighbor). It is easy to see that in this way we get all the minimum determining sets of a tree.

While the smallest set that determines a clique has $n - 1$ elements, some other interesting classes of graphs are determined by a subset of constant size. For instance paths are determined by only one vertex, and circuits by two. For classes of graphs that can be determined by a constant number of vertices the Subgraph Isometry problem can also be solved in polynomial time, because one can then define R to be a set of constant size that determines A, and one can check the existence of a subgraph isometry for all possible choices for an image of R. (There is a constant number of them.)

This raises the problem of finding the smallest determining set of a graph. Is this problem polynomially solvable or NP-hard? For arbitrary k can the structure of graphs that can be determined by at most k vertices be described? For dense graphs this problem is close to subgraph isomorphism.

Another generalization of the problem studied in this paper is the minimization of the number of components of a minimum T-join (or in a join). We do not know the complexity status of this problem.

References

1. Bandelt Hans-Jürgen, "Recognition of tree metrics", *SIAM Journal of Discrete Math.*, **3** (1990), 1-6.
2. Buneman Peter, "A note on the metric properties of trees", *Journal of Combinatorial Theory* (B), **17** (1974), 48-50.
3. Cai Leizhen and Corneil Derek, "Tree Spanners", *SIAM journal of Discrete Mathematics*, **8** (1995), 359-378.
4. Cunningham James, "Free Trees and Bidirectional Trees as Representations of Psychological distance", *Journal of mathematical psychology*, **17** (1978), 165-188.
5. Deza Michel Marie and Laurent Monique *Geometry of cuts and metrics*, Springer, 1991.
6. Edmonds Jack and Johnson Ellis, "Matching, Euler Tours and the Chinese Postman", *Mathematical Programming*, **5** (1973), 88-124.
7. Frank András, "A Survey on T-joins, T-cuts, and Conservative Weightings", *Combinatorics, Paul Erdős is eighty*, **2** (1996), 213-252.
8. Frank András and Szigeti Zoltán, "On packing T-cuts", *Journal of Combinatorial Theory* (B), **61** (1994), 263-271.
9. Garey Michael and Johnson David, *Computers and intractability, a Guide to the Theory of NP-Completeness*, Freeman, 1979.
10. Guan Mei Gu, "Graphic programming using odd and even points", *Chinese Journal of Mathematics*, **1** (1962), 273-277.
11. Hakimi S.L. and Yau S.S., "Distance matrix of a graph and its realizability", *Quarterly Applied Mathematics*, **22** (1964), 305-317.
12. Korach E. and Penn M., "Tight integral duality gap in the Chinese postman problem", *Mathematical Programming*, **55** (1992), 183-191.
13. Korte Bernhard, Prömel Hans Jürgen and Steger Angelika, "Steiner trees and VLSI-layout", *Paths, Flows and VLSI-layouts*, Korte, Lovász, Prömel, Schrijver, eds, Springer-Verlag, 1980.
14. Lovász László and Plummer M. D., *Matching Theory*, North-Holland, Amsterdam, 1986.
15. Middendorf M. and Pfeiffer F., "On the complexity of the edge-disjoint path problem", *Combinatorica*, **8** (1998), 103-116.
16. Penny David, Foulds and Hendy, "Testing the theory of evolution by comparing phylognenetic trees", *Nature*, **297** (1982), 197-200.
17. Sebő András, "A quick proof of Seymour's Theorem on t-joins", *Discrete Mathematics*, **64** (1987), 101-103.
18. Sebő András, "Undirected distances and the postman-structure of graphs", *Journal of Combinatorial Theory/B*, **49** (1990), No 1.
19. Seymour Paul, "On odd cuts and planar multicommodity flows", *Proc. London Math. Soc.*, **42** (1981), 178-192.
20. Winter Pawel, "Steiner Problem in Networks : A Survey", *Networks*, **17** (1987), 129-167.

Two NP-Hardness Results for Preemptive Minsum Scheduling of Unrelated Parallel Machines

René Sitters

Department of Mathematics and Computer Science
Technische Universiteit Eindhoven
P.O.Box 513, 5600 MB Eindhoven, The Netherlands
r.a.sitters@tue.nl

Abstract. We show that the problems of minimizing total completion time and of minimizing the number of late jobs on unrelated parallel machines, when preemption is allowed, are both NP-hard in the strong sense. The former result settles a long-standing open question.

1 Introduction

Suppose that m machines M_i $(i = 1, \ldots, m)$ have to process n jobs J_j $(j = 1, \ldots, n)$. Each job can be processed on any of the machines but it can only be worked on by one machine at a time. Each machine can process at most one job at a time. The time it takes to process job J_j completely on machine M_i is given by a positive integer p_{ij}. Preemption is allowed, i.e., we may interrupt the processing of a job, and continue it later on the same, or at any time on another machine. We are interested in finding a schedule for which a certain optimality criterion is minimized.

In this paper we consider two optimality criteria: the sum of completion times $\sum_{j=1}^n C_j$, and the sum of unit penalties $\sum_{j=1}^n U_j$. The completion time C_j of a job J_j is the last moment in time that it is processed. If a job J_j has a given due date d_j, that is, the moment in time by which it should be completed, then we say the job is *late* if $d_j < C_j$. The unit penalty U_j is 1 if job J_j is late, and 0 otherwise.

With the two optimality criteria we have defined two scheduling problems. In the notation introduced by Graham et al. [8] they are denoted by $R|pmtn|\sum U_j$ and $R|pmtn|\sum C_j$. The 'R' indicates we have unrelated parallel machines, i.e., no relation between the mn processing times p_{ij} is presumed. The number of machines m is defined as part of the problem instance. If the number of machines m is fixed, the notation 'Rm' is used. The acronym '$pmtn$' indicates that preemption is allowed. The third field indicates which optimality criterion is used.

Lawler [13] proved that the problem $P|pmtn|\sum U_j$, i.e., the problem of minimizing the number of late jobs, is already NP-hard in the ordinary sense on identical parallel machines . We say machines are *identical* if $p_{i1} = p_{i2} = \ldots = p_{im}$

K. Aardal, B. Gerards (Eds.): IPCO 2001, LNCS 2081, pp. 396–405, 2001.

for all i, $1 \le i \le n$. Hence $R|pmtn| \sum U_j$ is NP-hard in the ordinary sense as well. Du and Leung [4] show that the problem $R|pmtn, r_j| \sum U_j$, where the given jobs have release dates, is strongly NP-hard.

We show that $R|pmtn| \sum U_j$ is NP-hard in the strong sense.

Little is known about the preemptive minimization of the sum of completion times. If no preemption is allowed ($R|| \sum C_j$), an optimal schedule can be found in polynomial time by solving a weighted matching problem (Horn [9], Bruno et al. [3]). In the case of uniform parallel machines, Gonzalez [7] shows that an optimal preemptive schedule can be found in polynomial time. We say that the machines are uniform if $p_{ij} = p_j/s_i$ for given processing requirement p_j of job J_j, and speed s_i of machine M_i. The problem $P2|pmtn, r_j| \sum C_j$, is NP-hard in the ordinary sense [5], and the problem $P2|chains, pmtn| \sum C_j$, where chainlike precedence relations between the jobs are added, is NP-hard in the strong sense [6].

We show that the problem $R|pmtn| \sum C_j$ is NP-hard in the strong sense.

For almost all scheduling problems the preemptive version of the problem is not harder to solve than the non-preemptive version. For at least two scheduling problems this emperical law does not hold true.

Brucker, Svetlana, Kravchenko, and Sotskov [2] showed that the preemptive job shop scheduling problem with two machines and three jobs ($J2|n = 3, pmtn|C_{max}$) is NP-hard in the ordinary sense. However, Kravchenko and Sotskov [11] show that the non-preemptive version can be solved in $O(r^4)$ time, where r is the maximum number of operations of a job.

The other exception is that of finding an optimal preemptive schedule for equal length jobs on identical parallel machines. The optimality criterion is the sum of weighted lateness penalties. This problem ($P|p_j = p, pmtn| \sum w_j U_j$) was proven to be NP-hard in the ordinary sense by Brucker, Svetlana, and Kravchenko [1]. In the same paper they give a $O(n \log n)$ time algorithm for the non-preemptive version. Recently they even proved strong NP-hardness for the preemptive problem.

$R|pmtn| \sum C_j$ is the third problem type on this short list.

2   Minimizing the Number of Late Jobs

$R|pmtn| \sum U_j$

Instance: A number α, a set $\{M_1, M_2, \ldots, M_m\}$ of m unrelated machines, a set $\{J_1, J_2, \ldots, J_n\}$ of n independent jobs, a set $\{p_{ij} \mid i = 1 \ldots m, j = 1 \ldots n\}$ where p_{ij} is the processing time of job J_j on machine M_i, and a due date d_j for each job J_j.

Question: Does there exist a preemptive schedule for which $\sum_{j=1}^{n} U_j \le \alpha$?

It is not obvious this problem is in the class NP since we have to exclude the possibility that there is a schedule with a superpolynomially number of preemptions that has a strictly smaller objective value than any schedule with a polynomial number of preemptions. Lawler and Labetoulle [12] show that for any monotone, non-decreasing objective function $f(C_1, C_2, \ldots, C_n)$, an optimal schedule can be constructed by solving a linear program, if the completion times of the jobs in an optimal schedule are given. The number of preemptions for this schedule is bounded by a function of order $O(m^2 n)$. Verifying the feasibility of a schedule with $O(m^2 n)$ preemptions requires polynomial time, and so $R|pmtn|\sum U_j$ is in NP.

We present a reduction from the 3-Dimensional Matching problem to $R|pmtn|\sum U_j$. The former was proven to be NP-complete by Karp [10].

3-Dimensional Matching (3DM)

Instance: Three sets $U = \{u_1, \ldots, u_n\}$, $V = \{v_1, \ldots, v_n\}$, and $W = \{w_1, \ldots, w_n\}$, and a subset $S \subset U \times V \times W$ of size m.
Question: Does S contain a perfect matching, that is, a subset S' of cardinality n that covers every element in $U \cup V \cup W$?

As a preliminary we define one set of jobs and three sets of machines. For each element u_i of the set U, we define one machine which we denote by U_i. The set of these n machines is denoted by U as well. In the same way we define the sets of machines V and W. We use the following notation for the set S: $S = \{(u_{\alpha_j}, v_{\beta_j}, w_{\gamma_j})| \ j = 1, \ldots, m\}$. For each element $s_j = (u_{\alpha_j}, v_{\beta_j}, w_{\gamma_j})$ of S we define one job which we denote by J_j. The set of these m jobs is denoted by J. The processing time of a job is small on the three machines that correspond to the related triple, and is large on all other machines. To be specific: let $s_j = (u_{\alpha_j}, v_{\beta_j}, w_{\gamma_j})$ be an element of S, then the processing time of job J_j is $3p$ on machine U_{α_j}, $\frac{3}{2}p$ on machine V_{β_j}, and p on machine W_{γ_j}, where $p \in \mathbb{R}$, $p \geq 2$. The processing time is K on any of the other $3n - 3$ machines, where $K \in \mathbb{R}$, $K \geq 6p$. K and p will be chosen appropriately.

We use the following terminology. We say that a machine is a 'slow' machine for the job J_j if the processing time of job J_j is K on that machine. In any other case we say that the machine is 'fast' for the job.

Lemma 1. *Given an instance I of 3-Dimensional Matching with $S \subset U \times V \times W$ and $|S| = m$, $|U| = |V| = |W| = n$, define the sets of machines U, V, and W and the set of jobs J as described above. If we add the restriction that none of the V-machines can process a job before time $t = 1$ and none of the W-machines can be used for processing before time $t = 2$, then for every preemptive schedule the following holds:*

(i) $C_j \geq p + 1$ for any job $J_j \in J$,

(ii) if there are n jobs for which $C_j < (p + 1) + \frac{1}{6}$, then I contains a perfect matching.

Proof. (i) Schedule job J_j on machine U_{α_j} from $t = 0$ to $t = 1$, on machine V_{β_j} from $t = 1$ to $t = 2$, and on machine W_{γ_j} from time $t = 2$ onwards. Scheduled in this way $C_j = p + 1$. Any other schedule will give a strictly larger completion time.

(ii) Call the schedule of job J_j as described in (i), the optimal schedule of this job. If I does not contain a perfect matching, it is impossible to schedule a set of n jobs such that each job gets its optimal schedule. At least one job of any set of n jobs must diverge from its optimal schedule for a total time of at least 0.5. This will delay the best possible completion time for this job by at least $\frac{1}{6}$.   □

We will use this construction with the gradual schedule of the J-jobs to prove Theorem 1. In the lemma we made the restriction that sets of machines can only be used from some point in time onwards; in the NP-hardness proof below we will have to find a way to enforce this.

Theorem 1. $R|pmtn| \sum U_j$ *is strongly NP-hard.*

Proof. The problem is in the class NP as we already showed. To complete the proof we reduce the 3-Dimensional Matching problem to $R|pmtn| \sum U_j$. Given an instance of 3-DM of size m and n, we will define a scheduling instance containing a set of $3n$ machines and $2n + m$ jobs, and prove that a 3-dimensional matching exists if and only if there exists a preemptive schedule for which the number of late jobs does not exceed a certain value.

Let I be an instance of 3-DM with $S \subset U \times V \times W$ and $|S| = m$, $|U| = |V| = |W| = n$. Define the sets of machines U, V, and W as before. For each machine of the set V we introduce one job with processing time 1 on that specific machine, and with processing time K on any of the other $3n - 1$ machines. The due date is 1 for all these jobs. For each machine of W we define one job with processing time 2 on that specific machine, and with processing time K on any of the other $3n - 1$ machines. The due date is 2 for all these jobs. We denote the set of these $2n$ jobs as A. We define the set of jobs J related to S, with their processing times as before. The due dates of these jobs are $p + 1$. We choose the value $p = 3n + 3$.

We claim that for any schedule, the number of late jobs is less than or equal to $m - n$, if and only if a perfect matching exists. Notice that this is the same as claiming that the number of early jobs is at least $3n$ if and only if a perfect matching exists.

If there exists a perfect matching, then it is possible to schedule the jobs such that $3n$ jobs are early: Schedule the $2n$ A-jobs such that they are early (there is only one way to do this), and schedule the n J-jobs that correspond to the elements in the 3-dimensional matching, in the way described in the proof of Lemma 1.

Observe that it is impossible to have more than n early J-jobs. For a J-job to be early, it must be scheduled on its fast W-machine for a time of at least $p - 2$. So if there are at least $n + 1$ early J-jobs, then the total processing time required on the W-machines is at least $(n + 1)(p - 2) = (n + 1)(3n + 1)$. For any n this is strictly larger than $n(3n + 4)$, which is the total available processing time on the W-machines before the due date, $t = 3n + 1$.

We conclude that if at least $3n$ jobs are early, then these jobs are the $2n$ A-jobs and exactly n of the J-jobs. From Lemma 1 it follows that in this case a perfect matching exists.   □

3   Minimizing the Sum of Completion Times

$R|pmtn|\sum C_j$

Instance: A number α, a set $\{M_1, M_2, \ldots, M_m\}$ of m unrelated machines, a set $\{J_1, J_2, \ldots, J_n\}$ of n independent jobs, and a set $\{p_{ij}|i = 1 \ldots m, j = 1 \ldots n\}$ where p_{ij} is the processing time of job J_j on machine M_i.
Question: Does there exist a preemptive schedule for which $\sum_{j=1}^{n} C_j \leq \alpha$?

Theorem 2. $R|pmtn|\sum C_j$ *is strongly NP-hard.*

Proof. The membership of the class NP follows from the same arguments as given for the problem $R|pmtn|\sum U_j$.
To complete the proof we reduce 3-Dimensional Matching to $R|pmtn|\sum C_j$. Given an instance of the 3-DM problem we define an instance of a scheduling problem and prove that a perfect matching exists if and only if there exists a preemptive schedule for which the sum of completion times does not exceed a certain value.

Let I be an instance of 3-DM with $S \subset U \times V \times W$ and $|S| = m, |U| = |V| = |W| = n$. The set of machines consists of the sets U, V and W as defined before and, additionally, one machine which we denote by Z.

The set of jobs consists of three different sets. First we define the set of jobs J related to S, and use the same processing times for these jobs as defined before. The processing time on the Z-machine is p for any job from J. The value of p is set to $p = 2$. The value of K will be specified later.

The second set of jobs is the set A which contains many jobs with a small processing time. For each V-machine we define M A-jobs with processing time $\frac{1}{M}$ on that specific machine and processing time K on any other machine. The value of M will be specified later. For each W-machine we define $2M$ A-jobs with processing time $\frac{1}{M}$ on that specific machine and with processing time K on any other machine. For the Z-machine we define $3M$ A-jobs with processing time $\frac{1}{M}$ on the Z-machine and processing time K on any of the other $3n$ machines. The total number of jobs in A is $(3n + 3)M$. The A-jobs are meant to keep the V-machines busy until time 1, the W-machines until time 2, and the Z-machines until time 3.

The third set of jobs is B. For each U-machine we define $\frac{1}{3}(m - n)$ B-jobs with processing time $2m + 4$ on that specific machine, and processing time K on any other machine. (Without loss of generality we may assume that $\frac{1}{3}(m - n)$ is integer.) For each V-machine we define $\frac{2}{3}(m - n)$ B-jobs with processing time $2m + 4$ on that specific machine and processing time K on any other machine. For each W-machine we define $(m - n)$ B-jobs with processing time $2m + 4$ on that specific machine and processing time K on any other machine. The B-jobs are introduced to ensure that, in an optimal schedule, a limited part of the J-jobs is scheduled on the U-, V-, and W-machines.

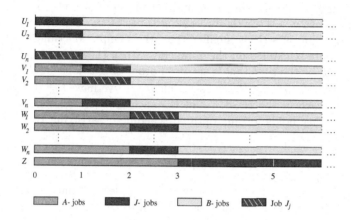

Fig. 1. Sketch of the schedule σ_{3DM}.

If a perfect matching exists then the jobs can be scheduled as shown in Fig. 1. All A-jobs are scheduled on their fast machines such that their sum of completion times is minimized. There is only one way to do this. The n jobs from J that correspond to the perfect matching are scheduled as in the proof of Lemma 1. The completion time of these jobs is 3. All other J-jobs are scheduled after the A-jobs on the Z-machine. Each B-job is scheduled on its (unique) fast machine. The B-jobs are placed directly after the other jobs. This schedule is denoted by σ_{3DM}. The sum of completion times of σ_{3DM} is denoted by $C_{\sigma_{3DM}}$. The value of $C_{\sigma_{3DM}}$ is clearly a polynomial in m,n, and M. The expression is omitted here.

We use the notation $C_{\widetilde{\sigma}}(\widetilde{J})$ for the sum of completion times of the jobs in a set $\widetilde{J}$ in a schedule $\widetilde{\sigma}$, and $C_{\widetilde{\sigma}}$ for the sum of completion times of all jobs in schedule $\widetilde{\sigma}$.

We now show that a perfect matching exists if there exists a preemptive schedule σ with $C_{\sigma} \leq C_{\sigma_{3DM}}$. Assume that I does not contain a perfect matching and that σ is an optimal preemptive schedule for which $C_{\sigma} \leq C_{\sigma_{3DM}}$. Three cases are distinguished, each next one allowing a more general schedule.

- *Case 1:* In σ no jobs are processed on slow machines, and all A-jobs are scheduled as in σ_{3DM}.
- *Case 2:* In σ no jobs are processed on slow machines.
- *Case 3:* Neither of these assumptions about σ is made.

The essence of the reduction is contained in Case 1. It is intuitively clear that the other two cases follow from Case 1 for appropriate values of M and K. In

Case 2 and Case 3 we give explicit values for the numbers M and K and prove that these values satisfy.

Case 1: We will prove that $C_\sigma \geq C_{\sigma 3DM} + \frac{1}{6}$. Notice that the J-jobs are the only jobs that can be scheduled on more than one machine. Also notice that, by the choice of $2m+4$ for the processing times of the B-jobs, the completion time of any J-job is strictly smaller than the completion time of any B-job. (If all J-jobs are scheduled on the Z-machine, then they are completed at time $2m+3$). Let T be the time that is spent on processing J-jobs on the Z-machine, and let $\widetilde{T} = T - 2(m-n)$. (In the schedule σ_{3DM} we have $\widetilde{T} = 0$.) The number of B-jobs on each of the U-, V-, and W-machines were chosen such that the following equality holds for any optimal schedule σ satisfying the assumptions we made for Case 1.

$$C_\sigma(B) = C_{\sigma 3DM}(B) - \widetilde{T}(m-n). \tag{1}$$

Now let $C_\sigma(J)_1$ be the sum of the n smallest completion times among the J-jobs in the schedule σ, and let $C_\sigma(J)_2$ be the sum of the $m-n$ largest completion times among the J-jobs. In a similar way we define $C_{\sigma 3DM}(J)_1$ and $C_{\sigma 3DM}(J)_2$.

Let $c_1 \leq c_2 \leq \ldots \leq c_m$ be an ordering of the completion times of the J-jobs in σ. The largest completion time c_m is at least $3+T$, and the last but one is at least $3+T-2$, and so on. In general we have $c_i \geq \max\{3+T-2(m-i)\,,\,3\}$. Therefore if $\widetilde{T} \geq 0$ it follows that

$$\begin{aligned}
C_\sigma(J)_2 = \sum_{i=n+1}^{m} c_i &\geq \sum_{i=n+1}^{m} 3+T-2(m-i) \\
&= \sum_{i=n+1}^{m} 3 + 2(m-n) - 2(m-i) + \widetilde{T} \\
&= C_{\sigma 3DM}(J)_2 + \widetilde{T}(m-n).
\end{aligned} \tag{2}$$

The same inequality holds if $\widetilde{T} < 0$. The arguments are more subtle and omitted here. Combining equality (1) and inequality (2), and using $C_{\sigma 3DM}(A) = C_\sigma(A)$ and $C_{\sigma 3DM}(J)_1 = 3n$ we obtain

$$\begin{aligned}
C_\sigma = C_\sigma(A) + C_\sigma(B) + C_\sigma(J)_1 + C_\sigma(J)_2 \\
\geq C_{\sigma 3DM}(A) + C_{\sigma 3DM}(B) + C_\sigma(J)_1 + C_{\sigma 3DM}(J)_2 \\
= C_{\sigma 3DM} - 3n + C_\sigma(J)_1.
\end{aligned}$$

From Lemma 1 we have $C_\sigma(J)_1 \geq 3n + \frac{1}{6}$, and therefore $C_\sigma \geq C_{\sigma 3DM} + \frac{1}{6}$.

Case 2: We will prove that $C_\sigma \geq C_{\sigma 3DM} + \frac{1}{12}$. The numbers M and K, introduced earlier, have not been specified yet. We choose $M = (2n+1)C^*(6C^*+1)^2$, where $C^* = C_{\sigma 3DM}(J) + C_{\sigma 3DM}(B)$, and $K = 48(2m+4)NC_{\sigma 3DM}$, where $N = (m+2n(m-n)+(3n+3)M)$, which is simply the total number of jobs. Notice that M and K are well-defined since C^* does not depend on M, and $C_{\sigma 3DM}$ does not depend on K. We could have chosen much smaller values for M and K. However, this will make it much harder to verify the correctness of the reduction.

The sum of completion times of the A-jobs in σ is $C_{\sigma 3DM}(A)$ if and only if these jobs are scheduled as in σ_{3DM}. Suppose now that some A-jobs are not scheduled according to σ_{3DM}. For example, assume that $1 - \delta(V_1)$ is the time that machine V_1 spends on processing A-jobs between time $t = 0$ and $t = 1$. Let $c_1 \leq c_2 \leq \ldots \leq c_M$ be an ordering of the completion times of the A-jobs on machine V_1. The largest completion time, c_M, is at least $1 + \delta(V_1)$, and the largest but one is at least $1 + \delta(V_1) - \frac{1}{M}$, and so on. Compared to the schedule σ_{3DM} this will increase the sum of completion times for the A-jobs by at least $\lceil \delta(V_1)M \rceil \delta(V_1) \geq \delta(V_1)^2 M$. Now let $1 - \delta(V_i)$, $1 - \delta(W_j)$, and $1 - \delta(Z)$ be the time that, respectively, machine V_i, W_j, and Z spends on processing A-jobs between time $t = 0$ and respectively $t = 1$, $t = 2$, and $t = 3$ ($1 \leq i, j \leq n$). Let $\delta = \delta(V_1) + \ldots + \delta(V_n) + \delta(W_1) + \ldots + \delta(W_n) + \delta(Z)$. Then we obtain

$$C_\sigma(A) \geq C_{\sigma_{3DM}}(A) + \sum_{i=1}^{n} \delta(V_i)^2 M + \sum_{i=1}^{n} \delta(W_i)^2 M + \delta(Z)^2 M$$
$$\geq C_{\sigma_{3DM}}(A) + \frac{\delta^2 M}{2n+1}.$$

If $\frac{\delta^2 M}{2n+1} > C^*$, then certainly $C_\sigma \geq C_{\sigma_{3DM}} + \frac{1}{12}$, and we are done. So assume the opposite and substitute the value of M. This gives us

$$\delta \leq \frac{1}{6C^* + 1}. \tag{3}$$

The δ time can be used to schedule at most a fraction $\delta/2$ of one J-job. For B-jobs this fraction is of course smaller. Now consider the problem in which all processing times of J- and B-jobs are multiplied by a factor $1 - \delta/2$, slow machines may not be used, and the machines from V, W and Z may not be used until time 1, 2 and 3 respectively. From Case 1 it follows that, for this problem, the sum of completion times of these scaled J- and B-jobs is at least $(1 - \frac{1}{2}\delta)(C^* + \frac{1}{6})$. We conclude that in the original problem

$$C_\sigma(J) + C_\sigma(B) \geq (1 - \frac{1}{2}\delta)(C^* + \frac{1}{6}).$$

From $C_\sigma(J) + C_\sigma(B) < C^* + \frac{1}{12}$ we obtain

$$\delta > \frac{1}{6C^* + 1}. \tag{4}$$

This contradicts the earlier obtained upper bound (3) on δ. We conclude that $C_\sigma \geq C_{\sigma_{3DM}} + \frac{1}{12}$.

Case 3: We will prove that $C_\sigma \geq C_{\sigma_{3DM}} + \frac{1}{24}$. Suppose that some parts of jobs are scheduled on slow machines. The sum of the fractions of all jobs scheduled on slow machines can not be more than $\frac{C_{\sigma_{3DM}}}{K}$ since the total processing time of these jobs would already exceed $C_{\sigma_{3DM}}$. From σ we define a new schedule in three steps. First, remove all the work that is scheduled on slow machines. Secondly, shift the remaining schedule to the right over a time $\frac{1}{24N}$. That is, all work is postponed by $\frac{1}{24N}$. Thirdly, reschedule the removed work on fast machines between $t = 0$ and $t = \frac{1}{24N}$. This is possible since the total processing time of this work is at most $\frac{C_{\sigma_{3DM}}}{K}(2n + 4) = \frac{1}{24N}$ if completely scheduled on fast machines. In this new schedule no job is scheduled on a slow machine, and the increase in the total sum of completion times is at most $\frac{1}{24}$. Using Case 2 we obtain

$$C_\sigma + \frac{1}{24} \geq C_{\sigma 3DM} + \frac{1}{12} \quad \Rightarrow \quad C_\sigma \geq C_{\sigma 3DM} + \frac{1}{24}.$$

Since Case 3 allows optimal schedules of any form we conclude that if $C_\sigma \leq C_{\sigma 3DM}$, then a perfect matching exists. $\square$

4 Discussion

The complexity of $R|pmtn|\sum C_j$ and $R|pmtn|\sum U_j$ is still open for a fixed number of machines, even for $m = 2$. Another open question is whether $P|pmtn|\sum U_j$ is solvable in pseudopolynomial time or NP-hard in the strong sense.

Acknowledgements. I would like to thank Jan Karel Lenstra and Leen Stougie for their comments on an earlier version of the paper.

References

1. P. Brucker and S.A. Kravchenko: Preemption can make parallel machine scheduling problems hard, Osnabruecker *Schriften zur Mathematik*, Reihe P (Preprints), Heft 211, (1999).
2. P. Brucker, S.A. Kravchenko and Y.N. Sotskov, Preemptive job-shop scheduling problems with a fixed number of jobs, *Mathematical Methods of Operations Research* 49 (1999), 41-76.
3. J. Bruno, E.G. Coffman Jr, R. Sethi, Scheduling independent tasks to reduce mean finishing time, *Communications of the ACM* 17 (1974), 382-387.
4. J. Du and J.Y.-T. Leung, Minimizing the number of late jobs on unrelated machines, *Operations Research Letters* 10 (1991), 153-158.
5. J. Du, J.Y.-T. Leung, Minimizing mean flow time with release time constraints, *Theoretical Computer Science* 75 (1990) 347-355.
6. J. Du, J.Y.-T. Leung, G.H. Young, Scheduling chain-structured tasks to minimize makespan and mean flow time, *Information and Computation* 92 (1991) 219-236.
7. Gonzalez, Optimal mean finish time preemptive schedules, Technical Report 220, Computer Science Department, Pennsylvania State University (1977).
8. R.L. Graham, E.L. Lawler, J.K. Lenstra, A.H.G. Rinnooy Kan, Optimization and approximation in deterministic sequencing and scheduling: a survey, *Annals of Discrete Mathematics* 4 (1979), 287-326.
9. W.A. Horn, Minimizing average flow time with parallel machines, *Operations Research* 21 (1973), 846-847.
10. R.M. Karp, Reducibility among combinatorial problems, in: R.E. Miller and J.W. Thatcher (eds.), *Complexity of Computer Computations*, Plenum Press, New York (1972), 85-103.
11. S.A. Kravchenko, Y.N. Sotskov, Optimal makespan schedule for three jobs on two machines, *Mathematical Methods of Operations Research* 43 (1996), 233-238.

12. E.L. Lawler, J. Labetoulle, On preemptive scheduling of unrelated parallel processors by linear programming, *Journal of the Association for Computing Machinery* 25 (1978), 612-619.

13. E.L. Lawler, Recent results in the theory of machine scheduling, in: A. Bachem, M. Grötchel and B. Korte (eds.), *Mathematical Programming: the State of the Art-Bonn 1982*, Springer, Berlin-New York (1983), 202-234.

Approximation Algorithms for the Minimum Bends Traveling Salesman Problem

Clifford Stein* and David P. Wagner*

Dartmouth College, Department of Computer Science, Hanover, NH 03755-3510
{cliff,dwagn}@cs.dartmouth.edu.
http://www.cs.dartmouth.edu/~cliff
http://www.cs.dartmouth.edu/~dwagn

Abstract. The problem of traversing a set of points in the order that minimizes the total distance traveled (traveling salesman problem) is one of the most famous and well-studied problems in combinatorial optimization. In this paper, we introduce the metric of minimizing the number of turns in the tour, given that the input points are in the Euclidean plane. We give approximation algorithms for several variants under this metric. For the general case we give a logarithmic approximation algorithm. For the case when the lines of the tour are restricted to being either horizontal or vertical, we give a 2-approximation algorithm. If we have the further restriction that no two points are allowed to have the same x- or y-coordinate, we give an algorithm that finds a tour which makes at most two turns more than the optimal tour. We also introduce several interesting algorithmic techniques for decomposing sets of points in the Euclidean plane that we believe to be of independent interest.

1 Introduction

The problem of traversing a set of points in the order that minimizes the total distance traveled (traveling salesman problem) is one of the most famous and well-studied problems in combinatorial optimization [21]. It has many applications [24,13,14,25,18], and has been a testbed for many of the most useful ideas in algorithm design and analysis. The usual metric, minimizing the total distance traveled, is an important one, but many other metrics including maximum distance [19,12,15,20], minimum latency(traveling repairman problem) [1, 7], minimizing the shortest edge length[21], maximum scatter TSP [4], and minimizing the total angle traversed (angular-metric TSP) [2] are also of interest.

In this paper, we introduce the metric of minimizing the number of turns in the tour, given that the input points are in the Euclidean plane. Equivalently, we wish to find a tour through the points, consisting of straight lines, so that the number of lines is minimized. To our knowledge this metric has not been studied previously. It is motivated by applications in robotics and in the movement of

---

* Research partially supported by NSF Career Award CCR-9624828, NSF Grant EIA-98-02068, a Dartmouth Fellowship, and an Alfred P. Sloane Foundation Fellowship.

K. Aardal, B. Gerards (Eds.): IPCO 2001, LNCS 2081, pp. 406–421, 2001.

other heavy machinery: for many such devices turning is an expensive operation. Imagine that a robot needs to visit a set of locations distributed over a relatively small physical space (room or building sized). If the locations are fairly dense throughout the region, there will be many tours whose total length is close to the minimum. If this robot turns slowly, the time spent visiting all the points is dominated by the number of turns made. In a recent paper, Arkin et al. consider the problem of carving out an area with a drill while minimizing the number of times of the drill must stop to make a turn[3]. This problem is closely related to our problem. Further applications appear in VLSI, especially with the 2-layer chip model[23].

Over larger geographic areas, both distance and number of turns can contribute to the time spent traversing a tour. We view our work in this paper as an important first step towards understanding this more general problem. Metrics involving minimizing the number of turns or some function of the number of turns and total distance are well studied for the shortest paths problem [28,23, 31,32,22,29,30].

For the regular traveling salesman problem, Christofedes' algorithm gives a 3/2-approximation [9], and the algorithms of Arora [5] and Mitchell [27] give a PTAS for the problem in the case when the points are in the Euclidean plane. Objectives such as the traveling repairman problem, maximum length tour and maximum scatter also have constant-factor approximation schemes [19,4,7]. For longest tour, in the case when the points are in R^d for some fixed d, the problem is actually solvable in polynomial time [6]. In contrast, for the angular-metric TSP, the best known approximation algorithm is $O(\log n)$ [2].

In this paper, we give approximation algorithms for several variants of the traveling salesman problem for which the metric is to minimize the number of turns. For the case of an arbitrary set of n points in the Euclidean plane, we give an $O(\log(\min(z, c)))$-approximation algorithm, where z is the maximum number of collinear points and c is the minumum number of lines that can be used to cover all n points. In the worst case $\min(z, c)$ can be as big as $n/3$, but it will often be much smaller. We call this problem the *minimum bends traveling salesman problem*.

We also study interesting restricted cases and find better approximation ratios. We introduce the *rectilinear minimum bends traveling salesman problem*, in which the lines of the tour are restricted to being either horizontal or vertical, and we give a 2-approximation algorithm for this problem. The algorithms for both cases involve forming a relaxation which we call the *line cover* problem, where the line cover is the minimum-sized set of lines covering all the input points. The differences in the two algorithms arise in the choice of potential lines to include in the cover and the ability to approximate the resulting line cover instance. For the general case, we obtain a set-cover problem, Our logarithmic approximation ratio results from the logarithmic approximation ratio of set-cover.[8,10,11]. For the rectilinear case, we obtain a bipartite vertex cover problem, which can be solved in polynomial time.

Finally, we consider a special case of the rectilinear minimum bends traveling salesman problem in which we have the restriction that no two points in the input are allowed to have the same x- or y-coordinate. This allows us to study carefully an aspect of the problem which the previous approximation algorithms ignore. Once they find a line cover, the other algorithms link the lines together in a fairly straightforward way. For this problem, however, the minimum sized line cover is equal to the number of input points and so the line cover gives us essentially no helpful information. Instead we focus on how to carefully link together non-collinear points into a tour. For this case, we give an algorithm that finds a tour which makes at most two turns more than the optimal tour. Thus we have an approximation algorithm with an additive, rather than a multiplicative error bound.

Beyond the additive error bound, our algorithm for this problem introduces several interesting algorithmic techniques. We introduce two different ways to decompose a set of points in the Euclidean plane. We call these decompositions a *9-division* and a *4-division*. We then show that any set of points can either be decomposed into a 9-division or a 4-division. Guided by these decompositions, we repartition the points into a set of points that are monotonically increasing, a set of points that are monotonically decreasing, and a set of points that fall on the perimeters of a set of nested boxes. Using this second decomposition, we are able to find a tour that uses at most two turns more than the optimal tour. We believe that these decompositions may be of independent interest.

We do not yet know whether the problems considered in this paper are NP-hard. We believe the general minimum bends traveling salesman problem is NP-hard, and we would not be surprised if the non-collinear rectilinear variant (c.f. Section 2) is actually solvable in polynomial time. We omit many of the proofs in this extended abstract.

2   Preliminaries

In this section we introduce three versions of the Minimum Bends TSP. We also define the Line Cover problem and its rectilinear variant, both of which will be useful subroutines in our algorithms.

Throughout this paper, we use the convention that a point p_i has x and y coordinates x_i and y_i respectively. When a point p_i falls on line l_j, we will say that line l_j *covers* point p_i.

We will be concerned with approximation algorithms, and will define a ρ-approximation algorithm for a minimization problem to be one which, in polynomial time, finds a solution of value $\rho\text{OPT}+O(1)$, where OPT is the value of the optimal solution to the problem.

In this paper, we will consider traveling salesman tours in the plane. Our tours will differ from conventional tours in that the endpoints of their segments need not be at input points.

Definition 1 (Segmented Tour (S-Tour)). *Given a set of points $P = \{p_1, p_2, \ldots, p_n\}$ in the Euclidean plane, define an S-Tour over P to be a se-*

quence of line segments $\pi = l_0, l_1, \ldots, l_{m-1}$ *such that: 1) there exists a set of points* $Q = \{q_0, q_1, \ldots, q_{m-1}\}$ *such that the endpoints of* l_i *are* q_i *and* $q_{i+1 \bmod m}$, *and 2) each point* $p_i \in P$ *falls along some line* $l_j \in \pi$.

Definition 2 (Rectilinear Segmented Tour ($\square$S-Tour)). *Given a set of points* $P = \{p_1, p_2, \ldots, p_n\}$ *in the Euclidean plane, define a* $\square$*S-Tour over* P *to be an S-Tour* $\pi = l_0, l_1, \ldots, l_{m-1}$ *over* P *with the following additional property: 3)* $l_j \in \pi$ *is a horizontal or vertical line segment depending on the parity of* j.

Although we define a tour as a sequence of line segments, given a set of points $Q = \{q_0, q_1, \ldots, q_{m-1}\}$, and a starting direction d, a natural tour associated with Q and d traverses the points in order. To go from point q_i to point q_{i+1}, we greedily travel in the path that minimizes the number of lines needed to travel from q_i to q_{i+1} (having approached q_i in a particular direction). Thus the output from our algorithms may consist of an ordering on the points together with a starting direction.

We now define the main problem of this paper along with two restricted versions. In the *Minimum Bends TSP (MBTSP)*, we are given a set of points $P = \{p_1, p_2, \ldots, p_n\}$ in the Euclidean Plane, and we wish to find an S-Tour $\pi = l_0, l_1, \ldots, l_{m-1}$ over P such that m is minimized. We will let MBTSP(P) denote the optimal value for m. Similarly, in the *Rectilinear Minimum Bends TSP ($\square$MBTSP)*, we wish to find a $\square$S-Tour $\pi = l_0, l_1, \ldots, l_{m-1}$ over P such that m is minimized. We will let $\square$MBTSP(P) denote the optimal value for m. In the *Non-Collinear Rectilinear Minimum Bends TSP (NC-$\square$MBTSP)*, our points have the further restriction that for any $p_i, p_j \in P$ with $i \neq j$ we have $x_i \neq x_j$ and $y_i \neq y_j$ (in the future we will call this the *non-collinearity property*). We wish to find a $\square$S-Tour $\pi = l_0, l_1, \ldots, l_{m-1}$ over P such that m is minimized. We will let NC-$\square$MBTSP(P) denote the optimal value for m. For all these problems, the number of bends in a tour is equivalent to the number of line segments. Thus our objective, minimizing bends, is characterized by minimizing m, the number of line segments in the tour.

Our algorithms will also use, as subroutines, several related problems. In the *Line Cover Problem (LC)*, we are given a set of points $P = \{p_1, p_2, \ldots, p_n\}$ in the Euclidean Plane. We wish to find a set of lines $L = \{l_1, l_2, \ldots, l_m\}$ such that each $p_i \in P$ falls along some $l_j \in L$ and such that m is minimized. We let LC(P) be the optimal value for m. The *Rectilinear Line Cover Problem ($\square$LC)* is a variant of the Line Cover Problem in which the lines are restricted to being either horizontal or vertical.

3 A 2-Approximation Algorithm for $\square$MBTSP

In this section we give a 2-approximation algorithm for the Rectilinear version of the Minimum Bends Traveling Salesman Problem ($\square$MBTSP). As described in Section 2, we are given an arbitrary set of points in the Euclidean plane, and we are looking for a rectilinear tour which minimizes the number of turns in the tour. The Rectilinear Line Cover problem will be used in our approximation algorithm for □MBTSP. Hassin and Meggido showed that this problem can be solved in $O(n^{1.5})$ time by reducing it to the Bipartite Vertex Cover problem[16].

Our algorithm first computes an optimal rectilinear line cover on the set of input points. We then convert this to a rectilinear tour having no more than twice as many turns as there are lines in the optimal rectilinear line cover. The resulting tour is a 2-approximation since the number of lines in the rectilinear line cover instance defines an obvious lower bound on the □MBTSP instance.

Lemma 1. *Given a set of points $P = \{p_1, p_2, \ldots, p_n\}$ in the Euclidean plane, if $\Box LC(P) = m$, then $\Box MBTSP(P) \leq 2m + O(1)$.*

Proof. Let $L = l_1, l_2, \ldots, l_m$ be an optimal rectilinear line cover over the points in P. Let v be the number of vertical lines in L and let h be the number of horizontal lines in L. Construct a tour as follows. Define a bounding box B such that all points in P are contained within the boundaries of B. Connect all horizontal lines in a path that travels in a S-like fashion through every horizontal line and along the boundaries of B. This adds $2h - 1$ lines to the tour. Repeating this process in the vertical direction adds $2v - 1$ lines. Joining the horizontal and vertical sections at both ends adds 4 more lines for a total of $2h-1+2v-1+4 = 2m+O(1)$ lines. Figure 1 shows a sample tour. □

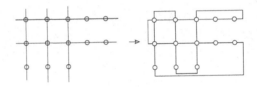

Fig. 1. A sample tour formed by □MBTSP

The lemma is tight in the sense that we can show that there exist sets of points P for which $\Box MBTSP(P) \geq 2\Box LC(P)$. Combining this lemma with Hassin and Meggido's algorithm we obtain:

Theorem 1. *A 2-approximation for the Rectilinear Minimum Bends Traveling Salesman Problem can found in $O(n^{1.5})$ time.*

Rectilinear Cycle Cover from Rectilinear Line Cover

For several TSP problems, a cycle cover relaxation is a useful algorithmic tool. We can define a Rectilinear Cycle Cover Problem analogous to the □MBTSP. We note that from a given rectilinear line cover, we can optimize the rectilinear cycle cover using those lines by finding an appropriate matching between vertical and horizontal endpoints. This, however, does not give us an algorithm for finding the optimal rectilinear cycle cover, as the optimal rectilinear cycle cover may not use the lines from the optimal rectilinear line cover. On the other hand, we are able to show that given a rectilinear cycle cover, we can convert it to a rectilinear tour, using no more than 5/4 times as many turns. Given the above mentioned difficulty with finding the optimal rectilinear cycle cover, we do not yet know how to make use of this fact.

4   An Approximation Algorithm for MBTSP

We now turn to the general minimum bends TSP problem. Consider what happens if we try to apply the algorithm of the previous section to this problem. Although the details are more involved, we can formulate a line cover problem, by constraining the candidate lines to be those which cover maximal collinear subsets of the input points, together with the degenerate "lines" formed by single points. We can then solve the line cover problem, and use this to obtain a tour, paying roughly a factor of 2 in the process. The only problem in this approach is that the resulting line cover problem is no longer equivalent to a bipartite vertex cover problem. Instead, it is now a set cover problem, and so our approximation bound will not be as good.

The details of our algorithm appear in Figure 2. In the first seven lines, we compute the set T, which contains all lines which might be in the optimal tour. We include in this set all lines going through two or more points. The optimal tour may also consist of lines that go through only one input point, hence we include singleton points as degenerate lines. The set T can easily be computed in polynomial time, and here we omit the discussion of the data structures needed to compute it efficiently.

Function FIND-MBTSP(P)

1) $T = \emptyset; k = 0$
2)   for all $(p_i \in P)$
3)   for all $(p_j \in P)$
4)   let S = the set of all points in P along the line through (p_i, p_j)
5) $T = T \cup \{S\}$
6)   for all $(p_i \in P)$
7)   let $T = T \cup \{\{p_i\}\}$
8)   let $T' = $ SET-COVER(T, P) // This returns a log-approximation of Set Cover
9)   for all $(S \in T')$
10)   let (q_{2k-1}, q_{2k}) = the two extremum points in S; $k{+}{+}$
          (if S is a singleton p, then $q_{2k-1} = q_{2k} = p$)
11) let $Q = \bigcup q_i$
12) for all $(q_i \in Q)$
13)   let l_i = the line segment $(q_i, q_{(i+1) \bmod |Q|})$ // l_i may have length 0.
14) return $\pi = l_1, l_2, \ldots, l_{|Q|}$

Fig. 2. An approximation algorithm for MBTSP

We now form a set-cover instance as follows. The elements of our set system are the initial input points, while the sets are the sets of points comprising lines in T. The optimal set cover is a lower bound on the optimal tour. Further, by arguments similar to those in Lemma 1, if we take the line segments in T', a line cover of the points P, order them, and connect the two endpoints of each two consecutive segments with an additional line segment, we do indeed get a tour. The code in lines 9 through 13 achieves this. Thus we have shown:

Theorem 2. *Algorithm* FIND-MBTSP, *given an input to the Minimum Bends TSP problem, computes a tour which is a $2\rho + 2$-approximation, where ρ is the approximation bound for Set Cover.*

In general, the best set cover approximation is $\ln n$ [10,17]. However, in the case when each set is of size no more than z, the approximation ratio is roughly $\ln z$, and tighter bounds are known for small values of z[11]. Further we note that the VC-Dimension of this set system is 2 which allows us to additionally bound the approximation ratio by $O(\log c)$ where c is the size of the minimum set cover[8]. The maximum set size corresponds to the maximum number of collinear points, and the minimum set cover corresponds to the minimum number of lines that can cover the points. Thus we have:

Corollary 1. *Given a set of points P among which no more than z are collinear, and which can be covered by c lines, Algorithm* FIND-MBTSP *is an $O(\log(\min(z, c)))$-approximation algorithm.*

5   An Approximation of NC-□MBTSP Using OPT+2 Bends

In this section we consider □MBTSP with the additional constraint that no two points share an x-coordinate or a y-coordinate. In this case, no two points may lie along the same line of the tour, and hence n is a lower bound on the number of bends in the tour. Also the number of lines in any rectilinear tour must be even, since the tour must have the same number of horizontal and vertical lines. Thus if n is odd, then $n + 1$ is a lower bound on the number of lines.

Our approximation algorithm finds a tour with $n + 2$ lines if n is even and $n + 3$ lines if n is odd. Thus, the algorithm finds a tour with at most OPT+2 bends.

5.1   Box Points and Diagonal Points

Our algorithm depends heavily on the division of the points into two categories: diagonal points, and box points. Here we define these two sets. In the following section we show that the input to NC-□MBTSP can always be partitioned into one set of box points and one set of diagonal points.

We say that a set of points $P = \{p_1, p_2, \ldots, p_n\}$ is *monotonically increasing* if, for any two points $p_i = (x_i, y_i)$, and $p_j = (x_j, y_j) \in P$ we have $x_i > x_j$

if and only if $y_i > y_j$. Similarly, P is *monotonically decreasing* if for any two points $p_i, p_j \in P$ we have $x_i > x_j$ if and only if $y_i < y_j$. A set of points may be considered *diagonal points* if they can be partitioned into two sets $\mathcal{I}$ and $\mathcal{D}$ such that the points in $\mathcal{I}$ are monotonically increasing and the points in $\mathcal{D}$ are monotonically decreasing. We define a *4-box* to be any set of four points $P = p_1, \ldots, p_4$, such that a single point falls along each of the four boundaries of the smallest enclosing rectangle around P, and no point falls at any corner of that rectangle. Finally, we say that a set of points may be considered *box points* if they can be partitioned into subsets of cardinality 4, such that each subset forms a 4-box, and such that given any two of these 4-boxes, one lies entirely within the other (See Figure 5).

Ultimately we want to show that for any given set of points with the non-collinearity property, all points can be partitioned into a single set of diagonal points and single set of box points as defined above.

5.2   Planar Subdivisions

We will classify the input points using a method we call the *Planar Subdivision Method*. We will define two different ways to divide the Euclidean plane, a 4-division and a 9-division. Then we will show that among any set of points with the non-collinearity property there exists a way to divide the plane into a 4-division or a 9-division. See Figure 3 for examples of the two divisions.

Definition 3 (4-division). *A 4-division is a division of the Euclidean plane via a single horizontal and a single vertical line into 4 quadrants, NE, NW, SE, and SW, such that: 1) The points in the NW region and the points in the SE region are monotonically decreasing. 2) The points in the NE region and the points in the SW region are monotonically increasing.*

Due to the relative positions of the quadrants, it is also the case that the union of the points in the NW and SE regions are monotonically decreasing. Likewise the union of the points in the NE and SW regions are monotonically increasing. In a tour this will allow us to cover these sets of points at a rate of one per line within each set. Note that in order to have a 4-division it is not necessary that a particular quadrant contain any points. Among any arrangement of 0, 1, 2, or 3 points, there exists a 4-division, but it is not not necessarily true that there exists a 4-division among 4 points. (A simple case analysis will show this.)

Definition 4 (9-division). *A 9-division is a division of the Euclidean plane into 9 regions, NE, NW, SE, SW, N, S, E, W, and C (the Center), defined by two horizontal and two vertical lines, satisfying the properties of the 4-division as well as: 3) The N, S, E, and W regions are all empty. 4) There exists exactly one point along each of the four boundaries of the center region, and none at any of the corners. The interior of the center region may contain any arrangement of points.*

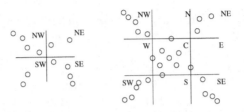

Fig. 3. A 4-division and a 9-division

We now define a function PLANAR-SUBDIVISION which, given a set of points with a 9-division, returns NW, NE, SW, SE, C, and B. These are the points in the NW, NE, SW, SE, and C regions, and the 4 points along the boundary of C respectively. If no 9-division exists, the algorithm finds a 4-division, and returns NW, NE, SW, and SE, the sets of points in the respective quadrants.

The algorithm repeatedly finds the smallest enclosing rectangle around the set of points. If there are 4 points along the border of that box, then a 9-division exists. If there is one point, then a 4-division exists. Otherwise, the box must have a corner point. We remove that corner point, classify it in the appropriate quadrant, and repeat the algorithm on the remaining points. Of the sets returned, B must contain exactly four points, but any of the other sets returned may be empty. Detailed pseudocode appears in Figure 4.

Function PLANAR-SUBDIVISION(P)
___
    let NW=NE=SW=SE=C=B=$\emptyset$ // The values to return
    while ($P \neq \emptyset$)
        let R=SMALLEST-ENCLOSING-RECTANGLE(P)
        if ($|R| == 4$)
            Division-Type=9 ; $C = P - R$; $B = R$
            return (Division-Type,NW,NE,SW,SE,C,B)
        else if ($|R| == 1$)
            Division-Type=4 ; NW = NW $\cup$ R
            return (Division-Type,NW,NE,SW,SE)
      else
            let maxx, minx, maxy, and miny be the points $p_i \in B$ such that
                x_i or y_i is minimized or maximized as appropriate
            if (maxx == maxy) NE = NE $\cup$ maxx ; P = P - maxx
            else if (minx == maxy) NW = NW $\cup$ minx ; P = P - minx
            else if (maxx == miny) SE = SE $\cup$ maxx ; P = P - maxx
            else if (minx == miny) SW = SW $\cup$ minx ; P = P - minx

Fig. 4. The Planar Subdivision Algorithm. It returns either a 4-division or a 9-division.

In order to analyze the planar subdivision algorithm, we will need several additional properties. Note that if some edge of a smallest enclosing rectangle did not contain any points, then the region would not properly be a smallest rectangle, as we could shrink it on that side. Thus the smallest enclosing rectangle about a set of points with the non-collinearity property must have exactly one point along each of its edges. In the event that fewer than four points fall along the union of all the edges of a smallest enclosing rectangle, then at least one point must lie at a corner of that rectangle. We state this as the corner lemma:

Lemma 2. *(Corner Lemma) Given a set of points $P = p_1, p_2, ..., p_n$ such that fewer than 4 points lie along the boundary of the smallest enclosing rectangle, R, at least one point in P must lie at a corner of R.*

We now know that given a set of four points P with the non-collinearity property, the following are equivalent: 1) The set P forms a 4-box. 2) The smallest enclosing rectangle about P has four points along its boundary. 3) No point lies at the corner of the boundary of the smallest enclosing rectangle about P.

The 4-box plays an important role in the classification of box points. Here we state the uniqueness of an enclosing 4-box.

Lemma 3. *Given a set of points $P = \{p_1, p_2, \ldots, p_n\}$ with the non-collinearity property, if there exists a 4-box in P, then there exists a unique 4-box which encloses all other 4-boxes in P.*

Lemma 4. *Given a set of points $P = \{p_1, p_2, \ldots, p_n\}$ with the non-collinearity property, if there exists a 9-division in P, then PLANAR-SUBDIVISION(P) finds the 9-division, and the points returned form the unique enclosing 4-box in P.*

Proof. First, we will show that the algorithm finds the unique enclosing 4-box if it exists. The set P initially contains all points. A point p' is only removed from P when it is on the corner of the smallest enclosing rectangle around P. Such a point cannot be in a 4-box, since it would have to be at the corner of any 4-box containing p'. Thus, we do not remove any points from P that could be in a 4-box. The algorithm only terminates when all points are removed from P, or when it finds that the smallest enclosing rectangle contains four points along its boundary. Thus it follows that either there does not exist a 4-box in P, or the algorithm finds the unique enclosing 4-box.

Next we show that if there exists a 4-box in P, then the algorithm finds a 9-division. Assume there exists a 4-box in P. Then we know that the algorithm finds the unique enclosing 4-box in P. Assume, by way of contradiction, that this 4-box does not define a 9-division. Then one of the properties of a 9-division would have to be violated. We will attempt to violate each property, and then contradict each violation.

Properties 1 and 2: Let there be two points in a corner region (NW, NE, SW, SE) which do not follow the region's monotonicity property. Then these two points, together with the two far points from the 4-box would form a larger 4-box, and the given 4-box would not be the unique enclosing 4-box.

Property 3: Let there be a point in a side region (N,S,E,W). Then this point together with the 3 far points of the given 4-box would form a larger 4-box.

Property 4 holds trivially and thus, our algorithm finds a 9-division if a 4-box exists. Since every 9-division contains a 4-box our algorithm finds a 9-division if one exists.   □

Lemma 5. *Given a set of points $P = \{p_1, p_2, \ldots, p_n\}$ with the non-collinearity property, if there does not exist a 9-division in P, then there exists a 4-division in P, and* PLANAR-SUBDIVISION *finds a 4-division.*

Proof Sketch. First let us show that if there does not exist a 9-division in P, then there exists a 4-division. We do this by induction on the cardinality of a set of points with no 9-division. Assume there does not exist a 9-division in P and consider the smallest bounding rectangle about P. There cannot be 4 points on this rectangle, or there would be a 4-box and thus a 9-division in P. Therefore, there must be a point, say p_c, at the corner of this rectangle. Now consider the set of points $P - \{p_c\}$. Assume inductively that all sets of $|P| - 1$ points which do not have a 9-division contain a 4-division. Thus $P - \{p_c\}$ contains a 4-division (since it contains no 9-division).

Adding p_c to this set can only eliminate the 4-division if it breaks the monotonicity property in some region. Since p_c was a corner point in the set P, it follows that it is extremal in both the x and y direction, among the points in P. We can show that for at least one region R, p_c will satisfy the monotonicity property for that region. Consider that p_c is not located in R. Then we can show that it is possible to reassign the boundaries so that p_c is in R without changing the region to which any other point in P is designated.

Thus, we have our inductive hypothesis. Let P be a set of points of cardinality n which has the non-collinearity property but no 9-division. If there exists a 4-division on any set Q of cardinality $n - 1$ which has no 9-division, then there exists a 4-division on P. For our base case we simply state that there is a 4-division on any set of points of cardinality 1, and we are done with the inductive proof.

It remains to show that our algorithm finds a 4-division, if there is no 9-division. Assume there is no 9-division on P. Then our algorithm can never find a 4-box in P. Thus at each iteration, it must find a corner point. As this point is extremal in two directions, it maintains a particular monotonicity with all remaining points in P. Thus, assigning it to the region which has that monotonicity property cannot violate the property in that region. Our algorithms does this. Furthermore, the regions remain properly defined, since, when a point is assigned to a region (and removed from P) it is extremal with respect to all remaining points in P in the proper direction. Thus for any two points in different regions, we were guaranteed the proper directional relationship when the first of these was removed from P. Therefore, the 4 regions the algorithm returns will have the proper monotonicity property, and thus they will form a 4-division.   □

Theorem 3. *(Planar Subdivision Theorem) Given any set of points $P = p_1, p_2, ..., p_n$ such that no two points share an x-coordinate or a y-coordinate, there must exist either a 4-division or a 9-division among those points, and* PLANAR-SUBDIVISION(P) *returns a proper 4-division or 9-division among P.*

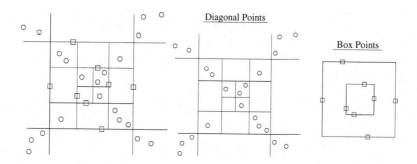

Fig. 5. Recursively finding a 9-division; finishing with a 4-division at the center. Classifying the points into Diagonal Points and Box Points.

5.3   The Algorithm

Figure 6 describes the approximation algorithm which finds tour of length at most OPT+2. In lines 2-7, it repeatedly applies the planar subdivision theorem to obtain a decomposition of the points. The loop will run for $O(n)$ iterations because each iteration, save the last, reduces the cardinality of C by at least 4. Recall that each decomposition partitions the points into sets NW, NE, SW, SE, B and C. The points in NE, SW, and SE are placed in the appropriate diagonals, either $\mathcal{I}$ or $\mathcal{D}' = \mathcal{D} \cap$ SE. The points in NW are indexed by the iteration in which they were found. The box points are indexed similarly.

The tour then consists of the points in $\mathcal{I}$, followed by the points in $\mathcal{D}'$, and then alternates between the points in $\mathcal{D}$ from NW and the various boxes. In order to get the exact bounds claimed, we have to be careful about the starting direction and we may also have to move points from one diagonal to another. These details are discussed in the proof of Theorem 5. Our algorithm may require that we move diagonal points from one diagonal to another. We call the algorithm that performs this SWITCH-DIAGONAL.

Theorem 4. *(Diagonal Switching Theorem) For any set of diagonal points derived using the Planar Subdivision method, there exists at least one point which may be swapped between $\mathcal{D}$ and $\mathcal{I}$, while retaining the monotonicity properties.*

Theorem 5. *Algorithm* FIND-NC-□MBTSP, *given an input to the Non-Collinear Minimum Bends TSP, finds a tour which contains at most 2 additional bends more than the optimal.*

Function FIND-NC-□MBTSP(P)

1) let $i = 1$
2) while ($P \neq \emptyset$)
3)   if ((4-division,NW,NE,SW,SE)=PLANAR-SUBDIVISION(P)) then
4) $\mathcal{I} = \mathcal{I} \cup$ NE $\cup$ SW ; $\mathcal{D}' = \mathcal{D}' \cup$ NW $\cup$ SE ; $P = \emptyset$
5)   else if ((9-division,NW,NE,SW,SE,B,C)=PLANAR-SUBDIVISION(P)) then
6) $\mathcal{I} = \mathcal{I} \cup$ NE $\cup$ SW ; $\mathcal{D}' = \mathcal{D}' \cup$ SE ; $\mathcal{D}_i =$ NW ; Box$_i = B$; P = C
7)   i++
8) $\mathcal{I}_{sort} = $ SORT($\mathcal{I}$) ; $\mathcal{D}'_{sort} = $ SORT($\mathcal{D}'$) // Along the y-coordinate
9) if ($|\mathcal{I}_{sort}|$) is odd then Starting-Direction = East
10) else if ($|\mathcal{I}_{sort}|$) is even then Starting-Direction = North
11) $\pi = $ (Starting-Direction,$\mathcal{I}_{sort}$,$\mathcal{D}'_{sort}$,Box$_{i-1}$,$\mathcal{D}_{i-1}$,...,Box$_1$,$\mathcal{D}_1$)
12) if (FINISHING-DIRECTION(π) == Starting-Direction) then
13)   return SWITCH-DIAGONAL(π,$\mathcal{I}$,$\mathcal{D}'$) // Apply Diagonal Switching Theorem
14) else return π

Fig. 6. An Approximation algorithm for NC-□MBTSP

Proof Sketch. As stated earlier, if n is even, then n is a lower bound on the number of lines in the optimal tour, by the non-collinearity property. Similarly, if n is odd, then $n+1$ is a lower bound. Now consider the number of lines in a tour returned by FIND-NC-□MBTSP. The return value consists of a series of point sequences, together with a starting direction. The total number of lines will be the sum of the number of lines in each sequence, plus the sum of the additional lines used in connecting adjoining sequences. An additional line will be necessary between two sequences if the preceding sequence finishes its path heading away from the start of the subsequent sequence. More than one additional line would be necessary only if a particular entrance direction were required at a point. Our definition of a tour defined by points will not require this, except to rejoin the end of a tour to its starting point.

The sequences $\mathcal{I}_{sort}$ and $\mathcal{D}'_{sort}$, because of their monotonicity, can be covered, starting from an extremal point, using a staircase-like path. This can be done at a rate of one line per point unless the starting direction requires that an extra turn be made in traveling from the first point to the second point. Transferring from $\mathcal{I}_{sort}$ to $\mathcal{D}'_{sort}$ will incur an extra turn, unless the tour can proceed directly to $\mathcal{D}'_{sort}$, in which case an extra line is needed after the first point of $\mathcal{D}'_{sort}$. Either way the total number of line segments needed to cover $\mathcal{I}_{sort}$ and $\mathcal{D}'_{sort}$ will be $|\mathcal{I}| + |\mathcal{D}'| + 1$ (See figure 7). Note that this algorithm does not handle the degenerate cases where the sets may have 0 or 1 points, but optimal paths can be found on a case-by-case basis.

Upon completion of $\mathcal{D}'_{sort}$ the path is inside the innermost uncovered Box$_i$, heading in either the North direction or the West direction, and all remaining points in $\mathcal{D}_i$ are to the North and West of the most recently covered point. These conditions are invariants to maintain after each subsequent point sequence, $\mathcal{D}_j$, is completed. The conditions are sufficient to cover all points of the inner most uncovered Box$_i$ using exactly 4 additional lines. The invariants remain true upon completion of Box_i. The invariants are also sufficient to cover a particular D_i using $|D_i|$ lines, and remain true after such a covering. Thus all points in $\bigcup_i$ $(Box_i \cup D_i)$ can be covered with $\sum_i(|Box_i| + |D_i|)$ lines (See figure 7).

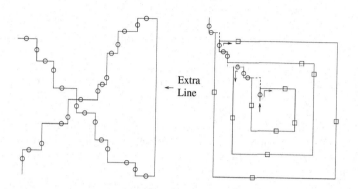

Fig. 7. The path through the diagonal points, and splicing in the box points

Joining the end of the tour back to its beginning will require 1 or 2 or 3 additional lines, depending on the starting and finishing directions. If 3 additional lines are required, then the starting and finishing directions must both be North. In this case we can apply the Diagonal Switching Theorem to modify the sizes of $\mathcal{D}'$ and $\mathcal{I}$ each by 1. This changes both the starting and finishing directions, allowing the tour to complete with 1 additional line. The resultant tour will have $n + 2$ total lines, or $n + 3$ total lines. By the parity argument, the tour can only have $n + 3$ lines if n is odd. Thus we achieve the desired optimality bound.   □

The total number of iterations of both the *while* loops in FIND-NC-□MBTSP and in PLANAR-SUBDIVISION is at most n, since an iteration of either loop reduces the size of P. The SMALLEST-ENCLOSING-RECTANGLE is thus run at most n times, and if run in a brute force fashion takes $O(n)$ time to run. However, by using 2 binary search trees, one keyed on the x-coordinate and one keyed on the y-coordinate, and an amortized analysis of the work needed per iteration, we can show that this portion of the algorithm runs in $O(n \log n)$ time overall. The remainder of the algorithm runs in $O(n)$ time.

Acknowledgements. We thank Neal Young for many helpful discussions on this work. Early work on this problem, particularly that in Section 3, was done jointly with Neal. We also thank Robert Fitch and Daniela Rus for their discussions on the Rectilinear Minimum Bend Path problem, and Michel Goemans and Neal Young for their advice regarding the use of the VC-Dimension.

References

1. F. Afrati, S. Cosmadakis, C. Papadimitriou, G. Papageorgiou, and N. Papakostantinou. The complexity of the travelling repairman problem. *Informatique Theoretique et Applications*, pages 79–87, 1986.
2. Alok Aggarwal, Don Coppersmith, Sanjeev Khanna, Rajeev Motwani, and Baruch Schieber. The angular-metric traveling salesman problem. *SIAM Journal on Computing*, 29(3):697–711, June 2000.
3. Esther M. Arkin, Michael A. Bender, Erik D. Demaine, Sandor P. Fekete, Joseph S. B. Mitchell, and Saurabh Sethia. Optimal covering tours with turn costs. In *Proceedings of the 12th ACM-SIAM Symposium on Discrete Algorithms*, pages 138–147, Washington, DC, 2001.
4. Esther M. Arkin, Yi-Jen Chiang, Joseph S. B. Mitchell, Steven S. Skiena, and Tae-Cheon Yang. On the maximum scatter TSP (extended abstract). In *Proceedings of the Eighth Annual ACM-SIAM Symposium on Discrete Algorithms*, pages 211–220, New Orleans, Louisiana, 5–7 January 1997.
5. S. Arora. Polynomial time approximation schemes for euclidean traveling salesman and other geometric problems. *JACM: Journal of the ACM*, 45, 1998.
6. A. Barvinok, Sandor P. Fekete, David S. Johnson, Arie Tamir, Gerhard J. Woeginger, and D. Woodroofe. The maximum traveling salesman problem. *submitted to Journal of Algorithms*, 1998.
7. A. Blum, P. Chalasani, D. Coppersmith, B. Pulleyblank, P. Raghavan, and M. Sudan. The minimum latency problem. In *Proceedings of the 26th Annual ACM Symposium on Theory of Computing*, pages 163–172, May 1994.
8. H. Bronnimann and M.T. Goodrich. Almost optimal set covers in finite VC-dimension. *Discrete Computat. Geom.*, 14:263–279, 1995.
9. N. Christofedes. Worst case analysis of a new heuristc for the traveling salesman problem. Technical report, Graduate School of Industrial Administration, Carnegie Mellon University, Pittsburgh, PA, 1976.
10. V. Chvátal. A greedy heuristic for the set-covering problem. *Mathematics of Operations Research*, 4(3):233–235, August 1979.
11. R. Duh and M. Furer. Approximation of k-set cover by semi-local optimization. In *Proceedings of the 29th Annual ACM Symposium on Theory of Computing*, pages 256–264, 1997.
12. M. L. Fisher, G. L. Nemhauser, and L. A. Wolsey. An analysis of approximations for finding a maximum weight hamiltonian circuit. *Operations Research*, 27(4):799–809, July–August 1979.
13. R.S. Garfinkel. Minimizing wallpaper waste, part I: a class of traveling salesman problems. *Operations Research*, 25:741–751, 1977.
14. P.C. Gilmore and R.E. Gomory. Sequencing a one state-variable machine: a solvable case of the traveling salesman problem. *Operations Research*, 12:655–679, 1964.
15. Rafael Hassin and Shlomi Rubinstein. Better approximations for max TSP. *Information Processing Letters*, 75:181–186, 2000.
16. R. Hasssin and N. Meggido. Approximation algorithms for hitting objects with straight lines. *Discrete Applied Mathematics*, 30:29–42, 1991.
17. Dorit Hochbaum, editor. *Approximation Algorithms*. PWS, 1997.
18. L.J. Hubert and F.B. Baker. Applications of combinatorial programming to data analysis: the traveling salesman and related problems. *Pyschometrika*, 43:81–91, 1978.

19. R. Kosaraju, J. Park, and C. Stein. Long tours and short superstrings. In *Proceedings of the 35th Annual Symposium on Foundations of Computer Science*, pages 166–177, 1994.

20. A. V. Kostochka and A. I. Serdyukov. Polynomial algorithms with the estimated 3/4 and 5/6 for the traveling salesman problem of the maximum (in russian). *Upravlyaemye Sistemy*, 26:55–59, 1985.

21. E.L. Lawler, J.K. Lenstra, A.H.G. Rinooy Kan, and D.B. Shmoys, editors. *The Traveling Salesman Problem*. John Wiley and Sons, 1985.

22. D.T. Lee, C.D. Yang, and C.K. Wong. Problem transformation for finding rectilinear paths among obstacles in two-layer interconnection model. Technical Report 92-AC-104, Dept. of EECS, Northwestern University, 1992.

23. D.T. Lee, C.D. Yang, and C.K. Wong. Rectilinear paths among rectilinear obstacles. *Discrete Applied Mathematics*, 70, 1996.

24. J.K. Lenstra and A.H.G. Rinnooy Kan. Some simple applications of the travelling salesman problem. *Operations Research Quarterly*, 26:717–733, 1975.

25. W.T. McCormick, P.J. Schweitzer, and T.W. White. Problem decomposition and datareorganization by a clustering technique. *Operations Research*, 20:993–1009, 1972.

26. N. Meggido and A. Tamir. On the complexity of locating linear facilities in the plane. *Operations Research Letters*, 1:194–197, 1982.

27. Joseph S. B. Mitchell. Guillotine subdivisions approximate polygonal subdivisions: A simple polynomial-time approximation scheme for geometric TSP, k-MST, and related problems. *SIAM Journal on Computing*, 28:1298–1309, 1999.

28. J.S.B. Mitchell, C. Piatko, and E.M. Arkin. Computing a shortest k-link path in a polygon. In *Proceedings of the 33rd Annual Symposium on Foundations of Computer Science*, pages 573–582, 1992.

29. C.D. Yang, D.T. Lee, and C.K. Wong. On bends and lengths of rectilinear paths: a graph-theoretic approach. *Internat. J. Comp. Geom. Appl.*, 2:61–74, 1992.

30. C.D. Yang, D.T. Lee, and C.K. Wong. On minimum-bend shortest recilinear path among weighted rectangles. Technical Report 92-AC-122, Dept. of EECS, Northwestern University, 1992.

31. C.D. Yang, D.T. Lee, and C.K. Wong. Rectilinear path problems among rectilinear obstacles revisited. *SIAM Journal on Computing*, 24:457–472, 1992.

32. C.D. Yang, D.T. Lee, and C.K. Wong. On bends and distance paths among obstacles in two-layer interconnection model. *IEEE Transactions on Computers*, 43:711–724, 1994.

Lecture Notes in Computer Science

For information about Vols. 1–1979
please contact your bookseller or Springer-Verlag

Vol. 2024: H. Kuchen, K. Ueda (Eds.), Functional and Logic Programming. Proceedings, 2001. X, 391 pages. 2001.

Vol. 2025: M. Kaufmann, D. Wagner (Eds.), Drawing Graphs. XIV, 312 pages. 2001.

Vol. 2026: F. Müller (Ed.), High-Level Parallel Programming Models and Supportive Environments. Proceedings, 2001. IX, 137 pages. 2001.

Vol. 2027: R. Wilhelm (Ed.), Compiler Construction. Proceedings, 2001. XI, 371 pages. 2001.

Vol. 2028: D. Sands (Ed.), Programming Languages and Systems. Proceedings, 2001. XIII, 433 pages. 2001.

Vol. 2029: H. Hussmann (Ed.), Fundamental Approaches to Software Engineering. Proceedings, 2001. XIII, 349 pages. 2001.

Vol. 2030: F. Honsell, M. Miculan (Eds.), Foundations of Software Science and Computation Structures. Proceedings, 2001. XII, 413 pages. 2001.

Vol. 2031: T. Margaria, W. Yi (Eds.), Tools and Algorithms for the Construction and Analysis of Systems. Proceedings, 2001. XIV, 588 pages. 2001.

Vol. 2032: R. Klette, T. Huang, G. Gimel'farb (Eds.), Multi-Image Analysis. Proceedings, 2000. VIII, 289 pages. 2001.

Vol. 2033: J. Liu, Y. Ye (Eds.), E-Commerce Agents. VI, 347 pages. 2001. (Subseries LNAI).

Vol. 2034: M.D. Di Benedetto, A. Sangiovanni-Vincentelli (Eds.), Hybrid Systems: Computation and Control. Proceedings, 2001. XIV, 516 pages. 2001.

Vol. 2035: D. Cheung, G.J. Williams, Q. Li (Eds.), Advances in Knowledge Discovery and Data Mining – PAKDD 2001. Proceedings, 2001. XVIII, 596 pages. 2001. (Subseries LNAI).

Vol. 2037: E.J.W. Boers et al. (Eds.), Applications of Evolutionary Computing. Proceedings, 2001. XIII, 516 pages. 2001.

Vol. 2038: J. Miller, M. Tomassini, P.L. Lanzi, C. Ryan, A.G.B. Tettamanzi, W.B. Langdon (Eds.), Genetic Programming. Proceedings, 2001. XI, 384 pages. 2001.

Vol. 2039: M. Schumacher, Objective Coordination in Multi-Agent System Engineering. XIV, 149 pages. 2001. (Subseries LNAI).

Vol. 2040: W. Kou, Y. Yesha, C.J. Tan (Eds.), Electronic Commerce Technologies. Proceedings, 2001. X, 187 pages. 2001.

Vol. 2041: I. Attali, T. Jensen (Eds.), Java on Smart Cards: Programming and Security. Proceedings, 2000. X, 163 pages. 2001.

Vol. 2042: K.-K. Lau (Ed.), Logic Based Program Synthesis and Transformation. Proceedings, 2000. VIII, 183 pages. 2001.

Vol. 2043: D. Craeynest, A. Strohmeier (Eds.), Reliable Software Technologies – Ada-Europe 2001. Proceedings, 2001. XV, 405 pages. 2001.

Vol. 2044: S. Abramsky (Ed.), Typed Lambda Calculi and Applications. Proceedings, 2001. XI, 431 pages. 2001.

Vol. 2045: B. Pfitzmann (Ed.), Advances in Cryptology – EUROCRYPT 2001. Proceedings, 2001. XII, 545 pages. 2001.

Vol. 2047: R. Dumke, C. Rautenstrauch, A. Schmietendorf, A. Scholz (Eds.), Performance Engineering. XIV, 349 pages. 2001.

Vol. 2048: J. Pauli, Learning Based Robot Vision. IX, 288 pages. 2001.

Vol. 2051: A. Middeldorp (Ed.), Rewriting Techniques and Applications. Proceedings, 2001. XII, 363 pages. 2001.

Vol. 2052: V.I. Gorodetski, V.A. Skormin, L.J. Popyack (Eds.), Information Assurance in Computer Networks. Proceedings, 2001. XIII, 313 pages. 2001.

Vol. 2053: O. Danvy, A. Filinski (Eds.), Programs as Data Objects. Proceedings, 2001. VIII, 279 pages. 2001.

Vol. 2054: A. Condon, G. Rozenberg (Eds.), DNA Computing. Proceedings, 2000. X, 271 pages. 2001.

Vol. 2055: M. Margenstern, Y. Rogozhin (Eds.), Machines, Computations, and Universality. Proceedings, 2001. VIII, 321 pages. 2001.

Vol. 2056: E. Stroulia, S. Matwin (Eds.), Advances in Artificial Intelligence. Proceedings, 2001. XII, 366 pages. 2001. (Subseries LNAI).

Vol. 2057: M. Dwyer (Ed.), Model Checking Software. Proceedings, 2001. X, 313 pages. 2001.

Vol. 2059: C. Arcelli, L.P. Cordella, G. Sanniti di Baja (Eds.), Visual Form 2001. Proceedings, 2001. XIV, 799 pages. 2001.

Vol. 2064: J. Blanck, V. Brattka, P. Hertling (Eds.), Computability and Complexity in Analysis. Proceedings, 2000. VIII, 395 pages. 2001.

Vol. 2066: O. Gascuel, M.-F. Sagot (Eds.), Computational Biology. Proceedings, 2000. X, 165 pages. 2001.

Vol. 2068: K.R. Dittrich, A. Geppert, M.C. Norrie (Eds.), Advanced Information Systems Engineering. Proceedings, 2001. XII, 484 pages. 2001.

Vol. 2070: L. Monostori, J. Váncza, M. Ali (Eds.), Engineering of Intelligent Systems. Proceedings, 2001. XVIII, 951 pages. 2001. (Subseries LNAI).

Vol. 2072: J. Lindskov Knudsen (Ed.), ECOOP 2001 – Object-Oriented Programming. Proceedings, 2001. XIII, 429 pages. 2001.

Vol. 2073: V.N. Alexandrov, J.J. Dongarra, B.A. Juliano, R.S. Renner, C.J.K. Tan (Eds.), Computational Science – ICCS 2001. Part I. Proceedings, 2001. XXVIII, 1306 pages. 2001.

Vol. 2074: V.N. Alexandrov, J.J. Dongarra, B.A. Juliano, R.S. Renner, C.J.K. Tan (Eds.), Computational Science – ICCS 2001. Part II. Proceedings, 2001. XXVIII, 1076 pages. 2001.

Vol. 2081: K. Aardal, B. Gerards (Eds.), Integer Programming and Combinatorial Optimization. Proceedings, 2001. XI, 423 pages. 2001.

Vol. 2082: M.F. Insana, R.M. Leahy (Eds.), Information Processing in Medical Imaging. Proceedings, 2001. XVI, 537 pages. 2001.

Vol. 2091: J. Bigun, F. Smeraldi (Eds.), Audio- and Video-Based Biometric Person Authentication. Proceedings, 2001. XIII, 374 pages. 2001.

Vol. 2092: L. Wolf, D. Hutchison, R. Steinmetz (Eds.), Quality of Service – IWQoS 2001. Proceedings, 2001. XII, 435 pages. 2001.